Current Topics in Developmental Biology

Volume 54

Cell Surface Proteases

Current Topics in Developmental Biology

Volume 54

Cell Surface Proteases

Edited by

Stanley Zucker

Professor of Medicine
Departments of Medicine/VA Medical Center
Northport, New York and State University
of New York at Stony Brook, Stony Brook, New York

and

Wen-Tien Chen

Professor of Medicine
Department of Medicine/Division of Neoplastic Diseases
State University of New York at Stony Brook
Stony Brook, New York

ACADEMIC PRESS

An imprint of Elsevier Science

Amsterdam   Boston   London   New York   Oxford   Paris
San Diego   San Francisco   Singapore   Sydney   Tokyo

This book is printed on acid-free paper. ∞

Academic Press
An Elsevier Science Imprint.
525 B Street, Suite 1900, San Diego, California 92101-4495, USA
http://www.academicpress.com

Academic Press
84 Theobald's Road, London WC1X 8RR, UK
http://www.academicpress.com

International Standard Book Number: 0-12-153154-6

PRINTED IN THE UNITED STATES OF AMERICA
03   04   05   06   07   08   9   8   7   6   5   4   3   2   1

Contents

1

Membrane Type-Matrix Metalloproteinases (MT-MMP)
Stanley Zucker, Duanqing Pei, Jian Cao, and Carlos Lopez-Otin

2

Surface Association of Secreted Matrix Metalloproteinases
Rafael Fridman

3

Biochemical Properties and Functions of Membrane-Anchored Metalloprotease-Disintegrin Proteins (ADAMs)
J. David Becherer and Carl P. Blobel

4

Shedding of Plasma Membrane Proteins
Joaquín Arribas and Anna Merlos-Suárez

5

Expression of Meprins in Health and Disease
Lourdes P. Norman, Gail L. Matters, Jacqueline M. Crisman, and Judith S. Bond

6

Type II Transmembrane Serine Proteases
Qingyu Wu

7

DPPIV, Seprase, and Related Serine Peptidases in Multiple Cellular Functions

Wen-Tien Chen, Thomas Kelly, and Giulio Ghersi

8

The Secretases of Alzheimer's Disease

Michael S. Wolfe

9

Plasminogen Activation at the Cell Surface

Vincent Ellis

10

Cell-Surface Cathepsin B: Understanding Its Functional Significance

Dora Cavallo-Medved and Bonnie F. Sloane

11

Protease-Activated Receptors

Wadie F. Bahou

12

Emmprin (CD147), a Cell Surface Regulator of Matrix Metalloproteinase Production and Function

Bryan P. Toole

13

The Evolving Roles of Cell Surface Proteases in Health and Disease: Implications for Developmental, Adaptive, Inflammatory, and Neoplastic Processes

Joseph A. Madri

14

Shed Membrane Vesicles and Clustering of Membrane-Bound Proteolytic Enzymes

M. Letizia Vittorelli

Contents

This book is dedicated to the vision of Fannie and Morris Zucker and the fulfillment of their dream to bring up their sons in a land of freedom and opportunity where education would open doors not available in their homeland.

Contributors

Numbers in parentheses indicate the pages on which authors' contributions begin.

Joaquín Arribas (125), Laboratori de Recerca Oncològica, Servei d'Oncologia Mèdica, Hospital Universitari Vall d'Hebron, Barcelona 08035, Spain

Wadie F. Bahou (343), Division of Hematology, State University of New York, Stony Brook, New York 11794-8151

J. David Becherer (101), Department of Biochemical and Analytical Pharmacology, GlaxoSmithKline Research Inc., Research Triangle Park, North Carolina 27709

Carl P. Blobel (101), Cellular Biochemistry and Biophysics Program, Sloan-Kettering Institute, Memorial Sloan-Kettering Cancer Center, New York, New York 10021

Judith S. Bond (145), Department of Biochemistry and Molecular Biology, The Pennsylvania State University College of Medicine, Hershey, Pennsylvania 17033-0850

Jian Cao (1), Department of Hematology, School of Medicine, State University of New York at Stony Brook, Stony Brook, New York 11794

Dora Cavallo-Medved (313), Department of Pharmacology, Wayne State University, School of Medicine, Detroit, Michigan 48201

Wen-Tien Chen (207), Department of Medicine/Division of Neoplastic Diseases, State University of New York at Stony Brook, Stony Brook, New York 11794-8154

Jacqueline M. Crisman (145), Department of Biochemistry and Molecular Biology, The Pennsylvania State University College of Medicine, Hershey, Pennsylvania 17033-0850

Vincent Ellis (263), School of Biological Sciences, University of East Anglia, Norwich, NR4 7TJ, United Kingdom

Rafael Fridman (77), Department of Pathology, School of Medicine, Wayne State University, Detroit, Michigan 48201

Giulio Ghersi (207), Department of Cellular and Developmental Biology, University of Palermo, Viale delle Scienze 90128 Palermo, Italy

Thomas Kelly (207), Department of Pathology, Arkansas Cancer Research Center, Little Rock, Arkansas 72205-7199

Carlos Lopez-Otin (1), Departamento de Bioquímica y Biología Molecular, Instituto Universitario de Oncologia, Universidad de Oviedo, 33071-Oviedo, Spain

Joseph A. Madri (391), Department of Pathology, Yale University School of Medicine, New Haven, Connecticut 06510

Gail L. Matters (145), Department of Biochemistry and Molecular Biology, The Pennsylvania State University College of Medicine, Hershey, Pennsylvania 17033-0850

Anna Merlos-Suárez (125), Laboratori de Recerca Oncològica, Servei d'Oncologia Mèdica, Hospital Universitari Vall d'Hebron, Barcelona 08035, Spain

Lourdes P. Norman (145), Department of Biochemistry and Molecular Biology, The Pennsylvania State University College of Medicine, Hershey, Pennsylvania 17033-0850

Duanqing Pei (1), Department of Pharmacology, University of Minnesota, Minneapolis, Minnesota 55455

Bonnie F. Sloane (313), Department of Pharmacology and Barbara Ann Karmanos Cancer Institute, Wayne State University, School of Medicine, Detroit, Michigan 48201

Bryan P. Toole (371), Department of Anatomy and Cellular Biology, Tufts University School of Medicine, Boston, Massachusetts 02111

M. Letizia Vittorelli (411), Dipartimento di Biologia cellulare e dello Sviluppo, Viale delle Scienze, Parco D'Orleans II, 90128-Palermo, Italy

   0

Michael S. Wolfe (233), Center for Neurologic Diseases, Brigham and Women's Hospital and Harvard Medical School, Boston, Massachusetts 02115

Qingyu Wu (167), Department of Cardiovascular Research, Berlex Biosciences, Richmond, California 94806

Stanley Zucker (1), VA Medical Center, Northport, New York 11768 and School of Medicine, State University of New York at Stony Brook, Stony Brook, New York 11794

Preface

This book represents the first book dedicated entirely to cell surface proteases, primarily of mammalian origin. The field has reached sufficient maturity to permit the presentation of 14 individual chapters. Membrane proteases that function primarily in other cell organelles beside the plasma membranes were not included in the book, i.e., pro-protein convertases that function in the Golgi apparatus. An attempt was made to describe interactions between different cell surface proteases and to avoid duplication wherever possible. In a few instances, the same topics were covered in more than one chapter, so as to present alternative points of views.

For anyone working in any single area related to cell surface proteolytic events, it is important to have an overview of the larger cast of characters that may be interacting on the cell surface. It is reasonable to speculate that none of the cell surface proteases functions alone and many important interactions remain to be elucidated. The importance of plasma membrane proteases in crucial aspects of cell function is only beginning to be appreciated. Not only are these enzymes involved in cleaving other membrane-bound and extracellular matrix proteins, but they are involved in processing physiologically important peptide substrates including growth factors, cytokines, chemokines, hormones, binding proteins, and apoptotic factors. As a result, membrane proteases are involved in a variety of cell functions, including metabolic regulation, immune response, cell growth, apoptosis, motility, and invasion. An important role for these proteases in embryonic development has been demonstrated throughout the phylogenetic tree. Furthermore, these proteases exhibit their multiple cell functions through interactions with other cell surface proteins such as receptors and integrins. The field is beginning to expand into new areas including protein localization, activation, inhibition, and three-dimensional structure. Inhibitors specific for plasma membrane proteases, including molecular inhibitors and naturalizing monoclonal antibodies, have been developed and tested *in vitro*. The involvement of cell surface proteases in disease represents an area of exciting new developments. The potential to treat disease with various categories of protease inhibitors is only beginning to be explored.

We thank the many authors for their thoughtful and timely contributions. The readers undoubtedly will appreciate the considerable amount of work and dedication required to put together such an extensive review of each of these subjects. Every one of these authors is an acknowledged expert in their field. As editors, we thank all of these contributors for their enthusiasm and patience. We thank our research colleagues for providing advice and editorial guidance (Michelle

Hymowitz) and the staff of Academic Press, especially Hilary Rowe, for their contributions to the development of this book. We extend our appreciation to the multitude of scientists who have made contributions to the field of cell surface proteases. This book is a tribute to their efforts. We apologize to those scientists whose work was not specifically mentioned or were misrepresented; space limitations prevented us from providing more extensive references.

We ask the readers to send along any corrections or missing information that can be corrected in future editions of this book.

We hope that this book will prove helpful for the large numbers of investigators and students who have an interest in cell surface proteolytic events and to the larger number of scientists interested in the plasma membrane.

Foreword

Recent years have seen an enormous expansion in the field of proteolytic enzymes, with a wealth of new knowledge of structures and mechanisms of the enzymes and the realization that a wide array of biological processes is controlled by proteolysis. Proteolytic events occur intracellularly, pericellularly, and extracellularly, but for many years the majority of studies have dealt with intracellular and extracellular proteolysis. Although the importance of pericellular proteolysis has long been recognized, it is only in the last several years that many cell surface anchored proteases have been discovered. These enzymes have rapidly become subjects of intense investigation as they play important roles in many biological processes. Some of the soluble proteases released from cells have shown to accumulate on the cell surface by binding to specific cell surface molecules. Localization of proteases on the cell surface either by a transmembrane domain or by other means serves to increase the local concentration of the enzymes, target proteolysis, enhance specificity, and also slow down reactions with endogenous inhibitors, thus regulating the protease action.

This is the first book that focuses on cell surface proteolysis. The Editors of this book, Stanley Zucker and Wen-Tien Chen, have long-standing interests in the role of pericellular proteolysis, especially in cell migration and cancer metastasis. The editors have chosen leaders in the field as contributors to the book. It covers metallo- (matrix metalloproteinases, ADAMs, meprins), serine (type II transmembrane serine proteases, dipeptidyl peptidase IV, seprase, plasminogen activators), cysteine (cathepsin B on the cell surface), and aspartic (memapsin 2/BACE, presenilins) proteases. Other subjects include protease-activated receptors, an MMP-inducing cell surface molecule of the Ig superfamily, emmprin, and shed membrane vesicles.

Many of the membrane-bound proteases are multidomain proteins and the noncatalytic domains often dictate the specificity of the enzymes as they allow the enzymes to interact with other cell surface or extracellular matrix molecules to express their biological activities. Such molecular assemblies on the cell surface may participate in cell–cell contacts and cell–matrix interactions. Each chapter, while describing biochemical and catalytic properties of a particular group of proteases, delineates biological and pathological implications of a specific group of cell surface associated proteases.

While the unified theme of this book is "Cell Surface Proteases," the functions of these proteases are diverse and therefore many avenues exist for medical applications. For example, membrane-bound MT1-MMP assists cells in migration by cleaving cell surface adhesion molecules such as CD44, while ADAMs, which have more restricted substrate specificity, are involved in shedding of cell surface molecules, development, inflammation, and release of signaling molecules. Memapsin 2 and presenilins are associated with Alzheimer's disease. Membrane vesicles shed from cancer cells contain a number of proteases, integrins, and HLA molecules and they may alter cellular behavior.

The book is timely and comprehensive, covering substantial amounts of recently reported information. The reader will value the excellent reviews of each subject. At the same time, one may realize that there are many key questions yet to be addressed as only a small fraction of this area of biology has been unravelled to date. "Cell Surface Proteases" will provide a foundation on which future knowledge in the field will expand. It will also be valuable for identifying future research paths.

Hideaki Nagase
Professor of Matrix Biology
Kennedy Institute of Rheumatology Division
Imperial College of Science, Technology and Medicine
London, UK

Introduction

Although the biological importance of membrane-bound serine proteases was recognized more than a hundred years ago, interest in the field of cell surface proteases was slow to develop in comparison to proteases identified initially in serum
(coagulation and complement factors), locally secreted by cells (plasminogen activators and matrix metalloproteinases), or intracellular compartments (cathepsins and calpains). It is of interest that the discovery of plasma-membrane-bound
metalloproteinases (Chapters 1–5), serine proteases (Chapters 7–9), and cathepsins (Chapter 10) occurred about the same time in the mid-1980s. From a biological point of view, the late appreciation of this important class of enzymes
may seem surprising. Technical difficulties in isolating and purifying proteases
from plasma membranes contributed considerably to the delay. In the past decade,
the field has rapidly progressed, as the reader will appreciate in reading this
book. Undoubtedly, the identification of an extensive number of serine-, cysteine-,
and metalloproteases that are displayed on the cell surface is not yet complete.
The importance of plasma membrane proteases in crucial aspects of cell function as well as in health and disease (Chapters 11–14) is only beginning to be
appreciated.

This book encompasses both intrinsic membrane proteases (Chapters 1–8) and
plasma membrane receptor related protease mechanisms (Chapters 9–12). The
figure on the book cover provides the reader with the sense of the close proximity
of these various proteins displayed on the cell surface. A constant theme noted
throughout this book is that the surface localization of proteases provides the cell
with the capability to process a vast spectrum of surface precursor molecules, that
once cleaved, play pivotal roles in normal cellular and tissue functions. Furthermore, co-localization of both the protease and the substrate as interacting molecule
within specific regions of the plasma membrane, such as lipid rafts (Chapters 1,
2, and 9), invadopodia (Chapters 1, 2, 7, 10, and 12), caveolae (Chapter 9), and
shed vesicles (Chapter 14), may increase their concentration, making regulatory
events more efficient *in vivo*. Surface binding of intrinsic and soluble proteases
also allows the cell to terminate proteolysis by rapid internalization of membrane-
bound enzymes, thereby allowing the cells to fine tune its external environment
(Chapters 1 and 2). These enzymes are involved in cleaving numerous extracellular matrix proteins (Chapters 1–10 and 14); they are also involved in processing
peptide substrates including growth factors, cytokines, chemokines, hormones,
and angiogenic factors (Chapters 1, 3–8, and 11). Cleavage of protease activated

receptors (PARs) at the cell surface by serine proteases generated by inflammatory or hemostatic pathways results in activation of signal transduction pathways that modulate numerous cellular activities (Chapter 11). Cleavage of proteins at the cell surface initiated by intrinsic plasma membrane proteases is also likely to activate intracellular signaling pathways; these events remain to be more thoroughly characterized.

The involvement of cell surface proteases in cell movement is an important new area of investigation (Chapters 1, 7, and 9–10). Cell migration requires constant modulation of the adhesive properties of a cell; this is achieved through numerous complex interactions between the extracellular matrix and the actin cytoskeleton mediated by transmembrane adhesion molecules. New clues have suggested that cleavage by membrane-type-MMPs (MT-MMPs), of adhesive-type proteins may provide a mechanism for enhancing cell migration and invasion (Chapter 1). Alternatively, association of seprase complexes with $\alpha 3\beta 1$ integrin localizes active molecules to invadopodia (Chapter 7). Chapter 9 deals with how migrating cells utilize the activity of the plasminogen activator system in the presence/absence of a transmembrane linkage between the glycolipid-anchored protein uPAR and the cytoskeleton. Considerable evidence indicates that serine peptidases and MT-MMPs accumulate at cell surface protrusions, termed invadopodia, that may have a prominent role in processing soluble factors (including growth factors, hormones, chemokines, bioactive peptides) in addition to the well-established role of invadopodia in degrading components of the extracellular matrix (Chapters 1 and 7). These membrane proteases may directly activate either themselves or soluble latent proteases such as MMP-2 and plasmin (Chapters 2 and 9). Different membrane proteases form complexes at invadopodia or other specialized locations that could provide distinct or overlapping functions (Chapters 10 and 14).

It has long been known that certain cell surface proteins can release their biologically active extracellular domains to the pericellular millieu through limited proteolysis. However, the broad recognition of this type of proteolysis (ectodomain shedding) as a general mechanism that regulates many functions of virtually all types of cell surface proteins has been relatively recent (Chapters 3–5, 11, and 14).

An important role for these proteases in embryonic development has been demonstrated throughout the phylogenetic tree. The cleavage of a cell surface protein (Notch) sequentially by ADAM family members, followed by the release of the cytoplasmic domain of Notch and transport to the nucleus, ultimately leading to control of gene expression, represents a fascinating effect of protease activity (Chapters 3 and 4). Knockout and transgenic mice have provided a fertile experimental model to delineate the function of various proteases.

Multiple interactions between cell surface proteases have complicated interpretation of the biological role of individual enzymes. A given membrane protease may have several functions and more than one protease or one protease family may mediate the same function (Chapter 7). For example, matriptase/MT-SP1, a recently identified type II serine membrane protease, has been proposed to have a biological role in activating pro-uPA, as has the cysteine protease cathepsin B (see Chapters 6 and 9). Cross talk between cell surface receptors and ligands through cell surface proteolysis represents another mechanism through which plasma membrane proteases are involved in various cell functions (see Chapter 3). Another mechanism that cells have evolved to enhance protease activity at the cell surface is to have low affinity docking sites. For example, EMMPRIN, secreted tissue plasminogen activator, and cathepsin B binding to annexin II, a protein peripherally associated with the plasma membrane, provide for protein complex formation at the cell surface; the effect of these interactions needs to be better understood (Chapters 9, 10, and 12).

Interactions between membrane proteases and the lipid bilayer have not been studied in detail. The turnover of plasma membrane proteases involving cell surface shedding, endocytosis in clathrin-coated pits, and intracellular degradation in lysosomes and proteosomes has received scant attention. In sharp contrast, the fate and control of endocytosed plasma membrane receptors has become an area of intense scientific interest. The potential role of caveolae in regulating protease function at the cell surface, involving the presence of multiple proteases and their surface receptors including cathepsin B, uPAR, MT1-MMP, and MMP-2, is discussed in Chapters 1, 9, and 10.

The potential to treat disease with various categories of protease inhibitors is a burgeoning new field. Little is known definitively about the role of cell surface proteases in most diseases with the exception of tumor necrosis factor-α (TNF-α) converting enzyme (TACE). Because TNF-α plays a pivotal role in rheumatoid arthritis, several pharmaceutical companies have developed potent inhibitors of TACE that are effective anti-inflammatory agents in animal models (Chapter 3). Of considerable surprise, TACE has also been implicated in diabetes, as has DDPIV (Chapter 7) and meprin B (Chapter 5). The involvement of multiple cell surface proteases represents an area of exciting new development, as represented by Alzheimer's disease, where different types of surface proteases cleave the β-amyloid protein at different sites, leading to release of pathogenic soluble peptides (see Chapter 8). The involvement of membrane-bound serine proteases in cancer (matriptase and hepsin) and in heart failure (corin, the endogenous pro-atrial natriuretic peptide convertase) has been a strong impetus for pharmaceutical development of specific inhibitors for this category of enzymes (Chapter 6). Targeted strategies focusing on design of specific inhibitors of PARs have been initiated for possible applications in thrombotic/vascular diseases (Chapter 11).

Clinical trials of MMP inhibitors that effectively abrogate MT-MMPs as well as several other MMPs have been reported in advanced cancer; the lack of positive results in this setting suggests that these drugs were employed too late in the course of the disease. An alternative explanation is that MMPs may not be pivotal in cancer progression or that more specific inhibitors need to be developed to avoid interfering with proteases that might protect the host from disease (Chapter 13).

It is fair to predict that our understanding of cell surface proteases will likely evolve considerably over the next decade. A three-dimensional structure of a complete plasma membrane protease remains to be presented. Interactions between membrane proteases and the lipid bilayer need to be better understood. The role of individual proteases in health and disease needs to be better defined. These scientific advances will be eagerly anticipated.

1

Membrane Type-Matrix Metalloproteinases (MT-MMP)

Stanley Zucker,[1] *Duanqing Pei,*[2] *Jian Cao,*[3]
and *Carlos Lopez-Otin*[4]
[1]VA Medical Center
Northport, New York 11768
and School of Medicine, State University of New York
at Stony Brook, New York 11794

[2]Department of Pharmacology, University of Minnesota
Minneapolis, Minnesota 55455

[3]Department of Hematology, School of Medicine, State University of New York
at Stony Brook, Stony Brook, New York 11794

[4]Departamento de Bioquímica y Biología Molecular, Instituto Universitario de Oncologia
Universidad de Oviedo
33071-Oviedo, Spain

Current Topics in Developmental Biology, Vol. 54

I. Introduction

Matrix metalloproteinases (MMPs) belong to the family of zinc endopeptidases collectively referred to as metzincins. The metzincin superfamily is distinguished by a highly conserved motif containing three histidines that bind zinc at the catalytic site and a conserved methionine that sits beneath the active site (Stoker and Bode, 1995). The metzincins are subdivided into four multigene families: serralysins, astacins, ADAMs/adamalysins, and MMPs (Stoker and Bode, 1995). Our knowledge of the field of metzincin biology is expanding at a rapid rate, yet we still do not fully understand how these enzymes regulate biologic functions.

Once activated, MMPs degrade a variety of extracellular matrix components and assorted other proteins including growth factors, growth factor binding proteins, and protease inhibitors (Birkedal-Hansen, 1995; Brinckerhoff and Matrisian, 2002; Egeblad and Werb, 2002; Nagase and Woessner, 1999; Parks *et al.*, 1998; Stamenkovic, 2000; Sternlicht and Werb, 2001; Stetler-Stevenson, 1999; Woessner, 1991). The ability to degrade extracellular matrix proteins is essential for any cell to interact properly with its immediate surroundings and for multicellular organisms to develop and function normally. MMPs also generate matrix protein fragments, which have functional activity of their own. These various functions of MMPs modulate cell invasion and metastasis, cell migration, apoptosis, and angiogenesis, to name just a few. A major limitation to our understanding of biologic function of human MMPs relates to difficulties in extrapolating from *in vitro* and animal models to human biology and disease. The belief that an MMP is involved in a given biologic/pathologic process is often based on the strength of its association with that process and the existence of plausible mechanisms that can be tested by experimental approaches (Nagase and Woessner, 1999; Sternlicht and Werb, 2001). Because of extensive redundancy of MMP function in animals, it has been difficult to appreciate the role of individual genes in various experimental models (Shapiro, 1998).

There have been 25 vertebrate MMPs and 22 human homologs identified to date, as well as several nonvertebrate MMPs; the list undoubtedly will grow. These enzymes have both a descriptive name typically based on a preferred substrate (e.g., interstitial collagenase) or on the tissue or cell types identified with the MMP (epilysin and leukolysin) and a sequential numerical nomenclature reserved for the vertebrate MMPs. Because of uncertainty concerning relevant substrates for many of these enzymes, the number system will be used in this chapter except for the membrane type-MMP subfamily, where the designation MT-MMP is almost universally employed.

The multiplicity of MMPs with distinct but somewhat overlapping functions appears to act as a safeguard against loss of regulatory control. The molecular weights of proMMPs vary between 28,000–92,000. The known MMP genes have been divided into four subfamilies based on gene structure. Group I consists of the collagenase subfamily. Group 2 consists of the gelatinases/modular alteration (fibronectin-like domains) subfamily. Group 3 consists of the variant hemopexin exon subfamily. Group 4 consists of the variant catalytic exon and unique 3′-end exon (plasma membrane and cytoplasmic domains) subfamily which consists of the MT-MMPs (Apte *et al.*, 1997). Alternative classification systems have been proposed.

II. Secreted MMPs

A. Structural Features of Secreted MMPs

All MMPs share several highly conserved domains, most importantly a latency locus [PRCG(V/N)PDV] in the amino-terminal "pro" domain consisting of about 80 amino acids and a zinc atom binding domain [VAAHExGHxxGxxH] in the active site (catalytic domain) with the three histidines coordinating the zinc (Birkedal-Hansen, 1995). A second structural zinc ion and calcium ions are also present elsewhere in the catalytic domain of MMPs (Woessner and Nagase, 2000). A "pre" region (signal peptide) is removed after it targets the protein for synthesis in the endoplasmic reticulum and subsequent secretion or insertion into the plasma membrane; the only exception is the signal peptide of MMP-23 which is retained as a plasma membrane anchor (Pei *et al.*, 2000). The concept that the cysteine residue in the conserved propeptide sequence PRCGVPD is positioned directly opposite the zinc atom at the active center of MMPs and is coordinated to it through the –SH group was proposed by Springman *et al.* (1990); this was termed the "cysteine switch" because displacement of the cysteine residue by a variety of means (proteolytic cleavage, oxidation, mercurial compounds, etc.) resulted in activation of the enzyme. The contact Cys-SH with zinc was later confirmed directly. MMP-23 is the only exception without a cysteine residue in its putative domain upstream of its catalytic domain (Pei *et al.*, 2000). Several MMPs also contain a furin sensitive sequence RXKR near the C-end of the prodomain and activation occurs immediately to the carboxyl side of the distal arginine. The C-end of the propeptide is not sharply defined; there may be two or three residues grouped together where cleavage can occur and the resulting differences of 1–3 residues can have a powerful effect on the enzyme's activity of the resultant catalytic domain (Woessner and Nagase, 2000). The catalytic domain, approximately 160–170 amino acids in length with the catalytic active center in the C-terminal 50–54 residues, dictates cleavage site specificity through its active site cleft through specificity subsite pockets that bind amino acid residues immediately adjacent to the scissile peptide bond, and through secondary substrate-binding exosites located outside the active site (Overall, 2001). With the exception of MMP-7, -23, and -26, all MMPs have a hemopexin/vitronectin domain that is connected to the catalytic domain by a proline-rich hinge or linker. When present, the hemopexin domain region, approximately 200 residues long and shaped like a beta propeller fold with pseudo four fold symmetry, influences TIMP binding, the binding of substrates, membrane activation events, and some proteolytic activities (Murphy *et al.*, 1992; Sanchez-Lopez *et al.,* 1988). The proline-rich hinge region varies in length among various MMPs and is important in determining substrate specificity (Knauper *et al.,* 1997; Parks *et al.,* 1998). MMP-2 and MMP-9 are further distinguished by the insertion of three head-to-tail cysteine-rich repeats within their catalytic domain. A type V collagen-like insert consisting of 54 residues rich in proline is found in MMP-9 where it appears between the active center and the hemopexin domain (Wilhelm *et al.,* 1989).

B. Regulation of Secreted MMPs

The biologic function of MMPs requires that the activated enzyme is present in the right concentration, in the right place, and at the proper time. This requires tightly regulated transcription and posttranslational control. The N-terminal prodomain of secreted MMPs primarily acts as an internal inhibitor of MMP activity and maintains the enzyme in an inactive state until proteinase activity is required. Classical activation of MMPs is achieved by removal of the N-terminal pro-sequence of approximately 80 amino acids (Birkedal-Hansen *et al.,* 1993). The propeptide domain of MMPs is cleaved sequentially, with the initial cleavage taking place in a "bait" region located in a readily accessible site between the first and second α-helix in the propeptide domain. This cleavage destabilizes the interaction of the propeptide with the catalytic domain. Subsequent cleavages are then possible with the final cleavage often occurring as a bimolecular, autoproteolytic event (Imai *et al.,* 1996; Knauper *et al.,* 1996; Nagase and Woessner, 1999). The amino acids surrounding the prodomain cysteine are also involved in maintaining latency.

Kotra *et al.* (2001) have described the critical importance of a protonation event at the coordinated cysteine thiolate as a prerequisite for the departure of the propeptide from the active site of MMPs. A catalytically active glutamate transiently moves toward the zinc ion to achieve coordination and "masks" the positive potential of the zinc ion, thereby lowering the energy barrier for dissociation of the protonated cysteine side chain from the zinc ion. A subtle conformational change by the propeptide is needed in the course of zymogen activation.

Although the physiologic activators of specific MMPs are uncertain, the initial cleavage event can be carried out *in vitro* by a variety of serine proteinases including trypsin, plasma kallikrein, and neutrophil elastase. ProMMP-2 is an exception to direct serine proteinase activation of MMPs. Plasmin generation at the cell surface is cited as a potential mechanism for the physiological activation of latent MMP-1, MMP-3, and MMP-9. However, data derived from uPA and tPA knockout mice have not provided supporting evidence for the physiologic significance of this pathway (Lijnen *et al.,* 1998). Some activated MMPs can further activate other proMMPs. For example, MMP-3 has been shown to activate proMMP-1 and proMMP-9.

Many of the secreted MMPs, including MMP-1, -3, -9, -10, -11, and -13, are expressed at very low or nondetectable levels in healthy resting tissue. Plasma levels of MMP-1, -3, -8, -9, and -13 ranging between 2–100 ng/ml have been identified in healthy individuals (Zucker *et al.,* 1999). In contrast, readily detectable tissue levels of MMP expression are seen in repair or remodeling processes and in disease or inflamed tissue; higher plasma levels of MMPs are noted in these conditions (Zucker *et al.,* 1999). Some MMPs, including MMP-2, -7, -19, -24, -25, and -28, are expressed in healthy tissues (Lohi *et al.,* 2001). Normal plasma levels of MMP-2 vary in a range between 500–800 ng/ml. The production of many matrixins is transcriptionally regulated; inflammatory cytokines, growth factors, hormones, cellular transformation, and interaction with extracellular matrix components can enhance the transcription of a number of matrixin genes, whereas glucocorticoids, retinoic acid, and ovarian steroids often exhibit suppressive effects on some matrixins (Birkedal-Hansen, 1995; Nagase and Woessner, 1999). The promoter regions of inducible MMP genes show remarkable conservation of regulatory elements (Vincenti *et al.,* 1996).

MMP proteins are transcribed and secreted by the constitutive secretory pathway, except in the case of neutrophils and macrophages where MMPs can be stored in and released from secretory granules (Birkedal-Hansen, 1995; Woessner, 1991). A role for mitogen-activated protein kinases, AP-1, and ETS transcription factors in regulation of MMP gene expression has been proposed (Westermarck and Kahari, 1999). In contrast, MMP-2 is reported to be constitutively produced and is not enhanced by most cytokines except for possible induction by TGF-β, IL-8, and insulin-like growth factor receptor (IGF-I0) (Long *et al.,* 1998). Modulation of mRNA stability of MMPs in response to growth factors and cytokines has been described (Westermarck and Kahari, 1999).

The MMPs are often sequestered as inactive zymogens in the extracellular matrix (ECM) after secretion, thereby providing a reservoir of latent enzyme positioned for activation and proteolytic attack at focal sites. Heparan sulfate proteoglycans has been demonstrated to be an important docking molecule for MMPs in proximity to epithelial cells and underlying basement membranes (Yu and Woessner, 2000). Proteolytic processing is required to release the catalytically active enzyme. Interstitial collagenases (MMP-1, -8, -13) and MMP-3 have C-terminal hemopexin-like domains that bind to collagen.

C. Naturally Occurring MMP Inhibitors

Once activated, MMPs are modulated by endogenous proteinase inhibitors, which include four TIMPs (20-29 kDa), α_2-macroglobulin (α_2-M) (Birkedal-Hansen *et al.,* 1993; Stetler-Stevenson, 1999), and the recently described RECK (Oh *et al.,* 2001). TIMPs form strong, reversible, noncovalent complexes with active MMPs. The TIMP amino-terminal group is thought to fill the fourth coordinated site of the catalytic Zn^{2+} that was previously occupied by cysteine in the propeptide of MMPs, resulting in inhibition of the enzyme's hydrolytic activity (Brew *et al.,* 2000).

In vivo, TIMPs can inhibit cell invasion, tumorigenesis, metastasis, and angiogenesis. TIMPs also possess other cellular functions that are independent of MMP inhibition. These include growth promoting functions, induction of apoptosis, and suppression of receptor tyrosine kinase activation of growth factors (Hoegy *et al.,* 2001). These negative regulators are important for control of MMP activity.

Following the initial purification and characterization of TIMP-1 as an inhibitor of MMP-1, TIMP-2, -3, and -4 were subsequently identified. These four TIMPs exhibit 44–52% sequence identity and are cross-connected by six intrachain disulfide bridges to yield a conserved six-loop, two domain structure.

All of the TIMPs are capable of inhibiting virtually all of the MMPs to varying degrees following formation of tight noncovalent 1:1 complexes. The exception to the general rule is that TIMP-1 appears to be a very poor inhibitor of MT-MMPs attached to the cell surface. All of the TIMPs are soluble proteins, widely distributed in body fluids, except for TIMP-3, which has the unique ability to bind via its C-terminal domain to heparan sulfate proteoglycans within the extracellular matrix, thereby concentrating it to specific regions within tissues and basement membranes (Langton *et al.,* 2000). In addition to binding to the active site of MMPs, the C-terminal domains of TIMP-1 and TIMP-2 form complexes with the C-terminal domains of proMMP-9 and proMMP-2, respectively. These complexes preserve the inhibitory activity of the TIMPs while dampening the activity of the bound MMP (Nagase and Woessner, 1999).

α_2-Macroglobulin (α_2M) is a large protein (750 kDa) produced by the liver and present in high concentrations in normal serum. α_2M inhibits all four classes of proteases (serine, cysteine, metallo, aspartyl) and produces irreversible clearance of MMPs. α_2M presumably functions primarily in blood where it probably is of less consequence to MMPs, because these enzymes are inactive in this location (Zucker *et al.,* 1999).

RECK, reversion-inducing cysteine-rich protein with kazal motifs, is a plasma membrane (glycosylphosphatidylinositol-modified anchor) glycoprotein containing serine protease inhibitor-like domains, which has been described as an inhibitor of MMPs involved in cancer progression (MMP-2, MMP-9, and MT1-MMP). Although the molecular basis underlying this inhibitory effect is unclear, recombinant soluble RECK can directly inhibit soluble MT1-MMP proteolytic activity. Mice lacking a functional RECK gene die *in utero* with defects in collagen fibrils, the basal lamina, and vascular development, thereby suggesting unimpeded MMP degradation of ECM proteins. Vascular sprouting is dramatically suppressed in tumors derived from RECK-expressing fibrosarcomas implanted in nude mice (Oh *et al.,* 2001). (See Chapter 2 by Fridman for detailed discussion of RECK.)

Another class of MMP inhibitors, protein subdomains, have structural similarities to TIMPs. For example, proteolytic processing of the procollagen C-terminal protease enhancer protein releases a C-terminal fragment with MMP inhibitory activity and structural similarity to the N-terminal domain of TIMPs (Mott *et al.,* 2000).

D. Catabolism and Clearance of Secreted MMPs

Relatively little is known about the further proteolysis of activated MMPs. Whereas some cleavages of activated MMPs lead to inactivation, others, such as those that specifically remove the hemopexin domain, can generate truncated enzymes that lose their ability to cleave some substrates (Woessner and Nagase, 2000). Another means of regulating extracellular MMP levels is by clearance of intact enzymes. Clearance of MMP-13 through a specific cell surface receptor has been demonstrated (Barmina *et al.,* 1999) (see Chapter 2 by Fridman for further discussion). Thrombospondin has been implicated in the clearance of MMP-2, perhaps involving endocytosis of the MMP-2 complex by the low-density lipoprotein receptor-related protein LRP (Yang *et al.,* 2001). However, uncertainty concerning the biologic activity of radiolabeled MMP-2 in these experiments raises concern about the interpretation of the data.

III. Membrane Type-Matrix Metalloproteinases (MT-MMPs)

For 20 years following the initial discovery of the collagenase family (Gross and Lapiere, 1962) (later renamed MMPs), the small number of vertebrate MMPs identified was considered to be rapidly secreted from cells following synthesis. In the early 1980s, evidence of collagenolytic and gelatinolytic enzymatic activity associated with the plasma membranes of endothelial cells and tumor cells appeared in the literature (Kalebic *et al.,* 1983; Moll *et al.,* 1990; Zucker *et al.,* 1985a,b, 1987, 1990). By the year 2000, cell surface localization of proteolytic enzymes was increasingly recognized to be important in the cell's ability to adapt and modify its environment (Sternlicht and Werb, 2001). In addition to MT-MMPs, other metalloprotease family members, serine proteases, cysteine proteases, and aspartic proteases have been demonstrated to be intrinsic plasma membrane enzymes. A variety of membrane receptor-related proteases have also been described. These different categories of proteases will be discussed in Chapters 2-12.

A. Identification of Membrane Type-Matrix Metalloproteinases

Six distinct members of the MT-MMP subfamily of MMPs have been identified in human tissues (Pei, 1999; Velasco *et al.,* 2000) (Table I). Sato *et al.* (1994), using reverse transcriptase–polymerase chain reaction (RT-PCR) based strategies with degenerate oligonucleotides encoding conserved regions in MMPs, cloned the first member of this family (Fig. 1). The isolated placental cDNA encoded a protein that was originally called MT-MMP and now is known as MT1-MMP (MMP-14). Almost simultaneously, Strongin *et al.* (1995) isolated and partially characterized this membrane protein as part of their studies directed to clarify the

Table I General Characteristics of Human MT-MMPs

Name	MMP Number	Accession Number	Number of Amino Acids	Chromosome Location	Substrates	Expression
MT1-MMP	MMP-14	D26512	583	14q11-12	proMMP-2, proMMP-13, types I, II, III collagen, gelatin, fibronectin, laminin, fibrin, proteoglycans, proTNFα	Ubiquitous
MT2-MMP	MMP-15	Z48482	670	16q12.2-21	proMMP-2, fibronectin, tenascin, laminin, aggrecan, perlecan	Ubiquitous
MT3-MMP	MMP-16	AB009303	608	8q21.3-22.1	proMMP-2, type II collagen, fibronectin	Brain, lung, heart, placenta
MT4-MMP	MMP-17	X89576	550	12q24	proTNFα, fibrinogen, fibrin	Brain, leukocytes, colon, ovary, testis
MT5-MMP	MMP-24	AF131284	646	20q11.2	proMMP-2	Brain, kidney, pancreas, lung
MT6-MMP	MMP-25	AJ239053	563	16p13.3	proMMP-2(very low efficiency)	Leukocytes, lung, spleen

mechanisms of activation of proMMP-2. Further studies established that this MT1-MMP activation process involves the formation of a trimolecular complex between proMMP-2, MT1-MMP, and TIMP-2 that acts as a concentration mechanism on the cell surface that is crucial for the efficiency of activation (Fig. 2; see color insert) (Strongin *et al.*, 1995; Zucker *et al.*, 1998a). On the other hand, purified and soluble ectodomains of MT1-MMP can process proMMP-2 efficiently without appreciable contribution from TIMP-2 (Pei and Weiss, 1996).

The demonstration that MT1-MMP was very efficient in activation of MMP-2 represented a very important step toward our understanding of the mechanisms underlying the generation of pericellular proteolytic activity in multiple normal and pathological processes. Anchorage of MT-MMPs on the plasma membrane facilitates focused substrate degradation limited to a small area and partial resistance to protease inhibitor action. Localization of MT-MMPs and MMP-2 to specialized cell surface protrusions called invadopodia, especially in cancer cells, increases the degradation of surrounding ECM proteins (Monsky *et al.*, 1993; Nakahara *et al.*, 1998).

Soon after the finding of MT1-MMP, studies on MT-MMPs acquired additional interest and complexity with the discovery of five more members of this family: MT2-, MT3-, MT4-, MT5-, and MT6-MMP, whose officially approved numbers

Transmembrane MT-MMPs

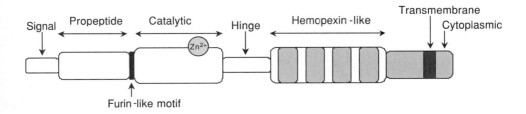

GPI-anchored MT-MMPs

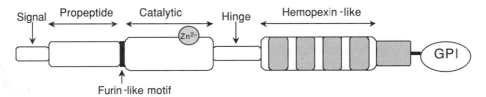

Figure 1 Schematic representation of the domain organization of human MT-MMPs. MT1-, MT2-, MT3-, and MT5-MMP belong to the transmembrane group of MT-MMPs and contain a type I transmembrane domain and a short cytoplasmic tail. MT4-MMP and MT6-MMP belong to the GPI group. They lack the cytoplasmic tail and are anchored to the plasma membrane via glycosylphosphatidylinositol.

following the MMP nomenclature system are MMP-15, MMP-16, MMP-17, MMP-24, and MMP-25, respectively (Table I). MT2-, MT3-, and MT4-MMP were cloned by RT-PCR with degenerate oligonucleotides deduced from cysteine-switch and Zn-binding sites of MMPs, and RNAs from different tumor tissues as templates: a kidney carcinoma for MT2-MMP, an oral melanoma for MT3-MMP, and a breast carcinoma for MT4-MMP. Full-length sequences for these enzymes were subsequently obtained by screening appropriate cDNA libraries (Puente *et al.*, 1996; Takino *et al.*, 1995a; Will and Hinzmann, 1995). The approach followed to identify mouse and human MT5-MMP was first based on a computer search of expressed sequence tag (EST) databases, looking for sequences with similarity to previously described MMPs (Llano *et al.*, 1999; Pei, 1999). Full-length mouse MT5-MMP was cloned by screening a mouse brain cDNA library using a human EST clone (Pei, 1999). For human MT5-MMP, further 5′-RACE experiments and analysis of PAC and BAC genomic libraries led to the cloning of a full-length cDNA for this membrane protease (Llano *et al.*, 1999). Finally, MT6-MMP was identified using the information generated during sequencing of a large genome region containing the gene responsible for familial Mediterranean fever, followed by RT-PCR and RACE-based strategies with RNA from human tissues (Pei, 1999; Velasco *et al.*, 2000).

Extensive genomic searching of the public and private versions of the human genome sequence has failed to reveal the presence of additional MT-MMPs, strongly suggesting that this subfamily of membrane-bound MMPs is just composed of six members. Nevertheless, it is remarkable that other MMPs unrelated to MT-MMPs can be also anchored to the cell surface. This is the case of CA-MMP (MMP-23), a type II transmembrane protease with an N-terminal signal anchor peptide, rather than the cleavable signal sequence present in all other MMPs both secreted and membrane-bound (Pei *et al.*, 2000). Finally, MMP family members lacking transmembrane domains, such as MMP-2 and MMP-9, may be also localized at the cell surface through the interaction with membrane proteins such as integrin $\alpha_v\beta_3$ and CD44 that act as docking receptors for these proteases (Brooks *et al.*, 1996; Yu and Stamenkovic, 2000) (see Chapter 2).

B. Expression of MT-MMP Genes in Tissues

The expression of MT-MMPs is differentially controlled at the transcriptional level as evident from the tissue distribution of their mRNAs (Puente *et al.*, 1996; Shofuda *et al.*, 1997; Takino *et al.*, 1995a,b; Will and Hinzmann, 1995). In contrast to most secreted members of the MMP family, the mRNAs for the six MT-MMPs are detectable in the extracts of many tissues and varies considerably (Table I). Northern blot analysis revealed MT1-MMP in adult human intestine, kidney, lung, ovary, placenta, prostate, and spleen (Will and Hinzmann, 1995). MT2-MMP is expressed in colon, heart, brain, intestines, pancreas, and testis, and less in the lung, kidney, spleen, liver, and muscle (Seiki, 1999). MT3-MMP is expressed in human placenta, ovary, brain, lung, and vascular smooth muscle cells (Shimada *et al.*, 1999). MT4-MMP is localized in brain, colon, ovary, testis, leukocytes, and in a variety of transformed cell types (Grant *et al.*, 1999; Puente *et al.*, 1996). MT5-MMP is expressed in brain, kidney, lung, and pancreas (Llano *et al.*, 1999; Romanic *et al.*, 2001). MT6-MMP is predominantly expressed in leukocytes, lung, and spleen and in high levels in colon carcinoma cells and anaplastic brain cancer cells, but not in normal colon or brain (Velasco *et al.*, 2000). Pei has shown that MT6-MMP appears to be a gene specifically expressed in neutrophils of the peripheral blood leukocytes and has named it leukolysin (Pei, 1999). By cellular fractionation, Pei's group also demonstrated that leukolysin is targeted to gelatinase granules (containing MMP-9), secretory vesicles and plasma membranes in resting neutrophils (Kang *et al.*, 2001), suggesting that leukolysin may function to facilitate the interactions between neutrophils and the endothelium, and the migration of leukocytes through the basement membrane.

C. Structural Chemistry and Biochemistry

1. Unique Structural Features of MT-MMP Proteins

Clues about the functional role of specific domains of transmembrane MT-MMPs have been deduced from structural analysis and mutagenesis studies. Structural comparisons among the six distinct MT-MMPs have revealed that they show a significant percentage of overall amino acid sequence identities (about 40–50%) and contain all protein domains characteristic of MMPs including signal peptide, propeptide, catalytic domain with the zinc-binding site, hinge region, and hemopexin domain (Fig. 1). However, MT-MMPs also contain unique structural features that confer them specific properties. Thus, they differ from the remaining MMP family members by the presence of a C-terminal extension rich in hydrophobic residues and involved in the membrane attachment of these proteases.

On the basis of their method of attachment to the plasma membrane, MT-MMPs may be classified into two groups: transmembrane-type and glycosylphosphatidylinositol (GPI)-type. MT1-, MT2-, MT3-, and MT5-MMP are type I transmembrane proteins with a short cytoplasmic tail involved in the regulation of intracellular trafficking and activity of these proteases (Jiang *et al.*, 2001; Lehti *et al.*, 2000; Uekita *et al.*, 2001). By contrast, MT4- and MT6-MMP are not transmembrane proteins and are bound to the cell surface by a GPI-mediated mechanism (Itoh *et al.*, 1999; Kojima *et al.*, 2000). In these proteins, the C-terminal hydrophobic domain is cleaved off in the ER lumen, and the resulting ectodomain is transferred to the GPI moiety. Consequently, the mature GPI-anchored MT4- and MT6-MMP lack transmembrane and intracellular domains (Fig. 1). MT4-MMP has the least degree of sequence identity to the other family members (Pei, 1999). Another distinction is that MT4-MMP is most potently inhibited by TIMP-1 rather than TIMP-2 or TIMP-3, like the proMMP-2 activating MT-MMPs.

a. Propeptide Domain of MT-MMPs. The propeptide domain of MT-MMPs has some characteristic features when compared with the equivalent region of other MMPs. It has been described that this domain functions in MT-MMPs as an intramolecular chaperone involved in trafficking these proteases to the cell surface, and in maintaining them in proper conformation for full enzymatic activity (Cao *et al.*, 2000). Recent structure–function relationship studies of the MT-MMP prodomains have revealed the presence of a conserved tetrapeptide sequence (Tyr^{42}-Gly^{43}-Tyr^{44}-Leu^{45}) within this region that is critical for the intramolecular chaperone function of the prodomain from these membrane proteases (Pavlaki *et al.*, 2002). Furthermore, mutations in this conserved region have been shown to be detrimental for MT1-MMP function in terms of activation of proMMP-2, binding of TIMP-2 to the cell surface, cell migration, and substrate degradation. Therefore, this four-residue amino acid sequence characteristic of MT-MMP propeptides contains important information for both the folding of these proteases and the intramolecular chaperone activity of this domain (Pavlaki *et al.*, 2002). MMP-28 also has this conserved tetrapeptide sequence within the propeptide domain (Lohi *et al.*, 2001). In a similar vein, the propeptide domain of other mammalian membrane-bound proteases, including ADAM-10 (Anders *et al.*, 2001) and β-secretase (Shi *et al.*, 2001), have been demonstrated to facilitate proper folding of their active protease domains and may serve as chaperones that thwart aggregation of unfolded protein.

All six MT-MMP family members have a furin recognition motif inserted between the propeptide and the catalytic domain containing the basic sequence motif Arg^{108}-Arg^{109}-Lys^{110}-Arg^{111}. In many types of cells (Pei and Weiss, 1996) this sequence is recognized and cleaved by at least four members of the proprotein convertase family (paired basic amino acid cleaving enzymes/Kex-like proteinases) including furin, PACE-4, PC6, and PC7. Cleavage of the propeptide of MT-MMP in the trans Golgi network between the Arg^{111}-Tyr^{112} bond releases the propeptide from the catalytic domain (Rozanov *et al.*, 2001; Yana and Weiss, 2000). These proprotein convertases are calcium-dependent transmembrane serine proteases of the subtilisin family and are prominently displayed in the trans Golgi network (Steiner, 1998). It has been speculated that the effect of actin binding-protein 280 in tethering of furin at the plasma membrane may serve to concentrate the furin protein for efficient MT1-MMP processing (Ellerbroek and Stack, 1999). Secreted forms of MT1-MMP transmembrane deletion mutants are also efficiently processed immediately downstream of the RRKR motif by a furin-dependent pathway (Pei and Weiss, 1996; Sato *et al.*, 1994, 1996). Whereas both furin and MT3-MMP are colocalized in the trans Golgi network, neither furin activity nor its recognition site in MT3-MMP is required for the observed colocalization (Kang *et al.*, 2002).

The occurrence of furin-independent alternative pathways of MT-MMP activation has been well documented in certain types of cells (Cao *et al.*, 1996; Li *et al.*, 1998; Rozanov *et al.*, 2001; Sato *et al.*, 1999). These alternative pathways include possible conformational effects induced by the plasma membrane localization of proMT1-MMP, autoproteolysis, or activation through the action of non-furin proprotein convertases (Yana and Weiss, 2000) and other proteases located at the plasma membrane. Co-transfection of mutant α_1-protease inhibitors cDNA [α_1-Pittsburgh (Cao *et al.*, 1996) and α_1 anti-trypsin Portland (Yana and Weiss, 2000)], which are known to inhibit furin/PC, with MT1-MMP cDNA into COS-1 cells resulted in loss of function of MT1-MMP in some cell types, but not others. An alternative overlapping "triplet" of minimal proprotein convertase recognition motifs is embedded in the [86]KAMRRPR peptide as [86]KXXR, [89]RXXR, or [89]RR and represents a second site for furin cleavage (Yana and Weiss, 2000). MT3-MMP and MT5-MMP have been shown to be processed in stably transfected MDCK cells via a proprotein convertase (PC) pathway (Kang *et al.*, 2002; Wang and Pei, 2001), further illustrating the generality of the PC-based intracellular activation mechanism proposed for MMP-11 (Pei and Weiss, 1995).

Furin/PACE cleave a wide range of precursor proteins at the consensus amino acid sequence RxK/RR (Cao *et al.*, 1996; Knauper and Murphy, 1999). Furins are routed through multiple cellular compartments and can be active at diverse processing sites including the trans Golgi network, cell surface, and early endosomes. MMP-11, MMP-23, and MMP-28 also contain a furin/PACE recognition domain in their propeptides (Lohi *et al.*, 2001; Velasco *et al.*, 1999) and are activated by this mechanism (Pei *et al.*, 2000). It has been proposed that the kinase and phosphatase-dependent subcellular localization of furin and the tethering of furin to the plasma membrane by actin binding protein 280 may be regulatory factors in proMT-MMP activation (Ellerbroek and Stack, 1999).

Plasmin has been reported to be able to activate proMT1-MMP by cleaving the molecule downstream of the furin consensus sequence; this effect was proposed to occur extracellularly after trafficking of latent enzyme to the cell surface (Okomura *et al.*, 1997). Although MT3-MMP containing an alternative spliced mRNA has been identified, the physiologic importance of soluble MT-MMPs is uncertain.

b. Catalytic Domain of MT-MMPs. The group of transmembrane MT-MMPs present a characteristic 8-amino acid insertion between strands β_{II} and β_{III} in the catalytic domain. In MT1-MMP this consists of the residues [163]PYAPIREG[170]. The function of this insertion, called the MT-loop (Fernandez-Catalan *et al.*, 1998), is not entirely clear. Inspection of the 3D structure of the complex between the catalytic domain of MT1-MMP and TIMP-2 shows that the MT-loop generates a pocket in the MMP catalytic domain fold that interacts with the AB-loop of TIMP-2 (Fernandez-Catalan *et al.*, 1998). Interestingly, studies with the MT-loop of MT1-MMP have revealed that neither mutation nor deletion of this region significantly impaired the catalytic activity of this enzyme toward synthetic peptides or fibrinogen (English *et al.*, 2001a). Therefore, these residues appear to have little influence on the conformation of the active-site cleft of these proteinases. Likewise, none of the MT1-MMP mutants in the MT-loop showed significant differences in K_i^{app} for the N-terminal inhibitory domain of TIMP-2, indicating that the net effect on affinity of the MT1-MMP:TIMP-2 complex is not influenced by this 8-residue loop (English *et al.*, 2001a). By contrast, analysis of the kinetics of activation of proMMP-2 by the MT-loop mutants showed a significant impairment in proMMP-2 processing when compared with wild-type MT1-MMP. Therefore, it is likely that the MT-loop of transmembrane MT-MMPs represents a structural adaptation of the MMP catalytic domain that facilitates the efficient activation of proMMP-2 (English *et al.*, 2001a). Consistent with this, MT4-MMP and MT6-MMP, the two MT-MMP family members lacking the MT-loop, have been shown to be either unable to activate, or very inefficient at activating, proMMP-2 (English *et al.*, 2001b). Finally, and also in agreement with these studies on MT1-MMP, studies with human and mouse MT2-MMP have revealed marked differences in the proMMP-2 activating properties of these orthologous enzymes, which have been ascribed to variations in the structure of their respective MT-loops (Miyamori *et al.*, 2000a). Thus, human MT2-MMP is somewhat defective in cell-mediated activation of proMMP-2, whereas mouse MT2-MMP is very efficient in this activity. Comparison of the amino acid sequence of their MT-loops revealed the presence of some residues (Pro[183] and Glu[185]) in the human enzyme that appeared to hamper its proMMP-2 activating function. Replacement of these residues by the equivalent ones of the mouse MT2-MMP (Ser[183] and Asp[185]) led to the recovery of efficient activation of proMMP-2.

c. Hinge Domain of MT-MMPs.   Similar to other MMPs, the MT-MMPs have a flexible hinge region linking the catalytic and hemopexin domains. The hinge domains in MT-MMPs are longer and more variable in length (34–65 amino acid residues) than in soluble MMPs; it is not clear whether this sequence motif is important for mediating substrate specificity as has been demonstrated for collagenases (Knauper and Murphy, 1999).

d. Hemopexin Domain of MT-MMPs.   The hemopexin-like region shows all the characteristics previously demonstrated for secreted MMPs, including conservation of the two cysteine residues that form a disulfide bond. The MT-MMPs contain an additional insertion at the end of the hemopexin domain just preceding the transmembrane domain. Ohuchi *et al.* (1997) suggested that the 27 amino acid segment in the C-terminus of the hemopexin domain of MT1-MMP (amino acids 509–535) is essential for collagenolytic activity probably because that sequence is necessary for intact conformation of the hemopexin domain. The role of the hemopexin domain in oligomerization of MT-MMPs will be discussed later in this chapter.

e. Transmembrane Domain of MT1-MMP.   After the hemopexin repeats that usually mark the end of the protein in most MMPs, all MT-MMPs contain a COOH-terminal hydrophobic extension rich in hydrophobic residues and involved in the attachment of these proteases to the cell surface (Cao *et al.,* 1995). There are clear differences in the structure of this region in the different MT-MMPs. Thus, MT1-, MT2-, MT-3, and MT5-MMP contain a long stretch of hydrophobic residues that is predicted to act as a transmembrane anchor in all of them. This sequence is followed by a cytoplasmic tail. By contrast, MT4- and MT6-MMP contain a shorter hydrophobic domain that is not capable of acting as a permanent anchoring sequence to the plasma membrane. This sequence is not followed by a cytoplasmic tail as is the case of bona fide type I transmembrane proteins. Itoh *et al.* (1999) and Kojima *et al.* (2000) have demonstrated that the hydrophobic sequence of MT4- and MT6-MMP is only mediating attachment of these proteases to the membrane via GPI-based mechanisms.

f. Cytoplasmic Domain of MT-MMPs.   Finally, the subgroup of transmembrane MT-MMPs has a cytoplasmic tail whose functional importance in different contexts has been addressed by a number of independent studies. The first experiments directed to analyze the role of the MT-MMP cytoplasmic tail in the function of these proteases were performed by Nakahara *et al.* (1998). By using deletion mutants, they demonstrated that MT1-MMP requires its cytoplasmic domain to localize to the invadopodia of tumor cells, a protruding membrane structure associated with invasive cancer cells (Nakahara *et al.,* 1998). Subsequently, Lehti *et al.* (2000) reported that a truncation of 10 amino acids, but not the 6 terminal amino acid residues in the MT1-MMP cytoplasmic tail, decreased the invasive activity of human melanoma cells, supporting a role for this MT-MMP domain in cell invasion. Ureña et al. (1999) and Nakahara et al. (1998) reported that the cytoplasmic tail is critically involved in trafficking of MT1-MMP to discrete regions of the cell surface and in communicating with putative intracellular components. In contrast, Hotary et al. (2000) observed that truncation of the MT1-MMP cytoplasmic tail did not affect the invasive phenotype of a cell line in a fibrin matrix. Hotary et al. concluded that either surface localization signals are no longer required when MT1-MMP is overexpressed or critical targeting information is encoded within the extracellular domain itself (Hotary et al., 2002). A single point mutation in a cysteine residue in the C-terminal domain of MT1-MMP has been reported to interfere with cell migration and invasion, possible by abolishing formation of MT1-MMP dimers (Rozanov et al., 2001). However, other studies disputed this observation (Gingras et al., 2001). Rozanov et al. (2001) concluded that the cytoplasmic tail is not required for locomotion of extremely migratory tumor cells and proposed that there may be two distinct mechanisms that affect cell locomotion: (1) the proteolytic activity of MT1-MMP facilitates cell motility, and (2) the cytoplasmic tail of MT1-MMP communicates with putative intracellular components leading to proteolytic-independent motility.

It has been reported that the cytoplasmic domain of MT1-MMP is essential for regulating the activity of this enzyme through a dynamin-mediated process of internalization in clathrin-coated vesicles (Jiang et al., 2001). Based on proMMP-2 activation, endocytosis downregulates MT1-MMP activity at the cell surface (Jiang et al., 2001) (Fig. 3, see color insert). These studies have been extended with the finding that the cytoplasmic tail-dependent internalization of MT1-MMP is important for its invasion-promoting activity (Uekita et al., 2001). Nakahara et al. (1998) and Ureña et al. (1999) have demonstrated that this region is critically involved in trafficking of MT-MMP to discrete regions of the cell surface. Further analysis of the putative internalization signals present in the cytoplasmic tail has shown that di-leucine and tyrosine residues from this region are essential for the MT1-MMP internalization process. Using the yeast two hybrid system to detect associated intracellular proteins, Uekita et al. demonstrated that the cytoplasmic domain of MT1-MMP (LLY[573]) was a binding site for a component of clathrin-coated pits, the μ_2 subunit of the adapter protein 2 (AP-2). Uekita et al. (2001) suggested that turnover of MT1-MMP at the adherent edge of the cell is an important step in regulating the enzyme during cell migration and invasion. Some of these sequence motifs are present in MT2- and MT3-MMP and are also effectively internalized from the cell surface, but are absent in MT5-MMP that is internalized less efficiently. Likewise, the GPI-anchored MT-MMPs (MT4-MMP and MT6-MMP), lacking a cytoplasmic tail, are not internalized efficiently (Uekita et al., 2001).

The cytoplasmic tail of MT1-MMP also appears to mediate specific interactions with intracellular proteins. Peptides derived from the cytoplasmic domain are capable of binding specifically to the p32/gC1q-R multifunctional (complement receptor); this interaction is proposed to result in directional trafficking of

MT1-MMP from the Golgi network to the plasma membrane (Rozanov *et al.*, 2001). Likewise, the cytoplasmic tail of MT1-MMP also appears to bind the Golgi protein p59/GRASP55, and this association may be important in the intracellular trafficking of the proteinase (Kuo *et al.*, 2000). It has been pointed out that the cytoplasmic tail of MT1-MMP contains potential phosphorylation sites that could be involved in the recruitment of intracellular proteins that drive the specific localization of MT1-MMP (Ellerbroek and Stack, 1999).

Homophilic complex formation of MT1-MMP on the cell surface has been demonstrated; this mechanism fits the requirement for two adjacent MT1-MMP molecules cooperating to cleave one proMMP-2 molecule. A cysteine residue (Cys^{574}) present in the cytoplasmic tail of MT1-MMP is apparently involved in an intermolecular disulfide bond linking monomers of this proteinase and generating stable covalent dimers of MT1-MMP on the cell surface (Kazes *et al.*, 2000; Rozanov *et al.*, 2001). In contrast, Itoh *et al.* (2001) proposed that the hemopexin-like domain of MT1-MMP is responsible for homophilic complex formation. Lehti *et al.* (2002) concluded that both the hemopexin-like and cytoplasmic domains of MT1-MMP are involved in formation of enzyme oligomers that function in intermolecular proteolytic events at the cell surface. The MT1-MMP fragment remaining on the cell surface after cleavage and deletion of the catalytic domain acts as an inhibitor of native MT1-MMP function on the cell surface by interfering with the oligomerization of wild-type MT1-MMP molecules (Lehti *et al.*, 2002). Overall *et al.* (2000), however, were unable to demonstrate homodimer formation of MT1-MMP C-terminal hemopexin domains.

Since MT1-MMP has been reported to be localized to lamellipodia structures, and a GTP-bound form of Rac1 stimulates the generation of lamellipodia in cells, the interaction between these molecules has been examined. The results suggest that the expression of Rac1DA generates ruffling membrane and forces MT1-MMP to localize at this site, thereby promoting homophilic complex formation and activation of proMMP-2 (Itoh *et al.*, 2001).

2. Crystal Structure of the MT1-MMP:TIMP-2 Complex

Fernandez-Catalan *et al.* (1998) have solved the 2.75-Å crystal structure of the complex between the catalytic domain of human MT1-MMP and bovine TIMP-2. Apart from exhibiting the classical MMP fold observed between TIMP-1 and soluble MMPs (Gomis-Ruth *et al.*, 1997), the catalytic domain of MT1-MMP displayed two large insertions remote from the active-site cleft that might be important for interactions with macromolecular substrates. The TIMP-2 proteolytic chain folds into a continuous wedge; the AB edge loop is much more elongated and tilted, wrapping around the S-loop and β-sheet rim of the MT1-MMP molecule. The C-terminal edge of TIMP-2 also makes many interactions with MT1-MMP. The substrate binding region of MT1-MMP looks similar to that of classical MMPs. Its voluminous S1$'$ pocket is much larger than that needed for accommodation of a P1$'$ Ile/Leu side chain as in MMP-3 which is in agreement with the observed cleavage preference of MT1-MMP for peptide substrates with long bulky P1'-analogs (Fernandez-Catalan *et al.,* 1998; Mucha *et al.,* 1998). Figure 4 (see color insert) displays a stereo view of the ribbon representation of the three dimensional X-ray structure of the catalytic domain of MT1-MMP. A computational model displaying different domains of MT1-MMP is displayed in Fig. 5 (see color insert). To date, information on the 3D structure of other members of the MT-MMP subfamily of MMPs has not been reported.

D. Evolutionary Analysis of MT-MMPs in the Context of the MMP Family

As discussed in the previous section, MT-MMPs exhibit many structural peculiarities when compared with the remaining MMP family members. Therefore, it is likely that these membrane proteinases diverged early in the evolutionary history of MMPs. In fact, dendrogram analysis of MMPs shows that MT-MMPs form a distinct evolutionary branch in the MMP phylogenetic tree (Fig. 6). This proposal on the early evolutionary divergence of MT-MMPs from MMPs is also supported by a number of observations derived from analysis of the structure, exon–intron organization, and chromosomal localization of these genes. In relation to the first aspect, it is widely accepted that all MMPs have been generated by several duplications of an ancestor gene. The presence of additional domains in MT-MMPs could be explained by an exon shuffling process as in the case of the MMP-2 and MMP-9 genes, where the novel domains are added by modular assembly of new exons (Huhtala *et al.,* 1990, 1991). However, gene structure analysis of MT-MMPs has revealed that the new domains characteristic of these membrane proteinases have not been added through this mechanism (Apte *et al.,* 1997; Lohi *et al.,* 2000; Velasco and López-Otín, unpublished data). Thus, the transmembrane and cytoplasmic domains of MT1-, MT2-, MT3-, and MT5-MMP are encoded in the same large exon that codes for the last hemopexin repeat of these enzymes. Likewise, the hydrophobic regions of the GPI-anchored MT-MMPs are also incorporated in the same exon coding for the fourth hemopexin repeat of MT4- and MT6-MMP. A more detailed analysis of the gene structure of MT-MMPs indicates that in addition to conserved exon–intron boundaries shared with MMPs, there are also some splicing sites specific for MT-MMPs. For example, in all of these genes an additional intron divides the prodomain coding exon in two parts. Furthermore, the fourth exon of MT1-MMP is very long and contains sequences usually encoded in two separate exons in other MMP genes.

The promoter regions of the diverse MT-MMP genes also have distinctive structural and functional features compared with other MMP genes, including the absence of the TATA box sequence and AP-1 binding site present in the proximal promoter of most MMPs (Lohi *et al.,* 2000; Pendás *et al.,* 1997). Finally, chromosomal localization studies have shown that these membrane-bound enzymes are encoded by genes widely dispersed in the human genome (Table I), and therefore,

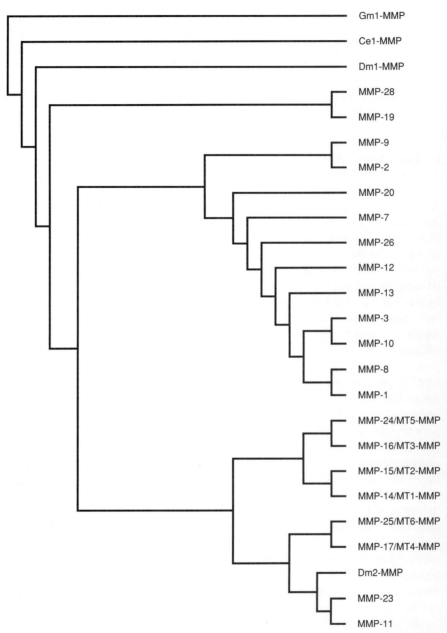

Figure 6 Dendrogram analysis of MMPs. MT-MMPs form a distinct evolutionary branch in the MMP phylogenic tree.

they are not part of the MMP gene cluster on chromosome 11q22 that contains at least eight members of this gene family (Pendas *et al.,* 1995). These data, together with the previously indicated differences in gene structure and organization, strongly support the proposal that MT-MMPs have undergone an early evolutionary divergence from the other MMPs.

An additional view of the evolutionary history of MT-MMPs can be obtained from the analysis of the presence of these membrane-anchored proteases in diverse organisms. In this regard, it is remarkable that orthologous enzymes for all six human MT-MMPs have been found in mouse (English *et al.,* 2000; Okada *et al.,* 1995; Pei, 1999; Tanaka *et al.,* 1997a; Velasco and López-Otín, unpublished results). This finding is interesting in light of studies showing that some MMPs have evolved somewhat differently in mice and humans (Balbín *et al.,* 2001). MT-MMPs have been also identified in many other mammalian species, as well as in birds and in fish (Kimura *et al.,* 2001; Yang *et al.,* 1996). However, in addition to these studies on vertebrate MMPs, the identification of putative MT-MMPs in more distantly related organisms in the phylogenetic tree, such as invertebrates and plants, should be very interesting and should aid in reconstructing the evolution of these enzymes. These studies would allow us to look for those functions that have been acquired or those that have been lost in the different organisms as well as to compare those MT-MMP functions which can be common in our heritage. The recent availability of the complete genome sequence of several eukaryotic organisms including *Caenorhabditis elegans, Drosophila melanogaster,* and *Arabidopsis thaliana* can be very useful in such studies. All these model organisms contain several MMPs in their genome structures, and although the presence of MT-MMP family members has not been formally demonstrated yet, our recent analysis of the MMPs identified in these organisms strongly suggests that MT-MMPs are present in both invertebrates and plants. This is the case with *Caenorhabditis elegans,* an extremely useful model organism because of its short life cycle, small genome, and ease of manipulation. In fact, because of its transparency all cells can be seen and followed during development, and a continuously growing number of mutant strains incorporating diverse phenotypes are being developed. Our analysis of the genome of this nematode has shown the presence of at least seven MMP-related sequences, including the three family members described by Wada *et al.* (1998). One of these sequences, called Y19-MMP, is likely a GPI-type MT-MMP because of the hydrophobicity pattern of the amino acid sequence of its C-terminal end. However, despite extensive searches, no transmembrane-type MT-MMP encoding genes were identified in the genome of this nematode. A similar situation can be found in the case of *Drosophila melanogaster,* a model organism that has been essential for the study of developmental biology. The fly MMP system is extremely simple. We have identified and characterized two *Drosophila* MMPs, *Dm1*-MMP and *Dm2*-MMP (Llano *et al.,* 2000, 2002, and unpublished data). Both enzymes have a domain organization identical to archetypal vertebrate MMPs including signal sequence, prodomain, catalytic domain, and hemopexin repeats.

Interestingly, both *Drosophila* MMPs contain a C-terminal extension that in the case of *Dm2*-MMP is predicted to act as a GPI-anchoring sequence. It is also remarkable that both *Dm2*-MMP and Y19-MMP contain a furin-activation sequence, another structural feature of MT-MMPs. Based on these observations, we can conclude that MT-MMPs are also present in invertebrates and their study can illuminate novel functions of human MT-MMPs. In addition, and because the only MT-MMPs found to date in invertebrates belong to the GPI-type, it is tempting to speculate that these enzymes may be ancestors of transmembrane-type MT-MMPs during evolution (Kojima *et al.,* 2000).

Nevertheless, the description of the complete genome sequence of *Arabidopsis thaliana* and the finding of MMPs in this organism (Maidment *et al.,* 1999) allow us to perform a comparative analysis of this proteolytic system across the plant and animal kingdoms. Despite the long independent evolution of these organisms from a common unicellular ancestor that they shared 1.6 billion years ago, at least five MMPs can be recognized in *Arabidopsis*. All show a simple domain structure lacking the C-terminal hemopexin domain, but four of them show a C-terminal hydrophobic region predicted to act as a transmembrane domain or to mediate GPI-based modifications. Therefore, it seems likely that these enzymes are anchored to the cell membrane. MMPs have been also found in other plant species, such as soybean and cucumber, but none of them appear to belong to the MT-MMP subfamily.

In summary, we can conclude that comparative studies of MMPs in vertebrates, invertebrates, and plants have revealed the presence of membrane-bound proteases belonging to this family in all of them. So, it is likely that there are some universal proteolytic functions in all these organisms which must be performed at the cell surface through the action of similar enzymes. The invertebrate MT-MMPs are closely related to vertebrate GPI-type MT-MMPs, whereas membrane-anchored MMPs from plants are only distantly related to MT-MMPs. However, it is also evident from these comparative studies that there is a strikingly increased complexity of the MT-MMP system in vertebrates, both in the number of proteins belonging to the family and in the domain structure and anchoring systems used by these enzymes for membrane binding. Presumably, this increased complexity of MT-MMPs in vertebrates is necessary to deal with a series of vertebrate innovations including those derived from the occurrence of more complex cell–cell and cell–matrix interactions in vertebrates, or from the prominent development of complex immunological, neural, and vascular systems in these organisms. In agreement with this possibility, some MT-MMPs have been reported to play important roles in the modulation of neovascularization processes (Hiraoka *et al.,* 1998; Zhou *et al.,* 2000). Functional analysis of the MT-MMP system in both human tissues and animal models, as described for MT1-MMP (Zhou *et al.,* 2000), will be essential for further clarifying the roles of the diverse MT-MMPs in multiple physiological processes, as well as to elucidate the basis of their altered expression and function in diseases such as cancer and arthritis.

E. Human Chromosomal Localization of MT-MMPs

The MT1-MMP gene is located on chromosome 14q11-q12; the T-cell receptor-α locus is located in the same region (Mignon *et al.*, 1995). The gene of MT2-MMP is localized to chromosome 16q13; MT3-MMP is localized to 8q21; MT4-MMP is localized to 12q24; MT5-MMP is localized to 20q11.2; MT6-MMP is localized to 16p13.3 (Llano *et al.*, 1999; Mattei *et al.*, 1997; Puente *et al.*, 1998; Velasco *et al.*, 2000). Other MMP genes have been mapped to chromosome 11 (MMP-1, MMP-3, MMP-7, MMP-8, MMP-10, MMP-12, MMP-13, MMP-20), chromosome 16 (MMP-2, MMP-9), and chromosome 22 (MMP-11). The diverse localization of MT-MMPs further supports the notion that they may have diverged early in the evolutionary process.

F. Transcriptional Regulation of MT-MMPs

Induction and regulation of the expression of MT1-MMP by growth factors and cytokines has been reported (Foda *et al.*, 1996; Lohi *et al.*, 1996; Yang *et al.*, 1996); these include IL1β, EGF, TNFα, bFGF, GM-CSF (Tomita *et al.*, 2000), scatter factor/hepatocyte growth factor (SF/HGF) (Hamasuna *et al.*, 1999; Kadono *et al.*, 1998; Wang and Keiser, 2000), and TGF-β. TNF-α, IL-1α, and IL-1β cause a time dependent increase in the steady-state MT1-MMP mRNA level within 4 h of exposure; levels remain elevated for 12 h in endothelial cells (Rajavashisth, *et al.*, 1999). The TNF-α effect on MT1-MMP only occurred when fibroblasts were embedded in type I collagen, indicating that the interaction between matrix and cytokine signals is required in a tissue environment (Han *et al.*, 2000). Some of these factors also function as mitogens, morphogens, and migration factors for a variety of cell types. Rapid tyrosine phosphorylation of c-met, the SF/HGF receptor, and MAPK/Erk serve as components of the signal transduction response to SF/HGF (Murakami *et al.*, 1999; Wang and Keiser, 2000). Recruitment and activation of PI-3 kinase and STAT pathways have also been implicated in the SF/HGF effect (Kadono *et al.*, 1998). A reduction in TIMP-2 production occurred as well (Hamasuna *et al.*, 1999).

Whereas all-*trans* retinoic acid is inhibitory for the expression of most MMPs, MMP-13, MT1-MMP and other MMPs involved in bone formation are induced by retinoic acid and its derivatives. The production of these MMPs is concomitant with the development of a retinoic acid-induced osteogenic differentiation program mediated by Cbfa1, a transcriptional factor of the runt gene family, which requires the participation of a signaling pathway involving the activity of p38 MAPK. Putative Cbfa1 elements in the MT1-MMP promoter were also observed (Jimenez *et al.*, 2001).

Knowledge of the regulatory elements of the MT-MMP promoter sequences is limited (Knauper and Murphy, 1999). Haas *et al.* (1999) have characterized the murine MT1-MMP promoter gene in which there are consensus binding sites for transcriptional factors including Sp1, Egr-1, AP-4, NF-kB, and c-ETS1, but not AP-1, AP-2, and stress response elements such as HSE. Analysis of the 5'upstream region of the MT1-MMP gene revealed that it lacks the typical TATA box and transcriptional factor binding sites that represent the target of cytokines and growth factors; in contrast, most other MMP genes contain these elements and are regulated strongly by AP-1, AP-2, and TGF-α responses. Increased binding of the transcription factor Egr-1 to the MT1-MMP promoter displaces Sp1 from this site, resulting in induction of MT1-MMP and initiating the invasive phenotype of endothelial cells in response to the ECM environment (3D collagen) (Haas *et al.*, 1999) (see Chapter 13 for additional details). ProMMP-2 and TIMP-2 genes also lack a TATA box immediately 5'upstream of their gene (Seiki, 1999). It has been proposed that activation of NF-kB is essential to induce MT1-MMP expression in fibroblasts after TNF-α exposure (Han *et al.*, 2000). A potential p65 NF-kB binding site in the 5'-flanking region of the human and mouse MT1-MMP gene has been identified (Haas and Madri, 1999; Han *et al.*, 2000). In contrast to this information on the mechanisms controlling MT1-MMP expression, no information regarding the regulation of expression of the remaining MT-MMP family members has yet been published.

G. Other Regulatory Mechanisms Controlling MT-MMPs

1. Identification of Genes Capable of Regulating MT1-MMP Function

Miyamori *et al.* (2001) used an expression cloning strategy to screen for genes that modulate the activation of proMMP-2 by MT1-MMP. Expression of claudin-5 (also claudin-1, -2, -3), an endothelial specific tight junction protein, was identified which replaced TIMP-2 for mediating proMMP-2 activation by all MT-MMPs, including mutants lacking the transmembrane domain. Direct interaction of claudin-1 with MT1-MMP and MT2-MMP was demonstrated and required that the ectodomain of claudin interact presumably with the catalytic domain of these MT-MMPs. It was suggested that claudin recruits all MT-MMPs and proMMP-2 on the cell surface to achieve elevated focal concentrations, and consequently enhanced activation of proMMP-2.

Nakada *et al.* (2001) screened a human fetal kidney library for genes that regulate MT1-MMP function. They demonstrated that proMMP-2 activation by MT1-MMP or MT3-MMP was inhibited by cotransfection of a cDNA derived from alternative splicing of the testican 3 gene that encodes the N-terminal 313 amino acid of a calcium binding proteoglycan with a three amino acid substitution at the C-terminal domain (designated N-testican). Expression of N-testican and testican 3 was detected in normal brain, but was downregulated in brain cancer. These observations, supported by cell transfection experiments, suggest that N-testican and testican 3 interfere with tumor invasion by inhibiting the function of MT-MMPs (Nakada *et al.,* 2001); the mechanism of inhibition remains to be determined.

2. Shedding/Release of MT-MMP from Cells

Proteolytic ectodomain processing of MT-MMP on the cell surface represents an important mechanism for regulating enzyme activity. Active MT1-MMP undergoes autocatalytic processing on the cell surface leading to the formation of an inactive 43/45 kDa fragment remaining on the cell surface and the release of soluble catalytic domain fragments; both autocatalytic (18 kDa) and nonautocatalytic (56, 50, 31–35 kDa) fragments have been identified. The shedding of the major 18 kDa fragment is initiated by cleavage at the G284-G285 site followed by cleavage between the conserved A255 and I256 residues near the conserved methionine turn, a structural feature of the catalytic domain of all MMPs. Thus, autocatalytic shedding evolves a mechanism to terminate MT-MMP activity on the cell surface, by disrupting enzyme integrity at a vital structural site (Toth *et al.,* 2002).

MT5-MMP tends to shed from the cell surface as a soluble enzyme (Pei, 1999; Wang and Pei, 2001). The shedding mechanism is evidently not mediated by the TACE or TACE-related ADAMs since the synthetic inhibitor of TACE, BB-94, failed to block the shedding process (Wang and Pei, 2001). Instead, a furin-type convertase is apparently involved in the shedding process through a cryptic furin recognition motif between the hemopexin domain and the transmembrane domain (Pei, 1999; Wang and Pei, 2001). This shedding mechanism appears to be unique to MT5-MMP since other MMPs lack such a cryptic furin motif in the analogous positions.

Rapid shedding of membrane vesicles containing proMT1-MMP and active and proMMP-2/MMP-9 from endothelial cells has been described. Serum and angiogenic factors (VEGF, FGF-2) stimulate shedding of MMPs as vesicle components. Addition of vesicles to endothelial cells resulted in autocrine stimulation of cell invasion through a reconstituted basement membrane (Tarboletti *et al.,* 2002).

Homology screening for human MT-MMPs has resulted in the identification of a minor cDNA encoding a soluble type of MT3-MMP lacking the transmembrane domain which is considered to be an alternatively spliced variant of MT3-MMP (Shofuda *et al.,* 1997). Soluble MT3-MMP has the same proteolytic effect on extracellular matrix substrates and proMMP-2 as native MT3-MMP and is more readily inhibited by TIMP-2 than TIMP-1; a physiologic role for the soluble enzyme was proposed (Matsumoto *et al.,* 1997).

3. Soluble Posttranscriptional Regulators of MT-MMPs

The role of growth factors and cytokines in transcriptional regulation of MT-MMPs has been described above. Zucker *et al.* (1995) and other investigators demonstrated that physiologic concentrations of thrombin activate proMMP-2 secreted by endothelial cells and smooth muscle cells; minimal activation occurred in the absence of cells. This effect requires the proteolytic site of thrombin, was inhibited by TIMP-2 (Zucker *et al.,* 1995), and involved thrombin induction of cellular expression of MT1-MMP (Lafleur *et al.,* 2001b). Pro-MMP-2 activation was not mediated by the classical thrombin receptors, PAR-1, PAR-2, or PAR-4 (Galis *et al.,* 1997; Zucker *et al.,* 1995) (refer to Chapter 11 by Bahou for further details of PAR mechanisms). Thrombin induction of synthesis of MMP-1 and MMP-3 was, however, mediated by the thrombin receptor PAR-1 (Duhamel-Clerin *et al.,* 1997). Coculture of smooth muscle cells and pericytes with endothelial cells results in suppression of endothelial cell activation of proMMP-2, presumably as a result of increased production of TIMP-2 and TIMP-3 by perivascular cells (Lafleur *et al.,* 2001a). These authors have also reported that the 63 kDa thrombin-activated MMP-2 is distinct from the 62 kDa species found following standard proMMP-2 activation protocols. Activation occurs in a two-step procedure with MT-MMP making the first cleavage in proMMP-2 to yield a 64 kDa intermediate followed by a second unique thrombin cleavage (Lafleur *et al.,* 2001b). Vascular endothelial growth factor (VEGF) affects the thrombin-MMP mechanism in endothelial cells by induction of tissue factor. Incubation of VEGF-treated endothelial cells in the presence of coagulation factors VIIa, Va, prothrombin, and lipids resulted in the generation of thrombin and subsequent activation of endothelial secreted proMMP-2 (Zucker *et al.,* 1998b).

The importance of thrombin enhancement of MMP function in tumor angiogenesis has been emphasized (Chiarugi *et al.,* 2000). Nguyen *et al.* (2000) proposed an interesting hypothesis wherein thrombin activation of proMMP-2 caused short-lived disruption of existing capillary beds, thus allowing endothelial cells to make contact with type I collagen, which then produced upregulation of MT1-MMP expression followed by proMMP-2 activation. Activated protein C, a serine protease that plays a physiologic role as an anticoagulant, has been reported to directly activate proMMP-2, even in the absence of cells (Nguyen *et al.,* 2000). It was proposed that thrombin-induced activation of proMMP-2 may be mediated in a three-step mechanism involving (1) binding of thrombin to thrombomodulin on the endothelial cell surface, (2) thrombin-induced activation of protein C, and (3) protein C activation of proMMP-2.

4. Alternative Mechanisms Involved in MT-MMP and MMP-2 Activation

Soluble latent MT1-MMP released by mesangial cells can be activated by urokinase, but not tissue plasminogen activator, leading to the activation of proMMP-2. In these experiments, the mechanism of shedding of MT1-MMP from the cell surface was not identified (Kazes *et al.,* 1998). It has also been reported that neutrophil serine proteases, elastase, cathepsin G, and proteinase-3, acting in concert with MT1-MMP, are capable of activating proMMMP-2 (Shamamian *et al.,* 2000).

Heat shock-mediated transient increase in intracellular cyclic AMP has been reported to result in tumor specific suppression of MT1-MMP production and proMMP-2 activation (Sawaji *et al.,* 2000).

H. Modulation of Functional Activities of MT-MMPs by Interaction with Other Molecules

1. MT-MMP Interactions with ECM Proteins

Proteins of the extracellular matrix such as collagen (Azzam and Thompson, 1992; Seltzer *et al.,* 1994; Tomasek *et al.,* 1997), laminin (Sang *et al.,* 1991), fibronectin (Stanton *et al.,* 1998), tenascin (Jones and Jones, 2000), bone matrix proteins such as osteopontin, bone sialoproteins (Gilles *et al.,* 2001; Teti *et al.,* 1998), and SPARC (Gilles *et al.,* 1998) regulate a number of cellular mechanisms, including cell adhesion, spreading, differentiation, migration, and apoptosis (Maquoi *et al.,* 2000b). Changes in the cell cytoskeleton affect MT-MMP synthesis. Certain cells grown in an anchored three-dimensional (3D) collagen matrix produce low levels of MT1-MMP; synthesis is upregulated when the collagen gel relaxes (Tomasek *et al.,* 1997). In contrast, endothelial cells produce MT1-MMP and activate proMMP-2 upon 3D culture in type I collagen; no effect was noted on relaxation of the collagen gel (Haas *et al.,* 1998). The degree of proMMP-2 activation has been related to the level of cell surface MT1-MMP and fibrillar collagen deposition (Ruangpanit *et al.,* 2002); a residual MT1-MMP-independent action of proMMP-2 was also noted, but the mechanism was unexplained (Ruangpanit *et al.,* 2002). 2D type IV collagen enhanced the activation of proMMP-2 when cells were plated on coated plastic, but in contrast to 3D type I collagen, transcriptional modulation of MT1-MMP, MMP-2, or TIMP-2 expression was not noted. In contrast to MT1-MMP, MT3-MMP and MT4-MMP are not upregulated by type I collagen.

Plating cells on fibronectin leads to MT1-MMP activation in fibrosarcoma cells without apparent change in MT1-MMP or TIMP-2 levels (Stanton *et al.,* 1998; Tomasek *et al.,* 1997). Fibronectin was only effective when the cells were plated onto a film, but not in solution. The 120 kDa binding fragment of fibronectin was also an effective stimulus for proMMP-2 activation. An activating antibody to the α_5 integrin that recognizes this region of fibronectin and mimics ligand occupancy, and an antibody to the β_1 integrin cause integrin clustering also caused activation of proMMP-2. The ligation of other integrins to other extracellular matrix components may have comparable effects including α_2 and α_3 in rhabdomyosarcoma cells, α_2 in dermal fibroblasts (Zigrino *et al.,* 2001), $\alpha_v\beta_3$ and $\alpha_6\beta_1$ in melanoma cells, and α_3 and α_5 in glioblastoma cells (Murphy *et al.,* 2000). Calreticulin, a ubiquitously expressed calcium binding intracellular protein, has been implicated in the signal transduction pathway of proMMP-2 activation involving α_3 integrin (Ito *et al.,*

2001). It should be emphasized that a number of secreted MMPs have also been associated with the consequences of signaling mediated by cell adhesion receptors. Potential interactions with MT-MMPs remain to be explored.

2. Role of Integrins and Other Adhesion Molecules in Function of MT-MMPs

Brooks *et al.* (1996) reported that intermediate-activated MMP-2 can further bind to $\alpha_v\beta_3$ integrin on the cell surface, thereby increasing the cell surface density of MMP-2 and facilitating full autocatalytic maturation of MMP-2. Bafetti *et al.* (1998), however, were unable to document direct binding of MMP-2 to $\alpha_v\beta_3$ through an RGD-mediated integrin region. It has been proposed that cross-talk between MT1-MMP and $\alpha_v\beta_3$ integrin clustered in discrete regions on the cell surface (invadopodia) enhances proMMP-2 activation at the cell surface and subsequent directional migration of cells. The collaboration between MT1-MMP and integrin $\alpha_v\beta_3$ occurs only if the two molecules are in close proximity on the same cell (Deryugina *et al.*, 2001). MT1-MMP dependent functional activation of $\alpha_v\beta_3$ correlated with modification of the β_3 subunit, resulting in a change in electrophoretic mobility and antibody binding (Deryugina *et al.*, 2000). Alternative processing of integrin α_v subunit in transfected tumor cells by MT1-MMP has been demonstrated (Ratnikov *et al.*, 2002). It is to be noted that intracellular α_v subunit processing at different cleavage sites is generally considered to be a function of furin. MT1-MMP specifically cleaves pro-α_v, generating a 115 kDa heavy chain with a truncated C-terminus and a 25 kDa light chain; MT1-MMP also cleaves α_3 and α_5 integrin chains, but not α_2. Cleavage specificity is largely defined by the presence of a hydrophobic residue at the P' position. Ratnikov *et al.* (2002) proposed that MT1-MMP-mediated integrin processing in the same cell may be a general mechanism by which cells selectively regulate the functionality of integrins in promoting cell adhesion, migration, and focal adhesion kinase phosphorylation. An alternative explanation is that $\alpha_v\beta_3$ leads to the release of mature MMP-2 from cells and the enzyme accumulates extracellularly before diffusing into the matrix. The hemopexin fragment derived from autocatalytic digestion of MMP-2 has been reported to act as a natural inhibitor of cancer cell proliferation, migration, and tumor angiogenesis (Bello *et al.*, 2001). Vitronectin, a specific ligand for $\alpha_v\beta_3$ integrin, competitively blocked the integrin-dependent maturation of MMP-2. The integrin binding effect on MMP-2 activation is independent of changes in cell shape, although both factors often occur simultaneously (Yan *et al.*, 2000). Synthetic MMP inhibitors alter this mechanism by interfering with cleavage and activation of the β_3 integrin subunit (Deryugina *et al.*, 2000). $\alpha_5\beta_3$ integrin demonstrates approximately 10% efficiency as compared to $\alpha_v\beta_3$ in the MMP-2 activation mechanism. Excess TIMP-2, but not TIMP-1, interferes with this activation mechanism.

Another potential mechanism of cell surface localization (indirect) is MMP-2 binding to $\alpha_2\beta_1$ integrin-associated collagen (Steffensen *et al.*, 1998; Theret *et al.,* 1999). Pro-MMP-2 associated with integrin-bound collagen may act as a store of enzyme that can feed into the MT1-MMP pathway upon release from the ECM (Ellerbroek *et al.,* 1999). The ability of integrin clustering events on the cell surface to enhance MT1-MMP activity raises the possibility that activation of the cell surface MMP proteolytic cascade in migratory cells may be influenced not only by the type of matrix encountered, but also by its proteolytic status. For example, if the cell encounters an intact interstitial collagen barrier, collagen binding integrins would be expected to cluster along the intact fibrils, thereby promoting an MT1-MMP mediated proteolytic response. In contrast, if a cell encounters a degraded matrix, occupied integrins may not be sufficiently clustered, thereby leading to failure to stimulate proteolysis (Ellerbroek and Stack, 1999).

The ligation of $\alpha_4\beta_1$ integrin on mesenchymal cells by its co-receptor vascular cell adhesion molecule (VCAM) selectively upregulates MT1-MMP, increases activation of proMMP-2, and stimulates cell migration and invasion (Pender *et al.,* 2000). High expression of MT1-MMP and MMP-3 mRNA may be a critical factor in the migration of mesenchymal cells into ulcer beds in inflammatory bowel disease (Pender *et al.,* 2000).

Binding of β_1 integrin to 3D-collagen followed by integrin aggregation provides another pathway for MT1-MMP induced activation of proMMP-2. Rac 1, a member of Rho-related GTPases, modulates MT1-MMP mRNA expression and possibly the distribution on the cell surface leading to the activation of proMMP-2 and the collagenolytic phenotype of cells (Zhuge and Xu, 2001). It remains unclear whether ECM proteolysis is controlled by direct integrin signaling, by associated changes in cell shape, or by both.

Cell–cell contact has been reported to downregulate expression of MT1-MMP. MT1-MMP protein appears to diminish more rapidly than mRNA from cells following achievement of cell confluency (Tanaka *et al.,* 1997). Cell expression of E-cadherin, which is involved in cell:cell adhesion, has also been demonstrated to exert an effect on MMPs. Transfection of E-cadherin cDNA into squamous carcinoma cells led to a decrease in MT1-MMP mRNA as well as decreased amounts of activated MMP-2 (posttranslational effect), presumably as a result of suppression of the mitogen-activated protein kinase cascade and/or cytoskeletal organization (Ara *et al.,* 2000).

I. Effect of Pharmacologic Agents on MT1-MMP Function

Whereas numerous pharmacologic agents have been demonstrated to induce proMMP-2 activation, their mechanism of action has been elusive. Concanavalin A, the most widely studied inducer of MMP-2 activation (Overall and Sodek, 1990), was initially reported to function by enhancing transcription of MT1-MMP (Lehti *et al.,* 1998; Lohi *et al.,* 1996; Peracchia *et al.,* 1997; Tomasek *et al.,* 1997). Other investigators did not find enhanced expression of MT1-MMP with concanavalin A, or reported that increased expression of MT1-MMP did not correlate with MMP-2 activation (Lohi *et al.*, 1996; Shofuda *et al.*, 1998). An early and critical effect of concanavalin A was to induce clustering of MT1-MMP on the plasma membrane (Yu *et al.*, 1998a). More recent evidence suggests that concanavalin A increases proMMP-2 activation in HT1080 cells by apparently inhibiting the endocytosis of MT1-MMP from the cell surface (Jiang *et al.*, 2001). Cyclic AMP has been reported to suppress expression of MT1-MMP (Peracchia *et al.*, 1997; Yu *et al.*, 1998b). Monensin, a sodium ionophore, and orthovanadate, an inhibitor of protein tyrosine phosphatases, have been reported to mediate the activation of proMMP-2 secreted by cells in monolayers (Li *et al.*, 1997). Other reports have implicated protein tyrosine phosphorylation in the signal transduction pathways leading to activation of proMMP-2 (Li *et al.*, 1997). c-ras and v-src induced expression of MT1-MMP and invasiveness have been described (Thant *et al.*, 1997).

Several studies have demonstrated that the organization of the cell cytoskeleton is able to regulate proMMP-2 activation. Treatment of cells with cytochalasin D (Ailenberg and Silverman, 1996), which causes disruption of actin stress fibers, resulted in increased expression of MT1-MMP and proMMP-2 activation (Tomasek *et al.*, 1997). Cytochalasin D also induces a rapid activation of proMMP-2 in brain cancer cells that was independent of protein synthesis and coincided with the appearance of 43 kDa processed MT1-MMP (Gingras *et al.*, 2000). Gingras *et al.* concluded that posttranslational changes in the actin cytoskeletal structure of cancer cells provide a rapid stimulus triggering the degradation of ECM proteins through the MT1-MMP/MMP-2 activation mechanism.

Phorbol ester induced activation of proMMP-2 is a protein kinase C-dependent mechanism which is plasma membrane-mediated and is accompanied by increased MT1-MMP expression (Foda *et al.*, 1999). Increasing intracellular calcium levels with ionomycin or thapsigargin inhibits concanavalin A, TNFα, or phorbol ester induced proMMP-2 activation, presumably indicating a role for calcium in the processing of MT1-MMP (Atkinson *et al.*, 2001; Murphy *et al.*, 2000). A calmodulin antagonist stimulates MT1-MMP expression, consistent with the role of calmodulin as an inhibitor of matrix hydrolysis and MMP expression (Ito *et al.*, 1998).

Activation of monocytes by lipopolysaccharide resulted in the induction of MT1-MMP mRNA and protein synthesis that was able to be suppressed by inhibitors of prostaglandin synthesis, adenyl cyclase, and protein kinase A. Reversal of suppression by prostaglandin E2 and dibutyryl cyclic AMP, respectively, indicates regulation through a prostacyclin-cAMP pathway (Shankavaram *et al.*, 2001). Resting monocytes lacked MT1-MMP function.

J. Substrate Specificity of MT-MMPs

1. MT-MMP Mediated Activation of ProMMPs

The importance of pericellular activation and function of MMPs in proximity to the cell surface has been recognized for a number of years (Kalebic *et al.*, 1983; Zucker *et al.,* 1985a,b). The first substrate identified for MT1-MMP was proMMP-2 (Sato *et al.,* 1994; Strongin *et al.,* 1995). Based on evidence presented so far, the likely mechanism involves stoichiometric binding of TIMP-2 to the catalytic domain of MT1-MMP on the cell surface followed by the binding of the C-terminal domain of proMMP-2 to the C-terminus of TIMP-2, resulting in a trimolecular complex (Seiki, 1999). A second active MT1-MMP molecule on the cell surface then cleaves proMMP-2, leaving highly focused active MMP-2 available for efficient substrate degradation and participation in other events (Butler *et al.,* 1998; Sato *et al.,* 1994; Strongin *et al.,* 1995; Zucker *et al.,* 1998a) (Fig. 2). TIMP-3, but not TIMP-1, can substitute for TIMP-2 in this process (Lafleur *et al.,* 2001a). Whereas small amounts of TIMP-2 or TIMP-3 are required for pericellular proMMP-2 activation, excess inhibitor abrogates activation. A strong interaction between the TIMP-2 C-terminal domain and the MMP-2 hemopexin domain leads to burying epitope sites within the TIMP-2 C-terminal tail. Overall *et al.* (2000) proposed that following the binding of TIMP-2 to the hemopexin domain of MMP-2, the N-terminal domain of TIMP-2 is aligned away from the active site of MMP-2 and therefore is available for binding MT1-MMP in the activation complex.

Activation of proMMP-2 has been correlated with the appearance of 57/60 kDa and 43/45 kDa processed forms of MT1-MMP on the surface of cells (the size of these proteins measured by SDS–PAGE has varied slightly in different reports) (Fig. 3). The 57/60 kDa (prodomain deleted) form represents activated MT1-MMP (Lehti *et al.,* 1998). Knauper and Murphy (1999) suggested that MT-MMP activation occurs intracellularly followed by the formation of TIMP-2 complexes. The 43/45 kDa form of MT-MMP is generated by MMP-dependent proteolysis of the 57/60 kDa form and lacks the catalytic domain function. Controversy exists as to whether the 43/45 kDa form is generated through MT1-MMP intermolecular cleavage or through the activity of another MMP, such as MMP-2 (Stanton *et al.,* 1998). Employing recombinant virus to regulate the delivery of cDNAs to cells, full-length TIMP-2 was demonstrated to regulate the amount of active MT1-MMP on the cell surface. Thus, TIMP-2 may act as a positive regulator of MT1-MMP on the cell surface and, consequently, may support pericellular proteolysis (Hernandez-Barrantes *et al.,* 2000). In the absence of TIMP-2, MT1-MMP undergoes rapid intermolecular autocatalytic cleavage to the 43/45 kDa form which lacks the entire catalytic domain; the 43/45 kDa form then undergoes further processing or accumulates on the cell surface as an inactive stable degradation product. TIMP-4, an efficient MT1-MMP inhibitor with the capacity to interfere with autocatalysis of active MT1-MMP, is not able to substitute for TIMP-2 in the activation of proMMP-2 (Toth *et al.,* 2000). Many studies performed *in vitro* showed that soluble forms of MT1-MMP were able to activate proMMP-2 in the absence of TIMP-2 (Butler *et al.,* 1998; Imai *et al.,* 1996; Jo *et al.,* 2000; Pei and Weiss, 1996; Sato *et al.,* 1996; Will *et al.,* 1996). This inactive fragment may compete with the formation of homooligomers of intact MT1-MMP molecules and inhibit proMMP-2 inactivation (Itoh *et al.,* 2001; Lehti *et al.,* 2002).

The initial propeptide cleavage in 72 kDa proMMP-2 is at the Asn^{37}-Leu^{38} peptide bond resulting in a 64 kDa intermediate.[1] The second propeptide cleavage to 62 kDa MMP-2 is autoproteolytic (Murphy *et al.,* 2000) and occurs in trans by intermolecular cleavage by an already activated, fully mature MMP-2 molecule, but not by an intramolecular cleavage or in cis (Deryugina *et al.,* 2000; Overall *et al.,* 2000). Although intracellular activation of proMMP-2 has been reported, the mechanism was not delineated (Lee *et al.,* 1997).

In some types of cells, two modes of TIMP-2 binding to the cell surface were identified: hydroxamate inhibitor-sensitive and -insensitive. TIMP-2 binding to the plasma membrane in an hydroxamate-insensitive manner specifically inhibits MMP-2 upon activation; the presence of collagen interferes with this effect (Itoh *et al.,* 1998). Based on these observations, it was proposed that the pericellular activity of MMP-2 is tightly regulated by membrane-bound TIMP-2 and surrounding ECM molecules.

It has been shown that in cells not producing detectable levels of TIMP-2, MT1-MMP can efficiently initiate activation of proMMP-2 to its intermediate form, but further processing to active MMP-2 does not occur without addition of TIMP-2 (Miyamori *et al.,* 2001). This suggests that the final autocatalytic activation step has a greater requirement for the MT1-MMP:TIMP-2:proMMP-2 triad rather than the initial attack by MT1-MMP. The initial attack appears to be through the promotion of direct ecto-site interactions between the MT1-MMP catalytic domain and proMMP-2 (English *et al.,* 2001a), thus raising the question whether TIMP-2 participates in the activation process for proMMP-2 binding or conversion of intermediates to final active products. Additional investigations are required to characterize the triplex on the cell surface and resolve the role of the triplex in MT1-MMP-mediated proMMP-2 activation.

The biochemical mechanism for the release of activated MMP-2 from the triplex is perplexing when one considers the dissociation kinetics; binding affinities, however, may shift following proMMP-2 activation (Ellerbroek *et al.,* 1999). ProMMP-2 is activated less efficiently by cells expressing MT2-MMP and MT3-MMP, as compared to MT1-MMP (Seiki, 1999). Although the levels of MT-MMPs translated in the cells were similar, the relative levels of cell surface associated enzyme were different with MT1-MMP > MT3-MMP ≫ MT2-MMP (Seiki, 1999). A second proMMP-2 binding site on the cell surface, independent of the MT2-MMP complex, has also been proposed (Jo *et al.,* 2000). An alternative TIMP-2 independent pathway for efficient MMP-2 activation involving MT2-MMP has been demonstrated in cells derived from TIMP-2 knockout mice; the C-terminal hemopexin domain of MMP-2 was required in this process (Morrison *et al.,* 2001). Despite being able to bind TIMP-2, MT4- and MT6-MMP are unable to initiate

---

[1]The common designation of molecular mass for MMPs with gelatinolytic activity (MMP-2, MMP-9) is based on the migration of the proteinase in a gelatin impregnated SDS–PAGE (gelatin zymography). The apparent kDa of gelatinases do not conform to the anticipated mass based on amino acid composition.

or are very inefficient in initiating proMMP-2 cleavage (English *et al.,* 2001a,b; Fernandez-Catalan *et al.,* 1998).

ProMMP-13 is also activated *in vitro* by soluble MT1-MMP in a two-step cleavage mechanism; TIMP-2 and TIMP-3, but not TIMP-1, regulate the process. Initial cleavage of proMMP-13 is observed at the Gly^{35}-Ile peptide bond, followed by a secondary cleavage event resulting in the release of the propeptide domain (Knauper and Murphy, 1999; Knauper *et al.,* 1996). In cells treated with concanavalin A, MMP-2 participates in the activation of proMMP-13 by native MT1-MMP in two sequential cleavage steps. The involvement of the hemopexin domain of MMP-13 in this activation mechanism is uncertain (Murphy *et al.,* 2000). Murphy *et al.* (2000) speculated that proMMP-9 may also be activated within the MT-MMP cascade; support for this hypothesis is awaited.

2. Cleavage of Extracellular Matrix Proteins (ECM) by MT-MMPs

Although MT-MMPs were originally described because of their ability to activate proMMP-2, these membrane-bound proteinases also participate in matrix remodeling processes through their capacity to directly hydrolyze ECM substrates.

ECM degradation involves more than merely disrupting and remodeling structural barriers and scaffolds. The extracellular matrices influence cell behavior by sequestering and serving as a reservoir for signaling molecules, such as growth factors and growth factor binding proteins, and by acting as contextual ligands for cellular adhesion receptors that transduce signals to the cell interior. Furthermore, these matrices regulate basic processes such as cell shape, movement, growth, differentiation, and survival by controlling cell adhesion and the cytoskeletal machinery (Egeblad and Werb, 2002; Sternlicht and Werb, 2001). As a result of these effects, MMPs influence these cellular processes by altering the organization of the ECM and by release of sequestered biologically active proteins. Matrix proteolysis also results in the release of modular breakdown products or alteration of ECM components with biological activity (Davis, 1992). Furthermore, disruption of subcellular matrices can induce apoptosis in anchorage dependent cells, thereby affecting normal physiologic cell death in involuting tissues (Sternlicht and Werb, 2001).

MT1-MMP and MT2-MMP are potent proteinases with the ability to cleave fibrillar collagens in the classical 3/4 and 1/4 manner; type I collagen is hydrolyzed more effectively than type II or III collagen (Ohuchi *et al.,* 1997). In contrast, MMP-1 digests type III collagen more efficiently than types I and II collagen. Quantitative comparison of the activity of soluble MT1-MMP and MMP-1 demonstrated that MT1-MMP is 5 to 7 fold less efficient at cleaving type I collagen. The possibility that these differences may be an artifact due to loss of Tyr^{112} at the N-terminus of MT1-MMP was considered (Ohuchi *et al.,* 1997). Another consideration is that the membrane-bound form of MT1-MMP is considerably more active than soluble recombinant MT1-MMP. D'Ortho *et al.* (1997) reported that the catalytic domain of MT1-MMP alone did not retain catalytic activity for collagens, but was active against most other substrates. The activity of MT1-MMP on type I collagen is synergistically increased by MMP-2. C-terminal deletion mutants of MT1-MMP lost the ability to hydrolyze the triple helical interstitial collagens, thus indicating that the mechanism of action of the enzyme involves the hemopexin-like domain, similar to findings with other interstitial collagenases (Knauper and Murphy, 1999). MT-MMPs also degrade aggrecan (Buttner *et al.,* 1997; Fosang *et al.,* 1998), dermatosulfate proteoglycan, nidogen, tenascin, perlecan, fibulins, large tenascin-C, fibronectin, fibrinogen, vitronectin, laminin-1, laminin-5, α_2-macroglobulin, and α_1-antiproteinase (d'Ortho *et al.,* 1997; Imai *et al.,* 1997; Kang *et al.,* 2001; Murphy *et al.,* 2000; Ohuchi *et al.,* 1997; Pei and Weiss, 1996; Wang *et al.,* 1999).

In the absence of plasminogen, endothelial cell surface MT1-MMP is capable of degrading fibrin and facilitating cell migration through a fibrin clot (Hiraoka *et al.,* 1998). Cleavage sites in both fibrinogen and cross-linked fibrin obtained by proteolysis with MT1-MMP (also MMP-3 and -7) were found to be in very close proximity to those obtained by plasmin on these substrates. Additional studies of fibroblasts obtained from MT1-MMP knockout mice revealed that MT1-MMP null fibroblasts exhibit a late compensatory fibrin-invasive activity. The MT1-MMP-independent proinvasive process was identified as due to MT2-MMP and MT3-MMP. However, MT4-MMP transfected cells lacked fibrin-invasive function. Thus, a triad of MT-MMPs working alone or in combination is sufficient to confer fibrin-invasive activity independent of the plasminogen activator–plasminogen axis (Hotary *et al.,* 2002). However, only MT1-MMP transfected cells, not MT2-MMP or MT3-MMP transfected cells, were able to accelerate invasion atop type I collagen gels (Hotary *et al.,* 2002). MT1-MMP, along with MMP-8, MMP-12, and MMP-13, is capable of rapidly degrading and inactivating factor XII (Hageman Factor) (Bini *et al.,* 1999; Hiller *et al.,* 2000). RGD integrin-binding sequences in fibrinogen chains which normally bind to platelet glycoproteins are cleaved by MT1-MMP, thus potentially leading to interference with platelet aggregation. The receptor of complement component 1q (gC1qr) is also susceptible to MT1-MMP proteolysis with the cleavage site localized within the structurally disordered loop connecting the β_3 and the β_4 strands of C1qR (Rozanov *et al.,* 2002). These combined data support the hypothesis that endothelial MT1-MMP may be important in an early phase of tumor angiogenesis, inflammation, arthritis, and atherosclerosis.

Recombinant MT2-MMP has activity against fibronectin, tenascin, laminin, aggrecan, and perlecan (d'Ortho *et al.,* 1997, 1998). Recombinant MT3-MMP degrades type III collagen and fibronectin (Matsumoto *et al.,* 1997). Truncated MT3-MMP, lacking the transmembrane and cytoplasmic domain, digested type III collagen into characteristic 3/4 and 1/4 fragments more efficiently than a similar deletion mutant of MT1-MMP. MT3-MMP also digested cartilage proteoglycan, gelatin, vitronectin, laminin-1, α_1-protease inhibitor, and α_2-M into almost identical fragments as MT1-MMP (Shimada *et al.,* 1999).

MT4-MMP is very poor at hydrolyzing components of the ECM in comparison with other MT-MMPs (English *et al.,* 2000). In common with MT1-MMP, the catalytic domain of MT4-MMP shows significant ability to degrade fibrinogen at the Aα and Bβ, but not the γ chain. MT6-MMP is closer in function to MMP-3 than MT1- and MT4-MMP in terms of substrate and inhibitor specificity, being able to cleave type IV collagen, gelatin, fibronectin, fibrin, and proteoglycans (English *et al.,* 2001b; Kang *et al.,* 2001; Pei, 1999). However, it differs from MMP-3 and MT1-MMP in its inability to cleave laminin-1, and unlike MMP-3 is not able to activate proMMP-9 (English *et al.,* 2001b).

Finally, it must be mentioned that most studies describing the enzymatic activity of MT-MMPs have employed soluble recombinant enzymes consisting of the isolated catalytic ($\pm$hemopexin) domains. It needs to be emphasized that membrane-inserted enzymes may have different substrate preferences from the full-length enzymes (Buttner *et al.,* 1997). In general, catalytic domains usually cleave fewer, rather than more, sites along the substrate. The determination of the particular site at which the actual cleavage might occur *in vivo* depends on the physical environment and substrate presentation of the molecule at the cell surface (Buttner *et al.,* 1997).

3. Cleavage of Cytokines and Growth Factors by MT-MMPs

Examples of soluble MMP-mediated release or activation of cytokines and growth factors include FAS ligand, tumor necrosis factor (TNFα), the ectodomain of FGF-1 receptor, heparin binding EGF, TGFβ, IL-1β, insulin-like growth factor binding proteins, and IL-8 (Parks *et al.,* 1998; Stamenkovic, 2000; Sternlicht and Werb, 2001). MMPs modulate the activity of the CC chemokine monocyte chemoattractant protein-3 by selective proteolysis to release an N-terminal tetrapeptide. N-terminal processing at positions 4-5 of stromal-cell-derived factor-1α and β (SDF-1) by MMP-2, other soluble MMPs, and MT1-MMP resulted in loss of binding to its cognate receptor CXCR-4. SDF-1α functions as an important chemokine for hematopoietic progenitor stem cell migration, for breast cancer metastasis, and as an inhibitor of HIV-1 entry into cells (McQuibban *et al.,* 2001). A common chemokine-binding exosite is present on the MMP-2 hemopexin C domain. Pro TNFα can be cleaved by MT4-MMP, but the physiologic relevance of this observation is unknown (English *et al.,* 2000; Murphy *et al.,* 2000).

4. Other Methods for Examining MT-MMP Substrates

Other methods for examining potential protease substrate specificity, including that of MT-MMPs, have emerged. These include substrate phage display libraries, positional-scanning peptide libraries, and mixture-based libraries. Although MMPs share many common features in their substrate consensus motifs including proline in P3, serine in P1, hydrophobic residues at P2, and leucine or methionine in P1$'$, the presence of subtle distinctions indicates that it might be possible to discriminate among MMPs with optimized peptide substrates (Turk *et al.*, 2001). It has been reported that MT1-MMP is more efficient in cleaving synthetic substrates containing unusual amino acids with extremely long side chains in their P1′ positions (Mucha *et al.*, 1998). Using a hexamer substrate phage library consisting of randomized amino acid sequences and then performing peptide cleavages, Ohkubo *et al.* (1999) demonstrated that MT1-MMP preferentially digested a synthetic substrate containing Pro in the P1 position. The consensus substrate sequence for MT1-MMP appeared to be P-X-G/P- at the P3-P1′ sites. Of interest, this sequence is present in human collagens and α2-M and is the site of substrate digestion by interstitial collagenases (Ohkubo *et al.*, 1999). In contrast, using mixture-based oriented octapeptide libraries, Turk *et al.* (2001) concluded that MT1-MMP preferred serine or alanine at P1, and leucine at P1′. They reported that the predicted optimal peptide was a significantly better substrate than the collagen peptide.

K. MT-MMP Sorting on the Cell Surface

Compartmentation of receptors and signaling intermediates into specialized membrane domains represents a mechanism by which cells expressing MT1-MMP can integrate extracellular signaling (Annabi *et al.*, 2001). Localization of MT-MMP at the cell surface in both clathrin-coated pits and caveoli has been reported; the mechanistic relationships between these different functional compartments remain to be described. Jiang *et al.* (2001) demonstrated that cell surface MT1-MMP is regulated by dynamin-dependent endocytosis in clathrin-coated pits; the cytoplasmic domain of MT1-MMP plays an important recognition role in the endocytic process. MT1-MMP was colocalized with clathrin on the plasma membrane and the early endosome antigen in the endosomes (Fig. 3). Wild-type MT1-MMP, not the cytosolic domain truncated mutant, internalized into intracellular vesicles in 40 min (Jiang *et al.*, 2001). Inhibition of endocytosis by concanavalin A treatment, cytosolic domain deletion of MT1-MMP, or the expression of a dominant negative dynamin mutant D44K all enhanced the expression of MT1-MMP on the cell surface and consequently proMMP-2 activation (Jiang *et al.*, 2001). Interestingly, the cytoplasmic domain of MT1-MMP is divergent at the amino acid level from those of MT2-, MT3-, and MT5-MMP, raising the possibility that MT1-MMP may be regulated differently. Intracellular processing of various forms of MT1-MMP has not been characterized as yet.

Using immunohistochemistry, Puyraimond *et al.* (2001) demonstrated that MT1-MMP, pro MMP-2, and $\alpha_v\beta_3$ integrin were localized with patches of caveolin-1 in caveoli at the basolateral side of endothelial cells. Using physical cell membrane separation techniques, MT1-MMP was shown to be preferentially compartmentalized and proteolytically processed in low density caveolin-enriched membrane domains of cancer cells. Inhibition of MT1-MMP dependent cell migration, but not proMMP-2 activation, by caveolin-1 cotransfection, suggested that the localization of MT1-MMP to caveolin-enriched domains functions as a negative regulator of enzymatic activity. A role for caveolin-1 in inhibition of lamellipod extension and cell migration has also been demonstrated (Annabi *et al.*, 2001).

1. Endocytosis and Degradation of TIMP-2: Role of MT1-MMP

Maquoi *et al.* (2000a) reported that prolonged treatment of cells with phorbol esters resulted in increased cell binding, internalization, and lysosomal/endosomal degradation of TIMP-2, thus tipping the balance for activation of proMMP-2. They similarly proposed that type IV collagen mediates proMMP-2 activation by: (1) bringing cell membrane-associated MT1-MMP/TIMP-2/MMP-2 complexes closer together, thereby promoting autolytic cleavage of intermediate MMP-2, and (2) enhancing lysosomal degradation of TIMP-2; no modulation of MMP-2, MT1-MMP, or TIMP-2 expression was demonstrated (Maquoi *et al.*, 2000b). In contrast to the Maquoi report, studies from the Zucker laboratory indicated that short-term treatment of cells with phorbol esters and other agents that lead to activation of proMMP-2 had only a small effect on degradation of cell surface bound TIMP-2; thus, the activation of proMMP-2 precedes the proposed degradation of TIMP-2 (Zucker *et al.*, 2002b). Zucker *et al.* hypothesized that presynthesized MT1-MMP is routinely sorted to a transient storage compartment in the trans-Golgi network/endosomes where it is available for rapid trafficking to the plasma membrane in response to physiologic stimulation. In the absence of a stimulus for matrix degradation, MT1-MMP activated by furin in the trans Golgi network appears to undergo rapid autolytic cleavage and degradation, thus accounting for the transient appearance of mature MT1-MMP on the plasma membrane.

L. Biologic and Pathologic Roles for MT-MMPs

1. Cell Migration and Invasion

The movement of cells within or through tissue barriers is an extremely complex process; the role of proteases appears to vary depending on the milieu of the extracellular matrix (Ellis and Murphy, 2001). Major questions include whether cells always use proteolytic mechanisms to modify the ECM in their path and whether the path is simply a clearing mechanism, or a way of reorganizing the matrix to facilitate cellular interactions (Murphy and Gavrilovic, 1999). Of additional interest is the role of MMPs in modulating cell adhesion interactions with the extracellular matrix to release cells and to provide traction to their movement, as well as to expose signals necessary to effect motor function and provide chemoattraction (Vu and Werb, 2000). In addition to the positive effects on migration exerted by most MMPs, MMP-9 has been reported to exert a negative control over cell migration by inhibiting cell replication in the migrating epithelial sheet (Mohan *et al.*, 2002).

A role for MMPs, especially MT1-MMP, in cell migration has been repeatedly demonstrated using both *in vitro* and *in vivo* techniques (El Fahime *et al.*, 2000; Nabeshima *et al.*, 2000). Soluble MMPs, especially MMP-1, interacting with the

α_2 integrin subunit in keratinocytes, exert a positive influence on cell migration (Parks *et al.*, 1998). When cohort migration was induced with scatter factor/HGF, MT1-MMP and MMP-2 were immunolocalized primarily to the leading edges of the cells at the front of the migrating sheet of cells, with the cells behind the edge being negative for these MMPs. This phenomenon was also associated with changes in tyrosine phosphorylation of the E-cadherin/catenin complex and production of fibronectin (Nabeshima *et al.*, 2000). The dominant expression of MMPs in the front cells might be related to the inhibitory effect of abundant cell contact on MMP expression or the enhancement effect of extracellular matrix contact on MMP expression (Nabeshima *et al.*, 2000). The role of promigratory neoepitopes generated during extracellular matrix degradation remains to be clarified (Murphy and Gavrilovic, 1999); aggregation of $\beta 1$ integrins appear to be involved in modulating these events. As discussed earlier, the role of the cytoplasmic domain of MT-MMPs in cell migration remains controversial. MT-MMP dependent cell migration is not only accompanied by, but may require, the internalization of MT1-MMP at the leading edge of the cell (Uekita *et al.*, 2001). These authors proposed that internalization of MT1-MMP occurs constitutively as a mechanism to "clean up all forms of MT1-MMP molecules at the surface" of cells. Mechanistically, it is not clear how internalization can propel the cells to migrate as observed by Ukeita and colleagues (Uekita *et al.*, 2001).

The role of soluble versus membrane-anchored MMPs in cell invasion and morphogenesis has been addressed by stably transfecting cells to overexpress each of 10 different MMPs that have been linked to matrix remodeling; cells were then stimulated with scatter factor/HGF and observed. Neither secreted collagenases (MMP-1, -13), gelatinases (MMP-2, -9), stromelysins (MMP-3, -11), nor matrilysin (MMP-7) affected migration responses. By contrast, MT1-MMP expression conferred invasion incompetent cells with the ability to penetrate and remodel type I collagen matrices by using membrane-anchored MMPs as pericellular collagenases (Hotary *et al.*, 2000). In contrast with other reports (Nakahara *et al.*, 1998; Urena *et al.*, 1999), not only was the MT1-MMP cytosolic tail-deleted mutant processed to its mature form in a normal fashion, but it also induced an invasion program indistinguishable from that of the wild-type enzyme. MT2-MMP transfected cells disrupted type I collagen by degrading the underlying substratum in a discoordinated fashion that prevented cells from establishing the adhesive interactions necessary for migration. MT3-MMP was less efficient than MT1-MMP and MT2-MMP in invasion/morphogenic properties.

MT1-MMP has been localized predominantly to invadopodia, specialized membrane extensions that are the sites of ECM degradation (Nakahara *et al.*, 1998); this localization appears to be essential for ECM degradation and cancer cell invasion. Overexpression of MT1-MMP in cells induces a marked increase in cell migration. The extracellular signal-related protein kinase (ERK) cascade has been proposed to be involved in the induction of cell migration. Osteoclasts are another cell type in which MT1-MMP was associated with lamellipodia and invadopodia with a "walking phenotype." Cells with a "sitting phenotype" have MT1-MMP localized to small areas on the cell periphery and colocalization with actin staining indicative of arrest and attachment of osteoclasts (Gilles *et al.*, 2001; Sato *et al.*, 1997). MMP-9 and MMP-13 have also been implicated in these functions in osteoblasts (Sato *et al.*, 1997). MT2-MMP appears to traffic in a different fashion than MT1-MMP and is primarily confined to the Golgi when overexpressed in CHO cells (Miyamori *et al.*, 2000b). Matrix-dependent proteolysis of cell surface transglutaminase by MT1-MMP has been reported to regulate cancer cell adhesion and locomotion (Belkin *et al.*, 2001). Reciprocally, fibronectin protects its cell surface receptor, tissue transglutaminase, from proteolysis by MT1-MMP, thereby supporting cell adhesion and locomotion. Transmigration of T lymphocytes into perivascular tissues involves cell movement across endothelial cell layers; $\alpha_4\beta_1$ integrin engagement and the coordinated induction of MT1-MMP and MMP-2 have been implicated in this process (Graesser *et al.*, 1998, 2000) (see Chapter13 for detailed discussion).

Numerous ECM proteins have been shown to contain "cryptic" information which affects bound ECM organization as well as cell-ECM interactions (Davis, 1992). Laminin-5, an important glycoprotein component of epithelial basement membranes, provides an important ECM substrate for cell adhesion and migration. Gianelli *et al.* (1997) initially reported that specific cleavage of the γ_2 chain of laminin-5 by MMP-2 exposed a cryptic promigratory site on this molecule and triggered cell motility. These observations emphasize that proteolysis of extracellular matrix components not only destroys matrix but also induces new biologic activity. Subsequently, a model was proposed whereby expression of MT1-MMP is the primary trigger for migration over laminin-5, whereas MMP-2, which is activated by MT1-MMP, may play an ancillary role, perhaps by amplifying the MT1-MMP effects. In contrast, cells not expressing functional MT1-MMP leave laminin-5 intact and use it for static adhesion (Koshikawa *et al.*, 2000). Recombinant soluble MT1-MMP cleaved the γ_2 subunit of laminin-5. The codistribution of MT1-MMP with laminin-5 in breast and colon carcinoma tissue specimens suggested a role for this mechanism in cancer invasion. Gilles *et al.* (2001) confirmed the role of MT1-MMP and laminin-5 γ_2 chain degradation in epithelial cell migration; the distribution of MT1-MMP in lamellipodia of migrating cells and in cells at the periphery of the outgrowth was emphasized. The main receptors for laminin-5, $\alpha_6\beta_4$ and $\alpha_3\beta_1$ integrins, have been implicated not only in cell adhesion, but also in cell migration (Gilles *et al.*, 2001).

The concept of proteolytic exposure of cryptic sites within extracellular matrix components has been extended with the demonstration of cryptic sites induced by MMP cleavage of type IV collagen; a monoclonal antibody directed to this cryptic site disrupted integrin-dependent cell interactions and inhibited angiogenesis and tumor growth *in vivo* (Xu *et al.*, 2001). Among the intriguing issues raised by these studies is the temporal sequence and directional effect of a migrating cell laying down extracellular components, then degrading these matrix components and releasing fragments that function as chemoattractants for the cells producing the fragments.

MMP processing of CD44, the multistructural and multifunctional cell adhesion molecule that binds hyaluronic acid, collagen type-I, and fibrin at the cell surface, has been reported to be involved in cell migration (Okamoto *et al.,* 1999). MT1-MMP induced shedding of CD44H from the cell surface results in stimulation of cell migration (Kajita *et al.,* 2001). Expression of mutant CD44H in the cells, as well as MMP inhibitor treatment, effectively inhibited cell migration, suggesting that these cells use CD44H and MT1-MMP as migratory devices. Based on these studies, Kajita *et al.* (2001) postulated that timely and spatial processing of CD44H at the adherent edge by MT1-MMP is critical for stimulation of migration, allowing cells to be detached from the ECM and move onto different sites.

2. Embryology and Development

MT-MMPs are widely, yet selectively distributed in both embryonic and adult tissues. MT1-MMP is coexpressed temporally and spatially with TIMP-2 during mouse embryogenesis, which suggests common regulatory pathways with functional implications for MMP-2 activation (Apte *et al.,* 1997; Kinoh *et al.,* 1996). At the cellular level in the mouse embryo, MT1-MMP mRNA is expressed mainly in mesenchymal cells including muscle cells, fibroblasts, chondrocytes, and neural cells; expression decreases with postnatal maturation (Apte *et al.,* 1997; Takino *et al.,* 1995b). ProMMP-2 and TIMP-2 mRNA were expressed in the same tissues as MT1-MMP.

Various approaches have been taken to examine the role of MMPs in organogenesis. MT1-MMP is involved in organogenesis of the metanephros and in the formation of kidney tubules (Kadono *et al.,* 1998). Paracrine/juxtacrine epithelial:mesenchymal interactions are involved. Other cell surface receptors and signaling pathways participate in this complicated process. Renal epithelial cells undergoing branching morphogenesis *in vivo* have been shown to transiently express high levels of MT1-MMP (Ota *et al.,* 1998; Tanney *et al.,* 1998). By *in situ* hybridization, MT1-MMP mRNA was confined to ureteric bud epithelia and was absent in the mesenchyme, while MMP-2 was confined to the mesenchyme. MT1-MMP protein expression as detected by immunohistochemistry was more extensive, also including induced mesenchyme and nascent nephrons, peaking during mid-gestation when organogenesis is maximum (Ota *et al.,* 1998).

Prominent expression of MT-MMPs and TIMP-2 mRNA in bone and perichondrium during mouse skeletal and muscle development have been noted; interesting differences in distribution sites of endochondrial calcification have been described (Kinoh *et al.,* 1996; Knauper and Murphy, 1999).

Abundant expression of mRNA for MT1-MMP, MMP-2, and TIMP-2 have been identified in mouse embryonic lungs from E11.5–18.5 days (Kheradmand *et al.,* 2002) and in rabbit embryonic lungs (Fukuda *et al.,* 2000). MMP-2 expression was confined to cells of mesenchymal origin and was absent from lung epithelium. By contrast, MT1-MMP and TIMP-2 were expressed in both the epithelium and the mesenchyme. Expression of MMP-3, -9, -10, and TIMP-1 mRNA was not detected in embryonic lung epithelium or mesenchyme at any stage of lung development (Kheradmand et al., 2002). Fukuda et al. (2000) proposed that MMP degradation of ECM proteins, especially elastin and collagens, is a necessary component for migration of epithelial cells into the mesenchyme during the invagination step in lung development.

MMP-mediated proteolysis of the ECM plays a major role in placental tro-phoblast invasion of the uterus. Evidence obtained by Northern blot analysis, *in situ* hybridization, and immunohistochemistry indicate that MT1-MMP is im-portant in tissue organization of early human placenta acting alone and as a cell surface activator of proMMP-2 (Hurskainen et al., 1998; Tanaka et al., 1998). The most abundant MT1-MMP mRNA signal was in the invasive trophoblastic cells, less in decidual cells, with lower signal in endothelial cells, and virtually no signal in fibroblasts (Tanaka et al., 1998).

A role for MT1-MMP and MMP-2 in cell migration through and invasion of type IV collagen during epithelial mesenchymal transformation in the embryonic heart has been demonstrated (Song et al., 2000). TGFβ-3 is able to regulate the expression of both MMP-2 and MT1-MMP in endocardial cells. It was suggested that coordinated specific cell–matrix interactions which facilitate cell migration only occur when the composition of the surrounding ECM is proteolytically altered. Type IV collagen presents an initial barrier which must be overcome by newly formed mesenchymal cells producing MMPs. Embryos injected with a hydroxamic acid-based MMP inhibitor BB-94 showed depressed mesenchyme formation (Song et al., 2000).

3. Animal Models Employed to Analyze the Function of MT-MMPs

a. MMP and TIMP Knockout Mice: Profound Effect of MT1-MMP

MT1-MMP knockout. To define the biologic role of MT1-MMP, two groups have generated MT1-MMP deficient mice by gene targeting in embryonic stem cells (Holmbeck et al., 1999; Zhou et al., 2000). Knockout mice exhibited se-vere defects in skeletal development and angiogenesis. Postnatal development of secondary ossification centers was markedly delayed. The basic defect was con-sidered to be due to impairment in the mechanism responsible for dissolution of cartilage matrix (Holmbeck et al., 1999). MT1-MMP deficient fibroblasts derived from mutant mice are incapable of degrading type I collagen matrices. However, osteoclast-mediated degradation of bone or calcified cartilage matrix was unim-paired since these functions are dependent on acidic proteases of the cathepsin family (Holmbeck et al., 1999). Lack of MT1-MMP in knockout mice was not sufficiently compensated by other MT-MMPs or secreted MMPs. In tissue extracts from MT1-MMP null mice, proMMP-2 was minimally activated as compared to wild-type mice. The craniofacial, axial, and appendicular skeletons were severely affected, leading to a short, smaller, and domed skull, marked deceleration of postnatal growth, and death within 3–16 weeks. Shortening of the bones is a consequence of decreased chondrocyte proliferation in the proliferative zones of the growth plates. The enlargement of the hypertrophic zones in null mice is attributed to delayed resorption of cartilage during endochondral ossification. Defective cell production in the growth plate is likely to be the cause of dwarfism; the regulatory function of MT1-MMP in cell growth is unknown. Defective vascular invasion of cartilage leads to enlargement of hypertrophic zones of growth plates and delayed formation of secondary ossification centers in long bones. Concurrently, MT1-MMP deficient mice developed generalized arthritis; all joints showed overgrowth of a hypercellular vascularized synovial tissue and destruction of articular cartilage, resulting in ankylosis. Increased cell proliferation and fibrosis occurred in skeletal tissue. Dermal fibrosis and hair loss followed. Postnatal angiogenesis in response to FGF-2 in MT1-MMP null mice was markedly deficient in an *in vivo* corneal implantation assay (Zhou *et al.*, 2000).

MMP-9 knockout.   Similar to MT1-MMP, MMP-9 null mice have a defect with hypertrophic zones in cartilage which is detected in the fetus. However, MMP-9 knockout mice do not exhibit problems with postnatal growth or craniofacial development (Holmbeck *et al.*, 1999; Zhou *et al.*, 2000). It is probable that the peak requirement for MMP-9 and MT1-MMP occurs at different ages so that one may partially compensate for deficiencies in the other. In contrast to the profound effect on embryonic development in MT1-MMP knockout mice, TIMP-2 deficient mice display no gross anatomical or microscopic abnormalities despite severe impairment of proMMP-2 activation *in vivo* or *in vitro* (Caterina *et al.*, 2000; Wang *et al.*, 2000). Wang *et al.* (2000) concluded that both TIMP-2 and activated proMMP-2 are dispensable for normal development. The fact that TIMP-3 can substitute for TIMP-2 in terms of MT1-MMP induced proMMP-2 activation may explain the lack of an abnormal skeletal phenotype in TIMP-2 deficient mice.

MMP-2 knockout.   MMP-2 knockout mice have smaller stature, diminished tumor-related angiogenesis (Itoh *et al.*, 1998), and abnormal lung alveolarization (Kheradmand *et al.*, 2002), but skeletal and joint defects have not been reported (Itoh *et al.*, 1998). In contrast, nonsense mutations in the human MMP-2 gene (16q12-21) have been associated with a disabling autosomal recessive multicentric osteolysis (carpal and tarsal resorption and marked osteoporosis) and arthritis syndrome associated with palmar and plantar subcutaneous nodules and distinctive faces (Torg's syndrome) in the kindred of several Saudi Arabian families (Martignetti *et al.*, 2001). As anticipated, no MMP-2 enzymatic activity was detected in the serum or fibroblasts of affected family members. Two family-specific homoallelic MMP-2 mutations were identified. The nonsense mutation effects a deletion of the substrate binding domain and catalytic sites of the fibronectin type II-like and hemopexin/TIMP-2 binding domains. Martignetti *et al.* (2001) pointed out that the phenotype of people deficient in MMP-2 more closely resembles the phenotype of MT1-MMP mice. The inherited osteolysis in these families is counterintuitive considering the function of MMP-2 in degrading extracellular matrix substrates. One possible explanation is that defects in the osteoblast bone deposition are caused by incomplete degradation of extracellular matrix. Vu (2001) proposed that features of familial osteolysis syndrome and MT1-MMP knockout mice resemble those caused by deregulated TGFβ signaling. MMP-2 activates TGFβ by proteolytic cleavage of the latent-associated peptide. Thus, lack of MMP-2 may affect bone formation and hemostasis through modulating the level of active TGFβ. An alternative explanation is that MMP-2 may help to recruit osteoblasts to sites of new bone formation (Vu, 2001).

b. Growth Factor Receptor Knockouts.   Studies of the Patch mouse, which carries a deletion of the entire coding region of the α subunit of the platelet de-rived growth factor receptor (PDGFRα) resulting in multiple connective tissue defects of the first brachial arch (cleft face, heart defects), have shed light on the developmental function of MMPs (Robbins *et al.,* 1999). In contrast to the normal embryo, tissue from the brachial arch of Patch mouse embryos exhibited a marked decrease in MMP-2 and MT-MMP. Treatment of mutant brachial arch cells *in vitro* with high dose recombinant PDGF-AA upregulated MMP-2 protein, leading to the conclusion that PDGF-A regulation of MMP-2 and MT-MMP during normal development is responsible, in part, for cell migration and development of the first brachial arch. Robbins *et al.* (1999) concluded that the balance between MMP-2 and MT-MMP may be more important to normal development than the absolute amount of each proteinase.

A more limited palatal defect has been reported in mice deficient in epidermal growth factor receptor (EGFR) (Brunet *et al.,* 1993). Delayed lung maturation characterized by abnormal branching and poor alveolarization is seen in mice deficient in EGFR and is associated with low levels of MT1-MMP expression and activated MMP-2. At birth, MMP-2 knockout mice had a lung phenotype that phenocopied that of EGFR knockout mice, albeit somewhat less severe. From these phenotypes, MT1-MMP was identified as a major downstream target of EGFR signaling in lungs *in vitro* and *in vivo* (Kheradmand *et al.,* 2002).

In transforming growth factor (TGF)-β-3 deficient mice, a cleft palate develops due to the inability of the medial edge epithelia to fuse. Fusion of palatal shelves can be restored in organ culture by adding the mature form of TGFβ-3 into the medium which is followed by increased MMP-13 production (Blavier *et al.,* 2001). Detailed analysis of MMP and TIMP production in the embryo implicated failure of coordinated production and function of MMP-13, MT1-MMP, MMP-2, and TIMP-2 leading to defective palatal fusion. Synthetic inhibitors of MMPs resulted in phenocopying of the defect in TGFβ-3 deficient mice. Blavier *et al.* (2001) linked MMP-13 expression to epithelial–mesenchymal trans differentiation. They further pointed out that functional redundancy of MMPs may explain why neither MMP-2 or MT1-MMP deficient mice develop a cleft palate.

c. Transgenic Mice Overexpressing MT-MMPs.   To investigate the role of MT1-MMP in the mammary gland, transgenic mice overexpressing MT1-MMP in the mammary gland were generated. Transgenic mice developed lymphocytic infiltration, fibrosis, hyperplasia, dysplasia, and adenocarcinomas. It was concluded that overexpression of MT1-MMP may play a key role in development of tumorigenesis in the breast as a result of aberrant degradation of the ECM (Ha *et al.*, 2001). MMP-3 overexpression in the mammary gland, resulting initially in alteration of the stromal phenotype, has also been associated with mammary carcinogenesis (Sternlicht *et al.*, 1999). These findings derived from studies with MT1-MMP and MMP-3 transgenic mice are of considerable relevance because they have provided additional evidence to support the concept that MMPs, including MT-MMPs, may play essential roles in early stages of tumor evolution.

4. Involvement of MT-MMPs in Physiologic Functions, Wound Injury, and Repair Processes

a. MT1-MMP Production in the Endometrium: Hormonal Control. MMPs have specific spatial and temporal expression patterns in human endometrium and are critical for menstruation; MMP-1 and MMP-3 are primarily implicated. MMP-2 is localized in all cellular compartments, but with greatest intensity in endometrial stromal cells, especially in degrading menstrual tissue. Intense focal labeling for MT1-MMP mRNA expression was observed in desquamating stromal cells around collapsed dispersing glands. MT1-MMP mRNA was present throughout the menstrual cycle (Zhang *et al.*, 2000) with the strongest staining in subsets of macrophages, extravascular neutrophils, and granular lymphocytes and during the menstrual, mid-proliferative, and mid-secretory phases. MT2-MMP staining was more widespread than MT1-MMP without cyclical variation and with maximal intensity in glandular epithelium. Of physiologic importance, progesterone inhibited the activation of proMMP-2 by attenuating cell expression of MT1-MMP mRNA (Zhang *et al.*, 2000). Maata *et al.* (2000) pointed out that MT1-MMP expression is involved most notably in the normal endothelium under progesterone control.

b. Role of MT-MMPs in Skin Wound Healing. Skin wound healing depends on cell migration and extracellular matrix remodeling. Both processes, which are necessary for reepithelialization and repair of the underlying connective tissue, involve the action of MMPs. MMP-3, -9, and -13 are expressed early (days 1–5), while MMP-2, -11, and MT1-MMP continue to be expressed late (after day 7) following a cutaneous incision. In wound repair models of the skin, Parks *et al.* (1998) have demonstrated that one group of MMPs produced by keratinocytes is required for early cell migration into the wound, whereas other MMPs are produced by other cell types at later stages in the repair process and presumably serve different functions (Parks *et al.*, 1998). Okada *et al.* (1997) proposed that MT1-MMP activates proMMP-2 at the stromal cell surface primarily in the granulation tissue, but not in the regenerating epidermis. Several MMPs have been identified in normal corneas and are upregulated during corneal wound healing (Mohan *et al.*, 2002). MT1-MMP, MMP-9, -2, and -13 were proposed to be important in reepithelialization after cornea wounding (Ye *et al.*, 2000).

c. Role of MT-MMPs in Inflammation.   The finding that MT6-MMP or leukolysin is specifically expressed by inflammatory granulocytes raises the possibility that it may be important during inflammation (Kang *et al.*, 2001; Pei, 1999). During the inflammatory response, neutrophils respond to chemokines such as IL-8, which leads to the influx of neutrophils into inflammatory sites where they discharge their intracellular contents to destroy invading microbes in infected tissues. Given the subcellular localization of leukolysin, Kang and colleagues (Kang *et al.*, 2001) have postulated that leukolysin is engaged in the interactions with endothelial cells and the migratory process through the underlying basement membranes. Interestingly, MT4-MMP, another GPI-anchored MT-MMP, was found to be expressed in monocytes, which might reflect comparable function to leukolysin in neutrophils (English *et al.*, 2000).

d. Role of MT-MMPs in Liver Injury.   It has been suggested that a mere expression of the component of the ternary complex consisting of MT1-MMP:TIMP-2:MMP-2 is not sufficient to account for the activation of proMMP-2 in several model systems and that additional factors might be necessary in the process (Young *et al.*, 1995; Yu *et al.*, 1995). In the liver, MT1-MMP mRNA is associated principally with stromal (stellate) cells which also express both MMP-2 and TIMP-2 mRNA, whereas MT2-MMP mRNA is found mostly in hepatocytes and bile duct epithelial cells. Theret *et al.* (1997, 1998, 1999) proposed that hepatocytes and bile duct epithelium may contribute to the membrane-mediated proMMP-2 activation by expressing MT2-MMP at the level of intimate hepatocyte–stellate cell cell contact.

5. MT-MMPs in Disease

Up until the present, the focus of research on MT-MMPs in disease has been primarily on cancer. This reflects a historic bias in which MMPs have been almost instinctively linked to cancer invasion. More recently, interest in the role of MT-MMPs in numerous other diseases has been awakened.

a. Cancer.   A close association between the expression of activated MMPs, including MT-MMPs, and tumor dissemination has been demonstrated in a plethora of studies (McKerrow *et al.*, 2000; Murphy *et al.*, 2000; Polette and Birembaut, 1998; Sato *et al.*, 1994; Seiki, 1999; Stetler-Stevenson, 1995). In contrast to classical oncogenes, MMPs are not upregulated by gene amplification or activating mutations. The increased MMP expression in tumors is probably due to transcriptional changes rather than genetic alterations (Egeblad and Werb, 2002). Polymorphisms in MMP promoters exist that affect gene transcription and may influence cancer susceptibility (Kanamon *et al.*, 1999).

The importance of cell surface MMPs in this process has been emphasized repeatedly (Young *et al.*, 1995). Serine, cysteine, and aspartic proteinase levels are also increased in cancer tissue, leading to uncertainty of the role of each proteinase

in cancer (McKerrow *et al.*, 2000; Zucker *et al.*, 2002a). *In vitro*, MT1-MMP transfection of cells enhances their invasive characteristics. The preferential localization
of MT1-MMP at the advancing border of human tumor cell nests (Harada *et al.*,
1998) and within the cell at the leading edge invadopodia has been emphasized.

(1) MT-MMPs in clinical human cancer. The expression of MT1-MMP has
been detected in or on tumor cells and adjacent stromal cells in a variety of human
tumors, including lung (Polette *et al.*, 1996; Sato *et al.*, 1994; Tokuraka *et al.*,
1995), gastric (Mori *et al.*, 1997; Nomura *et al.*, 1995; Ohtani *et al.*, 1996), breast
(Heppner *et al.*, 1996; Polette *et al.*, 1996; Ueno *et al.*, 1997), colon (Kikuchi *et al.*,
2000; Ohtani *et al.*, 1996; Okada *et al.*, 1995; Sardinhah *et al.*, 2000), liver (Seiki,
1999; Theret *et al.*, 1998), pancreatic (Ellenrieder *et al.*, 2000; Imamura *et al.*,
1998), thyroid (Nakamura *et al.*, 1999), head and neck (Imanashi *et al.*, 2000;
Ohtani *et al.*, 1996; Okada *et al.*, 1995; Shimada *et al.*, 2000), ovarian (Seiki,
1999), prostate (Upadhyay *et al.*, 1999), bladder (Kitagawa *et al.*, 1998), cervical
(Afzal *et al.*, 1998; Davidson *et al.*, 1999; Maata *et al.*, 2000; Seiki, 1999), and
brain tumors (Gingras *et al.*, 2000; Yamamoto *et al.*, 1996). Employing monoclonal antibodies to recombinant human MT1-MMP in a specific enzyme-linked
immunoassay, increased levels of MT1-MMP have been demonstrated in clinical
tumor samples from patients with cancer (head and neck and lung cancer), as compared to nonmalignant tissues. High levels of MT1-MMP were also detected in
cell lysates and conditioned media derived from cancer cell lines. However, MT1-
MMP was not detected in the serum of patients with cancer (Aoki *et al.*, 2002).

Disagreement about the cellular origin of MMPs in human tumors has presented
difficulties in extrapolating from animal and *in vitro* data to explain the role of MT-
MMPs in cancer progression (Zucker *et al.*, 2000). Results using *in situ* hybridization techniques seem to vary depending on the type of cancer and the technique
employed (Ellenrieder *et al.*, 2000; Nakada *et al.*, 1999; Okada *et al.*, 1995; Polette
et al., 1996; Ueno *et al.*, 1997). Nonetheless, these types of experiments have revealed important clues to help decipher the role of MT-MMPs in human cancer.
Using the technique of *in situ* hybridization to identify mRNA in human tumor
specimens, several authors have demonstrated MT1-MMP mostly in stromal cells
within tumors or have shown that both cancer cells and surrounding fibroblasts
and macrophages express MT1-MMP (Afzal *et al.*, 1998; Davidson *et al.*, 1999;
Ohtani *et al.*, 1996). In breast and lung tumors, MT1-MMP mRNA expression has
been reported in fibroblasts in close contact with cancer cells. MT1-MMP expression differences between adenocarcinomas and squamous cell lung cancer have
been noted (Polette *et al.*, 1996). The expression of MT1-MMP mRNA and the
activation of proMMP-2 are increased in hepatocellular carcinomas; MT1-MMP
is expressed in the stromal cells (Theret *et al.*, 1998). The MT1-MMP stroma
induction near preinvasive and invasive tumor clusters supports the concept of
release/shedding of tumor derived factors enhancing tumor invasion (Polette and
Birembaut, 1998). These same stromal cells express MMP-2 and TIMP-2 mRNA.
The explanation for enhanced MMP production in peritumoral stromal cells

became better appreciated with the demonstration that cancer cells display Extracellular Matrix Metalloproteinase Inducer (EMMPRIN) on their plasma membrane; this immunoglobulin superfamily member functions to induce the expression of MMP-1, -2, -3, and MT1-MMP in neighboring stromal cells (Biswas *et al.*, 1995; Sameshima *et al.*, 2000; Zucker *et al.*, 2001). However, numerous other *in situ* hybridization studies have claimed that cancer cells, rather than stromal cells, express the dominant amount of MT1-MMP and MT2-MMP in human cancers (human thyroid, brain, and head and neck cancers) (Nakada *et al.*, 1999).

Immunohistochemical studies have identified MT1-MMP in both tumor cells and stromal cells. In most reports, the specificity of the antibodies employed for MT-MMP identification was not clearly demonstrated. It has been pointed out that differences in antigen localization in tumors may be due to the nature of the antigen employed in the mouse immunizations (Chenard *et al.*, 1999). Using specific monoclonal antibodies produced against recombinant protein, MT1-MMP was identified in the stromal cells of human adenocarcinoma almost exclusively, and not in cancer cells (Chenard *et al.*, 1999). A correlation has been demonstrated between immunohistochemical detection of MT1-MMP/MMP-2 and both the depth of tumor invasion and vascular invasion in human colon cancer (Kikuchi *et al.*, 2000); no correlation with distant metastasis was shown. TIMP-2 levels in the stroma were found to be associated with lymphatic invasion and lymph node metastasis (Kikuchi *et al.*, 2000).

Activated MMP-2, as detected by gelatin zymography, has also been predominantly demonstrated in various tumors compared to the normal tissue counterparts (Davies *et al.*, 1993); increased activated MMP-2 generally correlated with MMP expression in the primary tumor (Ellenrieder *et al.*, 2000; Nakada *et al.*, 1999) and in metastasis (Ueno *et al.*, 1997). In a preliminary study, comparative RT-PCR (tumor versus normal tissue MT1-MMP/MMP-2 ratios) has been proposed as a preoperative molecular-level prognostic factor for overall survival in human gastric carcinoma (Caenazzo *et al.*, 1998).

Immuno-electron microscopy showed that MT1-MMP was localized along the plasma membrane of cancer cells, whereas it was detected in the rough endoplasmic reticulum in stromal cells. Ohtani *et al.* (1996) suggested that the expression of MT1-MMP in cancer cells may be related to the invasive growth and to tissue remodeling process in fibroblasts. Upadhyay *et al.* (1999) have investigated the expression and immunolocalization of MT1-MMP and MMP-2 in human prostate tissue from patients undergoing radical prostatectomy. They reported that the basal cells in benign glands were uniformly stained intensely for MT1-MMP, whereas secretory cells were rarely positive. Conversely, in high grade prostate intraepithelial neoplasm (PIN) showing patchy loss of basement membrane, secretory cells showed consistent cytoplasmic staining for MT1-MMP. In cancer, staining was heterogeneous. Using Western blotting, MT1-MMP was detected as both the 63 kDa latent and 57 kDa activated forms of MT1-MMP. A significant spatial association between the patterns of MMP-2 and MT1-MMP immunostaining was noted. It was concluded that the regulation of these enzymes is altered during the earliest stages of prostate cancer. A good correlation between MT1-MMP and MMP-2 expression has been reported in human pancreatic cancer; MMP-2 expression and activation were associated with the extent of desmoplastic reaction in pancreatic cancers (Ellenrieder *et al.*, 2000). Among seven different MMPs examined, the production of proMMP-2 and its MT1-MMP mediated activation in carcinoma cell nests were proposed to play an important role in lymph node metastasis of human thyroid and oral carcinomas (Shimada *et al.*, 2000). Of interest, overproduction of TIMP-1 seems to be a common feature of human carcinomas (Nakamura *et al.*, 1999).

MT2-MMP has been found in a variety of carcinomas; however, it has not been linked with disease progression (Sato and Seiki, 1996). Although Polette and Birembaut (1998) reported that MT3-MMP is not often observed in tumor tissue, other groups have reported the expression of MT3-MMP in melanoma, breast cancer, astrocytic tumors, and renal cell carcinoma (Kitagawa *et al.*, 1999; Nakada *et al.*, 1999; Takino *et al.*, 1995a; Ueno *et al.*, 1997). MT5-MMP and MT6-MMP have also been found to be overexpressed in human brain tumors, especially those with more aggressive behavior (Llano *et al.*, 1999; Velasco *et al.*, 2000).

(2) MT-MMPs in experimental cancer models. Numerous experimental observations have been made concerning the role of MT-MMPs in cancer progression. In contrast to human tissue specimens, cancer cell lines generally demonstrate enhanced expression of many MMPs, including MT-MMP, as compared to nonmalignant cells. MT1-MMP is also shed from tumor cell surfaces *in vitro* (Harayama *et al.*, 1999; Imai *et al.*, 1996). The relevance of these observations to clinical disease remains an enigma. High levels of MMP expression generally, but not invariably (Zucker *et al.*, 1992), correlate with the more invasive and metastatic phenotype in cancer cell lines (Liotta, 1992; Stetler-Stevenson, 1995). Not long after the identification of the MT-MMP subfamily, Tsunezuka *et al.* (1996) reported that injection of MT1-MMP transfected cancer cells into mice resulted in increased metastatic behavior of the cells in short-term experiments. It was also demonstrated that a cooperation between the plasminogen/plasmin system and MT1-MMP endowed cells with the ability to fully activate proMMP-2 and with enhanced invasive properties *in vitro;* further development of this concept is awaited.

MT1-MMP and TIMP-2 mRNA have been demonstrated to be present in many human melanoma cell lines. Cell surface expression of MT1-MMP, as determined by flow cytometry, was similar in all cell lines, as was the relative amounts of active MT1-MMP. However, the presence of functionally active MMP-2 was restricted to the most aggressive cell lines. Whereas MT1-MMP and TIMP-2 mRNA were detected by RT-PCR in subcutaneous xenografts from all tumors, MMP-2 mRNA was detected only in xenografts derived from highly metastatic tumors (Hofman *et al.*, 2000). Another report demonstrated that human melanoma cell line invasive potential was controlled not by the expression of MT-MMP, but by the expression of TIMP-2, which regulates proMMP-2 activation; production of TIMP-2 correlated inversely with cell invasion (Kurschat *et al.*, 1999). In experiments with HCAT epidermal tumor cell lines, following contact with fibrillar collagen type I, the more malignant cell line gained the capacity to activate proMMP-2, possibly due to downregulation of TIMP-2 levels (Baumann *et al.*, 2000).

Human brain cancer cell lines stably-transfected with MT1-MMP cDNA demonstrated increased cell proliferation, migration, production of vascular endothelial growth factor, and directional remodeling of a three-dimensional collagen matrix, associated with activation of proMMP-2 *in vitro*. It was suggested that MT1-MMP, independent of MMP-2, plays an important role *in situ* in cancer migration and invasion along white matter fiber tracts. Of special relevance to brain tumors, a human glioblastoma cell line and MT1-MMP mRNA transfected fibroblasts were able to degrade the inhibitory protein of the brain (bN1-220), as well as migrate and invade through brain tissue (Belien *et al.*, 1999). In other studies of brain cancer cells, MT1-MMP transfected cells displayed increased tumor growth rates and elevated levels of angiogenesis following tumor cell injections into nude mice.

Coculture of human squamous carcinoma cells with normal fibroblasts has been reported to mediate cell–cell contact, MT1-MMP dependent activation of proMMP-2 on the tumor cell surface, and increased invasiveness *in vitro*. EGF was proposed as a mediator of these events (Sato *et al.*, 1999). This experimental model is attractive in that it recapitulates the mixture of cells in human tumors. Clustering of collagen-binding integrins (β_1 integrins) on ovarian cancer cell membranes has been shown to induce proMMP-2 activation through a MT-MMP and signal transduction pathway (Ellerbroek *et al.*, 1999). High levels of lysophosphatidic acid in patients with ovarian cancer appear to contribute to metastatic dissemination of cancer cells, possibly via upregulation of proMMP-2 activation via enhanced trafficking of MT1-MMP to the plasma membrane (focalized to sites of cell–matrix contact) and subsequent changes in MMP-dependent cell migration and invasion. G_i protein-coupled receptors were proposed to participate in this mechanism (Fishman *et al.*, 2001). In some studies, however, the amount of the active form of MT1-MMP on the cell surface did not appear to be sufficient to explain the ability of cancer cells to activate proMMP-2.

Aznavoorian *et al.* (2001) examined three cell lines derived from human squamous cell carcinomas for MMP activation and digestion of collagen fibrils. They demonstrated that in the absence of exogenous growth factors or accessory stromal cells, degradation of interstitial collagen by cancer cells required a threshold level of active MT1-MMP on the cell surface. Invasive cells were capable of MT1-MMP mediated proMMP-2 activation which was induced by culturing cells on collagen matrix; this mechanism was lacking in noninvasive cells. The mechanism for displaying high levels of active MT1-MMP in more aggressive cancer cell lines was not identified in this study; no differences in collagen receptors (integrins) were found.

A correlation between furin expression and aggressiveness of cancer cell lines *in vitro* and *in vivo* has been demonstrated. Transfection of invasive human cancer cells with a furin-specific modified α_1-protease inhibitor resulted in decreased invasiveness demonstrated in tracheal xenotransplants and decreased tumorigenicity after transplantation into nude mice. As anticipated, furin inhibitor-transfected cells displayed decreased activation of proMT1-MMP and proMMP-2. Although it is tempting to speculate that the decreased tumorigenicity of furin-inhibited cells was due to inhibition of activation of proMT-MMPs, it must be kept in mind that other cancer-related proteins such as TGF-β_1 are activated by the furin pathway (Bassi *et al.*, 2001).

Increased expression of oncogene family members, that is, EIAF, appears to contribute to the invasive cancer phenotype, including MT1-MMP expression and enhanced cell migration, but is not sufficient for exhibiting high metastatic activity *in vivo* (Habelhah *et al.*, 1999).

Of interest, the α_3 chain of type IV collagen, which is prominently displayed in the basement membrane around bronchopulmonary carcinoma cells, but not in normal bronchial epithelium, downregulates MT1-MMP expression (both at the transcriptional and posttranscriptional levels) in lung cancer cell lines (Martinella-Catusse *et al.*, 2001). In contrast, elastin-derived hydrophobic peptides stimulate MT1-MMP mRNA expression (Brassart *et al.*, 1998); MMP-2 and TIMP-2 expression are also stimulated following peptide binding to a specific cell receptor.

Cyclooxygenase-1 or -2 is able to mediate the processing of proMMP-2 through the induction of MT1-MMP in cancer cells (Takahashi *et al.*, 1999). Tsujii *et al.* (1997) reported that increased cyclooxygenase expression in colon cancer cells resulted in increased MT1-MMP expression; higher tissue MT1-MMP levels were noted in moderately and poorly differentiated tumors. This finding is of significance in light of evidence that use of nonsteroidal antiinflammatory agents, which inhibit cyclooxygenase activity, results in a decrease in mortality from colorectal cancer.

MT1-MMP expression by tumor cells has been reported to promote tumor vascularization in nude mice (Sounni *et al.*, 2002). These experiments, however, did not take into consideration the effect of tumor size per se, on angiogenesis. Depending on the experimental tumor model employed, the contribution of MT1-MMP to cancer progression and angiogenesis may be independent/dependent on its capacity to activate proMMP-2 (Itoh *et al.*, 1998; Sounni *et al.*, 2002). Deryugina *et al.* (2002) proposed that overexpression of MT1-MMP involves stimulation of angiogenesis through the upregulation of VEGF production. One fascinating aspect of the MT1-MMP transfections was the increased growth rate of the transfected cells. This outcome suggests that MT1-MMP may be altering the phenotype of cells in a global fashion, thereby complicating analysis of the role of individual factors in enhanced tumorigenicity. Of relevance to these experiments, overexpression of TIMP-2 in carcinoma cells resulted in delayed *in vivo* tumor growth and vascularization by strongly reducing VEGF expression (Hajitou *et al.*, 2001). In experimental studies, antibodies directed against the catalytic domain of MT1-MMP blocked endothelial cell migration, invasion, and capillary tube formation (Galvez *et al.*, 2001).

b. Atherosclerosis.   Considerable recent evidence suggests that MMPs play a role in maintaining blood vessel integrity and in the pathogenesis of cardiovascular diseases. Synthesis and removal of blood vessel wall proteins by endothelial cells, smooth muscle cells, and inflammatory cells in both physiologic and pathologic conditions depend on the action of numerous neutral proteinases, especially MMPs, and their inhibitors. To avoid weakening from the normal mechanical stresses of blood pressure, a vessel must continuously remodel its connective tissue, primarily by breaking down the ECM while new matrix is synthesized. MMPs have been implicated in pathologic states in the coronary artery system (Celentano and Frishman, 1997; Libby and Lee, 2000), in congestive heart failure (Spinale et al., 2000), aortic aneurysms (Pyo et al., 2000), and aortic dissections (Dollery et al., 1995; Ishii and Asuwa, 2000). The critical issue in these studies is whether MMPs have a causal relationship in cardiovascular diseases or whether these enzymes are components of the tissue repair mechanism. In the blood vessel wall and heart, the outcome of the reconstructive process on organ function is difficult to discern.

MMPs can degrade the fibrous cap of an atherosclerotic plaque, thereby contributing to plaque rupture (Celentano and Frishman, 1997; Wang and Keiser, 1998). Collagen degradation and synthesis take place simultaneously within the same plaques (Rekhter et al., 2000). A mechanistic link between hypercholesterolemia and collagen degradation resulting from MMPs and other proteolytic enzymes produced by macrophages has been suggested (Rekhter et al., 2000). In atherosclerotic arteries, MT1-MMP expression was noted within the complex atheroma colocalized with smooth muscle cells and macrophages. Proinflammatory molecules (IL-1α, TNF-α, and oxidized LDL) upregulate MT1-MMP expression in vascular smooth muscle cells and macrophages, thus influencing extracellular matrix remodeling in atherosclerosis (Rajavashisth et al., 1999). An important role for MT1-MMP in the remodeling of balloon-injured arteries in myocardia ischemia and reperfusion injury involves upregulation and activation of MMP-2 and MMP-9 after vascular injury (Jenkins et al., 1998; Lu et al., 2000). Arterial furin and MT1-MMP expression preceded or coincided with increased arterial proMMP-2 activation during arterial remodeling, either enlargement or shrinkage (Kleijn et al., 2001). Regulation of proMMP-2 activation depends on the degree of shear stress and might be regulated by TGFβ_1. Stretch provides a mechanism for increasing MT1-MMP in cardiac fibroblasts (Tyagi et al., 1998). A role for MT1-MMP in cerebral aneurysm remodeling has also been proposed (Bruno et al., 1998).

Increased MT1-MMP expression in smooth muscle cells has been described at an early stage (Jenkins et al., 1998) following experimental balloon catheter deendothelialization of the aorta (Wang and Keiser, 1998). In both high and low blood flow models in rabbits, increased furin and MT1-MMP mRNA levels occurred prior to and at the onset of arterial remodeling and was followed by an increase in activated MMP-2 in approximately 1 week (Kleijn et al., 2001). The magnitude and rate of increase in furin and MT1-MMP transcription and proMMP-2 activation were considerably greater after blood flow increase than after decrease. Both smooth muscle cells in media and fibroblasts in the adventia of the blood vessel displayed increased MT1-MMP immunostaining. These cells are also the major collagen producing cells of the arterial wall.

Target specificity and selectivity is of primary importance in developing new treatments for atherosclerosis (Celentano and Frishman, 1997). Peptide-based inhibitors of MMP activity have been shown to block early vascular smooth muscle cell migration in both *in vivo* and *in vitro* models of atherosclerosis (Abbruzzese *et al.*, 1998; de Smet *et al.*, 2000; El Fahime *et al.*, 2000; Lu *et al.*, 2000); reports of long term outcomes have varied (Bendeck *et al.*, 1996). Pasterkamp *et al.* (2000) postulated that MMPs in the atherosclerotic plaque play a causal role not only in plaque vulnerability, but also in *de novo* atherosclerosis. Leville *et al.* (2000), however, emphasized that a cause–effect relationship was not established between MMPs and intimal hyperplasia. Based on informative preclinical studies of cardiovascular diseases (Lu *et al.*, 2000), clinical trials of MMP inhibitors (MMPIs) in patients with heart disease have been discussed, but uncertainty of outcome (Godin *et al.*, 2000) has hampered initiation of clinical trials to date (Carmeliet, 2000; Ducharme *et al.*, 2000; Galis, 1999; Libby, 2000).

c. Rheumatoid Disease.   A role for MMPs in the invasiveness and collagenolytic properties of the rheumatoid pannus has long been proposed; MMP-1, -8, and -13 have been implicated. MMP-2/MT1-MMP seem to be focalized to pannocytes at sites crucial for tissue destruction in arthritis (Konttinen *et al.*, 1998). MT1-MMP and MT3-MMP are highly expressed in both the lining and the sublining layers in rheumatoid synovial tissue localized to fibroblasts, macrophages, and osteoclast-like cells (Pap *et al.*, 2000). An important role in the activation of proMMP-2 in the rheumatoid lining cell layer and in cartilage destruction has been proposed (Yamanaka *et al.*, 2000). The level of expression of MT3-MMP in rheumatoid synovial tissue was markedly lower than MT1-MMP. It has also been proposed that MT1-MMP activation of proMMP-2 might contribute to aseptic loosening of polyethylene prostheses (Nawrocki *et al.*, 1999); cooperation between MT1-MMP expression in macrophages and MMP-2 expression in fibroblasts was proposed.

d. Neurologic Disease.   Astrocyte activation is a stereotype reaction to neuronal injury leading to tissue remodeling and the formation of a glial scar. Extracellular matrix remodeling after nerve injury involves expression of MMPs by neurons and astrocytes (Lukes *et al.*, 1999). mRNA levels and cellular distribution of MT1-MMP were examined in wobbler mutant mice affected by progressive neurodegeneration resembling amyotrophic lateral sclerosis; MT1-MMP mRNA levels were elevated in affected tissues, primarily in reactive astrocytes (Rathke-Hartlieb *et al.*, 2000). TIMP-1 and TIMP-3 mRNA levels were also increased. It was proposed that MT1-MMP may facilitate the breakdown of glia inhibitory proteins, thereby enabling astrocytes to have access to sites of motor neuron degeneration. Examination of brain tissue of patients with Alzheimer's disease demonstrated production of MT1-MMP by white matter microglial cells (Yamada *et al.*, 1995). An effect on activation of proMMP-2 and subsequent processing of beta-amyloid protein was proposed.

e. Lung Disease.   In a murine model of lung emphysema caused by transgenic overexpression of interleukin-13, MT1-MMP, MMP-2, -9, -12, and -13 were increased; the severity of the emphysema was attenuated by treatment with a synthetic MMP-inhibitor (Zheng *et al.,* 2000). In an *in vivo* rat model of ventilator-induced lung injury, an increase in MT1-MMP mRNA in vascular endothelial cells and bronchial epithelial cells of lungs injured by high volume ventilation was demonstrated. The injury was accompanied by upregulation and release of MMP-2, MMP-9, and EMMPRIN in the lungs when compared to noninjured lungs ventilated with low total volume. The ventilator-induced injury was attenuated by treatment with an MMP inhibitor (AG3340, Prinomastat) (Foda *et al.,* 2000). Furthermore, in an *in vitro* model examining the effect of cell stretching on MMPs, it was found that stress applied to pulmonary microvascular endothelial cells led to an increase in mRNA production, protein release, and activation of MT1-MMP and MMP-2, as well as upregulation of TIMP-2 (Haseneen *et al.,* 2002).

f. Kidney Disease.   A role for MT-MMPs in glomerular diseases, especially the severe form of glomerulonephritis accompanied by disruption of glomerular membranes, has been proposed. MT1-MMP has been demonstrated to be induced in infiltrating macrophages during the development of crescentic glomerulonephritis and may contribute to pathologic degradation of glomerular extracellular matrix through activation of proMMP-2 (Hayashi *et al.,* 2000).

M. Synthetic Inhibitors of MT-MMPs

Rational design of potent inhibitors of MMPs has resulted in production of numerous agents which inhibit MT-MMPs and secreted MMPs. Initially drugs were targeted to the chemical functional group that chelates the active site zinc (II) ion, which is a ubiquitous feature of MMPs. Biochemical studies using collagen-like substrates have shown that six amino acids are primarily responsible for the proteolytic activity of MMPs. These six subsites within the enzyme catalytic domain span the locations designated S3 to S3'. Peptide and peptide-like compounds have been designed that combine backbone features (P1, P1', P2', P3'regions) that would favorably interact with the enzyme subsites (S1, S1', S2', S3' pockets) and with functional capabilities of binding zinc in the catalytic site (Borkakoti, 1998; Gomis-Ruth *et al.,* 1997; Grams *et al.,* 1995). The S1' site is the most prominent pocket in the MT-MMP catalytic domain. Most drugs designed as MMP inhibitors essentially mimic the collagen substrate of MMPs and thereby work as highly competitive, but reversible inhibitors of enzyme activity.

Potent inhibitors have been developed that are selective for the deep-pocket enzymes (MMP-2, MMP-3, MMP-8, MMP-9, MMP-13, and MT-MMPs) over the shallow pocket enzymes (MMP-1 and MMP-7). Achieving drug inhibitor selectivity among MMP-2, MMP-9, and MT1-MMP has been difficult (Yamamoto *et al.,*

1998) [see Zucker *et al.* (2002a) for a review of the subject]. Several hydroxamic acid-derived inhibitors of MMPs have shown promising results in preclinical studies in cancer and arthritis. However, clinical trials of MMP inhibitors in advanced cancer have yielded disappointing results. It has been pointed out that the use of these MMP inhibitors in patients with preexisting metastasis may have precluded a more favorable outcome (Zucker *et al.,* 2000).

An exciting new approach in the development of more specific MMP inhibitors has involved the use of phage display random peptide libraries. Using this approach, Koivunen *et al.* (1999) reported that peptides containing the sequence cyclic HWGF are potent and selective inhibitors of MMP-2 and MMP-9, but not MMP-8, MMP-13, and MT1-MMP. Another approach involves the screening of combinatorial libraries to identify small molecule antagonists of MMP function (Boger *et al.,* 2001). Based on the prominent role of MT-MMPs in physiologic and pathologic processes, the development of specific inhibitors of MT-MMPs is anxiously awaited.

IV. Summary and Concluding Remarks

The original concept that MMPs function simply by excavating the structural groundwork that supports cells and separates tissue compartment has undergone extensive revision. Considerable evidence has accumulated over the past 15 years to indicate that pericellular effects of MMPs play an important role in normal development, normal physiologic processes, and homeostatic mechanisms in the body (Sternlicht and Werb, 2001). The function of these proteinases at the cell surface is consistent with their role in altering the local environment surrounding the cell. It is recognized that extracellular matrices convey considerable inherent information and that these structures act as profound modifiers of cell behavior. The function of MMPs in altering the extracellular matrix and thereby unearthing cryptic structural information and releasing sequestered signal molecules is only beginning to be understood.

The discovery of the MT-MMP subfamily has greatly expanded our understanding of mechanisms by which cells degrade extracellular matrices. MT-MMPs exert profound effects, not only on extracellular matrix degradation, but also on cell migration. The role of individual MT-MMPs on cell function varies depending on the cell type and the environment. MT1-MMP is the most prominent and appears to be the most active member of this MMP subfamily. Our knowledge of the functional roles of the remaining MT-MMPs is probably very limited. Much remains to be uncovered in our understanding of interaction between MT-MMPs and cell surface adhesion molecules, for example, integrins and activation of signal transduction pathways. Likewise, the positive regulatory role of claudins and the negative role of the testican gene family on MT-MMP function remain to be explored. Much excitement has been generated by reports that MT1-MMP cleavage of laminin 5 and

CD44 results in cell migration; the physiologic and pathologic relevance of these observations remains to be clarified. Concerns about cellular functional modifications that may result from transfection of MT-MMP cDNAs into cells, resulting in synthesis of nonphysiologic amounts of proteinases, need to be addressed.

The role of the propeptide domain of MT-MMPs as intramolecular chaperones and the formation of MT-MMP oligomers at the cell surface need to be further explored. Additional structural and enzymatic analyses of MT-MMPs are required in order to identify the *in vivo* substrates of the different proteinases as well as to design selective inhibitors for experimentation in pathological conditions. Further studies will be necessary to identify the regulatory elements present in the promoter regions of these genes, as well as to establish the determinants of specificity of expression of each of them in normal and pathological tissues. It will be also fascinating to examine putative SNPs that could be responsible for altered expression or activity of different variants of MT-MMPs in disease. With the exception of MT1-MMP, we know little concerning the molecular mechanisms controlling MT-MMP expression in health and disease.

Inappropriate proteolysis appears to be involved in numerous diseases. A major challenge is to determine which aspects of MT-MMP proteolysis are causative of disease and which represent a physiologic response to the underlying disease process resulting in tissue repair. It is necessary to determine exactly which of the many potential substrates and mechanism of enzyme function identified *in vitro* actually regulate biologic processes *in vivo*.

The generation of animal models of gain or loss of function for MT2-, 3-, 4-, 5-, and 6-MMPs, similarly to those available for MT1-MMP, will be crucial for a better understanding of the roles of these MT-MMPs. These studies may also help to clarify the basis of the putative redundancy among MT-MMP subfamily members.

Undoubtedly, future studies of MT-MMPs will provide new insights about the function and regulation of a group of important, widely distributed cell surface proteinases that have captured the fancy of numerous investigators from a variety of scientific disciplines.

References

Abbruzzese, T. A., Guzman, R. J., Yee, C., Zarins, C. K., and Dalman, R. L. (1998). Matrix metalloproteinase inhibition limits arterial enlargement in a rodent arteriovenous fistula model. *Surgery* **124,** 328–334.

Afzal, S., Lalani, E.-N., Poulsom, R., Stubbs, A., Rowlinson, G., Sato, H., and Seiki, M. (1998). MT1-MMP and MMP-2 mRNA expression in human ovarian tumors: possible implications for the role of desmoplastic fibroblasts. *Hum. Pathol.* **29,** 155–165.

Ailenberg, M., and Silverman, M. (1996). Cellular activation of mesangial gelatinase A by cytochalasin D is accompanied by enhanced mRNA expression of both gelatinase A and its membrane-associated gelatinase A activator (MT-MMP). *Biochem. J.* **313,** 879–884.

Anders, A., Gilbert, S., Garten, W., Postina, R., and Fahrenhoz, F. (2001). Regulation of the alpha-secretase ADAM10 by its prodomain and proprotein convertases. *FASEB J.* **15,** 1837–1839.

Annabi, B., Lachambre, M.-P., Bousquet, N., Page, M., Gingras, D., and Beliveau, R. (2001). Localization of membrane-type 1 matrix metalloproteinase in caveolae membrane domains. *Biochem J.* **353,** 547–553.

Aoki, T., Yonezawa, K., Ohuchi, E., Fujimoto, N., Iwata, K., Shimada, T., Shiomi, T., Okada, O., and Seiki, M. (2002). Two-step sandwich enzyme immunoassay using monoclonal antibodies for detection of soluble and membrane-associated human membrane type 1-matrix metalloproteinase. *J. Immunoassay Immunochem.* **23,** 49–68.

Apte, S. S., Fukai, N., Beier, D. R., and Olsen, B. R. (1997). The matrix metalloproteinase-14 (MMP-14) gene is structurally distinct from other MMP genes and is co-expressed with the TIMP-2 gene during mouse embryogenesis. *J. Biol. Chem.* **272,** 25511–25517.

Ara, T., Deyama, Y., Yoshimura, Y., Higashino, F., Shindoh, M., Matsumoto, A., and Fukuda, H. (2000). Membrane type 1-matrix metalloproteinase expression is regulated by E-cadherin through the suppression of mitogen-activated protein kinase cascade. *Cancer Lett.* **157,** 115–121.

Atkinson, S. J., Patterson, M. L., Butler, M. J., and Murphy, G. (2001). Membrane type 1 matrix metalloproteinase and gelatinase A synergistically degrade type 1 collagen in a cell model. *FEBS Lett.* **491,** 222–226.

Aznavoorian, S., Moore, B. A., Alexander-Lister, D., Hallit, S. L., Windsor, J., and Engler, J. A. (2001). Membrane type 1-matrix metalloproteinase-mediated degradation of type I collagen by oral squamous cell carcinoma cells. *Cancer Res.* **61,** 6264–6275.

Azzam, H. S., and Thompson, E. W. (1992). Collagen-induced activation of the Mr 72,000 type IV collagenase in normal and malignant human fibroblastoid cells. *Cancer Res.* **52,** 4540–4544.

Bafetti, L. M., Young, T. N., Itoh, Y., and Stack, M. S. (1998). Intact vitronectin induces matrix metalloproteinase-2 and tissue inhibitor of metalloproteinase-2 expression and enhanced cellular invasion by melanoma cells. *J. Biol. Chem.* **273,** 143–149.

Balbín, M., Fueyo, A., Knauper, V., López, J. M., Alvarez, J., Sanchez, L. M., Quesada, V., Bordallo, J., Murphy, G., and López-Otín, C. (2001). Identification and enzymatic characterization of two diverging murine counterparts of human interstitial collagenase (MMP-1) expressed at sites of embryo implantation. *J. Biol. Chem.* **276,** 10253–10262.

Barmina, O. Y., Walling, H. W., Fiacco, G. J., Freije, J. M., Lopez-Otin, C., Jeffrey, J. J., and Partridge, C. A. (1999). Collagenase-3 binds to a specific receptor and requires the low density lipoprotein receptor-related protein for internalization. *J. Biol. Chem.* **274,** 30087–30093.

Bassi, D., Lopez De Cicco, R., Mahloogi, H., Zucker, S., Thomas, G., and Klein-Szanto, A. J. P. (2001). Furin inhibition results in absent or decreased invasiveness and tumorigenicity of human cancer cells. *Proc. Natl. Acad. Sci. USA* **98,** 10326–10331.

Baumann, P., Zigrino, P., Mauch, C., Breitkreut, D., and Nischt, R. (2000). MT1-MMP mediated proMMP-2 activation in epidermal tumor cells. *Brit. J. Cancer* **83,** 1387–1393.

Belien, A. T. J., Paganetti, P. A., and Schwab, M. E. (1999). Membrane-type 1 matrix metalloproteinase (MT1-MMP) enables invasive migration of glioma cells in central nervous system white matter. *J. Cell Biol.* **144,** 373–384.

Belkin, A. M., Akimov, S. S., Zaritskayal, L. S., Ratnikov, B. I., Deryugina, E. I., and Strongin, A. Y. (2001). Matrix-dependent proteolysis of surface transglutaminase by membrane-type metalloproteinase has been reported to regulate cancer cell adhesion and locomotion. *J. Biol. Chem.* **276,** 18415–18422.

Bello, L., Lucini, V., Carrabba, G., Giussani, C., Machluf, M., Pluderi, M., Kikas, D., Zang, J., Tomei, G., Villani, R. M., Carroll, R. S., Bikfalvi, A., and Black, P. M. (2001). Simultaneous inhibition of glioma angiogenesis, cell proliferation, and invasion by a naturally occurring fragment of human metalloproteinase-2. *Cancer Res.* **61,** 8730–8736.

Bendeck, M. P., Irvin, C., and Reidy, M. A. (1996). Inhibition of matrix metalloproteinase activity inhibits smooth muscle cell migration but not neointimal thickening after arterial injury. *Circ. Res.* **78,** 38–43.

Bini, A., Wu, D., Schnuer, J., and Kudryk, B. J. (1999). Characterization of stromelysin 1 (MMP-3), matrilysin (MMP-7), and membrane type 1 matrix metalloproteinase (MT1-MMP) derived fibrin(ogen) fragments D-dimer and D-like monomer: NH2-terminal sequences of late stage digest fragments. *Biochemistry* **38,** 13928–13936.

Birkedal-Hansen, H. (1995). Proteolytic remodeling of extracellular matrix. *Curr. Opin. Cell Biol.* **7,** 728–735.

Birkedal-Hansen, H., Moore, W. G. I., Bodden, M. K., Windsor, L. J., Birkedal-Hansen, B., DeCarlo, A., and Engler, J. A. (1993). Matrix metalloproteinases: A review. *Crit. Rev. Oral Biol. Med.* **42,** 197–250.

Biswas, C., Zhang, Y., DeCastro, R., Guo, H., Nakamura, T., Kataoka, H., and Nabeshima, K. (1995). The human tumor cell-derived collagenase stimulating factor (renamed EMMPRIN) is a member of the immunoglobulin superfamily. *Cancer Res.* **55,** 434–439.

Blavier, L., Lazaryev, A., Groffen, J., Heisterkamp, N., DeClerck, Y. A., and Kaartinen, V. (2001). TGF-b3-induced palatogenesis requires matrix metalloproteinases. *Mol. Biol. Cell.* **12,** 1457–1466.

Boger, D., Goldberg, J., Siletti, S., Kessler, T., and Cheresh, D. A. (2001). Identification of a novel class of small-molecule antiangiogenic agents through the screening of combinatorial libraries which function by inhibiting the binding and localization of proteinase MMP-2 to integrin avb3. *J. Am. Chem. Soc.* **123,** 1280–1288.

Borkakoti, N. (1998). Matrix metalloproteases: variations on a theme. *Prog. Biophys. Mol. Biol.* **70,** 73–94.

Brassart, B., Randoux, A., Hornebeck, W., and Emonard, H. (1998). Regulation of matrix metalloproteinase-2 (gelatinase A, MMP-2), membrane-type matrix metalloproteinase-1 (MT1-MMP) and tissue inhibitor of metalloproteinase-2 (TIMP-2) expression by elastin-derived peptides in human HT-1080 fibrosarcoma cell line. *Clin. Exp. Metastasis* **16,** 489–500.

Brew, K., Dinakarpandian, D., and Nagase, H. (2000). Tissue inhibitors of metalloproteinases: evolution, structure and function. *Biochim. Biophys. Acta* **1477,** 267–283.

Brinckerhoff, C. E., and Matrisian, L. M. (2002). Matrix metalloproteinases: a tail of a frog that became a prince. *Nat. Rev. Mol. Cell. Biol.* **3,** 207–214.

Brooks, P. C., Stromblad, S., Sanders, L. C., von Schalscha, T. L., Aimes, R. T., Stetler-Stevenson, W. G., Quigley, J. P., and Cheresh, D. A. (1996). Localization of matrix metalloproteinase MMP-2 to the surface of invasive cells by interaction with integrin avb3. *Cell* **85,** 683–693.

Brunet, C. L., Sharpe, P. M., and Ferguson, M. W. (1993). The distribution of epidermal growth factor binding sites in the developing mouse palate. *Int. J. Dev. Biol.* **37,** 451–457.

Bruno, G., Todor, R., Lewis, I., and Chyatte, D. (1998). Vascular extracellular matrix remodeling in cerebral aneurysms. *J. Neurosurg.* **89,** 431–440.

Butler, G. S., Butler, M. J., Atkinson, S. J., Will, H., Tamura, T., van Westrum, S. S., Crabbe, T., Clemens, J., d'Ortho, M.-P., and Murphy, G. (1998). The TIMP-2 membrane type 1 metalloproteinase "receptor" regulates the concentration and efficiency of progelatinase A. A kinetic study. *J. Biol. Chem.* **273,** 871–880.

Buttner, F. H., Chubinskaya, S., Margerie, D., Huch, K., Flechtenmacher, J., Cole, A. A., Kuettner, K. E., and Bartnik, E. (1997). Expression of membrane type 1 matrix metalloproteinase in human articular cartilage. *Arthritis Rheum.* **40,** 704–709.

Caenazzo, C., Onisto, M., Sartor, L., Scalerta, R., Giraldo, A., Nitti, D., and Garbisa, S. (1998). Augmented membrane type 1 matrix metalloproteinase (MT1-MMP): MMP-2 messenger RNA ratio in gastric carcinomas with poor prognosis. *Clin. Cancer Res.* **4,** 2179–2186.

Cao, J., Sato, J., Takino, T., and Seiki, M. (1995). The C-terminal region of membrane type matrix metalloproteinase is a functional transmembrane domain required for progelatinase A activation. *J. Biol. Chem.* **270,** 801–805.

Cao, J., Rehemtulla, A., Bahou, W., and Zucker, S. (1996). Membrane type matrix metalloproteinase 1 (MT-MMP1) activates progelatinase A without furin cleavage of the N-terminal domain. *J. Biol. Chem.* **271,** 30174–30180.

Cao, J., Hymowitz, M., Conner, C., Bahou, W. F., and Zucker, S. (2000). The propeptide domain of membrane type 1-matrix metalloproteinase acts as an intramolecular chaperone when expressed in trans with the mature sequence in COS-1 cells. *J. Biol. Chem.* **275,** 29648–29653.

Carmeliet, P. (2000). Proteinases in cardiovascular aneurysms and rupture: targets for therapy? *J. Clin. Invest.* **105,** 1519–1520.

Caterina, J. J., Yamada, S., Caterina, C. M., Longenecker, G., Holmback, K., Shi, J., Yermovsky, A. E., Enger, J. A., and Birkedal-Hansen, H. (2000). Inactivating mutation of the mouse tissue inhibitor of metalloproteinases-2 (Timp-2) gene alters proMMP-2 activation. *J. Biol. Chem.* **275,** 26416–26422.

Celentano, D. C., and Frishman, W. H. (1997). Matrix metalloproteinases and coronary artery disease: A novel therapeutic target. *J. Clin. Pharmacol.* **37,** 991–1000.

Chenard, M.-P., Lutz, Y., Mechine-Neuville, A., Stoll, I., Bellocq, J.-B., Rio, M.-C., and Basset, P. (1999). Presence of high levels of MT1-MMP protein in fibroblastic cells of human invasive carcinomas. *Int. J. Cancer* **82,** 208–212.

Chiarugi, V., Magnelli, L., Dello Sbarba, P., and Ruggiero, M. (2000). Tumor angiogenesis: thrombin and metalloproteinases in focus. *Exp. Mol. Pathol.* **69,** 63–66.

d'Ortho, M. P., Will, H., Atkinson, S., Butler, G., Messent, A., Gavrilovic, J., Smith, B. R. T., Zardi, L., and Murphy, G. (1997). Membrane type metalloproteinases 1 and 2 exhibit broad-spectrum proteolytic capacities comparable to many matrix metalloproteinases. *Eur. J. Biochem.* **250,** 751–757.

d'Ortho, M.-P., Stanton, H., Butler, M., Atkinson, S. J., Murhpy, G., and Hembry, R. M. (1998). MT1-MMP on the cell surface causes focal degradation of gelatin films. *FEBS Lett.* **421,** 159–164.

Davidson, B., Goldberg, I., Kopolovic, J., Lerner-Geva, L., Gotlieb, W. H., Ben-Baruch, G., and Reich, R. (1999). MMP-2 and TIMP-2 expression correlates with poor prognosis in cervical carcinoma—a clinicopathologic study using immunohistochemistry and mRNA in situ hybridization. *Gyn. Oncol.* **73,** 372–382.

Davies, B., Miles, D. W., Haperfield, L. C., Naylor, M. S., Bobrow, L. G., Rubens, R. D., and Balkwill, F. R. (1993). Activity of type IV collagenases in benign and malignant breast disease. *Br. J. Cancer* **67,** 1126–1131.

Davis, G. E. (1992). Affinity of integrins for damaged extracellular-matrix: alpha-v-beta-3 binds to denatured collagen type-1 through RGD sites. *Biochem. Biophys. Res. Commun.* **183,** 1025–1031.

de Smet, B. J. G. L., de Kleijn, D., Hanemaaijer, R., Verheijen, J. H., Robertus, L., van der Helm, Y. J. M., Borst, C., and Post, M. J. (2000). Metalloproteinase inhibition reduces constrictive arterial remodeling after balloon angioplasty. A study in the atherosclerotic Yacatan micropig. *Circulation* **2000,** 2962–2967.

Deryugina, E. I., Bourdon, M., Jungwirth, K., Smith, J. W., and Strongin, A. Y. (2000). Functional activation of integrin avb3 in tumor cells expressing membrane-type 1 matrix metalloproteinase. *Int. J. Cancer* **86,** 15–23.

Deryugina, E. I., Ratnikov, B., Monosov, E., Postnova, T. I., DiScipio, R., Smith, J., and Strongin, A. Y. (2001). MT1-MMP initiates activation of pro-MMP-2 and integrin avb3 promotes maturation of MMP-2 in breast carcinoma cells. *Exp. Cell Res.* **263,** 209–223.

Deryugina, E. I., Soroceanu, L., and Strongin, A. Y. (2002). Upregulation of vascular endothelial growth factor by membrane-type 1 matrix metalloproteinase stimulates human glioma xenograft growth and angiogenesis. *Cancer Res.* **62,** 580–588.

Dollery, C. M., McEwan, J. R., and Henney, A. M. (1995). Matrix metalloproteinases and cardiovascular disease. *Circ. Res.* **77,** 863–868.

Ducharme, A., Frantz, S., Aikawa, M., Rabkin, E., Linsey, M., Rohde, L. E., Schoen, F. J., Kelly, R. A., Werb, W., Libby, P., and Lee, R. T. (2000). Targeted deletion of matrix metalloproteinase-9 attenuates left ventricular enlargement and collagen accumulation after experimental infarction. *J. Clin. Invest.* **106**, 55–62.

Duhamel-Clerin, E., Orvain, C., Lanza, F., Cazenave, J.-P., and Klein-Soyer, C. (1997). Thrombin receptor mediated increase of two matrix metalloproteinases, MMP-1 and MMP-3, in human endothelial cells. *Arterioscler. Thromb. Vasc. Biol.* **17**, 1931–1938.

Egeblad, M., and Werb, Z. (2002). New functions for the matrix metalloproteinases in cancer progression. *Nat. Rev. Cancer* **2**, 163–176.

El Fahime, E., Torrente, Y., Caron, N. J., Bresolin, M. D., and Tremblay, J. P. (2000). In vivo migration of transplanted myoblasts require matrix metalloproteinase activity. *Exp. Cell Res.* **258**, 279–287.

Ellenrieder, V., Alber, B., Lacmer, U., Hendler, S. F., Menke, A., Boeck, W., Wagner, M., Wilda, M., Friess, H., Buchler, M., Adler, G., and Gress, T. A. (2000). Role of MT-MMPs and MMP-2 in pancreatic cancer progression. *Int. J. Cancer* **85**, 14–20.

Ellerbroek, S., and Stack, M. S. (1999). Membrane associated matrix metalloproteinase in metastasis. *BioEssays* **21**, 940–949.

Ellerbroek, S., Fishman, D. A., Kearns, A. S., Bafetti, L. M., and Stack, M. S. (1999). Ovarian carcinoma regulation of matrix metalloproteinase-2 and membrane type 2 matrix metalloproteinase through b2 integrin. *Cancer Res.* **59**, 1635–1641.

Ellis, V., and Murphy, G. (2001). Cellular strategies for proteolytic targeting during migration and invasion. *FEBS Lett.* **506**, 1–5.

English, W. R., Puente, X. S., Freije, J. P., Knauper, V., Amour, A., Merryweather, A., Lopez-Otin, C., and Murphy, G. (2000). Membrane type 4 matrix metalloproteinase (MMP17) has tumor necrosis factor-a convertase activity but does not activate pro-MMP2. *J. Biol. Chem.* **275**, 14046–14055.

English, W. R., Holtz, B., Vogt, G., Knauper, V., and Murphy, G. (2001a). Characterization of the role of the "MT-loop." An eight amino acid insertion specific to progelatinase A (MMP2) activating membrane-type matrix metalloproteinase. *J. Biol. Chem.* **276**, 42018–42026.

English, W. R., Velasco, G., Sracke, J. O., Knauper, V., and Murphy, G. (2001b). Catalytic activities of membrane-type 6 matrix metalloproteinase (MMP25). *FEBS Lett.* **491**, 137–142.

Fernandez-Catalan, C., Bode, W., Huber, R., Turk, D., Calvete, J. J., Lichte, A., Tschesche, H., and Maskos, K. (1998). Crystal structure of the complex formed by the membrane type 1-matrix metalloproteinase with tissue inhibitor of metalloproteinase-2, the soluble progelatinase A receptor. *EMBO J.* **17**, 5238–5248.

Fishman, D. A., Liu, Y., Ellerbroek, S. M., and Stack, M. S. (2001). Lysophosphatidic acid promotes matrix metalloproteinase (MMP) activation and MMP-dependent invasion in ovarian cancer cells. *Cancer Res.* **61**, 3194–3199.

Foda, H. D., George, S., Conner, C., Drews, M., Tompkins, D. C., and Zucker, S. (1996). Activation of human umbilical vein endothelial cell progelatinase A by phorbol myristate acetate (PMA): A protein kinase C-dependent mechanism involving a membrane-type matrix metalloproteinase. *Lab. Invest.* **74**, 538–545.

Foda, H. D., George, S., Rollo, E., Drews, M., Conner, C., Cao, J., Panettieri, R. A., Jr., and Zucker, S. (1999). Regulation of gelatinases in human airway smooth muscle cells: Mechanism of progelatinase A activation. *Am. J. Physiol.* **227**, L174–182.

Foda, J. D., Rollo, E. E., Drews, M., Conner, C., Appelt, K., Shalinsky, D. R., and Zucker, S. (2000). Ventilator-induced lung injury upregulates and activates gelatinases and EMMPRIN: Attenuation by the synthetic matrix metalloproteinase inhibitor, Prinomastat. *Am. J. Respir. Cell Mol. Biol.* **25**, 717–724.

Fosang, A. J., Last, K., Fujii, Y., Seiki, M., and Okada, Y. (1998). Membrane-type 1 MMP (MMP-14) cleaves at three sites in the aggrecan integlobular domain. *FEBS Lett.* **430**, 186–190.

Fukuda, Y., Ishizaki, M., Okada, Y., Seiki, M., and Yamanaka, N. (2000). Matrix metalloproteinases and tissue inhibitor of metalloproteinase-2 in fetal rabbit lung. *Am. J. Physiol.* **279**, L555–L561.

Galis, Z. S. (1999). Metalloproteases in remodeling of vascular extracellular matrix. *Fibrinolysis Proteolysis* **13,** 54–63.

Galis, Z. S., Kranzhofer, R., Fenton, J. W., and Libby, P. (1997). Thrombin promotes activation of matrix metalloproteinase-2 produced by cultured vascular smooth muscle cells. *Arterioscler. Thromb. Vasc. Biol.* **17,** 483–489.

Galvez, B. G., Matias-Roman, S., Albar, J. P., Sanchez-Madrid, F., and Arroyo, A. G. (2001). Membrane type 2-matrix metalloproteinase is activated during migration of human endothelial cells and modulates endothelial motility and matrix remodeling. *J. Biol. Chem.* **276,** 37491–37500.

Giannelli, G., Falk-Marzillier, J., Schiraldi, O., Stetler-Stevenson, W. G., and Quaranta, V. (1997). Induction of cell migration by matrix metalloproteinase-2 cleavage of laminin-5. *Science* **277,** 225–228.

Gilles, C., Bassuk, J. A., Pulyaeva, H., Sage, E. H., Foidart, J. M., and Thompson, E. W. (1998). SPARC/osteonectin induces matrix metalloproteinase 2 activation in human breast cancer cell lines. *Cancer Res.* **58,** 5529–5536.

Gilles, C., Polette, M., Coraux, C., Tournier, J.-M., Meneguzzi, G., Munaut, C., Volders, L., Rousselle, P., Birembaut, P., and Foldart, J.-M. (2001). Contribution of MT1-MMP and of human laminin-5 g2 chain degradation to mammary epithelial cell migration. *J. Cell Sci.* **114,** 2967–2976.

Gingras, D., Page, M., Annabi, B., and Beliveau, R. (2000). Rapid activation of matrix metalloproteinase-2 by glioma cells occurs through a posttranslational MT1-MMP-dependent mechanism. *Biochim. Biophys. Acta* **1497,** 341–350.

Gingras, D., Bousquet, N., Langlois, S., Lamchambre, M.-P., Annabi, B., and Beliveau, R. (2001). Activation of the extracellular signal-regulated protein kinase (ERK) cascade by membrane-type I matrix metalloproteinase (MT1-MMP). *FEBS Lett.* **507,** 231–236.

Godin, D., Ivan, E., Johnson, C., Magid, R., and Galis, Z. S. (2000). Remodeling of carotid artery is associated with increased expression of matrix metalloproteinases in mouse blood flow cessation model. *Circulation* **102,** 2861–2866.

Gomis-Ruth, F. X., Maskos, K., Betz, M., Bergner, A., Huber, R., Suzuki, K., Yoshida, N., Nagase, H., Brew, K., Bourenkov, G. P., Bartunik, H., and Bode, W. (1997). Mechanism of inhibition of the human matrix metalloproteinase stromelysin-1 by TIMP-1. *Nature* **389,** 77–81.

Graesser, D., Mahooti, S., Haas, T. L., Davis, S. J., Clark, R., and Madri, J. A. (1998). The interrelationship of alpha4 integrin and matrix metalloproteinase-2 in the pathogenesis of experimental autoimmune encephalomyelitis. *Lab. Invest.* **78,** 1445–1458.

Graesser, D., Mahooti, S., and Madri, J. A. (2000). Distinct role for matrix metalloproteinase-2 and alpha 4 integrin in autoimmune T cell extravasation and residence in brain parenchyma during experimental autoimmune encephalitis. *J. Neuroimmunol.* **109,** 121–131.

Grams, F., Crimmin, M., Hinnes, L., Huxley, P., Tschesche, H., and Bode, W. (1995). Structure determination and analysis of human neutrophil collagenase complexed with a hydroxamate inhibitor. *Biochemistry* **34,** 14012–14020.

Grant, G. M., Giambernardi, T. A., Grant, A. M., and Klebe, R. J. (1999). Overview of expression of matrix metalloproteinases (MMP-17, MMP-18, and MMP-20) in cultured human cells. *Matrix Biol.* **18,** 145–148.

Gross, J., and Lapiere, C. M. (1962). Collagenolytic activity in amphibian tissues; a tissue culture assay. *Proc. Natl. Acad. Sci. USA* **48,** 1014–1022.

Ha, H.-Y., Moon, H.-B., Nam, M.-S., Lee, J.-W., Ryoo, Z.-Y., Lee, T.-H., Lee, K.-K., So, B.-J., Sato, H., Seiki, M., and Yu, D.-Y. (2001). Overexpression of membrane-type matrix metalloproteinase-1 gene induces mammary gland abnormalities and adenocarcinoma in transgenic mice. *Cancer Res.* **61,** 984–990.

Haas, T. L., and Madri, J. A. (1999). Extracellular matrix-driven matrix metalloproteinase production in endothelial cells: Implications for angiogenesis. *Trends Cardiovasc. Med.* **9,** 70–77.

Haas, T. L., Davis, S. J., and Madri, J. A. (1998). Three dimensional type I collagen lattices induce coordinate expression of matrix metalloproteinases MT1-MMP and MMP-2 in microvascular endothelial cells. *J. Biol. Chem.* **273,** 3604–3610.

Haas, T. L., Stitelman, D., Davis, S. J., Apte, S. S., and Madri, J. A. (1999). Egr-1 mediates extracellular matrix-driven transcription of membrane type 1 matrix metalloproteinase in endothelium. *J. Biol. Chem.* **274,** 22679–22685.

Habelhah, H., Okada, F., Kobayashi, M., Nakai, K., Choi, S., Hamada, J.-I., Moriuchi, T., Kaya, M., Yoshida, K., Fujinaga, K., and Hosokawa, M. (1999). Inflammatory E1Af expression in mouse fibrosarcoma promotes metastasis through induction of MT1-MMP expression. *Oncogene* **18,** 1771–1776.

Hajitou, A., Sounni, N. E., Devy, L., Grignet-Debrus, C., Lewalle, J. M., Li, H., Deroanne, C. F., Lu, H., Colige, A., Nusgens, B. V., Frankenne, F., Maron, A., Yeh, P., Perricaudet, M., Chang, Y., Soria, C., Calberg-Bacq, C. M., Foidart, J. M., and Noel, A. (2001). Down-regulation of vascular endothelial growth factor by tissue inhibitor of metalloproteinase-2: effect on in vivo mammary tumor growth and angiogenesis. *Cancer Res.* **61,** 3450–3457.

Hamasuna, R., Kataoka, H., Moriyama, T., Itoh, H., and Seiki, M. (1999). Regulation of matrix metalloproteinase-2 (MMP-2) by hepatocyte growth factors/scatter factor (HGF/SF) in human glioma cells. HGF/SF enhances MMP-2 expression and activation accompanying up regulation of membrane type-1 MMP. *Int. J. Cancer* **82,** 274–281.

Han, Y.-P., Tuan, T.-L., Wu, H., Hughes, M., and Garner, W. (2000). TNF-a stimulates activation of pro-MMP2 in human skin through NF-kB mediated induction of MT1-MMP. *J. Cell Sci.* **114,** 131–139.

Harada, T. S. A., Mise, M., Imamura, T., Higashitsuji, H., Furutani, M., Niwano, M., Ishigami, S., Fukumoto, M., Seiki, M., Sato, H., and Imamura, M. (1998). Membrane-type matrix metalloproteinase-1 (MT1-MMP) gene is overexpressed in highly invasive hepatocellular carcinomas. *J. Hepatol.* **28,** 231–239.

Harayama, T., Ohuchi, E., Aoki, T., Sato, H., Seiki, M., and Okada, Y. (1999). Shedding of membrane type I matrix metalloproteinase in a human breast carcinoma cell line. *Jpn. J. Cancer Res.* **90,** 942–950.

Haseneen, N., Vaday, G., Zucker, S., and Foda, H. D. (2002). Mechanical stretch induces an increase in MMP-2 release and activation in human lung microvascular endothelial cells: The potential role of EMMPRIN. *Am. J. Respir. Crit. Care Med.* **165** (abstract).

Hayashi, K., Hirokashi, S., Osada, S., Shofuda, K.-I., Shirato, I., and Tomino, Y. (2000). Macrophage-derived MT1-MMP and increased MMP-3 activity are associated with glomerulonephritis. *J. Pathol.* **191,** 299–305.

Heppner, K. J., Matrisian, L. M., and Jensen, R. A. (1996). Expression of most matrix metalloproteinase family members in breast cancer represents a tumor-induced host response. *Am. J. Pathol.* **149,** 273–282.

Hernandez-Barrantes, S., Toth, M., Bernardo, M., Yurkova, M., Gervasi, D. C., Raz, Y., Sang, Q. A., and Fridman, R. (2000). Binding of active (57 kDa) membrane type-1 matrix metalloproteinase (MT1-MMP) to tissue inhibitor of metalloproteinase (TIMP)-2 regulates MT1-MMP processing and pro-MMP-2 activation. *J. Biol. Chem.* **275,** 12080–12089.

Hiller, O., Lichte, A., Oberpichler, A., Kocourek, A., and Tchesche, H. (2000). Matrix metalloproteinases collagenase-2, macrophage elastase, collagenase-3, and membrane type 1-matrix metalloproteinase impair clotting by degradation of fibrinogen and factor XII. *J. Biol. Chem.* **275,** 33008–33013.

Hiraoka, N., Allen, E., Apel, I. J., Gyetko, M. R., and Weiss, S. J. (1998). Matrix metalloproteinases regulate neovascularization by acting as pericellular fibrinolysins. *Cell* **95,** 365–377.

Hoegy, S. E., Oh, H. R., Corcoran, M. L., and Stetler-Stevenson, W. G. (2001). Tissue inhibitor of metalloproteinases-2 (TIMP-2) suppresses TKR-growth factor signaling independent of metalloproteinase inhibition. *J. Biol. Chem.* **276,** 3203–3214.

Hofman, U. B., Westphal, J. R., Zendman, A. J. W., Becker, J. C., Ruiter, D. J., and van Muijen, G. N. P. (2000). Expression and activation of matrix metalloproteinase-2 (MMP-2) and its co-localization with membrane-type 1 matrix metalloproteinase (MT1-MMP) correlate with melanoma progression. *J. Pathol.* **191,** 245–256.

Holmbeck, K., Bianco, P., Caterina, J., Yamada, S., Kromer, M., Kuznetsov, S. A., Mankani, M., Robey, P. G., Poole, R., Pidoux, I., Ward, J. M., and Birkedal-Hansen, H. (1999). MT1-MMP deficient mice develop dwarfism, osteopenia, arthritis, and connective tissue disease due to inadequate collagen turnover. *Cell* **99,** 81–92.

Hotary, K., Allen, E., Punturieri, A., Yana, I., and Weiss, S. J. (2000). Regulation of cell invasion and morphogenesis in a three-dimensional type I collagen matrix by membrane-type matrix metalloproteinases 1, 2, and 3. *J. Cell Biol.* **149,** 1309–1323.

Hotary, K., Yana, I., Sabeh, F., Xiao-Y. L., Holmbeck, K., Birkedal-Hansen, H., Allen, E. D., Hiraoka, N., and Weiss, S. J. (2002). Matrix metalloproteinase (MMPs) regulate fibrin-invasive activity via MT1-MMP-dependent and -independent processes. *J. Exp. Med.* **195,** 295–308.

Huhtala, P., Chow, L., and Tryggvason, K. (1990). Structure of the human type IV collagenase gene. *J. Biol. Chem.* **265,** 11077–11082.

Huhtala, P., Tuuttila, I. A., Chow, L. T., Lohi, J., Keski-Oja, J., and Tryggvason, K. (1991). Complete structure of the human gene for 92-kDa Type IV collagenase. Divergent regulation of expression for the 92- and 72-kDa enzyme genes in HT-1080 cells. *J. Biol. Chem.* **266,** 16485–16490.

Hurskainen, T., Seiki, M., Apte, S. S., Syrjakallio-Ylitalol, M., Sorsa, T., Oikarinen, A., and Autio-Harmainen, H. (1998). Production of membrane-type matrix metalloproteinase-1 (MT-MMP-1) in early human placenta: a role in placental implantation. *J. Histochem. Cytochem.* **46,** 221–229.

Imai, K., Hiramatsu, A., Fukushima, D., Pierschbacher, M. D., and Okada, Y. (1997). Degradation of decorin by matrix metalloproteinase: identification of the cleavage sites, kinetic analysis and transforming growth factor beta-1 release. *Biochem. J.* **322,** 809–814.

Imai, K., Ohuchi, E., Aoki, T., Nomura, H., Fujii, Y., Sato, H., Seiki, M., and Okada, Y. (1996). Membrane-type matrix metalloproteinase 1 is a gelatinolytic enzyme and is secreted in a complex with tissue inhibitor of metalloproteinases 2. *Cancer Res.* **56,** 2707–2710.

Imamura, T., Oshio, G., Mise, M., Harada, T., Suwa, H., Okada, N., Wang, Z., Yoshitomi, S., Tanaka, T., Sato, H., Arii, S., Seiki, M., and Imamura, M. (1998). Expression of membrane-type matrix metalloproteinase-1 in human pancreatic adenocarcinomas. *J. Cancer Res. Clin.* **124,** 65–72.

Imanashi, Y., Fujii, M., Tokumaru, Y., Tomita, T., Kanke, M., Kanjaki, J., Kameyama, K., Otani, Y., and Sato, H. (2000). Clinical significance of expression of membrane type 1 matrix metalloproteinase and matrix metalloproteinase-2 in human head and neck squamous cell carcinoma. *Hum. Pathol.* **31,** 895–904.

Ishii, M., and Asuwa, N. (2000). Collagen and elastin degradation by matrix metalloproteinases and tissue inhibitors of matrix metalloproteinase in aortic dissection. *Hum. Pathol.* **31,** 640–646.

Ito, A., Yamada, M., Sato, H., Sanekata, K., Sato, H., Seiki, M., Nagase, H., and Mori, Y. (1998). Calmodulin antagonists increase the expression of membrane-type-1 matrix metalloproteinase in human uterine cervical fibroblasts. *Eur. J. Biochem.* **251,** 353–358.

Ito, H., Seyama, Y., and Kubota, S. (2001). Calreticulin is directly involved in anti-a3 integrin antibody-mediated secretion and activation of matrix metalloproteinase. *Biochem. Biophys. Res. Commun.* **283,** 297–302.

Itoh, T., Tanioka, M., Yoshida, H., Yoshioka, T., Nishimoto, H., and Itohara, S. (1998). Reduced angiogenesis and tumor progression in gelatinase A-deficient mice. *Cancer Res.* **58,** 1048–1051.

Itoh, Y., Ito, A., Iwata, K., Tanzawa, K., Mori, Y., and Nagase, H. (1998). Plasma membrane-bound tissue inhibitor of metalloproteinases (TIMP-2) specifically inhibits matrix metalloproteinase 2 (gelatinase A) activated on the cell surface. *J. Biol. Chem.* **273,** 24360–24367.

Itoh, Y., Kajita, M., Kinoh, H., Mori, H., Okada, A., and Seiki, M. (1999). Membrane type 4 matrix metalloproteinase (MT4-MMP, MMP-17) is a glycosylphosphatidylinositol-anchored proteinase. *J. Biol. Chem.* **274,** 34260–34266.

Itoh, Y., Takamura, A., Ito, N., Maru, Y., Sato, H., Suenaga, N., Aoki, T., and Seiki, M. (2001). Homophilic complex formation of MT1-MMP facilitates proMMP-2 activation on the cell surface and promotes tumor cell invasion. *EMBO J.* **20,** 4782–4793.

Jenkins, G. M., Crown, M. T., Bilato, C., Gluzband, Y., Ryu, W.-S., Li. Z., Stetler-Stevenson, W.,

Nater, C., Froehlich, J. P., Lakatta, E. G., and Cheng, L. (1998). Increased expression of membrane-type matrix metalloproteinase and preferential localization of matrix metalloproteinase-2 to the neointimal of balloon-injured rat carotid arteries. *Circulation* **97**, 82–90.

Jiang, A., Lehti, K., Wang, X., Weiss, S. J., Keski-Oja, J., and Pei, D. (2001). Regulation of membrane-type matrix metalloproteinase 1 activity by dynamin-mediated endocytosis. *Proc. Natl. Acad. Sci. USA* **98**, 13693–13698.

Jimenez, M. J. G., Balbin, M., Alvarez, J., Komori, T., Bianco, P., Holmbeck, K., Birkedal-Hansen, H. J. M. L., and Lopez-Otin, C. (2001). A regulatory cascade involving retinoic acid, Cbfa1, and matrix metalloproteinases is coupled to the development of a process of perichondrial invasion and osteogenic differentiation during bone formation. *J. Cell Biol.* **155**, 1333–1344.

Jo, Y., Yeon, J., Kim, H.-J., and Lee, S.-T. (2000). Analysis of tissue inhibitor of metalloproteinases-2 effect on pro-matrix metalloproteinase-2 activation by membrane-type 1 matrix metalloproteinase using baculovirus/insect-cell expression system. *Biochem. J.* **345**, 511–519.

Jones, P. L., and Jones, F. S. (2000). Tenascin-C in development and disease: gene regulation and cell function. *Matrix Biol.* **19**, 581–596.

Kadono, Y., Shibahara, K., Nakmiki, M., Watanabe, Y., Seiki, M., and Sato, H. (1998). Membrane type 1-matrix metalloproteinase is involved in the formation of hepatocyte growth factor/scatter factor-induced branching tubules in madin-darby canine kidney epithelial cells. *Biochem. Biophys. Res. Commun.* **251**, 681–687.

Kajita, M., Itoh, Y., Chiba, T., Mori, H., Okada, A., Kinoh, H., and Seiki, M. (2001). Membrane-type 1 matrix metalloproteinase cleaves CD44 and promotes cell migration. *J. Cell. Biol.* **153**, 893–904.

Kalebic, T., Gabrisa, S., Glaser, B., and Liotta, L. A. (1983). Basement membrane collagen: Degradation by migrating endothelial cells. *Science* **221**, 281–284.

Kanamon, Y., Matsushima, M., Minaguchi, T., Kobayashi, K., Sagae, S., Kuodo, R., Terakawa, N., and Nakamura, Y. (1999). Correlation between expression of the matrix metalloproteinase 2 gene in ovarian cancers and an insertion/deletion polymorphism in its promoter region. *Cancer Res.* **59**, 4225–4227.

Kang, T., Yi, J., Guo, A., Wang, X., Overall, C. M., Jiang, W., Elde, R., Borregaard, N., and Pei, D. (2001). Subcellular distribution and cytokine- and chemokine-regulated secretion of leukolysin/MT6-MMP/MMP-25 in neutrophils. *J. Biol. Chem.* **276**, 21960–21968.

Kang, T., Nagase, H., and Pei, D. (2002). Activation of membrane-type matrix metalloproteinase 3 zymogen by the proprotein convertase furin in the trans-Golgi network. *Cancer Res.* **62**, 675–681.

Kazes, I., Delarue, F., Hagege, J., Bouzhir-Sima, L., Rondeau, E., Sraer, J.-D., and Nguyen, G. (1998). Soluble latent membrane-type 1 matrix metalloproteinase secreted by human mesangial cells is activated by urokinase. *Kidney Int.* **54**, 1976–1984.

Kazes, I., Elalamy, I., Sraer, J. D., Hatmi, M., and Nguyen, G. (2000). Platelet release of trimolecular complex components MT1-MMP/TIMP2/MMP2: involvement in MMP2 activation and platelet aggregation. *Blood.* **96**: 3064–3069.

Kheradmand, F., Rishi, K., and Werb, Z. (2002). Signaling through the EGF receptor controls lung morphogenesis in part by regulating MT1-MMP-mediated activation of gelatinase A/MMP2. *J. Cell Sci.* **115**, 839–848.

Kikuchi, R., Noguchi, T., Takeno, S., Kubo, N., and Uchida, Y. (2000). Immunohistochemical detection of membrane-type-1-matrix metalloproteinase in colorectal carcinoma. *Br. J. Cancer* **83**, 215–218.

Kimura, A., Shinohara, M., Ohkura, R., and Takahashi, T. (2001). Expression and localization of transcripts of MT5-MMP and its related MMP in the ovary of the medaka fish *Oryzias latipes*. *Biochim. Biophys. Acta* **1518**, 115–123.

Kinoh, H., Sato, H., Tsunezuka, Y., Takino, T., Kawashima, A., Okada, Y., and Seiki, M. (1996). MT-MMP, the cell surface activator of proMMP-2 (progelatinase A) is expressed with its substrate in mouse tissue during embryogenesis. *J. Cell Sci.* **109**, 953–959.

Kitagawa, Y., Kunimi, K., Ito, H., Sato, H., Uchibayashi, Okada, Y., Seiki, M., and Namiki, M. (1998). Expression and tissue localization of membrane-types 1, 2, and 3 matrix metalloproteinases in human urothelial carcinomas. *J. Urol.* **160,** 1540–1545.

Kitagawa, Y., Kunimi, K., Uchibayashi, T., Sato, H., and Namiki, M. (1999). Expression of messenger RNAs for membrane-type 1, 2, and 3 matrix metalloproteinases in human renal cell carcinomas. *J Urol.* **162,** 905–909.

Kleijn, D. P. V., Sluijter, J. P. G., Smit, J., Velema, E., Richard, W., Schoneveld, A. H., Pasterkamp, G., and Borst, C. (2001). Furin and membrane type-1 metalloproteinase mRNA levels and activation of metalloproteinase-2 are associated with arterial modeling. *FEBS Lett.* **501,** 37–41.

Knauper, V., and Murphy, G. (1999). Membrane-type matrix metalloproteinases and cell surface-associated activation cascades for matrix metalloproteinases. *In* "Matrix Metalloproteinases" (W. C. Parks and R. P. Mecham, Eds.), pp. 199–218. Academic Press, San Diego.

Knauper, V., Will, H., Lopez-Otin, C., Atkinson, S. J., Stanton, H., Hembry, R. M., and Murphy, G. (1996). Cellular mechanism for human procollagenase-3 (MMP-13) activation. *J. Biol. Chem.* **271,** 17124–17131.

Knauper, V., Docherty, A. J. P., Smith, B., Tschesche, H., and Murphy, G. (1997). Analysis of the contribution of the hinge region of human neutrophil collagenase (HNC, MMP-8) to stability and collagenolytic activity by alanine scanning mutagenesis. *FEBS Lett.* **480,** 60–64.

Koivunen, E., Arap, W., Valtanen, H., Rainisalo, A., Medina, O. P., Heikkila, P., Kantor, C., Gahmberg, C. G., Salo, T., Konttinen, Y. T., Sorsa, T., Ruoslahti, E., and Pasqualini, R. (1999). Tumor targeting with a selective gelatinase inhibitor. *Nat. Biotechnol.* **17,** 768–774.

Kojima, S., Itoh, Y., Matsumoto, S., Masuho, Y., and Seiki, M. (2000). Membrane-type 6 matrix metalloproteinase (MT6-MMP, MMP-25) is the second glycosyl-phosphatidyl-inositol (GPI)-anchored MMP. *FEBS Lett.* **480,** 142–146.

Konttinen, Y. T., Ceponis, A., Takagi, M., Ainola, M., Sorsa, T., Sutinen, M., Salo, T., Ma, J., Santavirta, S., and Seiki, M. (1998). New collagenase enzymes/cascade identified at the panus–hard tissue junction in rheumatoid arthritis: destruction from above. *Matrix Biol.* **17,** 585–601.

Koshikawa, N., Giannelli, G., Curculli, V., Miyazaki, K., and Qaranta, V. (2000). Role of surface metalloproteinase MT1-MMP in epithelial migration over laminin-5. *J. Cell Biol.* **148,** 615–624.

Kotra, L. P., Dross, J. B., Shimura, Y., Fridman, R., Schlegel, H. B., and Mobashery, S. (2001). Insight into the complex and dynamic process of activation of matrix metalloproteinases. *J. Am. Chem. Soc.* **123,** 3108–3113.

Kuo, A., Zhong, C., Lane, W. S., and Derynck, R. (2000). The cytoplasmic tail of MT1-MMP has been shown to bind to the Golgi protein p59 (GRASP 55) as does TGF-a; the role of this association appears to be in intracellular trafficking of MT1-MMP. *EMBO J.* **19,** 6427–6439.

Kurschat, P., Zigrino, P., Nischt, R., Breitkreutz, D., Steurer, P., Klein, E. C., Krieg, T., and Mauch, C. (1999). Tissue inhibitor of matrix metalloproteinase-2 regulates matrix metalloproteinase-2 activation by modulation of membrane-type 1 matrix metalloproteinase (MT1-MMP) activity in high and low invasive melanoma cells. *J. Biol. Chem.* **274,** 21056–21062.

Lafleur, M. A., Forsyth, P. A., Atkinson, S. J., Murphy, G., and Edwards, D. R. (2001a). Perivascular cells regulate endothelial membrane type-1 matrix metalloproteinase activity. *Biochem. Biophys. Res. Commun.* **282,** 463–473.

Lafleur, M. A., Hollenberg, M. D., Atkinson, S. J., Kanauper, V., Murphy, G., and Edwards, D. R. (2001b). Activation of pro-(matrix metalloproteinase-2) (pro-MMP-2) by thrombin is membrane-type-MMP-dependent in human umbilical vein endothelial cells and generates a distinct 63 kDa active species. *Biochem. J.* **357,** 107–115.

Langton, K. P., McKie, N., Curtis, A., Bond, P. M., Barker, M. D., and Clarke, M. (2000). A novel tissue inhibitor of metalloproteinases-3 mutation reveals a common molecular phenotype in Sorby's fundus dystrophy. *J. Biol. Chem.* **275,** 27027–27031.

Lee, A. Y., Akers, K. T., Collier, L., Li, L., Eisen, J. L., and Seltzer, J. L. (1997). Intracellular activation of gelatinase A (72 kDa tpe IV collagenase) by normal fibroblasts. *Proc. Natl. Acad. Sci. USA* **90,** 4424–4429.

Lehti, K., Lohi, J., Valtanen, H., and Keski-Oja, J. (1998). Proteolytic processing of membrane-type-1 matrix metalloproteinase is associated with gelatinase activation at the cell surface. *Biochem. J.* **334,** 345–353.

Lehti, K., Valtanen, H., Wickstrom, S., Lohi, J., and Keski-Oja, J. (2000). Regulation of membrane-type-1 matrix metalloproteinase activity by its cytoplasmic domain. *J. Biol. Chem.* **275,** 15006–15013.

Lehti, K., Lohi, J., Juntunen, M. M., Pei, D., and Keski-Oja, J. (2002). Oligomerization through hemopexin and cytoplasmic domains regulates the activity and turnover of membrane-type I matrix metalloproteinase (MT1-MMP). *J. Biol. Chem.* **277,** 8440–8448.

Leville, C. D., Dassow, M. S., Seabrook, G. R., Jean-Claude, J. M., Towne, J. B., and Cambria, R. A. (2000). All-trans-retinoic acid decreases vein graft intimal hyperplasia and matrix metalloproteinase activity in vivo. *J. Surg. Res.* **90,** 183–190.

Li, L., Akers, K., Eisen, A. Z., and Seltzer, J. L. (1997). Activation of gelatinase A (72 kDa type IV collagenase) induced by monensin in normal human fibroblasts. *Exp. Cell Res.* **232,** 322–330.

Li, H., Bauzon, D. E., Xu, X., Tschesche, H., Cao, J., and Sang, Q. A. (1998). Immunological characterization of cell-surface and soluble forms of membrane type I matrix metalloproteinase in human breast cancer cells and in fibroblasts. *Mol. Carcinogen.* **22,** 84–94.

Libby, P. (2000). Changing concepts in atherogenesis. *J. Int. Med.* **247,** 349–358.

Libby, P., and Lee, R. T. (2000). Matrix matters. *Circulation* **102,** 1874–1876.

Lijnen, H. R., Van Hoef, B., Lupu, F., Moons, L., Carmeliet, P., and Collen, D. (1998). Function of the plasminogen/plasmin and matrix metalloproteinase systems after vascular injury in mice with targeted inactivation of fibrinolytic system genes. *Thromb. Vasc. Biol.* **18,** 1035– 1045.

Liotta, L. A. (1992). Cancer cell invasion and metastasis. *Sci. Am.,* **266:** 54–59, Feb.

Llano, E., Pendas, A. M., Freije, J. P., Nakano, A., Knauper, V., Murphy, G., and Lopez-Otin, C. (1999). Identification and characterization of human MT5-MMP, a new membrane-bound activator of progelatinase A overexpressed in brain tumors. *Cancer Res.* **59,** 2570–2576.

Llano, E., Pendás, A. M., Aza-Blanc, P., Kornberg, T. B., and López-Otín, C. (2000). Dm1-MMP, a matrix metalloproteinase from *Drosophila* with a potential role in extracellular matrix remodeling during neural development. *J. Biol. Chem.* **275,** 35978–35985.

Llano, E., Adams, G., Pendas, A. M., Quesada, V., Sanchez, L. M., Santamaria, I., Noseli, S., and Lopez-Otin, C. (2002). Structural and enzymatic characterization of *Drosphila* Dm2-MMP, a membrane-bound matrix metalloproteinase with tissue-specific expression. *J. Biol. Chem.* **277,** 23321–23329.

Lohi, J., Lehti, K., Westermarck, J., Kahari, V.-M., and Keski-Oja, J. (1996). Regulation of membrane-type matrix metalloproteinase-1 expression by growth factors and phorbol 12-myristate 13-acetate. *Eur. J. Biochem.* **239,** 239–247.

Lohi, J., Lehti, K., Valtanen, H., Parks, W., and Keski-Oja, J. (2000). Structural analysis and promoter characterization of the human membrane-type matrix metalloproteinase-1 (MT1-MMP) gene. *Gene* **242,** 75–86.

Lohi, J., Wilson, C. L., Roby, J. D., and Parks, W. C. (2001). Epilysin, a novel human matrix metalloproteinase (MMP-28) expressed in testis and keratinocytes and in response to injury. *J. Biol. Chem.* **276,** 10134–10144.

Long, L., Navab, R., and Brodt, P. (1998). Regulation of Mr 72,000 type IV collagenase by the type I insulin-like growth factor receptor. *Cancer Res.* **58,** 3243–3247.

Lu, L., Gunja-Smith, Z., Woessner, F., Ursell, P. C., Nissen, T., Galardy, R. E., Xu, Y., Zhu, P., and Schwartz, G. G. (2000). Matrix metalloproteinases and collagen ultrastructure in moderate myocardial ischemia and reperfusion in vivo. *Am. J. Physiol.* **279,** H601–609.

Lukes, A., Mun-Bruce, S., Lukes, M., and Rosenberg, G. A. (1999). Extracellular matrix degradation by metalloproteinases and central nervous system disease. *Mol. Neurobiol.* **19,** 267–284.

Maata, M., Soini, Y., Liakka, A., and Autio-Harmainen, H. (2000). Localization of MT1-MMP, TIMP-1, TIMP-2, and TIMP-3 messenger RNA in normal, hyperplastic, and neoplastic endometrium. *Am. J. Clin. Pathol.* **114,** 402–411.

Maidment, J., Moore, D., Murphy, G. P., Murphy, G., and Clark, I. M. (1999). Matrix metalloproteinase homologues from *Arabidopsis thaliana*—expression and activity. *J. Biol. Chem.* **274,** 34706–34710.

Maquoi, E., Frankenne, F., Baramova, E., Munaut, C., Sounni, N. E., Remacle, A., Noel, A., Murphy, G., and Foidart, J.-M. (2000a). Membrane type I matrix metalloproteinase-associated degradation of tissue inhibitor of metalloproteinase 2 in human tumor cell lines. *J. Biol. Chem.* **275,** 11368–11378.

Maquoi, E., Frankenne, F., Noel, A., Krell, H.-W., Grams, F., and Foidart, J.-M. (2000b). Type IV collagen induces matrix metalloproteinase 2 activation in HT1080 fibrosarcoma cells. *Exp. Cell Res.* **261,** 348–359.

Martignetti, J. A., Al Aqeel, A., Al Sewairi, W., Boumah, C. E., Kambouris, M., Al Mayouf, S., Sheth, K. V., Al. Eid, W., Dowling, O., Harris, J., Glucksman, M. J., Bahabri, S., Meyer, B. F., and Desnick, R. J. (2001). Mutation of the matrix metalloproteinase 2 gene (MMP2) causes a multicentric osteolysis and arthritis syndrome. *Nat. Genet.* **28,** 261–265.

Martinella-Catusse, C., Polette, M., Noel, A., Gilles, C., Dehan, P., Munaut, C., Colige, A., Volders, L., Monboisse, J.-C., Foidart, J.-M., and Birembaut, P. (2001). Down-regulation of MT1-MMP expression by the a3 chain of type IV collagen inhibits bronchial tumor cell line invasion. *Lab. Invest.* **81,** 167–174.

Matsumoto, S., Katoh, M., Saito, S., Watanabe, T., and Matsuhjo, Y. (1997). Identification of soluble type of membrane-type matrix metalloproteinase-3 formed by alternatively spiced mRNA. *Biochim. Biophys. Acta* **1354,** 159–170.

Mattei, M. G., Roeckel, N., Olsen, B. R., and Apte, S. S. (1997). Genes of the membrane-type matrix metalloproteinase (MT-MMP) gene family, MMP-14, MMP-15 and MMP-16, localize to human chromosomes 14, 16, and 8, respectively. *Genomics* **40,** 168–169.

McKerrow, J. H., Bhargava, V., Hansell, E., Huling, S., Kumahara, T., Matley, M., Coussens, L., and Warren, R. (2000). A functional proteomics screen of proteases in colorectal carcinoma. *Mol. Med.* **6,** 450–460.

McQuibban, G. A., Butler, G., Gong, J.-H., Bendall, L., Powers, C., Clark-Lewis, I., and Overall, C. M. (2001). Matrix metalloproteinase activity inactivates the CXC chemokine stromal cell-derived factor-1. *J. Biol. Chem.* **276,** 43503–43508.

Mignon, C., Okada, A., Mattei, M. G., and Bassett, P. (1995). Assignment of human membrane-type matrix metalloproteinase (MMP14) gene to 14q11-q12 by in situ hybridization. *Genomics* **28,** 360–361.

Miyamori, H., Takino, T., Seiki, M., and Sato, H. (2000a). Human membrane type-2 matrix metalloproteinase is defective in cell-associated activation of progelatinase A. *Biochem. Biophys. Res. Commun.* **267,** 796–800.

Miyamori, H., Takino, T., Seiki, M., and Sato, H. (2000b). MT2-MMP appears to traffic in a different fashion than MT1-MMP, and is primarily confirmed to the Golgi when overexpressed in CHO cells. *Biochem. Biophys. Res. Commun.* **267,** 796–800.

Miyamori, H., Takino, T., Kobayashi, Y., Tokai, H., Itoh, Y., Seiki, M., and Sato, H. (2001). Claudin promotes activation of pro-matrix metalloproteinase-2 mediated by membrane-type matrix metalloproteinase. *J. Biol. Chem.* **276,** 28204–28211.

Mohan, R., Chintala, S. K., Jung, J. C., Villar, W. V. L., McCabe, F., Russo, L. A., Lee, Y., McCarthy, B. E., Wollenberg, K. R., Jester, J. V., Wang, M., Welgus, H. G., Shipley, J. M., Senior, R. M., and Fini, M. E. (2002). Matrix metalloproteinase gelatinase B (MMP-9) coordinates and effects epithelial regeneration. *J. Biol. Chem.* **277,** 2065–2072.

Moll, U. M., Lane, B., Zucker, S., and Nagasi, H. (1990). Localization of collagenase at the basal plasma membrane of a human pancreatic cancer cell line. *Cancer Res.* **50,** 6995–7002.

Monsky, W. L., Kelly, T., Lin, C.-Y., Yeh, Y., Stetler-Stevenson, W. G., Mueller, S. C., and Chen, W. T. (1993). Binding and localization of Mr 72,000 matrix metalloproteinase at cell surface invadopodia. *Cancer Res.* **53,** 3159–3164.

Mori, M., Mimori, K., Shiraishi, T., Fujie, T., Baba, K., Kusumoto, H., Haraguchi, M., Ueo, H., and Akiyoshi, T. (1997). Analysis of MT1-MMP and MMP-2 expression in human gastric cancers. *Int. J. Cancer* **74,** 316–321.

Morrison, C. J., Butler, G. S., Bigg, H. F., Roberts, C. R., Soloway, P., and Overall, C. M. (2001). Cellular activation of MMP-2 (gelatinase A) by MT2-MMP occurs via a TIMP-2- independent pathway. *J. Biol. Chem.* **276,** 47402–47410.

Mott, J. D., Thomas, C. L., Rosenbach, M. T., Takaha, K., Greenspan, D. S., and Banda, M. J. (2000). Posttranslational proteolytic processing of procollagen C-terminal proteinase enhancer releases a metalloproteinase inhibitor. *J. Biol. Chem.* **275,** 1384–1390.

Mucha, A., Cuniasse, P., Kannan, R., Beau, F., Yiotakis, A., Basset, P., and Dive, V. (1998). Membrane type-1 matrix metalloproteinase and stromelysin-3 cleave more efficiently synthetic substrates containing unusual amino acids in their P1' positions. *J. Biol. Chem.* **273,** 2763–2768.

Murakami, K., Sakukawa, R., Ikeda, T., Matsuura, T., Hasumura, S., Nagamori, S., Yamada, Y., and Saiki, I. (1999). Invasiveness of hepatocellular carcinoma cell lines: contribution of membrane-type 1 matrix metalloproteinase. *Neoplasia* **1,** 424–430.

Murphy, G., and Gavrilovic, J. (1999). Proteolysis and cell migration: creating a path? *Curr. Opin. Cell Biol.* **11,** 614–621.

Murphy, G., Willenbrock, F., Ward, R. V., Cockett, M. I., Eaton, D., and Docherty, A. J. P. (1992). The C-terminal domain of 72 kDa gelatinase A is not required for catalysis, but is essential for membrane activation and modulates interactions with tissue inhibitors of metalloproteinases. *Biochem. J.* **283,** 637–641.

Murphy, G., Knauper, V., Atkinson, S., Gavrilov, J., and Edwards, D. (2000). Cellular mechanisms for focal proteolysis and the regulation of the microenvironment. *Fibrinolysis Proteolysis* **14,** 165–174.

Nabeshima, K., Inoue, T., Shimao, Y., Okada, Y., Itoh, Y., Seiki, M., and Koono, M. (2000). Front-cell-specific expression of membrane-type 1 matrix metalloproteinase and gelatinase A during cohort migration of colon carcinoma cells induced by hepatoctocyte growth factor/scatter factor. *Cancer Res.* **60,** 3364–3369.

Nagase, H., and Woessner, F. (1999). Matrix metalloproteinases. *J. Biol. Chem.* **274,** 21491–21494.

Nakada, M., Nakamura, H., Ikeda, E., Fujimoto, N., Yamashita, J., Sato, H., Seiki, M., and Okada, Y. (1999). Expression and tissue localization of membrane-type 1, 2, and 3 matrix metalloproteinases in human astrocytic tumors. *Am. J. Pathol.* **154,** 417–428.

Nakada, M., Yamada, A., Takino, T., Miyamori, H., Takahashi, T., Yamashita, J., and Sato, H. (2001). Suppression of membrane-type 1 matrix metalloproteinase (MMP)-mediated MMP-2 activation and tumor invasion by testican 3 and its splicing variant gene product, N-tes. *Cancer Res.* **61,** 8896–8902.

Nakahara, H., Howard, L., Thompson, E. W., Sato, H., Seiki, Y., Yeh, Y., and Chen, W. T. (1998). Transmembrane/cytoplasmic domain-mediated membrane type 1-matrix metalloproteinase docking to invadopodia is required for cell invasion. *Proc. Natl. Acad. Sci. USA* **94,** 7959–7964.

Nakamura, H., Ueno, H., Yamashita, K., Seiki, M., and Okada, Y. (1999). Enhanced production and activation of progelatinase A mediated by membrane-type 1 matrix metalloproteinase in human papillary thyroid carcinomas. *Cancer Res.* **59,** 467–473.

Nawrocki, B., Polette, M., Burlet, H., Birembaut, P., and Adnet, J.-J. (1999). Expression of gelatinase A and its activator MT1-MMP in the inflammatory periprosthetic response to polyethylene. *J. Bone Min. Res.* **14,** 288–294.

Nguyen, M., Arkell, J., and Jackson, C. J. (2000). Three-dimensional collagen matrices induce delayed but sustained activation of gelatinase A in human endothelial cells via MT1-MMP. *Int. J. Biochem. Cell Biol.* **32,** 621–631.

Nomura, H., Sato, H., Seiki, M., Mai, M., and Okada, Y. (1995). Expression of membrane-type matrix metalloproteinase in human gastric carcinomas. *Cancer Res.* **55,** 3263–3266.

Oh, J., Takahashi, R., Kondo, S., Mizoguchi, A., Adachi, E., Sasahara, R. M., Nishimura, S., Imamura, Y., Kitayama, H., Alexander, D. B., Ide, C., Horan, T. P., Arakawa, T., Yoshida, H., Nishikawa, S., Itoh, Y., Seiki, M., Itohara, S., Takahashi, C., and Noda, M. (2001). The membrane-anchore MMP inhibitor RECK is a key regulator of extracellular matrix integrity and angiogenesis. *Cell* **107,** 789–800.

Ohkubo, S., Miyadera, K., Sugimoto, Y., Matsuo, K.-I., Wierzba, K., and Yamada, Y. (1999). Identification of substrate sequences for membrane type-1 matrix metalloproteinase using bacteriophage peptide display library. *Biochem. Biophys. Res. Commun.* **266,** 308–313.

Ohtani, H., Motohashi, H., Sato, H. M. S., and Nagura, H. (1996). Dual over-expression pattern of membrane-type metalloproteinase-1 in cancer and stromal cells in human gastrointestinal carcinoma revealed by in situ hybridization and immunoelectron microscopy. *Int. J. Cancer* **68,** 565–570.

Ohuchi, E., Imai, K., Fuji, Y., Sato, H., Seiki, M., and Okada, Y. (1997). Membrane type 1 matrix metalloproteinase digests interstitial collagens and other matrix macromolecules. *J. Biol. Chem.* **272,** 2446–2451.

Okada, A., Bellocq, J.-B., Rouyer, N., Chenard, M.-P., Rio, M.-C., Chambon, P., and Basset, P. (1995). Membrane-type matrix metalloproteinase (MT-MMP) gene is expressed in stromal cells of human colon, breast, and head and neck carcinomas. *Proc. Natl. Acad. Sci. USA* **92,** 2730–2734.

Okada, A., Tomasetto, C., Lutz, Y., Bellocq, J.-P., Rio, M.-C., and Basset, P. (1997). Expression of matrix metalloproteinases during rat skin wound healing: Evidence that membrane type-1 matrix metalloproteinase is a stromal activator of pro-gelatinase A. *J. Cell Biol.* **137,** 67–77.

Okamoto, I., Kawano, Y., Tsuiki, H., Sasaki, J., Nakao, M., Matsumoto, M., Suga, M., Ando, M., Nakajima, M., and Saya, H. (1999). CD44 cleavage induced by a membrane-associated metalloproteinase plays a critical role in tumor cell migration. *Oncogene* **18,** 1435–1446.

Okomura, Y., Sato, H., Seiki, M., and Kido, H. (1997). Proteolytic activation of the precursor of the membrane type 1 matrix metalloproteinase by human plasmin. *FEBS Lett.* **402,** 181–184.

Ota, K., Stetler-Stevenson, W., Yang, Q., Kumar, A., Wada, J., Kashihara, N., Wallner, E., and Kanwar, Y. (1998). Cloning of murine membrane-type-1-matrix metalloproteinase (MT-1-MMP) and its metanephric developmental regulation with respect to MMP-2 and its inhibitor. *Kidney Int.* **54,** 131–142.

Overall, C. M. (2001). Matrix metalloproteinase substrate binding domains, modules and exosites. *In* "Matrix Metalloproteinase Protocols" (I. M. Clark, Ed.), pp. 79–120. Humana Press, Totowa, NJ.

Overall, C. M., and Sodek, J. (1990). Concanavalin A produces a matrix-degradative phenotype in human fibroblasts. Induction and endogenous activation of collagenase, 72-kDa gelatinase, and Pump-1 is accompanied by the suppression of tissue inhibitor of matrix metalloproteinases. *J. Biol. Chem.* **265,** 21141–21151.

Overall, C. M., Tam, E., McQuibban, G. A., Morrison, C., Wallon, U. M., Bigg, H. F., King, A. E., and Roberts, C. R. (2000). Domain interactions in the gelatinase A.TIMP-2. MT1-MMP activation complex: The ectodomain of the 44 kDa form of membrane type-1 does not modulate gelatinase A activation. *J. Biol. Chem.* **275,** 39497–39506.

Pap, T., Shigeyama, Y., Kuchen, S., Fernihough, J. K., Simmen, B., Gay, R. E., Bullingham, M., and Gay, S. (2000). Differential expression pattern of membrane-type matrix metalloproteinases in rheumatoid arthritis. *Arthritis Rheum.* **43,** 1226–1232.

Parks, W. C., Sudbeck, B. D., Doyle, G. R., and Saariahlo-Kere, U. K. (1998). Matrix metalloproteinases in tissue repair. *In* "Matrix Metalloproteinases" (W. C. Parks and R. P. Mecham, Eds.), pp. 263–299. Academic Press, San Diego.

Pasterkamp, G., Schoneveld, A. H., Hijnen, D. J., de Kleijnm, D. P. V., Teepen, H., van der Wal, A. C., and Borst, C. (2000). Atherosclerotic arterial remodeling and the localization of macrophages and matrix metalloproteinases 1, 2 and 9 in the human coronary artery. *Atherosclerosis* **150,** 245–253.

Pavlaki, M., Cao, J., Hymowitz, M., Chen, W.-T., Bahou, W., and Zucker, S. (2002). A conserved sequence within the propeptide domain of membrane type 1-matrix metalloproteinase (MT1-MMP) is critical for function as an intramolecular chaperone. *J. Biol. Chem.* **277,** 2740–2749.

Pei, D. (1999). Identification and characterization of the fifth membrane-type matrix metalloproteinase MT5-MMP. *J. Biol. Chem.* **274,** 8925–8932.

Pei, D., and Weiss, S. J. (1995). Furin-dependent intracellular activation of the human stromelysin-3 zymogen. *Nature (London)* **375,** 244–247.

Pei, D., and Weiss, S. J. (1996). Transmembrane-deletion mutants of the membrane-type matrix metalloproteinase process progelatinase A and express intrinsic matrix-degrading activity. *J. Biol. Chem.* **271,** 9135–9140.

Pei, D., Kang, T., and Qi, H. (2000). Cysteine array matrix metalloproteinase (CA-MMP)/MMP-23 is a type II transmembrane matrix metalloproteinase regulated by a single cleavage for both secretion and activation. *J. Biol. Chem.* **275,** 33988–33997.

Pendas, A. M., Matilla, T., Estivill, X., and Lopez-Otin, C. (1995). The human collagenase-3 (CLG3) gene is located on chromosome 11q22.3 clustered to other members of the matrix metalloproteinase gene family. *Genomics* **26,** 615–618.

Pendás, A. M., Balbín, M., Llano, E., Jiménez, M. G., and López-Otín, C. (1997). Structural analysis and promoter characterization of the human collagenase-3 gene (MMP13). *Genomics* **40,** 222–233.

Pender, S. L. F., Salmela, M. T., Monteleone, G., Schnapp, D., McKenzie, C., Spencer, J., Fong, S., Saariahlo-Kere, U. K., and MacDonald, T. T. (2000). Ligation of a4b1 integrin on human intestinal mucosal mesenchymal cells selectively up-regulates membrane type-1 matrix metalloproteinase and confers a migratory phenotype. *Am. J. Pathol.* **157,** 1955–1962.

Peracchia, F., Tamburro, A., Prontera, C., Mariani, B., and Rotilio, D. (1997). cAMP involvement in the expression of MMP-2 and MT-MMP1 metalloproteinases in human endothelial cells. *Arterioscl. Thromb. Vasc. Biol.* **17,** 3185–3190.

Polette, M., and Birembaut, P. (1998). Membrane-type metalloproteinases in tumor invasion. *Int. J. Biochem. Cell Biol.* **30,** 1195–1202.

Polette, M., Nawrocki, B., Gilles, C., Sato, H., Seiki, M., Tournier, J. M., and Birembaut, P. (1996). MT-MMP expression and localization in human lung and breast cancer. *Virchows Arch.* **428,** 29–35.

Puente, X. S., Pendas, A. M., Liano, E., Velasco, G., and Lopez-Otin, C. (1996). Molecular cloning of a novel membrane-type matrix metalloproteinase from human breast cancer. *Cancer Res.* **56,** 944–949.

Puente, X. S., Pendas, A. M., Llano, E., and Lopez-Otin, C. (1998). Localization of the human membrane type 4-matrix metalloproteinase gene (MMP17) to chromosome 12q24. *Genomics* **54,** 578–579.

Puyraimond, A., Fridman, R., Lemesle, M., Arbeille, B., and Menashi, S. (2001). Co-localization of MMP-2/MT1-MMP/TIMP-2/avb3 in caveolae has been demonstrated. *Exp. Cell Res.* **262,** 28–36.

Pyo, R., Lee, J. K., Shipley, J. M., Curci, J. A., Mao, D., Ziporin, S. J., Ennis, T. L., Shapiro, R. M., Senior, R. M., and Thompson, R. W. (2000). Targeted gene disruption of matrix metalloproteinase-9 (gelatinase B) suppresses development of experimental abdominal aortic aneurysms. *J. Clin. Invest.* **105,** 1641–1649.

Rajavashisth, T. B., Liao, L. K., Galis, Z. S., Tripathi, S., Laufs, U., Tripathi, J., Chai, N.-N., Xiao-X, P., Jovinge, S., and Libby, P. (1999). Inflammatory cytokines and oxidized low density lipoproteins increase endothelial cell expression of membrane type 1-matrix metalloproteinase. *J. Biol. Chem.* **274,** 11924–11929.

Rathke-Hartlieb, S., Budde, P., Ewert, S., Schlmann, U., Staege, M. S., Jockusch, H., Batsch, J. W., and Frey, J. (2000). Elevated expression of membrane type I metalloproteinase (MT1-MMP) in reactive astrocytes following neurodegeneration in mouse central nervous system. *FEBS Lett.* **481**, 227–234.

Ratnikov, B. I., Rozanov, D. V., Postnova, T. I., Baciu, P. G., Zhang, H., DiScipio, R. G., Chestukhina, G. G., Smith, J. W., Deyryugina, E. I., and Strongin, A. Y. (2002). An alternative processing of integrin av subunit in tumor cells by membrane type-1 matrix metalloproteinase. *J. Biol. Chem.* **277**, 7377–7385.

Rekhter, M. D., Hicks, G. W., Brammer, D. W., Hallak, H., Kindt, E., Chen, J., Rosebury, W. S., Anderson, M. K., Kuipers, P. J., and Ryan, M. J. (2000). Hypercholesterolemia causes mechanical weakening of rabbit atheroma: local collagen loss as a prerequisite of plaque rupture. *Circ. Res.* **86**, 101–108.

Robbins, J. R., McGuire, P. G., Wehrle-Haller, B., and Rogers, S. L. (1999). Diminished matrix metalloproteinase 2 (MMP-2) in ectomesenchyme-derived tissues of the Patch mutant mouse: regulation of MMP-2 by PDGF and effects on mesenchymal cell migration. *Dev. Biol.* **212**, 255–263.

Romanic, A. M., Burns-Kurtis, C. L., Ao, Z., Arleth, A. J., and Ohlstein, E. H. (2001). Upregulated expression of human membrane type-5 matrix metalloproteinase in kidneys from diabetic patients. *Am. J. Physiol.* **281**, F309–F317.

Rozanov, D. V., Deryugina, E. I., Ratnikov, B. I., Monosov, E. Z., Marchenko, G. N., Machenko, G. N., Quigley, J. P., and Strongin, A. Y. (2001). Mutation analysis of membrane type-1 matrix metalloproteinase (MT1-MMP). The role of the cytoplasmic tail Cys-574, the active site of Glu-240 and furin cleavage motifs in oligomerization, processing and self-proteolysis of MT1-MMP expressed in breast cancer. *J. Biol. Chem.* **276**, 25705–25714.

Rozanov, D. V., Ghebrehiwet, B., Postnova, T. I., Eichinger, A., Deryugina, E. I., and Strongin, A. Y. (2002). The hemopexin-like C-terminal domain of membrane type-1 matrix metalloproteinase (MT1-MMP) regulates proteolysis of a multifunctional protein gC1qR. *J. Biol. Chem.* **277**, 9318–9325.

Ruangpanit, N., Price, J. T., Holmbeck, K., Birkedal-Hansen, H., Guenzler, V., Huang, X. D. C., Bateman, J. F., and Thompson, E. W. (2002). MT1-MMP-dependent and -independent regulation of gelatinase A activation in long-term, ascorbate-treated fibroblast cultures: Regulation by fibrillar collagen. *Exp. Cell Res.* **272**, 109–118.

Sameshima, T., Nabeshima, K., Toole, B. P., Yokogami, K., Goya, T., Koono, M., and Wakisaka, S. (2000). Expression of EMMPRIN (CD147), a cell surface inducer of matrix metalloproteinases, in normal brain and gliomas. *Int. J. Cancer* **88**, 21–27.

Sanchez-Lopez, R. R., Nicholson, R. R., Gesnel, M. C., Matrisian, L. M., and Breathnach, R. (1988). Structure–function relationships in the collagenase family member transin. *J. Biol. Chem.* **263**, 11892–11899.

Sang, W.-X., Thompson, E. W., Grant, D., Stetler-Stevenson, W. G., and Byers, S. W. (1991). Soluble laminin and arginine–glycine–aspartic acid containing peptides differentially regulate type IV collagenase messenger RNA, activation, and localization in testicular cell culture. *Biol. Reprod.* **45**, 387–394.

Sardinhah, T. C., Nogueras, J. J., Xiong, H., Weiss, E. G., and Wexner, S. D. (2000). Membrane-type 1 matrix metalloproteinase mRNA expression in colorectal cancer. *Dis. Colon Rectum* **43**, 389–395.

Sato, H., and Seiki, M. (1996). Membrane-type matrix metalloproteinases (MT-MMPs) in tumor metastasis. *J. Biochem.* **119**, 209–215.

Sato, H., Takino, T., Okada, Y., Cao, J., Shinagawa, A., Yamamoto, E., and Seiki, M. (1994). A matrix metalloproteinase expressed on the surface of invasive tumor cells. *Nature* **370**, 61–65.

Sato, H., Takino, T., Kinoshita, T., Imai, K., Okada, Y., Stetler Stevenson, W. G., and Seiki, M. (1996). Cell surface binding and activation of gelatinase A induced by expression of membrane-type-1-matrix metalloproteinase (MT1-MMP). *FEBS Lett.* **385**, 238–240.

Sato, T., Ovejero, M., Hou, P., Heegaard, A. M., Kumegawa, M., Fogged, N. T., and Delaisse, J. M. (1997). Identfication of the membrane-type matrix metalloproteinase MT1-MMP in osteoclasts. *J. Cell Sci.* **110,** 589–596.

Sato, T., Sakal, T., Sato, H., Mori, Y., and Ito, A. (1999). Enhancement of membrane-type 1-matrix metalloproteinase (MT1-MMP) production and sequential activation of progelatinase A on human dermal fibroblasts. *Br. J. Cancer* **80,** 1137–1143.

Sawaji, Y., Sato, T., Seiki, M., and Ito, A. (2000). Heat shock-mediated transient increase in intracellular 3′,5′-cyclic AMP results in tumor specific supression of membrane type 1-matrix metalloproteinase production and progelatinase A activation. *Clin. Exp. Metastasis* **18,** 131–138.

Seiki, M. (1999). Membrane-type matrix metalloproteinases. *APMIS* **107,** 137–143.

Seltzer, J. L., Lee, A. Y., Akers, K. T., Subdeck, B., Southon, E. A., Wayner, E. A., and Eisen, A. Z. (1994). Activation of 72-kDa type IV collagen/gelatinase by normal fibroblasts in collagen lattices is mediated by integrin receptors but is not related to latice formation. *Exp. Cell Res.* **213,** 365–374.

Shamamian, P., Popock, B. J., Schwartz, J. D., Monea, S., Chuang, N., Whiting, D., Marcus, S. G., Galloway, A. C., and Mignatti, P. (2000). Neutrophil-derived serine proteinases enhance membrane type-1 matrix metalloproteinase-dependent tumor cell invasion. *Surgery* **126,** 142–147.

Shankavaram, U. T., Lai, W.-C., Netzel-Arnett, S., Mangan, P. R., Ardans, J. A., Caterina, N., Stetler-Stevenson, W. G., Bikedal-Hansen, H., and Wahl, L. M. (2001). Monocyte membrane type1-matrix metalloproteinase. Prostaglandin-dependent regulation and role in metalloproteinase-2 activation. *J. Biol. Chem.* **276,** 19027–19032.

Shapiro, S. D. (1998). Matrix metalloproteinase degradation of extracellular matrix: biological consequences. *Curr. Biol.* **10,** 602–608.

Shi, X.-P., Chen, E., Yin, K.-C., Na, S., Garsky, V. M., Lai, M.-T., Y.-M. L., Platchek, M., Register, R. B., Sardan, M. K., TAng, S.-J., Thiebeau, J., Wods, T., Shafer, J. A., and Gardell, S. J. (2001). The pro domain of B-secretase does not confer strict zymogen-like properties but does assist proper folding of the protease domain. *J. Biol. Chem.* **276,** 10366–10373.

Shimada, T., Nakamura, H., Ohuchi, E., Fujii, Y., Murakami, Y., Sato, H., Seiki, M., and Okada, Y. (1999). Characterization of a truncated recombinant form of human membrane type 3 matrix metalloproteinase. *Eur. J. Biochem.* **262,** 907–914.

Shimada, T., Nakamura, H., Yamashita, K., Kawata, R., Murakami, Y., Fujimoto, N., Sato, H., Seiki, M., and Okada, Y. (2000). Enhanced production and activation of progelatinase A mediated by membrane-type 1 matrix metalloproteinase in human oral squamous cell carcinomas: implication for lymph node metastasis. *Clin. Exp. Metastasis* **18,** 179–188.

Shofuda, K.-I., Yasamitsu, H., Nishihashi, A., Miki, K., and Miyazuki, K. (1997). Expression of three membrane-type matrix metalloproteinases (MT-MMPs) in rat smooth muscle cells and characterization of MT3-MMPs with and without transmembrane domain. *J. Biol. Chem.* **272,** 9749–9754.

Shofuda, K.-I., Moriyama, K., Kishihashi, A., Higashi, S., Mizushima, H., Yasumitsu, H., Miki, K., Sato, H., Seiki, M., and Miyazaki, K. (1998). Role of tissue inhibitor of metalloproteinase-2 (TIMP-2) in regulation of progelatinase A activation catalyzed by membrane-type matrix metalloproteinase-1 (MT1-MMP) in human cancer cells. *J. Biochem.* **124,** 462–470.

Song, W., Jackson, K., and McGuire, P. G. (2000). Degradation of Type IV collagen by matrix metalloproteinase is an important step in the epithelial–mesenchymal transformation of the endocardial cushions. *Dev. Biol.* **227,** 606–617.

Sounni, N. E., Baramova, E. N., Munalt, C., Maquoi, E., Frankenne, F., Foidart, J.-M., and Noel, A. (2002). Expression of membrane type 1 matrix metalloprfoteinase (MT1-MMP) in A2058 melanoma cells is associated with MMP-2 activation and increased tumor growth and vascularization. *Int. J. Cancer* **98,** 23–28.

Spinale, F. G., Coker, M. L., Bond, B. R., and Zellner, J. L. (2000). Myocardial matrix degradation and metalloproteinase activation in the failing heart: a potential herapeutic target. *Cardiovasc. Res.* **46,** 225–238.

Springman, E. B., Angleton, E. L., Birekedal-Hansen, H., and Van Wart, E. V. (1990). Multiple modes of activation of latent human fibroblast collagenase: evidence for the role of a Cys[73] active site zinc complex in latency and a "cysteine switch" mechanism for activation. *Proc. Natl. Acad. Sci. USA* **87**, 364–368.

Stamenkovic, I. (2000). Matrix metalloproteinases in tumor invasion and metastasis. *Cancer Biol.* **10**, 415–433.

Stanton, H., Gavrilovic, J., Atkinson, S., d'Ortho, M.-P., Yamada, K. M., Zardi, L., and Murphy, G. (1998). The activation of pro-MMP-2 (gelatinase A) by HT-1080 fibrosarcoma cells is promoted by culture on a fibronectin substrate and is concomitant with an increase in processing of MT1-MMP (MMP-14) to a 45 kDa form. *J. Cell Sci.* **111**, 2789–2798.

Steffensen, B., Bigg, H. F., and Overall, C. M. (1998). The involvement of the fibronectin type II-like modules of human gelatinase A in cell surface localization and activation. *J. Biol. Chem.* **273**, 20622–20628.

Steiner, D. F. (1998). The protein convertases. *Curr. Opin. Chem. Biol.* **2**, 31–39.

Sternlicht, M. D., and Werb, Z. (2001). How matrix metalloproteinases regulate cell behavior. *Annu. Rev. Cell Dev. Biol.* **17**, 463–516.

Sternlicht, M. D., Lochter, A., Sympson, C. J., Huey, B., Rougier, J. P., Gray, J. W., Pinkel, B., Bissell, M. J., and Werb, Z. (1999). The stromal proteinase MMP3/stromelysin-1 promotes mammary carcinogenesis. *Cell* **98**, 351–358.

Stetler-Stevenson, W. G. (1995). Proteinase A activation during tumor cell invasion. *Invasion Metastasis* **14**, 259–268.

Stetler-Stevenson, W. G. (1999). Matrix metalloproteinases in angiogenesis: a moving target for therapeutic intervention. *J. Clin. Invest.* **103**, 1237–1241.

Stoker, W., and Bode, W. (1995). Structural features of a superfamily of zinc-endopeptidases: the metzincins. *Curr. Opin. Str. Biol.* **5**, 383–390.

Strongin, A. Y., Collier, I., Bannicov, G., Marmer, B. L., Grant, G. Z., and Goldberg, G. I. (1995). Mechanism of cell surface activation of 72-kDa type IV collagenase. Isolation of the activated form of the membrane metalloproteinase. *J. Biol. Chem.* **270**, 5331–5338.

Takahashi, Y., Kawahara, F., Noguchi, M., Miwa, K., Sato, S., Seiki, M., Inoue, H., Tanabe, T., and Yoshimoto, T. (1999). Activation of matrix metalloproteinase-2 in human breast cancer cells overexpressing cyclooxygenase-1 or -2. *FEBS Lett.* **460**, 145–148.

Takino, T., Sato, H., Shinagawa, A., and Seiki, M. (1995a). Identification of the second membrane-type matrix metalloproteinase (MT-MMP-2) gene from a human placental cDNA library. *J. Biol. Chem.* **270**, 23013–23020.

Takino, T., Sato, H., Yamamoto, E., and Seiki, M. (1995b). Cloning of a human gene potentially encoding a novel matrix metalloproteinase having a C-terminal transmembrane domain. *Gene* **155**, 293–298.

Tanaka, M., Sato, H., Takino, T., Iwata, K., Inoue, M., and Seiki, M. (1997a). Isolation of a mouse MT2-MMP gene from a lung cDNA library and identification of its product. *FEBS Lett.* **402**, 219–222.

Tanaka, S. S., Mariko, Y., Mori, H., Ishijimo, J., Tachi, S., Sato, H., Seiki, M., Yamanouchi, K., Tojo, H., and Tachi, C. (1997b). Cell-cell contact down regulates expression of membrane-type matrix metalloproteinase-1 (MT1-MMP) in a mouse mammary gland epithelial cell line. *Zool Sci.* **14**, 95–99.

Tanaka, S. S., Tagooka, Y., Sato, H., Tojo, H., and Tachi, C. (1998). Expression and localization of membrane type matrix metalloproteinase-1 (MT1-MMP) in trophoblast cells of cultured mouse blastocyts and ectoplacental cones. *Placenta* **19**, 41–48.

Tanney, D. C., Feng, L., Pollock, A. S., and Lovett, D. H. (1998). Regulated expression of matrix metalloproteinases and TIMP in nephrogenis. *Dev. Dyn.* **213**, 121–129.

Tarboletti, G., D'Ascenzo, S., Porsotti, P., Flavazzi, R., Pavan, A., and Dolo, V. (2002). Shedding of matrix metalloproteinases MMP-2, MMP-9, and MT1-MMP as membrane vesicle-associated components by endothelial cells. *Am. J. Pathol.* **160**, 673–680.

Teti, A., Farina, A. R., Villanova, I., Tiberio, A., Tasconelli, A., Sciortino, G., Chambers, A. F., and MacKay, A. R. (1998). Activation of MMP-2 by human GCT23 giant cell tumor cells induced by osteopontin, bone sialoprotein and GRGDSP peptides is RGD and cell shape change dependent. *Int. J. Cancer* **77,** 82–93.

Thant, A. A., Serbulea, M., Kikkawa, F., Liu, E., Tomoda, Y., and Hamaguchi, M. (1997). c-Ras is required for the activaton of the matrix metalloproteinases by concanavalin A in 3Y1 cells. *FEBS Lett.* **406,** 28–30.

Theret, N., Musso, O., L'Helgoualc'h, A., and Clement, B. (1997). Activation of matrix metalloproteinase-2 from hepatic stellate cells requires interactions with hepatocytes. *Am J. Pathol.* **150,** 51–58.

Theret, N., Musso, O., L'Helgoualc'h, A., Campion, J.-P., and Clement, B. (1998). Differential expression and origin of membrane-type 1 and 2 matrix metalloproteinases (MT-MMPs) in association with MMP2 activation in injured human livers. *Am J. Pathol.* **153,** 945–954.

Theret, N., Lehti, K., Musso, O., and Clement, B. (1999). MMP2 activation by collagen I and concanavalin A in cultured human stellate cells. *Hepatology* **30,** 462–468.

Tokuraka, M., Sato, H., Murakami, S., Okada, Y., Watanabe, Y., and Seiki, M. (1995). Activation of the precursor of gelatinase A/72 kDa type IV collagenase/MMP-2 in lung carcinomas correlates with the expression of membrane-type matrix metalloproteinase (MT-MMP) and with lymph node metastasis. *Int. J. Cancer* **64,** 355–359.

Tomasek, J. J., Haliday, N. L., Updike, D. L., Ahern-Moore, J. S., Vu, T.-K., Liu, R. W., and Howard, E. W. (1997). Gelatinase A activation is regulated by the organization of polymerized actin cytoskeleton. *J. Biol. Chem.* **272,** 7482–7487.

Tomita, T., Fujii, M., Tokumara, Y., Imanishi, Y., Kanke, M., Yamashita, T., Ishiguro, R., Kanzaki, J., Kameyama, K., and Otani, Y. (2000). Granulocyte-machrophage colony-stimulating factor upregulates matrix metalloproteinase-2 (MMP-2) and membrane type-1 MMP (MT1-MMP) in human head and neck cancer cells. *Cancer Lett.* **156,** 83–91.

Toth, M., Bernardo, M. M., Gervasi, D. C., Soloway, P., Wang, Z., Bigg, H. F., Overall, C. M., DeClerck, Y. A., Tschesche, H., Cher, M. L., Brown, S., Mobashery, S., and Fridman, R. (2000). Tissue inhibitor of metalloproteinase (TIMP)-2 acts synergistically with synthetic matrix metalloproteinase (MMP) inhibitors but not TIMP-4 to enhance the (membrane type 1)-MMP-dependent activation of pro-MMP-2. *J. Biol. Chem.* **275,** 41415–41423.

Toth, M., Hernandez-Barrantes, S., Osenkowski, P., Bernardo, M. M., Gervasi, D. C., Shimura, Y., Merouch, O., Kotra, L. K., Galvez, B. G., Arroyo, A. G., Mobashery, S., and Fridman, R. (2002). Complex pattern of membrane type 1-matrix metalloproteinase shedding. Regulation by autocatalytic cell surface inactivation of active enzyme. *J. Biol. Chem.* **277:** 26340–26350.

Tsujii, M., Kawano, S., and DuBois, R. N. (1997). Cyclooxygenase-2 expression in human colon cancer cells increases metastatic potential. *Proc. Natl. Acad. Sci. USA* **94,** 3336–3340.

Tsunezuka, Y., Kinoh, H., Takin, T., Watanabe, Y., Okada, Y., Shinagawa, A., Sato, H., and Seiki, M. (1996). Expression of membrane type matrix metalloproteinase 1 (MT1-MMP) in tumor cells enhances pulmonary metastasis in an experimental metastasis assay. *Cancer Res.* **56,** 5678–5683.

Turk, B. E., Huang, L. L., Piro, E. T., and Cantley, L. C. (2001). Determination of protease cleavage site motifs using mixture-based oriented peptide libraries. *Nat. Biotechnol.* **19,** 651–667.

Tyagi, S. C., Lewis, K., Pikes, D., Marcello, A., Mujumdar, B. S., Smiley, L. M., and Moore, C. K. (1998). Stretch-induced membrane type matrix metalloproteinase and tissue plasminogen activator in cardiac fibrobast cells. *J. Cell Physiol.* **176,** 374–382.

Uekita, T., Itoh, Y., Yana, I., Ohno, H., and Seiki, M. (2001). Cytoplasmic tail-dependent internalization of membrane-type 1 matrix metalloproteinase is important for its invasion-promoting activity. *J. Biol. Chem.* **155,** 1345–1356.

Ueno, H., Nakamura, H., Inoue, M., Imai, K., Noguchi, M., Sato, H., Seiki, M., and Okata, Y. (1997). Expression and tissue localization of membrane-types 1, 2, and 3 matrix metalloproteinases in human invasive breast carcinomas. *Cancer Res.* **57,** 2055–2060.

Upadhyay, J., Shekarriz, B., Nemeth, J. A., Dong, Z., Cummings, G. D., Fridman, R., Sakr, W., Grignon, D. J., and Cher, M. L. (1999). Membrane type 1-matrix metalloproteinase (MT1-MMP) and MMP-2 immunolocalization in human prostate: change in cellular localization associated with high-grade prostatic intraepithelial neoplasia. *Clin. Cancer Res.* **5**, 4105–4110.

Ureña, J. M., Merlos-Suarez, A., Baselga, J., and Arribas, T. (1999). The cytoplasmic carboxyterminal amino acid determines the subcellular localization of proTGF-alpha and MT1-MMP. *J. Cell Sci.* **112**, 773.

Velasco, G., Pendas, A. M., Fueyo, A., Knauper, V., Murphy, G., and Lopez-Otin, C. (1999). Cloning and characterization of human MMP-23, a new matrix metalloproteinase predominantly expressed in reproductive tissues and lacking conserved domains in other family members. *J. Biol. Chem.* **274**, 4570–4576.

Velasco, G., Cal, S., Merlos-Suarez, A., Ferrando, A., Alvarez, S., Nakano, A., Arribas, J., and Lopez-Otin, C. (2000). Human MT6-matrix metalloproteinase: Identification, progelatinase A activation, and expression in brain tumors. *Cancer Res.* **60**, 877–882.

Vincenti, M. P., White, L. A., Schroen, D. J., Benbow, U., and Brinckerhoff, C. E. (1996). Regulating expression of the gene for matrix metalloproteinase-1 (collagenase): Mechanisms that control enzyme activity, transcription, and mRNA stability. *Crit. Revi. Euk. Gene Expr.* **6**, 391–411.

Vu, T. H. (2001). Don't mess with the matrix. *Nat. Genet.* **28**, 202–203.

Vu, T. H., and Werb, Z. (2000). Matrix metalloproteinases: effectors of development and normal physiology. *Genes Dev.* **14**, 2123–2133.

Wada, K., Sato, H., Kinoh, H., Kajita, M., Yamamoto, H., and Seiki, M. (1998). Cloning of three *Caenorhabditis elegans* genes potentially encoding novel matrix metalloproteinases. *Gene* **211**, 57–62.

Wang, H., and Keiser, J. A. (1998). Expression of membrane-type matrix metalloproteinase in rabbbit neointimal tissue and its correlation with matrix-metalloproteinase-2 activation. *J. Vasc. Res.* **35**, 45–54.

Wang, H., and Keiser, J. A. (2000). Hepatocyte growth factor enhances MMP activity in human endothelial cells. *Biochem. Biophys. Res. Commun.* **272**, 900–905.

Wang, X., and Pei, D. (2001). Shedding of membrane type matrix metalloproteinase 5 by a furin-type convertase: a potential mechanism for down-regulation. *J. Biol. Chem.* **276**, 35953–35960.

Wang, Y., Johnson, A. R., Ye, Q.-Z., and Dyer, R. D. (1999). Catalytic activities and substrate specificity of the human membrane type 4 matrix metalloproteinase catalytic domain. *J. Biol. Chem.* **274**, 33043–33049.

Wang, Z., Juttermann, R., and Soloway, P. (2000). TIMP-2 is required for efficient activation of proMMP-2 in vivo. *J. Biol. Chem.* **275**, 26411–26415.

Westermarck, J., and Kahari, V.-M. (1999). Regulation of matrix metalloproteinase expression in tumor invasion. *FASEB J.* **13**, 781–792.

Wilhelm, S. M., Collier, I. E., Marmer, B. L., Eisen, A. Z., Grant, G. A., and Goldberg, G. I. (1989). SV40-transformed human lung fibroblasts secrete a 92-kDa type IV collagenase which is identical to that secreted by normal macrophages. *J. Biol. Chem.* **264**, 17213–17222.

Will, H., and Hinzmann, B. (1995). cDNA sequence and mRNA tissue distribution of a novel matrix metalloproteinase with a potential transmembrane segment. *Eur. J. Biochem.* **231**, 602–608.

Will, H., Adkinson, S. J., Butler, G. S., Smith, B., and Murphy, G. (1996). The soluble catalytic domain of membrane type 1 matrix metalloproteinase cleaves the propeptide of progelatinase A and initiates autoproteolytic activation. *J. Biol. Chem.* **271**, 17119–17123.

Woessner, J. F. (1991). Matrix metalloproteinases and their inhibitors in connective tissue remodeling. *FASEB J.* **5**, 2145–2154.

Woessner, F., and Nagase, H. (2000). "Matrix Metalloproteinases and TIMPs." Oxford University Press, New York.

Xu, J., Rodriguez, D., Petitclerc, Kim, J. J., Hangai, M., Yuen, S. M., Davis, G. E., and Brooks, P. C.

(2001). Proteolytic exposure of a cryptic site within collagen type IV is required for angiogenesis and tumor growth in vivo. *J. Cell Biol.* **154**, 1069–1079.

Yamada, T., Yoshiyama, Y., Sato, H., Seiki, M., Shinagawa, A., and Takahashi, A. (1995). White matter microglia produce membrane-type matrix metalloprotease, an activator of gelatinase A, in human brain tissues. *Acta Neuropathol.* **90**, 421–424.

Yamamoto, M., Mohanam, S., Sawaya, R., Fuller, G. N., Seiki, M., Sato, H., Gokaslan, Z. L., Liotta, L. A., Nicolson, G. L., and Rao, J. S. (1996). Differential expression of membrane-type matrix metalloproteinase and its correlation with gelatinase A activation in human malignant brain tumors in vitro and in vivo. *Cancer Res.* **56**, 384–392.

Yamamoto, M., Tsujishita, H., Hori, N., Ohishi, Y., Inoue, S., Ikeda, S., and Okda, Y. (1998). Inhibition of membrane-type 1 matrix metalloproteinase by hydroxamate inhibitors: An examination of the subsite pocker. *J. Med. Chem.* **41**, 1209–1217.

Yamanaka, H., Makino, K.-I., Takizawa, M., Nakamura, H., Fujimoto, N., Moriya, H., Nemori, R., Sato, H., Seiki, M., and Okada, Y. (2000). Expression and tissue localization of membrane-types 1, 2, and 3 matrix metalloproteinases in rheumatoid synovium. *Lab. Invest.* **80**, 677–687.

Yan, L., Moses, M. A., Huang, S., and Ingber, D. E. (2000). Adhesion-dependent control of matrix metalloproteinase-2 activation in human capillary endothelial cells. *J. Cell Sci.* **113**, 3979–3987.

Yana, I., and Weiss, S. J. (2000). Regulation of membrane type-1 matrix metalloproteinase activation by proprotein convertases. *Mol. Biol. Cell* **11**, 2387–2401.

Yang, M., Hayashi, K., Hayashi, M., Fujii, J. T., and Kurkinen, M. (1996). Cloning and developmental expression of a membrane-type matrix metalloproteinase from chicken. *J. Biol. Chem.* **271**, 25548–25554.

Yang, Z., Strickland, D. K., and Bornstein, P. (2001). Extracellular MMP-2 levels are regulated by the low-density lipoprotein-related scavenger receptor and thrombospondin 2. *J. Biol. Chem.* **276**, 8403–8408.

Ye, H. Q., Maeda, M., Yu, F.-S. X., and Azar, D. (2000). Differential expression of MT1-MMP (MMP-14) and collagenase III (MMP-13) genes in normal and wounded rat corneas. *Invest. Ophthalm. Vis. Sci.* **41**, 2894–2899.

Young, T. N., Pizza, S. V., and Stark, M. S. (1995). A plasma membrane associated component of ovarian adenocarcinoma cells enhances the catalytic efficiency of matrix metalloproteinase-2. *J. Biol. Chem.* **270**, 999–1002.

Yu, M., Sato, H., Sieki, M., and Thompson, E. W. (1995). Complex regulation of membrane-type matrix metalloproteinase expression and matrix metalloproteinase-2 activation by concanavalin A in MD-MB-231 human breast cancer cells. *Cancer Res.* **55**, 3272–3277.

Yu, M., Bowden, E. T., Sitlani, J., Sato, H., Seiki, M., Mueller, S. C., and Thompson, E. W. (1998a). Tyrosine phosphorylation mediates conA-induced membrane type 1-matrix metalloproteinase expression and matrix metalloproteinase-2 activation in MDA-MB-231 human breast carcinoma cells. *Cancer Res.* **57**, 5028–5034.

Yu, M., Sato, H., Seiki, M., Spiegel, S., and Thompson, E. W. (1998b). Elevated cyclic AMP suppresses ConA-induced MT1-MMP expression in MDA-MB-231 human breast cancer cells. *Clin. Exp. Metastasis* **16**, 185–191.

Yu, Q., and Stamenkovic, I. (2000). Cell surface-localized matrix metalloproteinase-9 proteolytically activates TGF-b and promotes tumor invasion and angiogenesis. *Genes Dev.* **14**, 163–176.

Yu, W.-H., and Woessner, J. F., Jr. (2000). Heparan sulfate proteoglycans as extracellular docking molecules for matrilysin (matrix metalloproteinase 7). *J. Biol. Chem.* **275**, 4183–4191.

Zhang, J., Hampton, A., Nie, G., and Salamonsen, L. (2000). Progesterone inhibits activation of latent matrix metlloproteinase (MMP)-2 by membrane-type 1 MMP: enzymes coordinately expressed in human endometrium. *Biol. Reprod.* **62**, 85–94.

Zheng, T., Zhu, Z., Wang, R. J., Homer, B., Ma, R. J., Riese, J. H. A., Chapman, J., Shapiro, S. D., and Elias, J. A. (2000). Inducible targeting of IL-13 to the adult lung causes matrix metallproteinase- and cathepsin-dependent emphysema. *J. Clin. Invest.* **106**, 1081–1093.

Zhou, A., Apte, S. S., Soininen, R., Cao, R., Baaklin, G. Y., Rauser, R. W., Wang, J., Cao, Y., and Tryggvason, K. (2000). Impaired endochondral ossification and angiogenesis in mice deficient in membrane-type matrix metalloproteinase-1. *Proc. Natl. Acad. Sci. USA* **97**, 4052–4057.

Zhuge, Y., and Xu, J. (2001). Rac1 mediates type I collagen-dependent MMP-2 activation. Role in cell invasion across collagen barrier. *J. Biol. Chem.* **276**, 16248–16256.

Zigrino, P., Drescher, C., and Mauch, C. (2001). Collagen-induced proMMP-2 activation by MT1-MMP in human dermal fibroblasts and the possible role of a2b1 integrins. *Eur. J. Cell Biol.* **89**, 68–77.

Zucker, S., Lysik, R. M., Ramamurthy, N. S., Golub, J., Wieman, J. M., and Wilkie, D. P. (1985a). Diversity of melanoma plasma membrane proteinases. Inhibition of collagenolysis and cytolytic activity by minocycline. *J. Natl. Cancer Inst.* **75**, 517–525.

Zucker, S., Lysik, R. M., Wieman, J., Wilkie, D., and Lane, B. (1985b). Diversity of human pancreatic cancer cell proteinases. Role of cell membrane metalloproteinases in collagenolysis and cytolysis. *Cancer Res.* **45**, 6168–6178.

Zucker, S., Wieman, J., and Lysik, R. (1987). Enrichment of collagen and gelatin degrading activities in the plasma membranes of cancer cells. *Biochim. Biophys. Acta* **924**, 225–237.

Zucker, S., Moll, U. T., Lysik, R. M., DiMassimo, E. D., Stetler-Stevenson, W. G., Liotta, L. A., and Schwedes, J. M. (1990). Extraction of type IV collagenase/gelatinase from plasma membranes of human cancer cells. *Int. J. Cancer* **45**, 1137–1142.

Zucker, S., Lysik, R. M., Malik, M., Bauer, B. A., Caamano, J., and Klein-Szanto, A. J. P. (1992). Secretion of gelatinases and tissue inhibitors of metalloproteinases by human lung cancer cell lines and revertant cell lines: Not an invariant correlation with metastasis. *Int. J. Cancer* **52**, 1–6.

Zucker, S., Conner, C., DiMassimo, B. I., Ende, H., Drews, M., Seiki, M., and Bahou, W. F. (1995). Thrombin induces the activation of progelatinase A in vascular endothelial cells: Physiologic regulation of angiogenesis. *J. Biol. Chem.* **270**, 23730–23738.

Zucker, S., Drews, M., Conner, C., Foda, H. D., DeCLerck, A., Langley, K. E., Bahou, W. F., Docherty, A. J. P., and Cao, J. (1998a). Tissue inhibitor of metalloproteinase-2 (TIMP-2) binds to the catalytic domain of the surface receptor, membrane type 1-matrix metalloproteinase 1 (MT1-MMP). *J. Biol. Chem.* **273**, 1216–1222.

Zucker, S., Mirza, H., Conner, C., Lorenz, A., Drews, M., Bahou, W. F., and Jesty, J. (1998b). Vascular endothelial growth factor and matrix metalloproteinase production in endothelial cells: Conversion of prothrombin to thrombin results in progelatinase A activation and cell proliferation. *Int. J. Cancer* **75**, 780–786.

Zucker, S., Hymowitz, M., Conner, C., Zarrabi, H. M., Hurewitz, A. N., Matrisian, L., Boyd, D., Nicholson, G., and Montana, S. (1999). Measurement of matrix metalloproteinases (MMPs) and tissue inhibitors of metalloproteinases (TIMPs) in blood and tissues. *Ann. NY Acad. Sci.* **878**, 212–227.

Zucker, S., Cao, J., and Chen, W.-T. (2000). Critical appraisal of the use of matrix metalloproteinase inhibitors in cancer treatment. *Oncogene* **19**, 6642–6650.

Zucker, S., Hymowitz, M., Rollo, E. E., Mann, R., Conner, C. E., Cao, J., Foda, H., Tompkins, D. C., and Toole, B. (2001). Tumorigenic potential of extracellular matrix metalloproteinase induce (EMMPRIN). *Am. J. Pathol.* **158**, 1921–1928.

Zucker, S., Cao, J., and Molloy, C. J. (2002a). Role of matrix metalloproteinases and plasminogen activators in cancer and metastaasis: Therapeutic strategies. *In* "Anticancer Drug Development" (B. C. Baguley and D. J. Kerr, Eds.), pp. 91–122. Academic Press, San Diego.

Zucker, S., Hymowitz, M., Conner, C. E., DiYanni, E. A., and Cao, J. (2002b). Rapid trafficking of membrane type 1-matrix metalloproteinase (MT1-MMP) to the cell surface regulates progelatinase A activation. *Lab Investig.* **82**, 1673–1684.

2

Surface Association of Secreted Matrix Metalloproteinases

Rafael Fridman
Department of Pathology
School of Medicine
Wayne State University
Detroit, Michigan 48201

I. Introduction

Proteolysis of the pericellular milieu is a fundamental physiological process by which cells modify their immediate microenvironment to achieve homeostasis and fulfill their biological destiny. The pericellular environment is a complex ensemble of surface components and extracellular matrix (ECM) proteins and any change in its composition and structure has a profound impact on cell behavior and survival. The task of adjusting the nature and composition of the pericellular milieu in the organism relies on specific proteases produced by a variety of cells in the tissue. The tight control of proteolytic activity produces optimal cellular responses while uncontrolled activity causes significant tissue damage. Various proteolytic systems are at the cell's command to carry out surface proteolysis.

Current Topics in Developmental Biology, Vol. 54

One of the major groups of proteases responsible for pericellular proteolysis is the matrix metalloproteinases (MMPs), a family of zinc-dependent endopeptidases. The MMP family comprises at least 27 members in vertebrates (Massova *et al.*, 1998; Nagase and Woessner 1999; Sternlicht and Werb, 2001), which are divided into five major subgroups based on their substrate specificity and structural organization: collagenases, stromelysins, matrilysins, gelatinases, and membrane type-MMPs (Massova *et al.*, 1998). MMP-19 and MMP-23 do not fall into these categories and thus may represent new subfamilies.

The members of the MMP family are key mediators of pericellular proteolysis in many physiological and pathological conditions, and in the past 20 years a considerable effort has been put forth to understand their mechanism of action and inhibition. Targeting the MMPs may interfere with the pathogenesis of various human diseases characterized by uncontrolled degradation of the ECM such as cardiovascular diseases, arthritis, and cancer. Therefore, a vast number of synthetic MMP inhibitors were developed and tested in animal models and in human clinical trials for effectiveness. Although some MMP inhibitors have shown promise, the targeting of MMPs is hampered by the structural similarities among the members of the MMP family, which reduces selectivity. The MMPs possess considerable structural similarities in their individual domain structures, in particular in the catalytic and hemopexin-like domain (Massova *et al.*, 1998), and with few exceptions, they have a considerable overlap in substrate specificity (Sternlicht and Werb, 2001). Biologically, the members of the MMP family also exhibit a high degree of redundancy in their cellular function. For example, studies investigating the role of MMPs in tumor cell migration and invasion have shown that these processes can be equally mediated by a variety of both secreted and membrane-anchored MMPs. Furthermore, knockout mice deficient in MMP genes, with the exception of MT1-MMP (Holmbeck *et al.*, 1999; Zhou *et al.*, 2000), showed no significant phenotypes when unchallenged (Parks and Shapiro, 2001). Also, the evidence shows that in most physiological and pathological conditions there is a consistent coexpression of various MMPs with overlapping properties. From the drug design perspective, these properties of the MMPs represent a major challenge. From the biological standpoint, the similarities between MMPs raise a fundamental dilemma on regulation of pericellular proteolysis: how do cells execute their proteolytic programs given the vast number of MMPs at the cell's command? Obviously, differential tissue expression and presence of tissue inhibitors of metalloproteinases (TIMPs), a family of natural MMP inhibitors, afford the cells the ability to control MMP-dependent proteolysis in tissues. However, these regulatory mechanisms, albeit important, do not explain how do cells can control the activity of MMPs at the pericellular space. New evidence indicates that focusing the activity of MMPs at the cell surface represents an additional level of regulation that permits cells to strictly control the initiation, inhibition, and termination of proteolytic activity on the cell surface. Furthermore, surface proteolysis affords the cells the ability to hydrolyze biologically relevant molecules in their immediate milieu.

To concentrate MMP activity at the cell surface, the MMP family developed unique structural (domain additions) and non structural features (recruitment of binding proteins) and hence today the MMPs can be subdivided in two major groups: the secreted and the membrane-anchored MMPs. A subgroup of MMPs incorporated membrane-anchoring motifs and thus evolved as membrane type-MMPs (MT-MMPs), which are the topic of Chapter 1. On the other hand, the secreted MMPs recruited various cell membrane proteins to act as docking sites on the cell surface. Many studies reported the association of secreted MMPs with the cell surface in various cell types including primary, nonmalignant, and malignant cells (Table I). The presence of secreted MMPs on the cell surface was investigated using a variety of techniques including isolation of purified plasma membranes and flow cytometry. These findings led to the search for specific surface components responsible for the MMP binding, which culminated with the identification of

Table I Reported Cell Surface Association of Secreted MMPs

Enzyme	Cell Type	Reference
MMP-9	Human pancreatic cancer RWP-I cells	Zucker et al., 1990
	Human breast epithelial MCF10A cells	Toth et al., 1997
	Bovine microvascular endothelial cells	Partridge et al. 1997
	Human fibrosarcoma HT1080 cells[a]	Ginestra et al., 1997
	Human neutrophils	Gaudin et al., 1997
	Human breast carcinoma 8701-BC cells[a]	Dolo et al., 1998
	Rat vascular endothelial cells	Olson et al., 1998
	Murine mammary carcinoma Met-1 cells	Bourguignon et al., 1998; Yu and Stamenkovic, 1999
	Human ovarian carcinoma CABA I cells[a]	Dolo et al., 1999
	Prostate cancer DU-145 cells	Manes et al., 1999
	Breast cancer MCF7 cells	Mira et al., 1999
	Murine mammary carcinoma TA3 cells	Yu and Stamenkovic, 1999
	Prostate cancer PC3 cells	Festuccia et al., 2000
	Normal mouse keratinocytes	Yu and Stamenkovic, 2000
	Human ovarian carcinoma OVCA 429 cells	Ellerbroek et al., 2001
MMP-13	Rat osteosarcoma UMR 106-1	Barmina et al., 1999
	Rat osteosarcoma ROS 17/2.8	Barmina et al., 1999
	Normal rat osteoblasts	Barmina et al., 1999
	Normal rat embryo fibroblasts	Barmina et al., 1999
MMP-1	Rat mammary carcinoma BC1-5 cells	Whitelock et al., 1991
	Human lung carcinoma LX-1	Guo et al., 2000
	Primary human keratynocyttes	Dumin et al., 2001
MMP-7	Human neuroblastoma SH-SY5Y	Yu et al., 2002
MMP-19	Human peripheral blood mononuclear cells	Sedlacek et al., 1998
	Human myeloid THP-1 cells	Mauch et al., 2002
	Human myeloid HL-60 cells	Mauch et al., 2002

[a]Presence in shed vesicles.

Table II   Identified Proteins Associated with Binding of Secreted MMPs

Enzyme	Binding Protein	Function
MMP-9	α2 chain of collagen IV (Olson *et al.*, 1998)	Surface association and matrix localization
	CD44 hyaluronan receptor (Bourguignon *et al.*, 1998; Yu and Stamenkovic, 1999)	Surface association
	RECK (Takahashi *et al.*, 1998)	Inhibition of expression and activity
	LRP (Hahn-Dantona *et al.*, 2001)	Internalization of zymogen
MMP-1	$\alpha_2\beta_1$ integrin (Dumin *et al.*, 2001)	Localization at cell-collagen binding sites
	EMMPRIN (Guo *et al.*, 2000)	Surface association
MMP-13	Endo180? (Barmina *et al.*, 1999)	Internalization
MMP-7	Heparan sulfate proteoglycan (Yu and Woessner, 2000)	Matrix localization
	CD44/HSPG (Yu *et al.*, 2002)	Surface association

several MMP-binding proteins, each acting to control various aspects of enzyme function (Table II). Today, the surface association of MMPs is no longer viewed as a mere physical docking of the protease on the cell surface but a complex array of cell–protease interactions that regulate enzymatic activity and substrate preference. Indeed, new evidence shows that surface binding of MMPs provides the cells with the capability to process a vast spectrum of surface precursor molecules that, once activated, play pivotal roles in normal cellular and tissue functions (Sternlicht and Werb, 2001). On the other hand, surface binding also allows cells to terminate proteolysis by either inhibition (Takahashi *et al.*, 1998) or internalization of extracellular proteases (Hahn-Dantona *et al.*, 2001; Walling *et al.*, 1998). As later discussed in this review, these mechanisms allow the cells to fine-tune the control of closely homologous proteases such as the MMPs by adding an extra level of regulation involving subcellular localization. This review summarizes our current knowledge of the surface association of those secreted MMPs on which there is information available. MMP-2 (gelatinase A) is not included in this chapter, and its surface binding is discussed in Chapter 1.

II. Surface Binding of MMP-9

A. Overview

MMP-9 (gelatinase B), together with MMP-2 (gelatinase A), belongs to the gelatinase subfamily of MMPs, which comprise a distinct subgroup of MMPs characterized by unique structural features, interactions with TIMPs, and substrate specificity (Collier *et al.*, 1988; Wilhelm *et al.*, 1989). At the domain organization level, MMP-9, in addition to the basic domains of all MMPs, contains a so-called gelatin-binding domain (GBD) inserted within the catalytic domain, which consists of three tandem copies of 58 amino acid residues, each homologous to the fibronectin type II-like module (Murphy and Crabbe, 1995). The GBD mediates the binding of the latent and active gelatinases to denatured collagen, also known as gelatin (Strongin et al., 1993). Consistently, the substrate profile of the gelatinases demonstrates that these enzymes are efficient gelatin-degrading proteases and thus are thought to accomplish the later steps of collagen degradation by carrying out hydrolysis of collagen molecules after the attack of interstitial collagenases. However, MMP-2 was also shown to exhibit classical collagenase activity (Aimes and Quigley, 1995). The gelatinolytic activity of the gelatinases has been exploited as a means to identify these enzymes in biological samples including cell surfaces using the gelatin zymography technique (for a detailed review of zymography, see Toth and Fridman, 2001). Several unique features confer MMP-9 with distinct characteristics. For example, MMP-9 contains an additional 54-amino-acid long, proline-rich extension with homology to the $\alpha2(V)$ chain of collagen V that is located between the hinge region and the hemopexin-like domain (Wilhelm et al., 1989). Furthermore, MMP-9, as opposed to MMP-2, is heavily glycosylated with both N- and O-linked glycosylation sites, which comprise a significant portion of the total molecular mass (Opdenakker et al., 2001a). The function of the oligosaccharide moieties of MMP-9 remains unknown. In biological samples, MMP-9 is found in three distinct forms: a monomer, a disulfide-linked homodimer (Olson et al., 2000; Strongin et al., 1993), and as a covalent complex with lipocalin (Kjeldsen et al., 1993; Yan et al., 2001). Although both the monomeric and dimeric forms of MMP-9 are fully competent proteases, the zymogenic form of the dimer exhibits a significantly slower rate of activation by MMP-3 when compared to the monomeric form (Olson et al., 2000). The MMP-9–lipocalin complex has been shown to play a role in potentiation of zymogen activation (Tschesche et al., 2001) and to protect MMP-9 from degradation (Yan et al., 2001).

Like all MMPs, active MMP-9 is inhibited by the TIMPs, which bind to the catalytic site with high affinity resulting in complete inhibition of enzymatic activity (Olson et al., 1997). The latent form (referred to here as pro-MMP or progelatinase) of MMP-9 can form noncovalent complexes with TIMP-1, which are mediated by specific molecular interactions between the hemopexin-like domain of the enzyme and the C-terminal region of the TIMP (Strongin et al., 1993). Pro-MMP-9 can also bind TIMP-3 (Butler et al., 1999). The ability to generate zymogen/inhibitor complexes is a unique feature of the gelatinases among the members of the MMP family. The zymogen/inhibitor complex of pro-MMP-2 with TIMP-2 is involved in surface association and activation (Butler et al., 1998; Strongin et al., 1995), whereas the role of the pro-MMP-9/TIMPcomplexes remains undefined.

MMP-9 has been extensively studied in the context of its role in tumor metastasis and angiogenesis (Coussens et al., 2000). MMP-9 has also been shown to play a critical role in the degradation of ECM in cardiovascular diseases (Ducharme et al., 2000). Mice deficient in MMP-9 display an abnormal pattern of skeletal growth plate vascularization and ossification (Vu *et al.*, 1998). MMP-9 is expressed by various types of cells including endothelial, epithelial, fibroblast, and immune cells, but the major producers of MMP-9 in tissues are the inflammatory cells (Opdenakker *et al.*, 2001b).

Most of the studies describing the association of MMP-9 with the cell surface were carried out with cell lines exposed to physiological and nonphysiological MMP-9 inducers such as growth factors (Ellerbroek *et al.*, 2001; Manes *et al.*, 1999; Mira *et al.*, 1999) and phorbol ester (Ginestra *et al.*, 1997; Mazzieri *et al.*, 1997; Olson *et al.*, 1998; Toth *et al.*, 1997). Upon secretion, MMP-9 is usually released into the culture media in a latent form. However, a fraction of the enzyme pool can be detected on the cell surface (Ellerbroek *et al.*, 2001; Festuccia *et al.*, 2000; Manes *et al.*, 1999; Mira *et al.*, 1999; Toth *et al.*, 1997) and in purified plasma membrane fractions (Toth *et al.*, 1997; Zucker *et al.*, 1990). Surface-bound MMP-9 can be readily extracted with aqueous solutions consistent with the fact that MMP-9 is a peripheral plasma membrane-associated protein (Toth *et al.*, 1997). MMP-9 was identified on the surface of a variety of cultured cells including endothelial (Olson *et al.*, 1998; Partridge *et al.*, 1997), breast epithelial (Olson *et al.*, 1998; Toth *et al.*, 1997), breast cancer (Mira *et al.*, 1999), pancreatic cancer (Zucker *et al.*, 1990), ovarian cancer (Ellerbroek *et al.*, 2001), prostate cancer (Festuccia *et al.*, 2000), fibrosarcoma (Mazzieri *et al.*, 1997), and mouse mammary carcinoma (Yu *et al.*, 1999) cells. Other studies showed the presence of MMP-9 in shed plasma membrane vesicles of human fibrosarcoma (Ginestra *et al.*, 1997) and endothelial (Taraboletti *et al.*, 2002) cells. These studies helped to establish the concept that MMP-9 is also a surface-associated protease and led to the search for MMP-9 binding proteins.

B. Binding of MMP-9 to the $\alpha2$(IV) Chain

The $\alpha2$(IV) chain of collagen IV was the first protein to be identified as an MMP-9 binding surface protein (Olson *et al.*, 1998). Ligand binding studies of iodinated pro-MMP-9 in breast epithelial MCF10A cells showed that MMP-9 binds to the cell surface with high affinity ($K_d \sim 22$ nM) (Olson *et al.*, 1998), which suggested the presence of a specific MMP-9 binding molecule. To identify the putative MMP-9 binding molecule, cell extracts of surface biotinylated-MCF10A breast epithelial cells were subjected to an MMP-9-affinity purification and a major biotinylated protein of 190 kDa was identified (Olson *et al.*, 1998). Sequencing data and specific antibodies demonstrated that the 190-kDa protein was the $\alpha2$(IV) chain of basement membrane collagen IV. Coimmunoprecipitation experiments further confirmed the $\alpha2$(IV) chain as the major MMP-9 binding protein in various cell types including breast epithelial MCF10A, breast carcinoma MDA-MB-231, fibrosarcoma HT1080, and rat vascular endothelial cells (Olson *et al.*, 1998; Toth *et al.*, 1999). These finding were consistent with immunohistochemical studies showing the presence of MMP-9 on the surface of cancer cells in breast carcinomas (Visscher *et al.,* 1994) and in the basement membrane of skin tumors (Coussens *et al.,* 1999; Karelina *et al.,* 1993). Formation of the pro-MMP-9/TIMP-1 complex did not preclude binding of pro-MMP-9 to the $\alpha 2(IV)$ chain, suggesting that the site of interaction is not the hemopexin-like domain, which is responsible for binding TIMP-1 (Olson *et al.,* 1998). The gelatinases bind to denatured collagens via the gelatin-binding domain, which exhibits a high degree of homology between the two enzymes (Murphy and Crabbe, 1995). However, pro-MMP-2 showed a significantly lower affinity for the $\alpha 2(IV)$ chain when compared to pro-MMP-9 (Olson *et al.,* 1998). Possibly, other sites and/or the enzyme conformation regulate the interactions with the $\alpha 2(IV)$ chain.

Functionally, the binding of MMP-9 to the $\alpha 2(IV)$ chain does not appear to play a direct role in the regulation of zymogen activation and/or enzyme inhibition (Olson *et al.,* 1998). It is also unknown whether the activity of MMP-9 is altered in any way by this interaction. However, the fact that MMP-9 is an efficient gelatinolytic enzyme suggests that the binding of MMP-9 to the $\alpha 2(IV)$ chain plays a role in the degradation of the collagen IV network at the cell surface. If so, the discovery of the $\alpha 2(IV)$ chain is not surprising given the well-known ability of the gelatinases to bind with high affinity to denatured collagens by means of the gelatin binding domain (Allan *et al.,* 1995). Consistently, the gelatinases are very efficient gelatinolytic enzymes (Murphy and Crabbe, 1995), which are considered to participate in the degradation of the collagen matrix subsequent to the action of collagenolytic enzymes such as MMP-1, MMP-13, or MT1-MMP. Consistent with this view, the gelatinases exhibit a weak affinity for native (trimeric) collagens (Olson *et al.,* 1998; Steffensen *et al.,* 1998). It is conceivable that the binding of MMP-9 to the $\alpha 2(IV)$ chain is mediated by sites that are cryptic in native collagen IV, which are only exposed after partial denaturation and/or degradation of the collagen IV molecule. If so, the binding of pro-MMP-9 to the $\alpha 2(IV)$ chain uncovers a potential proteolytic mechanism that regulates the remodeling of basement membrane collagen IV by localizing the zymogen at a precise location where upon activation completes the breakdown of the collagen network.

Although both gelatinases can cleave denatured collagen IV with similar efficiencies, MMP-9 exhibits a higher affinity toward the $\alpha 2(IV)$ chain (Olson *et al.,* 1998), suggesting that the association of MMP-9 with $\alpha 2(IV)$ may serve a different purpose. Thus, beyond the potential role of the pro-MMP-9/$\alpha 2(IV)$ complex in degradation of denatured collagen IV, the existence of single $\alpha 2(IV)$ chains in the extracellular space raises the interesting possibility that $\alpha 2(IV)$ chains may serve as pro-MMP-9 binding proteins on the cell surface. However, the stability of monomeric $\alpha 2(IV)$ chains remains controversial (Toth *et al.,* 1999; Yoshikawa *et al.,* 2001). This fact does not minimize the potential significance of the high-affinity binding of MMP-9 to $\alpha 2(IV)$ and consequently its association with collagen IV. Because collagen IV molecules are very closely associated with the cell plasma membrane, most likely bound via integrins, a clear distinction between cell surface and ECM cannot be made. It is therefore permissible to consider the site of MMP-9 interaction with collagen IV as pericellular rather than extracellular. This pericellular milieu constitutes the biological front where relevant proteolysis must take place.

C. Binding of MMP-9 to CD44

A series of studies (Bourguignon *et al.*, 1998; Yu and Stamenkovic, 1999) showed the association of MMP-9 with the hyaluronan receptor CD44. The adhesion receptor CD44 is a heavily glycosylated transmembrane protein that as a consequence of extensive alternative splicing exists in multiple variant forms. Numerous studies have implicated CD44 in tumor growth, invasion, and metastasis (Goodison *et al.*, 1999) and thus, the association of MMP-9 with CD44 has been suggested to link cellular adhesion to ECM and pericellular proteolysis (Yu and Stamenkovic, 1999). Studies with murine mammary carcinoma and human melanoma cells expressing CD44 reported the presence of MMP-9 on the cell surface (Bourguignon *et al.*, 1998; Yu and Stamenkovic, 1999). The association of MMP-9 with CD44 was demonstrated by coimmunoprecipitation experiments and colocalization of proteins on the cell surface by immunofluorescence (Bourguignon *et al.*, 1998; Yu and Stamenkovic, 1999), but the relative binding affinity of MMP-9 toward CD44 remains unknown. Analyses of the surface-bound MMP-9 by gelatin zymography revealed presence of the active form as determined by molecular mass (Yu and Stamenkovic, 1999), suggesting that CD44 interacts specifically with the active species of MMP-9. However, it is unclear whether binding to CD44 plays a role in pro-MMP-9 activation. The binding of MMP-9 to CD44 appears to be associated with the ability of CD44 receptors to aggregate on the cell surface and binding of CD44 to hyaluronan is not necessary for the association of MMP-9 to CD44. However, hyaluronan induces coclustering of CD44 receptors with active MMP-9 in the mouse mammary cell line.

Functionally, the association of MMP-9 with CD44 has been shown to promote tumor cell invasion *in vitro* and in experimental metastasis assays (Yu and Stamenkovic, 1999). Furthermore, expression of a soluble CD44 receptor abrogated invasion *in vitro* and *in vivo* and inhibition of MMP-9 activity by an inhibitory antibody or antisense technology obliterated the invasive ability of the cells expressing CD44 and surface-bound MMP-9 (Yu and Stamenkovic, 1999). Although surface-associated MMP-9 may focus ECM degradation on the immediate cellular environment, active MMP-9 bound to the CD-44 receptor was also shown to process latent tumor growth factor-β (TGF-β) to the biologically active form (Yu and Stamenkovic, 2000). Generation of active TGF-β on the cell surface may enhance tumor growth and metastasis by promoting a degradative phenotype (Overall *et al.*, 1989) and by inducing angiogenesis (Yu and Stamenkovic, 2000).

D. Binding of MMP-9 to RECK

A negative mechanism of MMP-9 expression and activity has been reported to involve the action of the membrane-anchored glycoprotein RECK (Takahashi et al., 1998). The RECK gene was identified after screening genes that could suppress the transformed phenotype induced by the ras oncongene in mouse fibroblasts (Takahashi et al., 1996). RECK codes for a glycosylphosphatidylinositol (GPI)-anchored cysteine-rich glycoprotein containing serine protease inhibitor-like domains and regions with weak homology to epidermal growth factor (EGF) (Takahashi et al., 1998) and its expression is downregulated in a variety of tumor cell lines and in cells transformed by a variety of oncogenes. Expression of the recombinant RECK protein in various cell lines inhibited in vitro invasion and metastasis in vivo without significant effects on the proliferative capacity of the cells. Subsequently, it was found that the expression of RECK was associated with specific reduction in the amounts of secreted pro-MMP-9 while no differences were observed in pro-MMP-2 secretion. Although RECK has no homology to TIMPs, a soluble RECK protein was shown to inhibit the enzymatic activity of MMP-9 (Takahashi et al., 1998), MMP-2, and MT1-MMP (Oh et al., 2001), albeit with a lower affinity than TIMPs. The mechanism by which the RECK protein inhibits MMP activity is unclear given the homology of RECK to serine protease inhibitors, but the inhibitory effects are not specific since various MMPs, both soluble and membrane-anchored, were equally inhibited (Oh et al., 2001). In contrast, there appears to be specificity in the recognition of pro-MMP-9 by a soluble recombinant RECK (Takahashi et al., 1998). However, whether membrane-anchored RECK can bind pro-MMP-9 on the cell surface and regulate its activation remains unclear. The inhibitory effect of RECK is believed to be the cause of the tumor-suppressing effects of RECK and also for the lethal consequence of RECK gene ablation in knockout mice (Oh et al., 2001). The RECK null mice fail to develop beyond the embryonic stage and exhibit severe malformation of mesenchymal and vascular tissues, which were ascribed to reduced MMP activity when compared to the wild-type RECK embryos. Proteins such as RECK have been proposed to represent a new class of protease inhibitors that act specifically on the cell surface to control enzymatic activity in the pericellular space, as opposed to soluble MMP inhibitors, which are secreted (Oh et al., 2001). Thus, surface-bound molecules that can act as protease inhibitors may add an extra level of regulation during pericellular proteolysis by trapping and inhibiting secreted MMPs on the cell surface.

E. Binding of MMP-9 to LRP

The control of pericellular proteolysis is also mediated by an active process of internalization that eliminates proteolytic enzymes and inhibitors from the cell surface. Low-density lipoprotein receptor-related protein (LRP) is a member of the LDL receptor family known to mediate the endocytotic intake of ligands as diverse as lipoproteins, protease–inhibitor complexes, proteases, growth factors, ECM components, viruses, and bacterial toxins (Herz and Strickland, 2001). More than 30 ligands were found to bind LRP with high affinity indicating the broad role that LRP plays in the regulation of protein function by internalization. LRP-mediated internalization of protease–inhibitor complexes has been shown to regulate the activity of urokinase-type plasminogen activator (uPA) and its receptor (uPAr) after inhibition by the plasminogen activator inhibitor (PAI)-1 (Nykjaer *et al.,* 1997). Internalization of the uPA/uPAr/PAI-1 complex permits recycling of uPAr molecules to the cell surface to maintain plasmin-dependent activity (Nykjaer *et al.,* 1997). Evidence suggests that LRP also plays a role in the control of MMP activity at the cell surface by mediating the internalization of various members of the MMP family. MMP-9 was shown to bind with high affinity to purified LRP either as a free enzyme or in complex with TIMP-1 (Hahn-Dantona *et al.,* 2001). Binding of MMP-9 to LRP was inhibited by RAP (receptor associated protein), a protein that inhibits ligand binding to LRP (Herz and Strickland, 2001). Upon binding to embryonic fibroblasts expressing LRP, the MMP-9/TIMP-1 complex was shown to be internalized as a function of time, a process that was followed by degradation in a chloroquine-dependent mechanism (Hahn-Dantona *et al.,* 2001). Based on these studies a role for LRP has been proposed in which LRP controls the level of MMP-9 in the pericellular environment by promoting the clearance and subsequent catabolism of the latent enzyme and the complex of pro-MMP-9 with TIMP-1. Thus, LRP may act as a negative regulator of MMP-9 action. In this regard, it is interesting that LRP internalizes the complex of pro-MMP-9 with TIMP-1, which is catalytically inactive. There is not yet evidence that LRP can promote the internalization of the active enzyme or the active enzyme in complex with TIMPs.

III. Surface Binding of the Soluble Interstitial Collagenases (MMP-1 and MMP-13)

A. Overview

The degradation of the connective tissue matrix is mediated by a limited number of specialized MMPs, which acquired the capacity of hydrolyzing native triple helical interstitial collagen molecules. These enzymes are efficient in promoting the degradation of fibrilar collagen molecules at a single locus in both the $\alpha 1$ and $\alpha 2$ chains and thus are known to generate the classical 3/4 and 1/4 degradation fragments. Cleavage of collagen I molecules causes instability of the triple helical structure and the denatured collagen becomes sensitive to the proteolytic attack of gelatinases. Fibrilar collagen degradation is a fundamental process during the development and formation of connective tissues such as bone and cartilage and for the remodeling and turnover of the collagenous matrix in normal and pathological conditions that necessitate collagen degradation. Both membrane-anchored and soluble MMPs can cleave interstitial native collagen (I to III) molecules including MMP-1 (fibroblast collagenase, collagenase-1), MMP-13 (collagenase-3), MMP-14 (MT1-MMP), MMP-8 (neutrophil collagenase, collagenase-2), and MMP-2 (gelatinase A), albeit with various degrees of efficiency (Krane, 2001).

Degradation of interstitial collagens must be strictly regulated to prevent unnecessary tissue damage. Localization of collagenolytic activity at the cell–matrix interface would permit turnover and remodeling of the collagen matrix where it is really needed. It is evident that by being anchored to the plasma membrane MT1-MMP, a potent collagenolytic enzyme, can focus its collagenase activity at the cell surface. However, how do the soluble interstitial collagenases achieve localized degradation of fibrilar collagens? Recent studies have attempted to answer this question in regard to the surface regulation of MMP-1 and MMP-13.

B. Surface Binding of MMP-1

1. Binding to $\alpha_2\beta_1$ Integrin

MMP-1 is produced by a variety of cell types and its major role is to promote the degradation of native collagen I molecules in the interstitial connective tissue matrix. MMP-1 is particularly active in normal and pathological processes that require turnover and remodeling of fibrilar collagen, particularly during development and wound healing, but it has also been associated with tumor invasion and metastasis (Brinckerhoff et al., 2000). The activity of MMP-1 during wound healing must be efficiently and tightly regulated so that the proteolytic events produce the appropriate cellular responses and accomplish effective tissue repair. A series of studies in human keratinocytes, the major cell of the skin, elegantly demonstrated that the latent form of MMP-1 associates with the cell surface via the $\alpha_2\beta_1$ integrin (Dumin et al., 2001; Stricker et al., 2001). The $\alpha_2\beta_1$ integrin is the major collagen-binding receptor and is known to mediate the adhesion of keratinocytes to collagen I. Coimmunoprecipitation experiments, cellular colocalization, and solid-phase binding studies showed that pro-MMP-1 interacts with and binds specifically to $\alpha_2\beta_1$ integrin. Indeed, $\alpha_3\beta_1$, another integrin receptor for collagen I, was not detected in complex with MMP-1. MMP-1 binding to $\alpha_2\beta_1$ integrin was shown to be mediated by the I domain of the α_2 integrin subunit and the hinge region and hemopexin-like domain of pro-MMP-1 (Dumin et al., 2001; Stricker et al., 2001). Solid-phase assay studies with purified α_2 I domain and MMP-1 (active and latent) showed a high-affinity (around 10–40 nM) binding for both enzyme forms, although the zymogen appeared to bind preferentially (Stricker et al., 2001). Under the experimental conditions used, the interaction of pro-MMP-1 with $\alpha_2\beta_1$ was evident when the cells were plated on native collagen I and was significantly reduced on gelatin. This observation suggested that the nature of the matrix influences the ability of the $\alpha_2\beta_1$ integrin to cluster pro-MMP-1 on the cell surface.

The significance of the binding of pro-MMP-1 to $\alpha_2\beta_1$ has been associated with the ability of keratinocytes to migrate on collagen I during the process of wound healing in the skin. In response to injury, keratinocytes on the epidermal layer migrate to the underlying collagen I-rich connective tissue matrix to initiate the process of tissue repair. Expression of $\alpha_2\beta_1$ at the basal surface of the migrating keratinocytes confers on the cells the ability to form focal contacts with the collagen matrix, which in turn transduce extracellular signals that stimulate the repair process by inducing the expression of proteolytic enzyme such as MMP-1, among other factors (Sudbeck *et al.*, 1997). Dumin *et al.* (2001) proposed that binding of MMP-1 to $\alpha_2\beta_1$ would promote the generation of denatured collagen I molecules at areas of focal contacts. This process would cause a reduction in the affinity of $\alpha_2\beta_1$ toward collagen resulting in the dislodging of the integrin from its binding to the now-denatured collagen. This would allow the cells to detach and generate new contacts at the leading front, in the uncleaved collagen I matrix. In addition, hydrolysis of collagen I at focal contacts by MMP-1 would elicit a negative feedback on enzyme gene expression and binding to $\alpha_2\beta_1$ (Dumin *et al.*, 2001). This hypothesis was based on the observation that induction of pro-MMP-1 gene expression and binding to $\alpha_2\beta_1$ were more efficient when the cells were in contact with native collagen I (Dumin *et al.*, 2001). Paradoxically, Dumin *et al.* (2001) found that the zymogen form of MMP-1 was the major enzyme species that was consistently bound to $\alpha_2\beta_1$ and that pro-MMP-1 was the major form detected in the supernatant of keratinocytes plated on collagen I. These observations raise the question as to whether the binding of pro-MMP-1 to $\alpha_2\beta_1$ can regulate zymogen activation and whether the enzyme remains associated with the integrin after activation. Considering that the zymogen form is likely to be catalytically inactive, the detection of surface bound pro-MMP-1 (Dumin *et al.*, 2001) or pro-MMP-9 as described earlier (Ellerbroek *et al.*, 2001; Toth *et al.*, 1997) represents a major dilemma in our understanding of the relevance of cell surface association of the zymogen for cellular function. Recent evidence suggests that pro-MMP-9 exhibits limited enzymatic activity without removal of the propeptide if the enzyme is bound to gelatin (Bannikov *et al.*, 2002). Whether a similar process exists in pro-MMP-1 upon binding to $\alpha_2\beta_1$ and/or collagen I is still unknown. Further studies are required to support a functional role for the binding of pro-MMP-1 to $\alpha_2\beta_1$ on cell behavior or on enzymatic activity.

2. Binding to EMMPRIN/CD147

Invasive tumor cells migrate through the interstitial collagen matrix of the stroma to reach the circulation and thus they must possess the necessary proteolytic machinery to degrade various collagen types during cell migration. There is ample evidence indicating that MMP-1 expression is elevated in various invasive tumors (Brinckerhoff *et al.*, 2000). Similar to the findings with human keratinocytes and MMP-1 (Dumin *et al.*, 2001), the concept of surface localization as a means to improve the effectiveness of collagen degradation can also be applied for MMP-1 and tumor cells. Supporting this view, early studies showed the presence of MMP-1 on the plasma membranes of tumor cells (Moll *et al.*, 1990). In recent years, however, production of proteolytic enzymes, including MMPs, has been shown to take place mostly in the stromal cells that are closely associated with the cancer cells (Crawford and Matrisian, 1994; Hewitt and Dano, 1996). The expression of MMPs by the stromal cells is a consequence of specific tumor–stromal interactions mediated by a variety of factors. One of this factors is the transmembrane protein EMMPRIN (extracellular matrix metalloproteinase inducer) or CD147/basigin (Biswas *et al.*, 1995), a member of the immunoglobulin superfamily that is found on the surface of many tumor cells and is known to induce the expression of several MMPs in fibroblasts (Guo *et al.*, 1997). For instance, tumor-bound EMMPRIN has been shown to stimulate MMP-1 expression in fibroblasts and thus may contribute to the degradation of the interstitial collagen matrix in tumor tissues. However, this degradative process, if limited to the fibroblast pericellular space, may have little impact on the immediate environment of the tumor cells and their ability to migrate. The tumor cells, to take advantage of the MMPs produced by the stromal cells, must be able to retain the released enzymes on their surface. A study has provided a possible answer to this dilemma by showing that EMMPRIN itself can act as "receptor" for pro-MMP-1 on the surface of the tumor cells (Guo *et al.*, 2000). In that study, phage display screening was used to search for EMMPRIN binding proteins. This resulted in the isolation of clones containing DNA sequences of human MMP-1 suggesting that EMMPRIN may act as a surface binding protein for MMP-1. The ability of EMMPRIN to bind MMP-1 was verified using various approaches and the zymogen form of MMP-1 was found to bind preferentially to EMMPRIN (Guo *et al.*, 2000). Evidence suggests that EMMPRIN-mediated induction of MMP expression also involves both heterotypic and homotypic cell–cell interactions resulting in MMP expression in both fibroblasts and tumor cells (Sun and Hemler, 2001). Thus, EMMPRIN may not only act to induce expression of MMPs in the neighboring fibroblasts but also to serve as a docking protein for pro-MMP-1 on the surface of the invasive tumor cells in either a paracrine or an autocrine fashion. The functional consequences of binding the MMP-1 zymogen to EMMPRIN for both enzyme and cellular functions remain to be elucidated.

C. Surface Binding of MMP-13

MMP-13, or collagenase-3, was originally identified in breast carcinomas using a genetic approach and has been associated with the invasive and metastatic properties of various human tumors (Pendas *et al.*, 2000). MMP-13 has also been shown to play a major role in bone formation (Jimenez *et al.*, 2001) and in the pathogenesis of osteoarthritis (Neuhold *et al.*, 2001).

A series of studies demonstrated the binding of MMP-13 to the cell surface suggesting the existence of a specific receptor (Omura *et al.*, 1994). Studies in rat osteoblastic cells showed that secreted or exogenously added MMP-13 was gradually removed from the extracellular space by an active process of internalization of the enzyme. Ligand binding studies further demonstrated the existence of single high-affinity (nanomolar range) receptor (Omura *et al.*, 1994; Walling *et al.*, 1998). However, it turned out that the initial binding and subsequent internalization of pro-MMP-13 was mediated by two distinct surface components. The protein responsible for pro-MMP-13 internalization was identified as LRP. This was supported by the observation that LRP null fibroblasts, albeit able to bind pro-MMP-13, were unable to internalize the enzyme. Furthermore, RAP inhibited MMP-13 internalization consistent with a role for LRP in this process (Barmina *et al.*, 1999). A pro-MMP-13-binding protein with a relative mass of 170 kDa was isolated and identified as the transmembrane glycoprotein Endo180 (Barmina *et al.*, 1999). However, this study did not show evidence for the interaction (Barmina *et al.*, 1999). Later studies using various approaches and similar cell lines failed to confirm the role of Endo180 in pro-MMP-13 surface binding (Bailey *et al.*, 2002) and thus the nature of the putative pro-MMP-13 receptor remains elusive. The possible role of Endo180 as mediator of pro-MMP-13 cell surface binding is nevertheless interesting. Endo180 is a member of the C-type lectins of endocytic receptors, which presently include the macrophage mannose receptor, the phospholipase A_2 receptor, and the DEC-205/MR6 receptor (Sheikh *et al.*, 2000). Structurally, these receptors are type I transmembrane glycoproteins containing within the extracellular portion a cysteine-rich domain at the aminoterminal region followed by a collagen binding domain (fibronectin type II) and 8 to 10 consecutive carbohydrate recognition domains (Sheikh *et al.*, 2000). This family of proteins serves as endocytic receptors for extracellular glycoproteins in a process mediated by clathrin-coated vesicles, which is followed by lysosomal degradation (Sheikh *et al.*, 2000). Endo180 was also found to bind the urokinase-type plasminogen activator receptor (uPAr) in a process dependent on the presence of pro-uPA, and thus Endo180 is also known as the urokinase-type plasminogen activator receptor-associated protein (uPARAP) (Behrendt *et al.*, 2000). In addition, Endo180/uPARAP was shown to bind collagen V via its collagen-binding domain (Behrendt *et al.*, 2000). Although the precise role of Endo180/uPARAP in uPAr function remains unknown, the biological properties of the C type lectin suggest that it may play a role in the internalization of the uPA/uPAr complex similar to the role of LRP in this process (Engelholm *et al.*, 2001). The multiple interactions mediated by Endo180/uPARAP raise the interesting possibility that this lectin may serve as a link between serine proteases, MMPs, and collagen degradation on the cell surface. Although the proposed interaction of Endo180/uPARAP with pro-MMP-13 has been disproved (Bailey *et al.,* 2002), it is worth noting that pro-MMP-13 is heavily glycosylated and thus may have the potential to interact with lectin type proteins on the cell surface to mediate its internalization. Interestingly, pro-MMP-9, which is also heavily glycosylated, is also internalized by LRP (Hahn-Dantona *et al.,* 2001) suggesting the possibility that glycosylation of soluble MMPs may serve as a recognition signal for endocytic receptors to deplete the pericellular space of excess zymogens before they undergo activation. For pro-MMP-13, surface binding may also be associated with activation (Bailey *et al.,* 2002; Knauper *et al.,* 1997). However, whether this process necessitates the presence of a specific cell surface receptor to bring about the zymogen in close association with its activator remains unknown.

IV. Surface Binding of MMP-7

A. Overview

MMP-7 (matrilysin-1) (Wilson and Matrisian, 1996; Woessner, 1995), together with MMP-26 (matrilysin-2) (de Coignac *et al.,* 2000; Park *et al.,* 2000; Uria and Lopez-Otin, 2000), is the shortest member of the MMP family containing only a signal sequence, a propeptide, and a catalytic domain. Structural studies suggested that MMP-7's minimal domain composition represents a retrograde process in MMP evolution and thus MMP-7 may be a relatively new MMP in which deletion of the hemopexin-like domain occurred late in evolution (Massova *et al.,* 1998). MMP-7 exhibits broad substrate specificity, which includes ECM and non-ECM proteins (Wilson and Matrisian, 1996; Woessner, 1995). Accumulating evidence indicates that MMP-7 cleaves various proteins known to be associated with the cell surface including E-cadherin (Noe *et al.,* 2001), tumor necrosis factor-α precursor (Gearing *et al.,* 1994; Haro *et al.,* 2000), and Fas ligand (Powell *et al.,* 1999). MMP-7 is involved in a variety of physiological and pathological processes including uterus involution (Woessner, 1996), tumor progression (Fingleton *et al.,* 1999; Wilson and Matrisian, 1996), apoptosis (Powell *et al.,* 1999), and microbial defense (Wilson *et al.,* 1999). Mice deficient in MMP-7 lack efficient antimicrobial defenses in the lung (Wilson *et al.,* 1999) and do not support intestinal tumorigenesis (Wilson *et al.,* 1997).

B. Binding of MMP-7 to Heparan Sulfate

Several studies demonstrated the ability of MMP-7 to bind heparan sulfate (HS), a major glycosaminoglycan of the ECM (Yu and Woessner, 2000). HS moieties associate with a protein core forming heparan sulfate proteolglycans (HSPGs), which are heterogeneous and ubiquitous components of the ECM, in particular basement membranes, but are also expressed on cell surfaces in transmembrane proteins. The HS side chains are linear polysaccharides composed of alternating disaccharide units of either uronic acid or L-iduronic acid and D-glucosamine and are negatively charged because of the high content of sulfate groups (Esko and Lindahl, 2001). HSPGs interact with a large number of proteins and thus play key roles in many biological processes as diverse as tissue development, growth factor signaling, angiogenesis, and tumor metastasis (Forsberg and Kjellen, 2001; Iozzo and San Antonio, 2001; Sanderson, 2001), to name just a few. Thus, the association of MMP-7 with HS may link proteolytic activity with HSPG-mediated processes.

The association of MMP-7 with sulfated proteoglycans was inferred from the fact that the enzyme was extracted from rat uterus with various sulfated compounds and/or after heparitinase treatment. Further studies determined that MMP-7 colocalizes with HS in rat uterus and that they are similarly regulated during the estrous cycle (Yu and Woessner, 2000). MMP-7 and HS were detected on the surface of the epithelial cells lining the uterus glands. Although the relative affinity of MMP-7 toward the cell surface was not directly determined, affinity measurements were carried out with immobilized heparin. These studies showed a high-affinity binding of both the latent and active forms of MMP-7, although the latent form exhibited a stronger affinity. Structural analyses of the catalytic domain of rat MMP-7 revealed the existence of a putative heparin-binding site represented by a contiguous line of positive residues, which may be responsible for the tight binding of MMP-7 to HSPG (Yu and Woessner, 2000).

An isoform of the multifunctional CD44 hyaluronan receptor containing the alternatively spliced exon v3 can incorporate HS side chains and thus is considered a nonclassical HSPG (Iozzo, 2001). Functionally, CD44/HS exhibits some of the functions of classical surface-associated HSPG including the ability to bind fibroblast growth factor (FGF) and heparin-binding epidermal growth factor (HB-EGF) (Bennett et al., 1995). One study reported the association of CD44/HS with MMP-7 on the cell surface by colocalization, coimmunoprecipitation, and heparin elution studies (Yu et al., 2002). The active form of MMP-7 appears to be preferentially associated with the heparan sulfate–containing CD44 isoform. The significance of this finding was then explained by the ability of MMP-7 to cleave the precursor form of heparin-binding epidermal growth factor (pro-HB-EGF) to generate active EGF. This growth factor can bind to CD44/HS and thus becomes accessible to the proteolytic processing of MMP-7, as demonstrated after coprecipitation of CD44/HS with pro-HB-EGF and exposure of the precipitated growth factor to MMP-7. Furthermore, it was shown that the mature HB-EGF could engage its receptor ErbB4 and directly affect organ remodeling in uterus and mammary glands (Yu et al., 2002). These studies demonstrated that the association of active MMP-7 with the cell surface via CD44/HS can promote the hydrolysis of cell surface molecules and thus directly influence cell behavior. The findings with CD44/HS and MMP-7 raise interesting possibilities for other secreted MMPs, as they have also been shown to bind to heparin (Butler *et al.*, 1998). Furthermore, TIMPs are also known to exhibit tight binding to heparin, which has been used for their isolation and purification. TIMP-3 has been shown to bind to HSPG in rat uterus in a similar fashion as MMP-7 (Yu *et al.*, 2000), suggesting the possibility that their close interaction may serve to regulate MMP-7 activity in the ECM and on the cell surface. The ability of HS chains to bind MMPs and TIMPs suggests the possibility that other HSPGs such as syndecans and glypicans could also act as potential cell surface binding proteins for these molecules and thus regulate pericellular proteolysis.

V. Surface Binding of MMP-19

MMP-19 is a secreted new member of the MMP family that was identified by searching databases containing expressed sequence tags (Cossins *et al.*, 1996; Pendas *et al.*, 1997) and by cDNA library screening (Kolb *et al.*, 1997) and was originally designated MMP-18 (Cossins *et al.*, 1996) and RASI-1 (Kolb *et al.*, 1997). Structurally, MMP-19 contains the basic domains of all MMPs including a signal peptide, a prodomain, a catalytic domain, and a hemopexin-like domain (Pendas *et al.*, 1997). However, the absence of several common structural features present in other MMP subfamilies makes MMP-19 a unique enzyme (Pendas *et al.*, 1997). MMP-19 contains two unique features including a hinge rich in acidic residues and a threonine-rich region of 36 residues at the end of the hemopexin-like domain downstream of the terminal cysteine (Cossins *et al.*, 1996; Pendas *et al.*, 1997). The catalytic competence of MMP-19 has been characterized with a recombinant catalytic domain, which was found to cleave a variety of ECM components, in particular collagen IV, from the basement membrane, laminin, and nidogen but shows no activity against triple helical collagens (Stracke *et al.*, 2000). MMP-19 is highly expressed in the synovial blood vessels of rheumatoid arthritis (Kolb *et al.*, 1997) and in myeloid cells (Mauch *et al.*, 2002). MMP-19 is expressed in normal breast tissues in myoepithelial, endothelial, and smooth muscle cells. Interestingly, compared to normal and noninvasive breast tumors, invasive breast carcinomas contain lower levels of MMP-19, suggesting that this enzyme may play a role in early stages of tumor progression.

Evidence indicates that MMP-19 associates with the cell surface of myeloid cells including activated blood mononuclear cells and various myeloid cell lines as determined by flow cytometry and surface biotinylation (Mauch *et al.*, 2002). The surface association of MMP-19 appears not to be mediated by a putative GPI anchor, and deletion of the hemopexin-like domain impaired surface association (Mauch *et al.*, 2002). The precise mechanism of MMP-19 surface association remains unclear but the presence of the C-terminal threonine-rich region may play a role in surface binding. Furthermore, the significance of the surface association remains to be determined.

VI. Concluding Remarks. Surface Binding: A Balance between Positive and Negative Effects on Pericellular Proteolysis

It is now evident that the surface association of secreted MMPs is a multifunctional process mediated by diverse surface proteins and involving, so far, only a limited number of secreted MMPs. It is also evident that binding of the secreted MMPs to the cell surface to be functionally meaningful for the cells does not require binding to a classical membrane-anchored molecule analogous to signaling or adhesion receptors. Instead, it may include surface molecules that facilitate clustering of the enzyme in the immediate pericellular microenvironment and surface proteins that can terminate proteolysis by removing the protease from the extracellular space. The differential effects of surface association on MMP function suggest that the fate of a surface-bound enzyme is determined by a balance between positive and negative elements. As shown in Fig. 1, binding of MMPs permits surface localization as demonstrated with the findings with MMP-9 (Olson *et al.*, 1998; Yu and Stamenkovic, 1999), MMP-1 (Dumin *et al.*, 2001), and MMP-7 (Yu *et al.*, 2002), and thus it may serve the cells to concentrate the secreted MMP at areas where

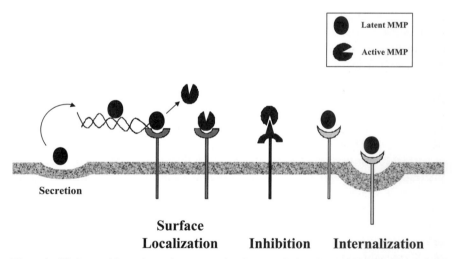

Figure 1 Distinct positive and negative aspects of surface association of secreted MMPs. The released MMPs can be retained on the cell surface by a variety of surface molecules including cell adhesion receptors or ECM components associated with the cell surface. This permits surface localization at areas of cell–matrix contact. Upon binding of the zymogen to these molecules, activation may ensue and the active enzyme may be released or remain bound. Alternatively, active enzyme may bind to a surface component. Binding to a surface-anchored protein with a protease inhibitory domain will result in inhibition of catalytic activity. Finally, surface binding may lead to enzyme internalization by endocytic receptors and thus terminate surface proteolysis. Interactions between these mechanisms and the action of TIMPs (not shown here) will determine the net proteolytic activity on the cell surface.

proteolysis is required. To achieve surface association, each of the secreted MMPs appears to utilize a different set of proteins suggesting a certain degree of specificity. If so, unique structural features in the secreted MMP molecules may target the enzyme to a specific surface-binding component. The presence of a gelatin-binding domain in the gelatinases is an example of adaptation for pericellular ECM binding. On the cells, however, the evidence so far indicates that there are not specific "receptors" for MMPs. Instead, surface proteins as different as cell adhesion receptors and ECM proteins have the ability to cluster secreted MMPs in areas of cell–matrix contacts. Although this association may facilitate substrate degradation at focal points, there is not yet evidence to indicate that such interactions regulate enzymatic activity. Further studies are required to determine the effects of surface binding on regulation of activation and catalytic activity of the secreted MMPs. If surface binding is required for efficient and localized control of enzymatic activity, a mechanism of inhibition that is also accessible to the pericellular environment must also exist to avoid excessive proteolysis. Membrane-anchored proteins with inhibitory domains such as RECK (Section II,E) (Takahashi et al., 1998) may provide a tighter control of MMP activity on the cell surface (Fig. 1) by directly binding the active MMP or by inhibiting active enzyme bound to an MMP surface "receptor." On the other hand, surface binding of active enzyme may disrupt enzyme–TIMP binding and thus act as a sort of safe haven for catalysis at the cell's doorstep. Although this is an interesting concept, there is no experimental evidence to support such a possibility with surface-bound secreted MMPs. In fact, the MT-MMPs, which are tethered to the plasma membrane, are readily accessible and inhibited by TIMP-2, a secreted inhibitor. Thus, at this juncture, the effects of surface binding on secreted MMP-TIMP interactions cannot be predicted. Enzyme internalization is another component of the overall mechanism aimed at terminating enzymatic activity on the cell surface. This process can be mediated by proteins such as LRP as shown for MMP-9 (Hahn-Dantona et al., 2001) and MMP-13 (Barmina et al., 1999) but may include other surface molecules. The studies with MMP-13 suggest that cell surface binding and internalization are mediated by different components and thus a complex process may regulate binding and subsequent internalization. The balance between positive and negative elements determines the level of net enzymatic activity in the pericellular milieu and permits the cells to tailor the appropriate proteolytic response. Further studies into the mechanisms of surface regulation and the molecules involved in these processes may provide new avenues for the development of specific inhibitors of MMP action in pathological conditions.

Acknowledgments

I am indebted to the past and present members of my laboratory and colleagues, for they have made the most important contributions to my understanding of MMP function. This work has been supported by a grant from the NIH CA-81222 to R.F.

References

Aimes, R. T., and Quigley, J. P. (1995). Matrix metalloproteinase-2 is an interstitial collagenase. Inhibitor-free enzyme catalyzes the cleavage of collagen fibrils and soluble native type I collagen generating the specific 3/4- and 1/4-length fragments. *J. Biol. Chem.* **270**, 5872–5876.

Allan, J. A., Docherty, A. J., Barker, P. J., Huskisson, N. S., Reynolds, J. J., and Murphy, G. (1995). Binding of gelatinases A and B to type-I collagen and other matrix components. *Biochem. J.* **309**(Pt 1), 299–306.

Bailey, L., Wienke, D., Howard, M., Knauper, V., Isacke, C. M., and Murphy, G. (2002). Investigation of the role of Endo180/urokinase-type plasminogen activator receptor-associated protein as a collagenase 3 (matrix metalloproteinase 13) receptor. *Biochem. J.* **363**, 67–72.

Bannikov, G. A., Karelina, T. V., Collier, I. E., Marmer, B. L., and Goldberg, G. I. (2002). Substrate binding of gelatinase B induces its enzymatic activity in the presence of intact propeptide. *J. Biol. Chem.* **277**, 16022–16027.

Barmina, O. Y., Walling, H. W., Fiacco, G. J., Freije, J. M., Lopez-Otin, C., Jeffrey, J. J., and Partridge, N. C. (1999). Collagenase-3 binds to a specific receptor and requires the low density lipoprotein receptor-related protein for internalization. *J. Biol. Chem.* **274**, 30087–30093.

Behrendt, N., Jensen, O. N., Engelholm, L. H., Mortz, E., Mann, M., and Dano, K. (2000). A urokinase receptor-associated protein with specific collagen binding properties. *J. Biol. Chem.* **275**, 1993–2002.

Bennett, K. L., Jackson, D. G., Simon, J. C., Tanczos, E., Peach, R., Modrell, B., Stamenkovic, I., Plowman, G., and Aruffo, A. (1995). CD44 isoforms containing exon V3 are responsible for the presentation of heparin-binding growth factor. *J. Cell Biol.* **128**, 687–698.

Biswas, C., Zhang, Y., DeCastro, R., Guo, H., Nakamura, T., Kataoka, H., and Nabeshima, K. (1995). The human tumor cell-derived collagenase stimulatory factor (renamed EMMPRIN) is a member of the immunoglobulin superfamily. *Cancer Res.* **55**, 434–439.

Bourguignon, L. Y., Gunja-Smith, Z., Iida, N., Zhu, H. B., Young, L. J., Muller, W. J., and Cardiff, R. D. (1998). CD44v(3,8-10) is involved in cytoskeleton-mediated tumor cell migration and matrix metalloproteinase (MMP-9) association in metastatic breast cancer cells. *J. Cell Physiol.* **176**, 206–215.

Brinckerhoff, C. E., Rutter, J. L., and Benbow, U. (2000). Interstitial collagenases as markers of tumor progression. *Clin. Cancer Res.* **6**, 4823–4830.

Butler, G. S., Apte, S. S., Willenbrock, F., and Murphy, G. (1999). Human tissue inhibitor of metalloproteinases 3 interacts with both the N- and C-terminal domains of gelatinases A and B. Regulation by polyanions. *J. Biol. Chem.* **274**, 10846–10851.

Butler, G. S., Butler, M. J., Atkinson, S. J., Will, H., Tamura, T., van Westrum, S. S., Crabbe, T., Clements, J., d'Ortho, M. P., and Murphy, G. (1998). The TIMP2 membrane type 1 metalloproteinase "receptor" regulates the concentration and efficient activation of progelatinase A. A kinetic study. *J. Biol. Chem.* **273**, 871–880.

Collier, I. E., Wilhelm, S. M., Eisen, A. Z., Marmer, B. L., Grant, G. A., Seltzer, J. L., Kronberger, A., He, C. S., Bauer, E. A., and Goldberg, G. I. (1988). H-ras oncogene-transformed human bronchial epithelial cells (TBE-1) secrete a single metalloprotease capable of degrading basement membrane collagen. *J. Biol. Chem.* **263**, 6579–6587.

Cossins, J., Dudgeon, T. J., Catlin, G., Gearing, A. J., and Clements, J. M. (1996). Identification of MMP-18, a putative novel human matrix metalloproteinase. *Biochem. Biophys. Res. Commun.* **228**, 494–498.

Coussens, L. M., Raymond, W. W., Bergers, G., Laig-Webster, M., Behrendtsen, O., Werb, Z., Caughey, G. H., and Hanahan, D. (1999). Inflammatory mast cells up-regulate angiogenesis during squamous epithelial carcinogenesis. *Genes Dev.* **13**, 1382–1397.

Coussens, L. M., Tinkle, C. L., Hanahan, D., and Werb, Z. (2000). MMP-9 supplied by bone marrow-derived cells contributes to skin carcinogenesis. *Cell* **103**, 481–490.

Crawford, H. C., and Matrisian, L. M. (1994). Tumor and stromal expression of matrix metalloproteinases and their role in tumor progression. *Invasion Metastasis* **14**, 234–245.

de Coignac, A. B., Elson, G., Delneste, Y., Magistrelli, G., Jeannin, P., Aubry, J. P., Berthier, O., Schmitt, D., Bonnefoy, J. Y., and Gauchat, J. F. (2000). Cloning of MMP-26. A novel matrilysin-like proteinase. *Eur. J. Biochem.* **267**, 3323–3329.

Dolo, V., D'Ascenzo, S., Violini, S., Pompucci, L., Festuccia, C., Ginestra, A., Vittorelli, M. L., Canevari, S., and Pavan, A. (1999). Matrix-degrading proteinases are shed in membrane vesicles by ovarian cancer cells in vivo and in vitro. *Clin. Exp. Metastasis* **17**, 131–140.

Dolo, V., Ginestra, A., Cassara, D., Violini, S., Lucania, G., Torrisi, M. R., Nagase, H., Canevari, S., Pavan, A., and Vittorelli, M. L. (1998). Selective localization of matrix metalloproteinase 9, beta1 integrins, and human lymphocyte antigen class I molecules on membrane vesicles shed by 8701-BC breast carcinoma cells. *Cancer Res.* **58**, 4468–4474.

Ducharme, A., Frantz, S., Aikawa, M., Rabkin, E., Lindsey, M., Rohde, L. E., Schoen, F. J., Kelly, R. A., Werb, Z., Libby, P., and Lee, R. T. (2000). Targeted deletion of matrix metalloproteinase-9 attenuates left ventricular enlargement and collagen accumulation after experimental myocardial infarction. *J. Clin. Invest.* **106**, 55–62.

Dumin, J. A., Dickeson, S. K., Stricker, T. P., Bhattacharyya-Pakrasi, M., Roby, J. D., Santoro, S. A., and Parks, W. C. (2001). Pro-collagenase-1 (matrix metalloproteinase-1) binds the alpha(2)beta(1) integrin upon release from keratinocytes migrating on type I collagen. *J. Biol. Chem.* **276**, 29368–29374.

Ellerbroek, S. M., Halbleib, J. M., Benavidez, M., Warmka, J. K., Wattenberg, E. V., Stack, M. S., and Hudson, L. G. (2001). Phosphatidylinositol 3-kinase activity in epidermal growth factor-stimulated matrix metalloproteinase-9 production and cell surface association. *Cancer Res.* **61**, 1855–1861.

Engelholm, L. H., Nielsen, B. S., Dano, K., and Behrendt, N. (2001). The urokinase receptor associated protein (uPARAP/endo180): a novel internalization receptor connected to the plasminogen activation system. *Trends Cardiovasc. Med.* **11**, 7–13.

Esko, J. D., and Lindahl, U. (2001). Molecular diversity of heparan sulfate. *J. Clin. Invest.* **108**, 169–173.

Festuccia, C., Angelucci, A., Gravina, G. L., Villanova, I., Teti, A., Albini, A., Bologna, M., and Abini, A. (2000). Osteoblast-derived TGF-beta1 modulates matrix degrading protease expression and activity in prostate cancer cells. *Int. J. Cancer* **85**, 407–415.

Fingleton, B. M., Heppner Goss, K. J., Crawford, H. C., and Matrisian, L. M. (1999). Matrilysin in early stage intestinal tumorigenesis. *Apmis* **107**, 102–110.

Forsberg, E., and Kjellen, L. (2001). Heparan sulfate: lessons from knockout mice. *J. Clin. Invest.* **108**, 175–180.

Gaudin, P., Berthier, S., Barro, C., Zaoui, P., and Morel, F. (1997). Proteolytic potential of human neutrophil membranes. *Eur. J. Cell Biol.* **72**, 345–351.

Gearing, A. J., Beckett, P., Christodoulou, M., Churchill, M., Clements, J., Davidson, A. H., Drummond, A. H., Galloway, W. A., Gilbert, R., and Gordon, J. L. (1994). Processing of tumour necrosis factor-alpha precursor by metalloproteinases. *Nature* **370**, 555–557.

Ginestra, A., Monea, S., Seghezzi, G., Dolo, V., Nagase, H., Mignatti, P., and Vittorelli, M. L. (1997). Urokinase plasminogen activator and gelatinases are associated with membrane vesicles shed by human HT1080 fibrosarcoma cells. *J. Biol. Chem.* **272**, 17216–17222.

Goodison, S., Urquidi, V., and Tarin, D. (1999). CD44 cell adhesion molecules. *Mol. Pathol.* **52**, 189–196.

Guo, H., Li, R., Zucker, S., and Toole, B. P. (2000). EMMPRIN (CD 147), an inducer of matrix metalloproteinase synthesis, also binds interstitial collagenase to the tumor cell surface. *Cancer Res.* **60**, 888–891.

Guo, H., Zucker, S., Gordon, M. K., Toole, B. P., and Biswas, C. (1997). Stimulation of matrix metalloproteinase production by recombinant extracellular matrix metalloproteinase inducer from transfected Chinese hamster ovary cells. *J. Biol. Chem.* **272**, 24–27.

Hahn-Dantona, E., Ruiz, J. F., Bornstein, P., and Strickland, D. K. (2001). The low density lipoprotein receptor-related protein modulates levels of matrix metalloproteinase 9 (MMP-9) by mediating its cellular catabolism. *J. Biol. Chem.* **276,** 15498–15503.

Haro, H., Crawford, H. C., Fingleton, B., Shinomiya, K., Spengler, D. M., and Matrisian, L. M. (2000). Matrix metalloproteinase-7-dependent release of tumor necrosis factor-alpha in a model of herniated disc resorption. *J. Clin. Invest.* **105,** 143–150.

Herz, J., and Strickland, D. K. (2001). LRP: a multifunctional scavenger and signaling receptor. *J. Clin. Invest.* **108,** 779–784.

Hewitt, R., and Dano, K. (1996). Stromal cell expression of components of matrix-degrading protease systems in human cancer. *Enzyme Protein* **49,** 163–173.

Holmbeck, K., Bianco, P., Caterina, J., Yamada, S., Kromer, M., Kuznetsov, S. A., Mankani, M., Robey, P. G., Poole, A. R., Pidoux, I., Ward, J. M., and Birkedal-Hansen, H. (1999). MT1-MMP-deficient mice develop dwarfism, osteopenia, arthritis, and connective tissue disease due to inadequate collagen turnover [In Process Citation] *Cell* **99,** 81–92.

Iozzo, R. V. (2001). Heparan sulfate proteoglycans: intricate molecules with intriguing functions. *J. Clin. Invest.* **108,** 165–167.

Iozzo, R. V., and San Antonio, J. D. (2001). Heparan sulfate proteoglycans: heavy hitters in the angiogenesis arena. *J. Clin. Invest.* **108,** 349–355.

Jimenez, M. J., Balbin, M., Alvarez, J., Komori, T., Bianco, P., Holmbeck, K., Birkedal-Hansen, H., Lopez, J. M., and Lopez-Otin, C. (2001). A regulatory cascade involving retinoic acid, Cbfa1, and matrix metalloproteinases is coupled to the development of a process of perichondrial invasion and osteogenic differentiation during bone formation. *J. Cell Biol.* **155,** 1333–1344.

Karelina, T. V., Hruza, G. J., Goldberg, G. I., and Eisen, A. Z. (1993). Localization of 92-kDa type IV collagenase in human skin tumors: comparison with normal human fetal and adult skin. *J. Invest. Dermatol.* **100,** 159–165.

Kjeldsen, L., Johnsen, A. H., Sengelov, H., and Borregaard, N. (1993). Isolation and primary structure of, NGAL, a novel protein associated with human neutrophil gelatinase. *J. Biol. Chem.* **268,** 10425–10432.

Knauper, V., Cowell, S., Smith, B., Lopez-Otin, C., O'Shea, M., Morris, H., Zardi, L., and Murphy, G. (1997). The role of the C-terminal domain of human collagenase-3 (MMP-13) in the activation of procollagenase-3, substrate specificity, and tissue inhibitor of metalloproteinase interaction. *J. Biol. Chem.* **272,** 7608–7616.

Kolb, C., Mauch, S., Peter, H. H., Krawinkel, U., and Sedlacek, R. (1997). The matrix metalloproteinase, RASI-1 is expressed in synovial blood vessels of a rheumatoid arthritis patient. *Immunol. Lett.* **57,** 83–88.

Krane, S. M. (2001). Petulant cellular acts: destroying the, ECM rather than creating it. *J. Clin. Invest.* **107,** 31–32.

Manes, S., Llorente, M., Lacalle, R. A., Gomez-Mouton, C., Kremer, L., Mira, E., and Martinez, A. C. (1999). The matrix metalloproteinase-9 regulates the insulin-like growth factor-triggered autocrine response in, DU-145 carcinoma cells. *J. Biol. Chem.* **274,** 6935–6945.

Massova, I., Kotra, L. P., Fridman, R., and Mobashery, S. (1998). Matrix metalloproteinases: structures, evolution, and diversification. *FASEB J.* **12,** 1075–1095.

Mauch, S., Kolb, C., Kolb, B., Sadowski, T., and Sedlacek, R. (2002). Matrix metalloproteinase-19 is expressed in myeloid cells in an adhesion-dependent manner and associates with the cell surface. *J. Immunol.* **168,** 1244–1251.

Mazzieri, R., Masiero, L., Zanetta, L., Monea, S., Onisto, M., Garbisa, S., and Mignatti, P. (1997). Control of type IV collagenase activity by components of the urokinase-plasmin system: a regulatory mechanism with cell-bound reactants. *EMBO J.* **16,** 2319–2332.

Mira, E., Manes, S., Lacalle, R. A., Marquez, G., and Martinez, A. C. (1999). Insulin-like growth factor I-triggered cell migration and invasion are mediated by matrix metalloproteinase-9. *Endocrinology* **140,** 1657–1664.

Moll, U. M., Lane, B., Zucker, S., Suzuki, K., and Nagase, H. (1990). Localization of collagenase at the basal plasma membrane of a human pancreatic carcinoma cell line. *Cancer Res.* **50,** 6995–7002.

Murphy, G., and Crabbe, T. (1995). Gelatinases A and B. *Methods Enzymol.* **248,** 470–484.

Nagase, H., and Woessner, J. F., Jr. (1999). Matrix metalloproteinases. *J. Biol. Chem.* **274,** 21491–21494.

Neuhold, L. A., Killar, L., Zhao, W., Sung, M. L., Warner, L., Kulik, J., Turner, J., Wu, W., Billinghurst, C., Meijers, T., Poole, A. R., Babij, P., and DeGennaro, L. J. (2001). Postnatal expression in hyaline cartilage of constitutively active human collagenase-3 (MMP-13) induces osteoarthritis in mice. *J. Clin. Invest.* **107,** 35–44.

Noe, V., Fingleton, B., Jacobs, K., Crawford, H. C., Vermeulen, S., Steelant, W., Bruyneel, E., Matrisian, L. M., and Mareel, M. (2001). Release of an invasion promoter E-cadherin fragment by matrilysin and stromelysin-1. *J. Cell Sci.* **114,** 111–118.

Nykjaer, A., Conese, M., Christensen, E. I., Olson, D., Cremona, O., Gliemann, J., and Blasi, F. (1997). Recycling of the urokinase receptor upon internalization of the uPA:serpin complexes. *EMBO J.* **16,** 2610–2620.

Oh, J., Takahashi, R., Kondo, S., Mizoguchi, A., Adachi, E., Sasahara, R. M., Nishimura, S., Imamura, Y., Kitayama, H., Alexander, D. B., Ide, C., Horan, T. P., Arakawa, T., Yoshida, H., Nishikawa, S., Itoh, Y., Seiki, M., Itohara, S., Takahashi, C., and Noda, M. (2001). The membrane-anchored, MMP inhibitor RECK is a key regulator of extracellular matrix integrity and angiogenesis. *Cell* **107,** 789–800.

Olson, M. W., Bernardo, M. M., Pietila, M., Gervasi, D. C., Toth, M., Kotra, L. P., Massova, I., Mobashery, S., and Fridman, R. (2000). Characterization of the monomeric and dimeric forms of latent and active matrix metalloproteinase-9. Differential rates for activation by stromelysin 1. *J. Biol. Chem.* **275,** 2661–2668.

Olson, M. W., Gervasi, D. C., Mobashery, S., and Fridman, R. (1997). Kinetic analysis of the binding of human matrix metalloproteinase-2 and -9 to tissue inhibitor of metalloproteinase (TIMP)-1 and TIMP-2. *J. Biol. Chem.* **272,** 29975–29983.

Olson, M. W., Toth, M., Gervasi, D. C., Sado, Y., Ninomiya, Y., and Fridman, R. (1998). High affinity binding of latent matrix metalloproteinase-9 to the alpha2(IV) chain of collagen IV. *J. Biol. Chem.* **273,** 10672–10681.

Omura, T. H., Noguchi, A., Johanns, C. A., Jeffrey, J. J., and Partridge, N. C. (1994). Identification of a specific receptor for interstitial collagenase on osteoblastic cells. *J. Biol. Chem.* **269,** 24994–24998.

Opdenakker, G., Van den Steen, P. E., Dubois, B., Nelissen, I., Van Coillie, E., Masure, S., Proost, P., and Van Damme, J. (2001a). Gelatinase B functions as regulator and effector in leukocyte biology. *J. Leukoc. Biol.* **69,** 851–859.

Opdenakker, G., Van den Steen, P. E., and Van Damme, J. (2001b). Gelatinase B: a tuner and amplifier of immune functions. *Trends Immunol.* **22,** 571–579.

Overall, C. M., Wrana, J. L., and Sodek, J. (1989). Independent regulation of collagenase, 72-kDa progelatinase, and metalloendoproteinase inhibitor expression in human fibroblasts by transforming growth factor-beta. *J. Biol. Chem.* **264,** 1860–1869.

Park, H. I., Ni, J., Gerkema, F. E., Liu, D., Belozerov, V. E., and Sang, Q. X. (2000). Identification and characterization of human endometase (Matrix metalloproteinase-26) from endometrial tumor. *J. Biol. Chem.* **275,** 20540–20544.

Parks, W. C., and Shapiro, S. D. (2001). Matrix metalloproteinases in lung biology. *Respir. Res.* **2,** 10–19.

Partridge, C. A., Phillips, P. G., Niedbala, M. J., and Jeffrey, J. J. (1997). Localization and activation of type IV collagenase/gelatinase at endothelial focal contacts. *Am. J. Physiol.* **272,** L813–L822.

Pendas, A. M., Knauper, V., Puente, X. S., Llano, E., Mattei, M. G., Apte, S., Murphy, G., and

Lopez-Otin, C. (1997). Identification and characterization of a novel human matrix metalloproteinase with unique structural characteristics, chromosomal location, and tissue distribution. *J. Biol. Chem.* **272,** 4281–4286.

Pendas, A. M., Uria, J. A., Jimenez, M. G., Balbin, M., Freije, J. P., and Lopez-Otin, C. (2000). An overview of collagenase-3 expression in malignant tumors and analysis of its potential value as a target in antitumor therapies. *Clin. Chim. Acta* **291,** 137–155.

Powell, W. C., Fingleton, B., Wilson, C. L., Boothby, M., and Matrisian, L. M. (1999). The metalloproteinase matrilysin proteolytically generates active soluble Fas ligand and potentiates epithelial cell apoptosis. *Curr. Biol.* **9,** 1441–1447.

Sanderson, R. D. (2001). Heparan sulfate proteoglycans in invasion and metastasis. *Semin. Cell. Dev. Biol.* **12,** 89–98.

Sedlacek, R., Mauch, S., Kolb, B., Schatzlein, C., Eibel, H., Peter, H. H., Schmitt, J., and Krawinkel, U. (1998). Matrix metalloproteinase MMP-19 (RASI-1) is expressed on the surface of activated peripheral blood mononuclear cells and is detected as an autoantigen in rheumatoid arthritis. *Immunobiology* **198,** 408–423.

Sheikh, H., Yarwood, H., Ashworth, A., and Isacke, C. M. (2000). Endo180, an endocytic recycling glycoprotein related to the macrophage mannose receptor is expressed on fibroblasts, endothelial cells and macrophages and functions as a lectin receptor. *J. Cell Sci.* **113**(Pt 6), 1021–1032.

Steffensen, B., Bigg, H. F., and Overall, C. M. (1998). The involvement of the fibronectin type II-like modules of human gelatinase A in cell surface localization and activation. *J. Biol. Chem.* **273,** 20622–20628.

Sternlicht, M. D., and Werb, Z. (2001). How matrix metalloproteinases regulate cell behavior. *Annu. Rev. Cell. Dev. Biol.* **17,** 463–516.

Stracke, J. O., Hutton, M., Stewart, M., Pendas, A. M., Smith, B., Lopez-Otin, C., Murphy, G., and Knauper, V. (2000). Biochemical characterization of the catalytic domain of human matrix metalloproteinase 19. Evidence for a role as a potent basement membrane degrading enzyme. *J. Biol. Chem.* **275,** 14809–14816.

Stricker, T. P., Dumin, J. A., Dickeson, S. K., Chung, L., Nagase, H., Parks, W. C., and Santoro, S. A. (2001). Structural analysis of the alpha(2) integrin I domain/procollagenase-1 (matrix metalloproteinase-1) interaction. *J. Biol. Chem.* **276,** 29375–29381.

Strongin, A. Y., Collier, I., Bannikov, G., Marmer, B. L., Grant, G. A., and Goldberg, G. I. (1995). Mechanism of cell surface activation of 72-kDa type IV collagenase. Isolation of the activated form of the membrane metalloprotease. *J. Biol. Chem.* **270,** 5331–5338.

Strongin, A. Y., Collier, I. E., Krasnov, P. A., Genrich, L. T., Marmer, B. L., and Goldberg, G. I. (1993). Human 92 kDa type IV collagenase: functional analysis of fibronectin and carboxyl-end domains. *Kidney Int.* **43,** 158–162.

Sudbeck, B. D., Pilcher, B. K., Welgus, H. G., and Parks, W. C. (1997). Induction and repression of collagenase-1 by keratinocytes is controlled by distinct components of different extracellular matrix compartments. *J. Biol. Chem.* **272,** 22103–22110.

Sun, J., and Hemler, M. E. (2001). Regulation of MMP-1 and MMP-2 production through CD147/extracellular matrix metalloproteinase inducer interactions. *Cancer Res.* **61,** 2276–2281.

Takahashi, C., Akiyama, N., Matsuzaki, T., Takai, S., Kitayama, H., and Noda, M. (1996). Characterization of a human MSX-2 cDNA and its fragment isolated as a transformation suppressor gene against v-Ki-ras oncogene. *Oncogene* **12,** 2137–2146.

Takahashi, C., Sheng, Z., Horan, T. P., Kitayama, H., Maki, M., Hitomi, K., Kitaura, Y., Takai, S., Sasahara, R. M., Horimoto, A., Ikawa, Y., Ratzkin, B. J., Arakawa, T., and Noda, M. (1998). Regulation of matrix metalloproteinase-9 and inhibition of tumor invasion by the membrane-anchored glycoprotein RECK. *Proc. Natl. Acad. Sci. USA* **95,** 13221–13226.

Taraboletti, G., D'Ascenzo, S., Borsotti, P., Giavazzi, R., Pavan, A., and Dolo, V. (2002). Shedding of the matrix metalloproteinases, MMP-2, MMP-9, and MT1-MMP as membrane vesicle-associated components by endothelial cells. *Am. J. Pathol.* **160,** 673–680.

Toth, M., and Fridman, R. (2001). Assesment of gelatinases by gelatin zymography. *In* "Metastasis Research Protocols" (S. A. Brooks and U. Schumacher, ed.), pp. 163–174. Humana Press, Totowa, New Jersey.

Toth, M., Gervasi, D. C., and Fridman, R. (1997). Phorbol ester-induced cell surface association of matrix metalloproteinase-9 in human MCF10A breast epithelial cells. *Cancer Res.* **57,** 3159–3167.

Toth, M., Sado, Y., Ninomiya, Y., and Fridman, R. (1999). Biosynthesis of alpha2(IV) and alpha1(IV) chains of collagen IV and interactions with matrix metalloproteinase-9. *J. Cell Physiol.* **180,** 131–139.

Tschesche, H., Zolzer, V., Triebel, S., and Bartsch, S. (2001). The human neutrophil lipocalin supports the allosteric activation of matrix metalloproteinases. *Eur. J. Biochem.* **268,** 1918–1928.

Uria, J. A., and Lopez-Otin, C. (2000). Matrilysin-2, a new matrix metalloproteinase expressed in human tumors and showing the minimal domain organization required for secretion, latency, and activity. *Cancer Res.* **60,** 4745–4751.

Visscher, D. W., Hoyhtya, M., Ottosen, S. K., Liang, C. M., Sarkar, F. H., Crissman, J. D., and Fridman, R. (1994). Enhanced expression of tissue inhibitor of metalloproteinase-2 (TIMP-2) in the stroma of breast carcinomas correlates with tumor recurrence. *Int. J. Cancer* **59,** 339–344.

Vu, T. H., Shipley, J. M., Bergers, G., Berger, J. E., Helms, J. A., Hanahan, D., Shapiro, S. D., Senior, R. M., and Werb, Z. (1998). MMP-9/gelatinase B is a key regulator of growth plate angiogenesis and apoptosis of hypertrophic chondrocytes. *Cell* **93,** 411–422.

Walling, H. W., Chan, P. T., Omura, T. H., Barmina, O. Y., Fiacco, G. J., Jeffrey, J. J., and Partridge, N. C. (1998). Regulation of the collagenase-3 receptor and its role in intracellular ligand processing in rat osteoblastic cells. *J. Cell Physiol.* **177,** 563–574.

Whitelock, J. M., O'Grady, R. L., and Gibbins, J. R. (1991). Interstitial collagenase (matrix metalloproteinase 1) associated with the plasma membrane of both neoplastic and nonneoplastic cells. *Invasion Metastasis* **11,** 139–148.

Wilhelm, S. M., Collier, I. E., Marmer, B. L., Eisen, A. Z., Grant, G. A., and Goldberg, G. I. (1989). SV40-transformed human lung fibroblasts secrete a 92-kDa type IV collagenase which is identical to that secreted by normal human macrophages. *J. Biol. Chem.* **264,** 17213–17221.

Wilson, C. L., and Matrisian, L. M. (1996). Matrilysin: an epithelial matrix metalloproteinase with potentially novel functions. *Int. J. Biochem. Cell Biol.* **28,** 123–136.

Wilson, C. L., Heppner, K. J., Labosky, P. A., Hogan, B. L., and Matrisian, L. M. (1997). Intestinal tumorigenesis is suppressed in mice lacking the metalloproteinase matrilysin. *Proc. Natl. Acad. Sci. USA* **94,** 1402–1407.

Wilson, C. L., Ouellette, A. J., Satchell, D. P., Ayabe, T., Lopez-Boado, Y. S., Stratman, J. L., Hultgren, S. J., Matrisian, L. M., and Parks, W. C. (1999). Regulation of intestinal alpha-defensin activation by the metalloproteinase matrilysin in innate host defense. *Science* **286,** 113–117.

Woessner, J. F., Jr. (1995). Matrilysin. *Methods Enzymol.* **248,** 485–495.

Woessner, J. F., Jr. (1996). Regulation of matrilysin in the rat uterus. *Biochem. Cell Biol.* **74,** 777–784.

Yan, L., Borregaard, N., Kjeldsen, L., and Moses, M. A. (2001). The high molecular weight urinary matrix metalloproteinase (MMP) activity is a complex of gelatinase B/MMP-9 and neutrophil gelatinase-associated lipocalin (NGAL). Modulation of MMP-9 activity by NGAL. *J. Biol. Chem.* **276,** 37258–37265.

Yoshikawa, K., Takahashi, S., Imamura, Y., Sado, Y., and Hayashi, T. (2001). Secretion of non-helical collagenous polypeptides of alpha1(IV) and alpha2(IV) chains upon depletion of ascorbate by cultured human cells. *J. Biochem. (Tokyo)* **129,** 929–936.

Yu, Q., and Stamenkovic, I. (1999). Localization of matrix metalloproteinase 9 to the cell surface provides a mechanism for CD44-mediated tumor invasion. *Genes Dev.* **13,** 35–48.

Yu, Q., and Stamenkovic, I. (2000). Cell surface-localized matrix metalloproteinase-9 proteolytically activates, TGF-beta and promotes tumor invasion and angiogenesis. *Genes Dev.* **14,** 163–176.

Yu, W. H., and Woessner, J. F., Jr. (2000). Heparan sulfate proteoglycans as extracellular docking molecules for matrilysin (matrix metalloproteinase 7). *J. Biol. Chem.* **275,** 4183–4191.

Yu, W. H., Woessner, J. F., Jr., McNeish, J. D., and Stamenkovic, I. (2002). CD44 anchors the assembly of matrilysin/MMP-7 with heparin-binding epidermal growth factor precursor and ErbB4 and regulates female reproductive organ remodeling. *Genes Dev.* **16,** 307–323.

Yu, W. H., Yu, S., Meng, Q., Brew, K., and Woessner, J. F., Jr. (2000). TIMP-3 binds to sulfated glycosaminoglycans of the extracellular matrix. *J. Biol. Chem.* **275,** 31226–31232.

Zhou, Z., Apte, S. S., Soininen, R., Cao, R., Baaklini, G. Y., Rauser, R. W., Wang, J., Cao, Y., and Tryggvason, K. (2000). Impaired endochondral ossification and angiogenesis in mice deficient in membrane-type matrix metalloproteinase I. *Proc. Natl. Acad. Sci. USA* **97,** 4052–4057.

Zucker, S., Moll, U. M., Lysik, R. M., DiMassimo, E. I., Stetler-Stevenson, W. G., Liotta, L. A., and Schwedes, J. W. (1990). Extraction of type-IV collagenase/gelatinase from plasma membranes of human cancer cells. *Int. J. Cancer* **45,** 1137–1142.

3

Biochemical Properties and Functions of Membrane-Anchored Metalloprotease-Disintegrin Proteins (ADAMs)

J. David Becherer[1] *and Carl P. Blobel*[2]
[1]Department of Biochemical and Analytical Pharmacology
GlaxoSmithKline Research Inc.
Research Triangle Park, North Carolina 27709

[2]Cellular Biochemistry and Biophysics Program
Sloan-Kettering Institute
Memorial Sloan-Kettering Cancer Center
New York, New York 10021

I. Introduction

Proteolysis has emerged as a key posttranslational regulator of the function of molecules on the cell surface and the extracellular matrix. In principle, proteolysis can activate or inactivate a substrate protein or can change its functional properties. Metalloprotease-disintegrin proteins [also referred to as ADAMs (a disintegrin and metalloprotease domain)] are a family of membrane anchored glycoproteins

Current Topics in Developmental Biology, Vol. 54

S   P   MP   D   CRR   EGF   CT
                                                                TM

Figure 1   A typical domain organization of an ADAM (a disintegrin and metalloprotease) consists of
an N-terminal signal sequence (SS), followed by a prodomain (P), a metalloprotease domain (MP), a
disintegrin domain (D), a cysteine-rich region (CRR), an EGF repeat (EGF), a transmembrane domain
(TM), and a cytoplasmic domain (CT).

that have been implicated in the processing of proteins on the cell surface and
the extracellular matrix. A typical domain organization of an ADAM is shown
in Fig. 1. Substrates of ADAMs are thought to include proteins with key roles in
development and in disease, such as cytokines, growth factors, receptors, adhesion
proteins, Notch, the amyloid precursor protein, and others. (Please see Chapter 4
by Arribas and Merlos-Suárez for details on protein ectodomain shedding. For
other reviews on ADAMs, please see Black and White, 1998; Blobel, 1997, 1999,
2000; Evans, 1999; Huovila *et al.,* 1996; McLane *et al.,* 1998; Mullberg *et al.,*
2000; Myles and Primakoff, 1997; Primakoff and Myles, 2000; Schlöndorff and
Blobel, 1999; Stone *et al.,* 1999; Turner and Hooper, 1999; Wolfsberg and White,
1996).

Of the 34 ADAMs that have been identified to date, only about half contain a
zinc binding catalytic site consensus sequence (HE*XX*H), a hallmark of the metzincin superfamily of zinc-peptidases (Blundell, 1994; Bode *et al.,* 1993; Jongeneel
et al., 1989; Stocker *et al.,* 1995), in their otherwise conserved metalloprotease
domain. Thus, only half of the known ADAMs are predicted to be catalytically
active. Individual ADAMs are referred to by numbers (ADAM1, ADAM2, etc.)
that are assigned by Dr. Judy White in the order in which full-length sequences
become available. Frequently updated tables of currently known ADAMs can be
found under two URLs (http://www.uta.fi/%7Eloiika/ADAMs/MMADAMs.htm
or http://www.people.Virginia.EDU/%7Ejag6n/Table_of_the_ADAMs.html). The
rate of discovery of new ADAMs has slowed considerably in the past few years,
suggesting that most of the more abundant or widely expressed ADAMs have now
been identified. Unfortunately, the annotation of the human genome sequence is not
yet complete, and therefore no conclusions about the total number of ADAMs can
be drawn. The list of 34 ADAMs that was available while this review was written
includes one that has only been described in *Caenorhabditis elegans* (ADAM14),
two that were only found in *Xenopus laevis* (ADAM13, 16), and two that have a
literature alias (ADAMs 27 and 31). Therefore the number of ADAMs that have
been identified in a given species will be smaller than 34. In mice, for example, 24
ADAMs have been identified. Of these, 9 do not have a catalytic-site consensus sequence, whereas the remaining 15 carry such a site and are therefore predicted to be
catalytically active. Of the 15 catalytically active ADAMs, 6 are expressed mainly

in the testis, whereas the remaining 9 (ADAMs 8, 9, 10, 12, 15, 17, 19, 28, 33) are expressed in a variety of somatic tissues. In this review we will focus mainly on the biochemical properties and functions of ADAMs that are catalytically active and expressed in somatic tissues.

II. Structural Chemistry and Biochemistry

A. Domain Structure

All ADAMs are membrane anchored glycoproteins that are characterized by a conserved domain structure: an amino-terminal signal sequence followed by pro-, metalloprotease, and disintegrin domains, a cysteine-rich region, usually containing an EGF repeat, and finally a transmembrane domain and cytoplasmic tail (see Fig. 1) (Black, 1998; Blobel and White, 1992; Blobel, 1997; Wolfsberg, 1996). ADAMs that contain a catalytic site consensus sequence usually also have an odd-numbered cysteine residue in their prodomain that is not found in ADAMs without a catalytic site. This odd-numbered cysteine inhibits the catalytic activity via a cysteine switch mechanism, thus presumably ensuring that the metalloprotease domain remains inactive until the prodomain is removed (Van Wart and Birkedal-Hansen, 1990). The disintegrin domain and cysteine-rich region of ADAMs are thought to have a role in cell–cell interaction and potentially also in substrate recognition. The amino acid sequence in these two protein domains is rich in cysteine residues, all of which are presumably engaged in intramolecular disulfide bonds. As a result, ADAMs migrate faster during sodium dodecyl sulfate–polyacrylamide gel electrophoresis (SDS–PAGE) under nonreducing conditions (which leaves the disulfide bonds intact) compared to reducing conditions (which breaks disulfide bonds and allows the protein to unfold more completely). The cytoplasmic domains of catalytically active ADAMs usually contain several signaling motifs such as potential phosphorylation sites and proline-rich SH3 ligand domains. In some cases, splice variants of membrane-anchored ADAMs have been found to give rise to soluble secreted proteins whose function would not be restricted to the cell surface (Gilpin et al., 1998; Howard et al., 2000; Jury et al., 1999). The same would be true for ADAM modules that are released from their membrane anchor through proteolysis, such as the metalloprotease domain of ADAM1 (Lum and Blobel, 1997). However, in most cases the metalloprotease domain of catalytically active ADAMs remains membrane anchored.

In addition to the membrane-anchored ADAMs there is a growing family of related molecules that are referred to as ADAMTS proteins (A disintegrin-like and metalloprotease with thrombospondin motifs; Hurskainen et al., 1999; Rodriguez-Manzaneque et al., 2000). These proteins, which will not be covered in this review, contain a conserved amino terminal metalloprotease domain and disintegrin-like domain that has a quite different pattern and spacing of cysteine residues compared to the disintegrin domain of membrane anchored ADAMs. In contrast to ADAMs, ADAMTS proteins lack a transmembrane domain, and instead harbor throm-bospondin motifs as well as other extracellular protein modules (see Tang and Hong, 1999, for recent reviews).

B. Processing, Activation, Trafficking, and Degradation

All ADAMs contain an N-terminal signal sequence, which directs translocation across the membrane of the endoplasmic reticulum and is immediately removed by signal peptidase. During translocation across the ER membrane, N-linked carbo-hydrates are attached and the proper disulfide bonds form. Following translocation, the core glycosylated membrane anchored ADAM has lost its signal sequence, but contains all other domains mentioned above, including the prodomain. A properly folded ADAM molecule will then emerge from the endoplasmic reticulum and migrate through the secretory pathway. Most N-linked carbohydrate residues are modified during transit through the secretory pathway. These modifications result in resistance to endoglycosidase H after the ADAM has moved through the medial Golgi apparatus. Folding of ADAMs in the ER and their subsequent transport to the surface can easily be disrupted by deletions within any of its protein domains, or by certain point mutations, such as mutating a cysteine residue or any one of the three Zn-coordinating histidines in the catalytic site. An unfolded ADAM is expected to be retained by intracellular chaperones and subsequently degraded. Therefore acquisition of endo H resistance provides a convenient assay for exit from the ER, and is thus indicative of proper protein folding (Cao *et al.*, 2002; Lum and Blobel, 1997; Lum *et al.*, 1998; Roghani *et al.*, 1999; Schlöndorff *et al.*, 2000).

Most ADAMs contain a proprotein convertase cleavage site between their pro- and metalloprotease domains. This is usually cleaved after acquisition of endoH resistance, suggesting that the prodomain is removed shortly after transit through the medial Golgi (Cao *et al.*, 2002; Kang *et al.*, 2002; Loechel *et al.*, 1999; Lum *et al.*, 1998; Roghani *et al.*, 1999; Schlöndorff *et al.*, 2000). This is consistent with the well-established function of proprotein convertases such as furin in the trans-Golgi network. However, at least one ADAM, fertilinα/ADAM1, contains a furin cleavage site between the metalloprotease and disintegrin domain (Lum, 1997). Cleavage at this site releases a soluble metalloprotease domain. Further-more, at least two ADAMs (ADAMs 8 and 28) do not have an evident proprotein convertase cleavage site between the pro- and metalloprotease domain. In the case of ADAM28, a point mutation that inactivates the catalytic site without affect-ing transport to the cell surface also prevents prodomain removal, suggesting that prodomain removal is autocatalytic (Howard, 2000). Following cleavage between the pro- and metalloprotease domains the prodomain or parts of the prodomain may still remain noncovalently attached to the ADAM (Lum, 1998; Milla *et al.*, 1999). Indeed, in the case of the TNFα convertase/ADAM17, the disintegrin domain and cysteine-rich region appear to be necessary for releasing the prodomain after it has been cleaved (Milla *et al.*, 1999).

During biosynthesis and transit through the secretory pathway, the prodomain of ADAMs is thought to keep the protease domain inactive via a cysteine-switch mechanism (Loechel *et al.*, 1999; Roghani *et al.*, 1999). When coexpressed in trans as separate polypeptides the prodomain inhibited the α-secretase activity of wild-type ADAM10 (Anders *et al.*, 2001). Following prodomain removal, it is likely that additional mechanisms regulate the catalytic activity of ADAMs. The prodomain of ADAM 10 has a second important function. The prodomain acted as a intramolecular chaperone and functionally rescued the α-secretase activity of a former inactive ADAM10 mutant lacking the prodomain. This rescue appears to be achieved by assisting the correct folding of the catalytically active site (Anders *et al.*, 2001). A similar chaperone function for the prodomain of MT1-MMP has been reported (see Chapter 1 by Zucker, Pei, Cao, and Lopez-Otin).

Interestingly, TIMPs 1 and 2, which together are known to inhibit all currently known MMPs, do not inhibit several ADAMs, with the exception of ADAM10, which is inhibited by TIMP1, but not TIMP2 (Amour *et al.*, 1998, 2000, 2002; Schwettmann and Tschesche, 2001). On the other hand, TIMP3 inhibits ADAM17 as well as ADAM10, but not, for example, ADAM9. It will be interesting to determine whether other TIMPs or other yet to be discovered inhibitors have a role in regulating the function of ADAMs once the prodomain is removed.

Following prodomain cleavage, mature ADAMs may be targeted to different subcellular localizations. This can be assessed using simple criteria such as cell surface biotinylation, accessibility to trypsin on the cell surface, and immunofluorescence microscopy (Cao *et al.*, 2002; Howard *et al.*, 2000; Lum *et al.*, 1998; Schlöndorff *et al.*, 2000; Weskamp *et al.*, 1996). Mature ADAM28, for example, was found predominantly on the cell surface of transiently transfected Cos-7 cells. ADAM9 also seems to reside mainly on the cell surface (Howard *et al.*, 2000; Weskamp *et al.*, 1996). In contrast, mature ADAM15 and ADAM17/TACE are mainly localized intracellularly, in a perinuclear compartment that resembles the trans-Golgi network/endosome (Lum *et al.*, 1998; Schlöndorff *et al.*, 2000). These results suggests that some ADAMs, such as ADAMs 28 and 9, may function predominantly on the cell surface, whereas others, such as ADAMs 15 and 17, are mainly active in the lumen of an intracellular compartment. Further studies will be necessary to more accurately define the subcellular localization of different ADAMs and to confirm that the localization of endogenous ADAMs can be compared to that of transiently transfected ADAMs.

Little is currently known about the degradation of ADAMs. Pulse chase analysis suggests that the half-life of some ADAMs is relatively long (on the order of 12–36 h; Lum *et al.*, 1998; Schlöndorff *et al.*, 2000). Treatment of cells with phorbol esters is known to activate TACE, while high concentrations lead to a rapid downregulation of mature TACE, presumably by increasing degradation in the lysosome (Doedens and Black, 2000). However, TACE activity can also be stimulated with lower doses of PMA that do not result in its downregulation and degradation (Zheng *et al.*, 2002), suggesting that activation is not necessarily linked to downregulation.

C. Protein Preparation and Catalytic Features

The first ADAM that was shown to have catalytic activity *in vitro* was ADAM10, which was initially purified as an enzyme that cleaves the myelin basic protein (Chantry *et al.*, 1989; Howard and Glynn, 1995; Howard *et al.*, 1996). Subsequently, the long-sought-after TNFα converting enzyme (TACE) was isolated by two groups and found to be an ADAM (ADAM17) (Black *et al.*, 1997; Moss *et al.*, 1997). Thus both TACE and ADAM10 were first purified from primary tissue or cell lysates based on their specific activity toward a substrate of interest. Once the genes for these two ADAMs were cloned, recombinantly expressed ectodomains were used for further studies of substrate specificity and to determine their inhibitor profiles.

Additional information on the catalytic activity of other ADAMs has since emerged from studies with recombinantly expressed soluble proteins. In addition to TACE/ADAM17 and ADAM10 (Rosendahl *et al.*, 1997), two other ADAMs have been shown to have catalytic activity in assays in which peptide or protein substrates are turned over. The recombinantly expressed metalloprotease domain of ADAM9 can cleave the insulin B chain in two positions, whereas a catalytically inactive control does not (Roghani *et al.*, 1999). Active ADAM9 also can process a number of other peptide substrates. In contrast, recombinantly expressed ADAM28 can cleave myelin basic protein, but not several peptides that are cleaved by other ADAMs, or extracellular matrix proteins that can be cleaved by certain MMPs (Howard *et al.*, 2001). One important conclusion from these studies is that ADAMs apparently have distinct cleavage specificities *in vitro*, even though no clear consensus cleavage site has emerged yet for any ADAM. This suggests that the substrate specificity of ADAMs *in vivo* may at least in part be dictated by their preference for different cleavage sites.

The first evidence for catalytic activity of ADAMs 12 and 19 was derived from assays with the general protease inhibitor α2 macroglobulin (Loechel *et al.*, 1998; Wei *et al.*, 2001). This inhibitor becomes covalently attached to the active site of a protease after it is cleaved. The complex of α2M and an ADAM can then be detected on an SDS gel, providing a convenient assay for catalytic activity. An advantage of this assay is that it does not require purification of the protease so long as specific antibodies against the protease are available for Western blot analysis. However, since α2M is a "suicide inhibitor" that irreversibly binds to the protease after it is cleaved, this assay is not suitable to determine inhibitor profiles of competitive inhibitors, or the cleavage site and specificity constant of an ADAM for biologically relevant peptides. In order to gather information on the inhibition of ADAMs by TIMPs or hydroxamic acid derivatives or to study cleavage of peptides it is necessary to identify substrates that can be turned over by a purified ADAM. It will be interesting to learn more about the inhibitor profile and substrate specificity of other catalytically active ADAMs in order to gain a better understanding of the potential selectivity of these enzymes *in vivo*.

D. Substrate Specificity

The substrate specificity of TACE has been explored via multiple strategies. Analysis of cells from $TACE^{Zn-/Zn-}$ mice reveals that TACE is involved in production not only of soluble TNF but also of L-selectin, p75 TNFR, and TGFα (Peschon *et al.*, 1998). Immunoprecipitated TACE substrates are cut poorly by TACE, presumably because both enzyme and substrate must be membrane anchored. A somewhat more physiological approach takes advantage of TACE-deficient fibroblasts. Using these cells in which the endogenous TACE gene is defective, putative sheddase substrates are evaluated by cotransfection with catalytically active TACE. Using this method, the type II IL-1 receptor and the p55 TNF receptor also are shed by TACE (Reddy *et al.*, 2000).

A third approach utilizes synthetic substrates with purified TACE to determine cleavage site specificity and k_{cat}/K_m values. The substrates are usually 8–14 amino acids and represent the sequences spanning the cleavage sites of the ectodomain of shed proteins (Table 1). Only TNF and TGFα peptides are processed efficiently by TACE, with k_{cat}/K_m values $>1 \times 10^5$ $M^{-1}s^{-1}$. Other peptide substrates, such as βAPP, KL, Rank-L, TNFR p55 and TNFr p75, IL-6 R, and Notch, are correctly processed by TACE but at least 50- to 1000-fold less efficiently (Mohan *et al.*, 2002). These data raise the possibility that TACE does not function as a broad-spectrum sheddase and that other ADAMs or metalloproteases may also be responsible for the ectodomain shedding of the aforementioned proteins.

Finally, inhibitors can be used to further define the nature of the enzyme(s) involved in shedding various surface molecules in cellular assays. A report profiles ~20 inhibitors in their capacity to inhibit shedding of CD23, βAPP, ACE, and TNF. Consequently, several inhibitors can selectively discriminate among these shedding events, yielding tools to specifically identify the corresponding sheddase and to probe the role these sheddases play in various experimental systems. For example, using a compound that discriminates between TACE and the βAPP sheddase, Parkin *et al.* (2002) demonstrate that TACE is not the α-secretase involved in βAPP shedding from neuronal cells.

Similar to TACE, ADAM9 cleaves several peptide substrates but at sites distinct from that reported for the native protein (Roghani *et al.*, 1999). Intriguing is the cleavage of βAPP by ADAM9. Although ADAM9 does not target the major α-secretase site, it does cleave efficiently at a site two residues upstream, with a k_{cat}/K_m of 2×10^5 $M^{-1}s^{-1}$. Perhaps in certain cell types ADAM9 functions as the

Table I   Membrane Proximal Cleavage Sites of Proteins Known to be Shed by Metalloproteinases[a]

Sheddase Substrate	Predicted Cleavage Sequence	Peptide Sequence Cleaved by:	
		TACE	ADAM-9
TNFα	SPLAQAVRSSSR	SPLAQA*VRSSSR	SPLA*QAVRSSSR
			SPLAQAVRS*SSR
TNFp55	LPQLEN*VKGTED	LPQLEN*VKGTED	NP
TNFp75	SMAPGA*VHLPQP	SMAPGAVH*LPQP	SMAPGAVH*LPQP
		SMAPGA*VHLPQP	
		SMAPG*AVHLPQP	
TGFα (N terminus)	SPVAAA*VVSHFN	SPVAAA*VVSHFN	ND
TGFα (C terminus)	HADLLA*VVAASQ	HADLLA*VVAASQ	NP
βAPP	EVHHQK*LVFFAE	EVHHQK*LVFFAE	EVHH*QKLVFFAE
C-Kit ligand	LPPVAA*SSLRND	LPPVAA*SSLRND	LPPVAAS*SLRND
		LPPVAAS*SLRND	LPPVAA*SSLRND
		LPPVAASS*LRND	LPPVA*ASSLRND
Trance/Rank-L	IVGPQR*FSGAPA	IVGPQR*FSGAPA	ND
IL-6R	TSLPVQ*DSSSVP	TSLPVQ*DSSSVP	NP
Notch	PYKIEA*VKSEPV	PYKIEA*VKSEPV	ND
HER4	HGLSLPVENRLYTYDH	HGLSLPVE*NRLYTYDH	ND
		HGLSLPV*ENRLYTYDH	
L-Selectin	KLDK*SFSKIKEGDYN	KLDK*SFSKIKEGDYN	NP

[a] Peptides mimicking the predicted cleavage sites were incubated with TACE, or ADAM9, and products formed were characterized by HPLC and mass spectrometry. All peptides were based on the human sequence except for L-selectin, which is from the mouse sequence. Data are derived from the following publications: Lum et al. (1999): Mohan et al. (2002): Peschon et al. (1998): Roghani et al. (1999). ND, Not determined: NP, no proteolysis.

βAPP α-secretase (Giulian et al., 1998). Further studies are required to determine if there is any physiological significance to this activity of ADAM9.

E. Natural and Synthetic Inhibitors

Several ADAMs are inhibited by tissue inhibitors of matrix metalloproteinases (TIMPs). In fact, inhibitor profiling was initially used to demonstrate that the "TNF sheddase" belonged to a subfamily of metalloproteases distinct form the MMPs. Whereas small molecular weight MMP hydroxamates are potent inhibitors of TNF shedding, the endogenous MMP inhibitors TIMP-1, TIMP-2 and TIMP-4 do not significantly inhibit TNF shedding. TIMP-3, however, is a potent inhibitor of both TNF and L-selectin release from cells, with an IC_{50} of $\sim$100 nM and $\sim$400 nM, respectively, in cellular assays (Borland et al., 1999). In enzyme assays using murine TACE and a synthetic substrate, TIMP-3 inhibits the enzyme with an apparent K_i of 180 pM. In addition to TACE, TIMP-3 also inhibits ADAM12

(Loechel *et al.*, 2000) and both TIMP-1 and TIMP-3 inhibit ADAM10 (Amour *et al.*, 2000). The physiological relevance of these observations remains to be demonstrated; however, it is intriguing that TIMP-3 is the only TIMP that binds to heparan sulfate proteoglycans and can therefore be localized on the cell surface in proximity to cell surface ADAMs.

As mentioned previously, small-molecular-weight MMP inhibitors also inhibit ADAM17/TACE (Gearing *et al.*, 1994; McGeehan *et al.*, 1994; Mohler *et al.*, 1994). To date, the majority of ADAM inhibitor development is due to the role TNF plays in arthritis. Early TACE inhibitors such as TAPI and GW 9471 (Becherer *et al.*, 2000; Beckett and Whittaker, 1998) are potent against TACE and block TNF release from cells both *in vitro* and *in vivo*. A growing collection of ADAM17/TACE inhibitors exists in the literature today, driven by the desire to reproduce the clinical efficacy seen with such biologicals as Enbrel (a soluble TNF receptor) and Remicade (an anti-TNF antibody), but with a small-molecule TNF inhibitor in pill form. For the most part, these inhibitors possess the characteristic hydroxamate functionality, which is critical to their mechanism of inhibition. This evolution of the TACE inhibitor is characterized by increasing selectivity for TACE relative to the MMPs (>100-fold selectivity is now common), improved potency in cellular TNF assays (IC50 < 100 nM), and potent inhibition of TNF *in vivo* (Barlaam *et al.*, 1999; Holms *et al.*, 2001; Kottirsch *et al.*, 2002; Letavic *et al.*, 2002; Levin *et al.*, 2002; Rabinowitz *et al.*, 2001; Xue *et al.*, 2001). However, very limited pharmacokinetic data are reported on these inhibitors and the challenge remains to develop an orally bioavailable TACE inhibitor that shows efficacy in the clinic. To date, the only published study of a TACE inhibitor in the clinic is that for GI 5402, a broad-spectrum TACE and MMP inhibitor (Dekkers *et al.*, 1999). This compound effectively blocks TNF release in response to LPS *without* a concomitant increase in the levels of membrane-bound TNF.

III. Expression in Different Tissues and Regulation of Gene Expression

As outlined in the introduction, some ADAMs with a catalytic site are mainly expressed in the testis, whereas others have a more ubiquitous expression pattern in somatic tissues (ADAMs 8, 9, 10, 12, 15, 17, 19, 28, 33). Northern blot analysis indicates that the expression pattern of several widely expressed ADAMs is at least partially overlapping, although individual ADAMs have distinct expression profiles (Black *et al.*, 1997; Howard *et al.*, 2000; Inoue *et al.*, 1998; Krätzschmar *et al.*, 1996; Lum *et al.*, 1998; Moss *et al.*, 1997; Rosendahl *et al.*, 1997; Weskamp *et al.*, 1996; Yagami-Hiromasa *et al.*, 1995). This conclusion is further supported by RNA *in situ* hybridization experiments. For example, ADAM9 is found in all cells and tissues evaluated to date, but is highly expressed in the developing heart, in mesenchymal tissues, and in certain areas of the brain, such as the hippocampus and hypothalamus (Weskamp *et al.*, 2002). Expression of ADAMs 12 and 19 is high in the bone, although both proteins are also quite widely expressed (Kurisaki *et al.*, 1998). *Xenopus laevis* ADAM13 is highly expressed in the cranial neural crest and somitic mesoderm (Alfandari *et al.*, 1997). Further studies will be necessary to understand whether ADAMs have different functions in different tissues, or whether their roles in different tissues is comparable and conserved.

Little is currently known about the regulation of ADAMs at the transcriptional level, although anecdotal evidence suggests that the expression of most catalytically active ADAMs can be regulated. ADAM8, for example, is upregulated in affected brain areas of the wobbler mouse, a mouse model for neuronal degeneration (Schlomann *et al.*, 2000). ADAM9 is upregulated in cells derived from a prostate tumor (LnCAP cells) that are treated with androgen (DHT), suggesting that ADAM9 may have a role in the androgen-dependent stage of prostate cancer (McCulloch *et al.*, 2000). ADAM19 expression is enhanced during dendritic cell maturation as well as during maturation of double positive to single positive T cells, suggesting an important role in the immune system (Deftos *et al.*, 2000; Fritsche *et al.*, 2000). We expect that a much better understanding of the regulation of ADAMs will result from increasing use of gene chips, which allow a rapid analysis of gene expression in cells treated with different stimuli, or in different tissues or in normal versus diseased tissues.

IV. Interactions of ADAMs with Other Extracellular and Intracellular Proteins

A. Evidence for Roles in Cell–Cell Interactions

ADAMs are related to snake venom integrin ligands termed disintegrins, and therefore they are predicted to function as membrane-anchored integrin ligands in addition to their predicted roles as metalloproteases. A number of studies now support this prediction. Specifically, ADAM9 has been shown to interact with $\alpha6\beta1$ integrins (Nath *et al.*, 2000), and human ADAM15, which contains the integrin ligands consensus sequence RGD in its disintegrin domain, has been shown to interact with the RGD-binding integrin $\alpha5\beta1$ or $\alpha v\beta3$ (Nath *et al.*, 1999; Zhang *et al.*, 1998). In contrast to human ADAM15, mouse ADAM15 does not contain an RGD sequence, and therefore does not bind to these two integrins (Lum *et al.*, 1998). Both mouse and human ADAM15 as well as ADAM12 also interact with integrin $\alpha9\beta1$(Eto *et al.*, 2000). Finally, ADAM28 has been shown to interact with the $\alpha4\beta1$ integrin (Bridges *et al.*, 2002).

Since ADAMs are membrane anchored (with the exception of soluble splice variants), one would predict that they should interact with integrins on adjacent cells, and thus promote cell–cell interactions. In this scenario, an ADAM may also function as counterreceptor and transmit a signal back to the cell it is associated with. However, it is also possible that an ADAM could interact with an integrin on the same cell, although there is no experimental evidence for this. An interaction of an ADAM with an integrin on the same cell in cis could, for example, attract the ADAM to focal contacts or other structures where integrins are localized. This could, in turn, direct the protease activity toward substrates that are colocalized with an integrin. Now that the role of ADAMs as integrin ligands is well supported, it will be important to determine what the functional relevance of this interaction is. Does binding of an ADAM to an integrin, for example, have a role in mediating cell–cell contact, or signaling, or both, or is it necessary for targeting the ADAM to the vicinity of relevant substrates?

In addition to their predicted role as integrin ligands, ADAMs are also capable of interacting with other extracellular proteins. ADAM12, for example, binds to syndecans, which in turn triggers activation of beta1 integrins and thus promotes cell spreading (Iba *et al.,* 2000). At least two ADAMs can also interact directly with specific substrates. ADAM10 binds ephrin 2A, a protein with a role in neuronal repulsion (Hattori *et al.,* 2000), as well as Notch (Lieber *et al.,* 2002). Indeed, truncated forms of ADAM10 that lack the metalloprotease domain function as dominant negative inhibitors of ADAM10 function, presumably because they can bind and sequester the ligand (Pan and Rubin, 1997). The related ADAM17/TACE reportedly also interacts with its substrate TNFα (Itai *et al.,* 2001). Although it remains to be determined whether or not other ADAMs also interact with their substrates, and whether TACE and ADAM10 bind to all relevant substrates, it is tempting to speculate that an interaction between an ADAM and its substrate may contribute to substrate targeting and specificity of an ADAM.

B. Evidence for Regulation of the Catalytic Activity of ADAMs and for Interactions with Cytoplasmic Proteins

All catalytically active and widely expressed ADAMs have potential signaling motifs, including SH3 ligand domains, in their cytoplasmic domains, suggesting that cytoplasmic proteins affect the function of ADAMs. In principle, this could be via an inside-out activation of an ADAM, or alternatively, binding of an integrin or other cell-surface protein to the ectodomain of an ADAM may trigger an outside-in signal. The best evidence for inside-out activation of the catalytic activity of an ADAM has come from studies of TACE. Shedding of the TACE substrate TGFalpha can be induced by receptor tyrosine kinases and activators of the MAP kinase pathway, suggesting that the catalytic activity of TACE is regulated by this pathway (Fan and Derynck, 1999). Shedding of TACE substrates such as TNFα and TGFα can also be activated by other means, such as by inhibiting tyrosine phosphatases or activating protein kinase C with phorbol esters such as

PMA (Hooper *et al.,* 1997). Remarkably, the cytoplasmic domain of TACE is not essential for PMA-dependent stimulation of its catalytic activity (Reddy *et al.,* 2000). The mechanism underlying this activation thus may involve other proteins, perhaps even another transmembrane protein which responds to PMA and can activate TACE. Cholesterol apparently also has a role in regulating the function of ADAMs, since depletion of cholesterol enhances the metalloprotease-dependent α-secretase cleavage of the amyloid precursor protein, which is thought to depend on ADAM10 (Kojro *et al.,* 2001). The proposed mechanism of this activation of α-secretase activity include increased membrane fluidity, decreased internalization of APP that would remove APP from the cell surface, and increased ADAM10 production.

Several cytoplasmic proteins have been found to interact with the cytoplasmic domain of different ADAMs. Two SH3-domain-containing proteins with a role in intracellular protein transport, endophilin and sorting nexin 9 (SNX9, initially named SH3PX1), were found to bind the different proline-rich sequences in the cytoplasmic domains of ADAM9 and ADAM15 (Howard *et al.,* 1999). Overexpression of endophilin or SNX9 leads to a relative increase in the immature pro form of ADAMs 9 or 15, consistent with a role in regulating the intracellular maturation or transport of these ADAMs. In separate studies, ADAMs 12 and 15 were found to interact with Grb2 and with members of the Src family of tyrosine kinases (Lck, Fyn, Abl, Src for ADAM15, p85α and Src for ADAM12) (Poghosyan *et al.,* 2002; Suzuki *et al.,* 2000). These results suggest the possibility of an integration of the function of ADAMs 15 and 12 and cellular signaling. The cytoplasmic domain of ADAM12 has also been shown to bind α-actinin-1 and α-actinin-2 and may therefore have a role in myoblast fusion (Cao *et al.,* 2001; Galliano *et al.,* 2000). TACE has been shown to bind the cell-cycle checkpoint gene MAD2, whereas ADAM9 binds to the related MAD2β (Nelson *et al.,* 1999). While the functional relevance of these interactions remains to be determined, it is tempting to speculate that they provide a link between the function of ADAMs 9 and TACE and the cell cycle. Finally, some of the most compelling evidence for a functional regulation of an ADAM through a cytoplasmic protein *in vivo* has come from studies of ADAM13, a protein with a role in neural crest development in *Xenopus laevis.* ADAM13 interacts with the SH3-domain-containing protein X-pacsin2 via a cytoplasmic SH3 ligand domain (Cousin *et al.,* 2000). Overexpression of X-pacsin2 reduces ADAM13 function in neural crest migration, but can also rescue developmental alterations induced by overexpression of ADAM13 (see later discussion), suggesting a functionally relevant interaction between these two proteins. Taken together, these results provide ample evidence for interactions of ADAM cytoplasmic domains with intracellular proteins, including ones that have clear links to signaling pathways or to the cytoskeleton. The next challenge will be to elucidate functional relevance of these interactions *in vivo* and to learn how exactly they are integrated in these signaling pathways.

V. Roles in Embryogenesis and Development

The first evidence for a role of an ADAM in development came from studies of a *Drosophila* ADAM named kuzbanian (KUZ/ADAM10), which was identified as a critical component of the Notch signaling pathway (Rooke *et al.*, 1996). Flies carrying a mutation in KUZ have a neurogenic phenotype that resembles the phenotype of flies lacking functional Notch. It has since become clear that KUZ is important for signaling via Notch, either by cleaving and activating a ligand such as Delta (Qi *et al.*, 1999), or by cleaving Notch itself (Lieber *et al.*, 2002) (see Blobel, 2000; Hartmann *et al.*, 2001; Kopan and Cagan, 1997; Nye, 1997; Schlöndorff and Blobel, 1999, for a more detailed discussion of the role of proteolysis in Notch signaling). The mammalian ortholog of KUZ, ADAM10, has been implicated in neurogenesis and cleaves the cell surface molecule ephrin 2A (Hattori *et al.*, 2000). This finding provides an intriguing explanation for a previously unsolved dilemma regarding the function of ephrin 2A. On the one hand, ephrin 2A was known to function as a repellent for axons by activating the Ephrin receptor. However, both ephrin 2A and its receptor are membrane anchored, and furthermore, they interact with high affinity with one another. How then could a high-affinity interaction between two cells result in repulsion? The answer is that ADAM10 evidently is activated by the interaction between ephrin 2A and its receptor, and then cleaves ephrin 2A close to the plasma membrane, allowing the axon carrying the ephrin receptor to retract.

Shortly after the discovery of TACE, the phenotype of mice lacking functional TACE was reported (Peschon *et al.*, 1998). Mice lacking TACE usually die in the first days after birth, and have eye, hair, and skin defects that resemble those seen in mice lacking TGFα. Indeed, in addition to being deficient in their ability to secrete TNF, mice lacking TACE are also unable to process TGFα properly, which provides a plausible explanation for why the phenotypes of mice lacking TACE and TGFα resemble one another. Further evidence that the lethality of this mutation appears to be related to TACE's role in TGFα shedding was provided by TNF-/- mice and TNFRp55-/- and TNFRp75-/, which are overtly normal. Furthermore, crossing the TACE deficient mice with the TNFRp55-/- and TNFRp75-/- mice had no effect on lethality. Taken together, these results provide compelling evidence for an essential role of ectodomain shedding in mammalian development and in activation of TGFα and potentially also other EGFR-ligands (Peschon *et al.*, 1998).

Additional information on the role of ADAMs in development has come from studies of the African clawed frog *Xenopus laevis*, which is considered to be a good experimental system to study early vertebrate development. Several *Xenopus* ADAMs have been identified and their expression during development has been analyzed (Alfandari *et al.*, 1997; Cai *et al.*, 1998). The role of one of these proteins, ADAM13, in development has been characterized in more detail. ADAM13 was initially reported as a gene that is highly expressed in the neural crest region of the developing brain and in somitic mesoderm (Alfandari *et al.,* 1997). In subsequent functional studies, ectopically overexpressed wild-type ADAM13 was found to result in abnormal positioning of neural crest cells, whereas overexpressed catalytically inactive ADAM13 decreased migration of neural crest cells (Alfandari *et al.,* 2001). These findings suggest that ADAM13 indeed has a critical role in cranial neural crest migration.

Little information has been published about the role of other ADAMs besides ADAM10, ADAM13, and ADAM17 in development. ADAM9 is clearly not essential for embryogenesis, as mice lacking ADAM9 develop normally, are viable and fertile, and do not display any evident defects (Weskamp *et al.,* 2002). The same is true for mice lacking ADAM15 (Weskamp and Blobel, manuscript in preparation). In contrast, most mice lacking functional ADAM19 die in the first few days after birth, most likely due to a defect in heart development (Zhou and Blobel, manuscript in preparation; Fujisawa-Sehara, personal communication). It will now be interesting to learn about the functions of the remaining widely expressed and catalytically active ADAMs in development.

VI. Roles in Disease

Little is known about the role of most ADAMs in disease. The exception to this is TACE, which is responsible for the release of $TNF\alpha$ from cells. Because TNF plays a pivotal role in rheumatoid arthritis, many pharmaceutical companies have developed potent inhibitors of TACE. Studies have confirmed that inhibition of TACE is antiinflammatory and suppresses paw swelling in the PG-PS and adjuvant-induced models of arthritis (Conway *et al.,* 2001; DiMartino *et al.,* 1997). The pharmacologic affects of nonselective TACE inhibitors are often subject to the criticism that many of their antiinflammatory properties are due to inhibition of the collagenases and gelatinases. However, a selective TACE inhibitor is as effective as Enbrel in the collagen-induced model of arthritis (Newton *et al.,* 2001). These data support the supposition that soluble and not membrane-bound TNF drives the autoimmune inflammation in this disease model (Letavic *et al.,* 2002; Ruuls *et al.,* 2001).

In addition to arthritis, TACE inhibitors are efficacious in other models of inflammation where TNF is implicated. These include acute models of airway inflammation induced by LPS or ovalbumin (Fiorucci *et al.,* 1998; Trifilieff *et al.,* 2002), and pneumococcal meningitis (Lieb *et al.,* 2001). However, in an experimental model of pulmonary tuberculosis, i.p. treatment with the hydroxamate BB-94 over a 30-day course delays granuloma formation and enhances the rate of disease progression and mortality. This data is accompanied by an increase in IL-4 in the lung, a diminished DTH response, and a general bias toward a type 2 cytokine response (Hernandez-Pando *et al.,* 2000).

TACE inhibitors (GW 9471) are effective at blocking both lethality and liver injury in the D-galactosamine/LPS model of endotoxemia (Murakami *et al.*, 1998). These data are consistent with those observed when transgenic mice expressing only the membrane bound form of TNF are challenged with D-galactosamine/LPS (Nowak *et al.*, 2000). However, unlike the D-galactosamine/LPS model of liver injury, a TACE/MMP inhibitor (GM 6001) exacerbates the hepatic necrosis seen when concanavalin A is used to induce hepatitis (Solorzano *et al.*, 1997). This increase in hepatic injury in the presence of GM 6001 is presumably due to the broad inhibitory spectrum of this metalloprotease inhibitor, thereby stabilizing the surface expression of the FAS-L and/or TNFRp55.

TNFα is one of the factors linked to obesity-induced insulin resistance, a characteristic of type 2 diabetes. This is based on the observation that TNF levels are increased in adipose tissues in rodent models for obesity (the fa/fa rat, db/db rat, and the KKAy mouse, Hotamisligil *et al.*, 1993). The TACE/MMP inhibitor KB-R7785 reduces plasma glucose and insulin levels when administered orally to insulin-resistant KKAy mice (Morimoto *et al.*, 1997). Furthermore, studies with the same inhibitor demonstrate that TNFα affects insulin resistance in nonobese insulin-resistant hypertensive fructose-fed rats (Togashi *et al.*, 2002). Neutralizing TNF improves the metabolic profile in obese mice and it appears that TNF exerts its affects in an autocrine or paracrine manner at the level of the adipocyte (Uysal *et al.*, 1997). Additional studies are required to better understand the role TACE and TNF play in modulating insulin sensitivity and glucose metabolism, both locally in the adipocyte as well as systemically (Xu *et al.*, 2002).

In addition to releasing TNF, TACE is also involved in the shedding of TGFα (see earlier discussion; Peschon *et al.*, 1998). Polycystic kidney disease is characterized by massive kidney enlargement due to rapidly expanding cysts and by abnormal expression of TGFα in the cyst-lining epithelial cells of the kidney. In a mouse model of PKD, inhibiting TACE results in a significant decrease in kidney size, BUN and serum creatine levels and an improvement in cystic index (Dell *et al.*, 2001). This study provides good evidence that inhibiting TACE will block TGFα shedding and suggests TACE inhibitors should be further evaluated in cancer models where EGFR ligands are known to play a key role in driving tumor growth.

VII. Concluding Comments and Future Directions

Since their discovery about a decade ago, some very intriguing insights into the function of a few ADAMs have emerged. They have been linked to a role in cell–cell interactions as potential integrin ligands, and in the cleavage and release and, in several cases, activation of membrane-anchored substrates. Yet much still remains to be discovered about the role of other ADAMs in cell–cell interaction, in protein ectodomain shedding, in processing of other molecules on the cell surface or in the extracellular matrix, or in signaling. From a medical point of view, it will be particularly interesting to learn more about the role of ADAMs in diseases such as cancer and autoimmune disorders. It is tempting to speculate that ADAMs could both enhance or suppress tumor growth by cleaving growth factors or their receptors. The exact consequences of this would depend on the stage of tumorigenesis and the nature of the substrate. One could imagine that release of an EGFR ligand might promote tumor-cell proliferation and migration. In this scenario, it might be beneficial to inhibit the catalytic activity of the appropriate ADAM, for example by administering a selective hydroxamic acid derivative. Alternatively, it could be that an ADAM may have a role in suppressing tumor growth, for example by cleaving an inhibitory factor, or by inactivating a receptor that is required for tumor growth. If this were to be the case, then care should be taken to not inhibit the relevant protease. Other potential functions in cell adhesion and signaling must also be taken into consideration. Given the insights that have already emerged from studies of TACE as a prototypical ADAM, we expect that similar approaches to characterizing other less well defined ADAMs will yield interesting information about the roles of these proteins in normal homeostasis and in pathophysiology.

References

Alfandari, D., Wolfsberg, T. G., White, J. M., and DeSimone, D. W. (1997). ADAM13: A novel ADAM expressed in somitic mesoderm and neural crest cells during *Xenopus laevis* development. *Dev. Biol.* **182,** 314–330.

Alfandari, D., Cousin, H., Gaultier, A., Smith, K., White, J. M., Darribere, T., and DeSimone, D. W. (2001). Xenopus ADAM 13 is a metalloprotease required for cranial neural crest-cell migration. *Curr. Biol.* **11**(12), 918–930.

Amour, A., Slocombe, P. M., Webster, A., Butler, M., Knight, C. G., Smith, B. J., Stephens, P. E., Shelley, C., Hutton, M., Knauper, V., Docherty, A. J., and Murphy, G. (1998). TNF-alpha converting enzyme (TACE) is inhibited by TIMP-3. *FEBS Lett.* **435**(1), 39–44.

Amour, A., Knight, C. G., Webster, A., Slocombe, P. M., Stephens, P. E., Knauper, V., Docherty, A. J., and Murphy, G. (2000). The in vitro activity of ADAM-10 is inhibited by TIMP-1 and TIMP-3. *FEBS Lett.,* **473**(3), 275–279.

Amour, A., Knight, C., English, W., Webster, A., Slocombe, P., Knauper, V., Docherty, A., Becherer, J., Blobel, C., and Murphy, G. (2002). The enzymatic activity of ADAM8 and ADAM9 is not regulated by TIMPs. *FEBS Lett.,* **524**(1-3), 154–158.

Anders, A., Gilbert, S., Garten, W., Postina, R., and Fahrenholz, F. (2001). Regulation of the alpha-secretase ADAM10 by its prodomain and proprotein convertases. *FASEB J.* **15**(10), 1837–1839.

Barlaam, B., Bird, T. G., Lambert-Van Der Brempt, C., Campbell, D., Foster, S. J., and Maciewicz, R. (1999). New alpha-substituted succinate-based hydroxamic acids as TNFalpha convertase inhibitors. *J. Med. Chem.* **42**(23), 4890–4908.

Becherer, J., Lamber, M., and Andrews, R. (2000). The TNF converting enzyme. *In* "Proteases as Targets for Therapy" (K. Von der Helm, B. Korant, and J. Cheronis, Eds.), pp. 235–258. Springer Berlin.

Beckett, R., and Whittaker, M. (1998). Matrix metalloproteinase inhibitors. *Expert Opin. Ther. Pat.* **8,** 259–282.

Black, R., Rauch, C. T., Kozlosky, C. J., Peschon, J. J., Slack, J. L., Wolfson, M. F., Castner, B. J., Stocking, K. L., Reddy, P., Srinivasan, S., Nelson, N., Boiani, N., Schooley, K. A., Gerhart, M., Davis, R., Fitzner, J. N., Johnson, R. S., Paxton, R. J., March, C. J., and Cerretti, D. P. (1997). A metalloprotease disintegrin that releases tumour-necrosis factor-α from cells. *Nature* **385,** 729–733.

Black, R. A., and White, J. M. (1998). ADAMs: focus on the protease domain. *Curr. Opin. Cell Biol.* **10**(5), 654–659.

Blobel, C. P., and White, J. M. (1992). Structure, function and evolutionary relationship of proteins containing a disintegrin domain. *Curr. Opin. Cell Biol.* **4,** 760–765.

Blobel, C. P. (1997). Metalloprotease-disintegrins: Links to cell adhesion and cleavage of TNFα and Notch. *Cell* **90,** 589–592.

Blobel, C. P. (1999). Roles of metalloprotease-disintegrins in cell–cell interactions, in neurogenesis, and in the cleavage of TNFalpha. *In* "Advances in Developmental Biochemistry" (P. M. Wassarman, Ed.), pp. 165–198. JAI Press Inc., Stamford, Connecticut.

Blobel, C. P. (2000). Remarkable roles of proteolysis on and beyond the cell surface. *Curr. Opin. Cell Biol.* **12**(5), 606–612.

Blundell, T. L. (1994). Metalloproteinase superfamily and drug design. *Struct. Biol.* **1,** 73–75.

Bode, W., Gomis-Ruth, F. X., and Stockler, W. (1993). Astacins, serralysins, snake venom and matrix metalloproteinases exhibit identical zinc-binding environments (HEXXHXXGXXH and Met-turn) and topologies and should be grouped into a common family, the "metzincins." *FEBS Lett.* **331**(1–2), 134–140.

Borland, G., Murphy, G., and Ager, A. (1999). Tissue inhibitor of metalloproteinases-3 inhibits shedding of L-selectin from leukocytes. *J. Biol. Chem.* **274**(5), 2810–2815.

Bridges, L. C., Tani, P. H., Hanson, K. R., Roberts, C. M., Judkins, M. B., and Bowditch, R. D. (2002). The lymphocyte metalloprotease MDC-L (ADAM 28) is a ligand for the integrin alpha4beta1. *J. Biol. Chem.* **277**(5), 3784–3792.

Cai, H., Krätzschmar, J., Alfandari, D., Hunnicutt, G., and Blobel, C. P. (1998). Neural crest-specific and general expression of distinct metalloprotease-disintegrins in early *Xenopus laevis* development. *Dev. Biol.* **204,** 508–524.

Cao, Y., Kang, Q., and Zolkiewska, A. (2001). Metalloprotease-disintegrin ADAM 12 interacts with alpha-actinin-1. *Biochem. J.* **357**(Pt 2), 353–361.

Cao, Y., Kang, Q., Zhao, Z., and Zolkiewska, A. (2002). Intracellular processing of metalloprotease disintegrin ADAM12. *J. Biol. Chem.,* **277**(29), 26403–26411.

Chantry, A., Gregson, N. A., and Glynn, P. (1989). A novel metalloproteinase associated with brain myelin membranes. Isolation and characterization. *J. Biol. Chem.* **264**(36), 21603–21607.

Conway, J. G., Andrews, R. C., Beaudet, B., Bickett, D. M., Boncek, V., Brodie, T. A., Clark, R. L., Crumrine, R. C., Leenitzer, M. A., McDougald, D. L., Han, B., Hedeen, K., Lin, P., Milla, M., Moss, M., Pink, H., Rabinowitz, M. H., Tippin, T., Scates, P. W., Selph, J., Stimpson, S. A., Warner, J., and Becherer, J. D. (2001). Inhibition of tumor necrosis factor-alpha (TNF-alpha) production and arthritis in the rat by GW3333, a dual inhibitor of TNF-alpha-converting enzyme and matrix metalloproteinases. *J. Pharmacol. Exp. Ther.* **298**(3), 900–908.

Cousin, H., Gaultier, A., Bleux, C., Darribere, T., and Alfandari, D. (2000). PACSIN2 is a regulator of the metalloprotease/disintegrin ADAM13. *Dev. Biol.* **227**(1), 197–210.

Deftos, M. L., Huang, E., Ojala, E. W., Forbush, K. A., and Bevan, M. J. (2000). Notch1 signaling promotes the maturation of CD4 and CD8 SP thymocytes. *Immunity* **13**(1), 73–84.

Dekkers, P. E., Lauw, F. N., ten Hove, T., te Velde, A. A., Lumley, P., Becherer, D., van Deventer, S. J., and van der Poll, T. (1999). The effect of a metalloproteinase inhibitor (GI5402) on tumor necrosis factor-alpha (TNF-alpha) and TNF-alpha receptors during human endotoxemia. *Blood* **94**(7), 2252–2258.

Dell, K. M., Nemo, R., Sweeney, W. E., Jr., Levin, J. I., Frost, P., and Avner, E. D. (2001). A novel inhibitor of tumor necrosis factor-alpha converting enzyme ameliorates polycystic kidney disease. *Kidney Int.* **60**(4), 1240–1248.

DiMartino, M., Wolff, C., High, W., Stroup, G., Hoffman, S., Laydon, J., Lee, J. C., Bertolini, D., Galloway, W. A., Crimmin, M. J., Davis, M., and Davies, S. (1997). Anti-arthritic activity of hydroxamic acid-based pseudopeptide inhibitors of matrix metalloproteinases and TNF alpha processing. *Inflamm. Res.* **46**(6), 211–215.

Doedens, J. R., and Black, R. A. (2000). Stimulation-induced down-regulation of tumor necrosis factor-alpha converting enzyme. *J. Biol. Chem.* **275**(19), 14598–14607.

Eto, K., Puzon-McLaughlin, W., Sheppard, D., Sehara-Fujisawa, A., Zhang, X. P., and Takada, Y. (2000). RGD-independent binding of integrin a9b1 to the ADAM-12 and -15 disintegrin domains mediates cell-cell interaction. *J. Biol. Chem.* **275**(45), 34922–34930.

Evans, J. P. (1999). Sperm disintegrins, egg integrins, and other cell adhesion molecules of mammalian gamete plasma membrane interactions. *Front. Biosci.* **4**, D114–D131.

Fan, H., and Derynck, R. (1999). Ectodomain shedding of TGF-alpha and other transmembrane proteins is induced by receptor tyrosine kinase activation and MAP kinase signaling cascades. *EMBO J.* **18**(24), 6962–6972.

Fiorucci, S., Antonelli, E., Migliorati, G., Santucci, L., Morelli, O., Federici, B., and Morelli, A. (1998). TNFalpha processing enzyme inhibitors prevent aspirin-induced TNFalpha release and protect against gastric mucosal injury in rats. *Aliment. Pharmacol. Ther.* **12**(11), 1139–1153.

Fritsche, J., Moser, M., Faust, S., Peuker, A., Buttner, R., Andreesen, R., and Kreutz, M. (2000). Molecular cloning and characterization of a human metalloprotease disintegrin—a novel marker for dendritic cell differentiation. *Blood* **96**(2), 732–739.

Galliano, M. F., Huet, C., Frygelius, J., Polgren, A., Wewer, U. M., and Engvall, E. (2000). Binding of ADAM12, a marker of skeletal muscle regeneration, to the muscle-specific actin-binding protein, alpha-actinin-2, is required for myoblast fusion. *J. Biol. Chem.* **275**(18), 13933–13999.

Gearing, A. J. H., Beckett, M., Christodoulou, M., Churchhill, M., Clements, J., Davidson, A. H., Drummond, A. H., Galloway, W. A., Gilbert, R., Gordon, J. L., Leber, T. M., Mangan, M., Miller, K., Nayee, P., Owen, K., Patel, S., Thomas, W., Wells, G., Wood, L. M., and Woolley, K. (1994). Processing of tumour necrosis factor-a precursor by metalloproteinases. *Nature* **370**, 555–558.

Gilpin, B. J., Loechel, F., Mattei, M. G., Engvall, E., Albrechtsen, R., and Wewer, U. M. (1998). A novel, secreted form of human ADAM 12 (meltrin alpha) provokes myogenesis in vivo. *J. Biol. Chem.* **273**(1), 157–166.

Giulian, D., Haverkamp, L. J., Yu, J., Karshin, W., Tom, D., Li, J., Kazanskaia, A., Kirkpatrick, J., and Rocher, A. E. (1998). The HHQK domain of beta-amyloid provides a structural basis for the immunopathology of Alzheimer's disease. *J. Biol. Chem.* **273**(45), 29719–29726.

Hartmann, D., Tournoy, J., Saftig, P., Annaert, W., and De Strooper, B. (2001). Implication of APP secretases in notch signaling. *J. Mol. Neurosci.* **17**(2), 171–181.

Hattori, M., Osterfield, M., and Flanagan, J. G. (2000). Regulated cleavage of a contact-mediated axon repellent. *Science* **289**(5483), 1360–1365.

Hernandez-Pando, R., Orozco, H., Arriaga, K., Pavon, L., and Rook, G. (2000). Treatment with BB-94, a broad spectrum inhibitor of zinc-dependent metalloproteinases, causes deviation of the cytokine profile towards type-2 in experimental pulmonary tuberculosis in Balb/c mice. *Int. J. Exp. Pathol.* **81**(3), 199–209.

Holms, J., Mast, K., Marcotte, P., Elmore, I., Li, J., Pease, L., Glaser, K., Morgan, D., Michaelides, M., and Davidsen, S. (2001). Discovery of selective hydroxamic acid inhibitors of tumor necrosis factor alpha converting enzyme. *Bioorg. Med. Chem. Lett.* **11**, 2907–2910.

Hooper, N. M., Karran, E. H., and Turner, A. J. (1997). Membrane protein secretases. *Biochem. J.* **321**, 265–279.

Hotamisligil, G. S., Shargill, N. S., and Spiegelman, B. M. (1993). Adipose expression of tumor necrosis factor-alpha: direct role in obesity-linked insulin resistance. *Science* **259**(5091), 87–91.

Howard, L., and Glynn, P. (1995). Membrane-associated metalloproteinase recognized by characterisitic cleavage of myelin basic protein: assay and isolation. *In* "Proteolytic Enzymes: Aspartic and Metalloproteases," (A. J. Barrett, Ed.), pp. 388–395. Academic Press, San Diego.

Howard, L., Lu, X., Mitchell, S., Griffiths, S., and Glynn, P. (1996). Molecular cloning of MADM: a catalytically active disintegrin-metalloprotease expressed in various cell types. *Biochem. J.* **317**, 45–50.

Howard, L., Nelson, K. K., Maciewizc, R. A., and Blobel, C. P. (1999). Interaction of the metalloprotase disintegrins MDC9 and MDC15 with two SH3 domain-containing proteins, endophilin I and SH3PX1. *J. Biol. Chem.* **274**(44), 31693–31699.

Howard, L., Maciewicz, R. A., and Blobel, C. P. (2000). Cloning and characterization of ADAM28: evidence for autocatalytic pro-domain removal and for cell surface localization of mature ADAM28. *Biochem. J.* **348 Pt 1**, 21–27.

Howard, L., *et al.* (2001). Catalytic activity of ADAM28. *FEBS Lett.* **498**(1), 82–86.

Huovila, A. P. J., Almeida, E. A., and White, J. M. (1996). ADAMs and cell fusion. *Curr. Opin. Cell Biol.* **8**(5), 692–699.

Hurskainen, T. L., Hirohata, S., Seldin, M. F., and Apte, S. S. (1999). ADAM-TS5, ADAM-TS6, and ADAM-TS7, novel members of a new family of zinc metalloproteases. General features and genomic distribution of the ADAM-TS family. *J. Biol. Chem.* **274**(36), 25555–25563.

Iba, K., Albrechtsen, R., Gilpin, B., Frohlich, C., Loechel, F., Zolkiewska, A., Ishiguro, K., Kojima, T., Liu, W., Langford, J. K., Sanderson, R. D., Brakebusch, C., Fassler, R., and Wewer, U. M. (2000). The cysteine-rich domain of human ADAM 12 supports cell adhesion through syndecans and triggers signaling events that lead to beta1 integrin-dependent cell spreading. *J. Cell Biol.* **149**(5), 1143–1156.

Inoue, D., Reid, M., Lum, L., Krätzschmar, J., Weskamp, G., Myung, Y. M., Baron, R., and Blobel, C. P. (1998). Cloning and initial characterization of mouse meltrin beta and analysis of the expression of four metalloprotease-disintegrins in bone cells. *J. Biol. Chem.* **273**, 4180–4187.

Itai, T., Tanaka, M., and Nagata, S. (2001). Processing of tumor necrosis factor by the membrane-bound TNF-alpha-converting enzyme, but not its truncated soluble form. *Eur. J. Biochem.* **268**(7), 2074–2082.

Jongeneel, C. V., Bouvier, J., and Bairoch, A. (1989). A unique signature identifies a family of zinc-dependent metallopeptidases. *FEBS Lett.* **242**(2), 211–214.

Jury, J. A., Perry, A. C., and Hall, L. (1999). Identification, sequence analysis and expression of transcripts encoding a putative metalloproteinase, eMDC II, in human and macaque epididymis. *Mol. Hum. Reprod.* **5**(12), 1127–1134.

Kang, T., Zhao, Y. G., Pei, D., Sucic, J. F., and Sang, Q. X. (2002). Intracellular activation of human Adamalysin 19/disintegrin and metalloproteinase 19 by furin occurs via one of the two consecutive recognition sites. *J. Biol. Chem.,* **277**(28), 25583–25591.

Kojro, E., Gimpl, G., Lammich, S., Marz, W., and Fahrenholz, F. (2001). Low cholesterol stimulates the nonamyloidogenic pathway by its effect on the alpha-secretase ADAM 10. *Proc. Natl. Acad. Sci. USA* **98**(10), 5815–5820.

Kopan, R., and Cagan, R. (1997). Notch on the cutting edge. *Trends Genet.* **13**(12), 465–467.

Kottirsch, G., Koch, G., Feifel, R., and Neumann, U. (2002). Beta-aryl-succinic acid hydroxamates as dual inhibitors of matrix metalloproteinases and tumor necrosis factor alpha converting enzyme. *J. Med. Chem.* **45**(11), 2289–2293.

Krätzschmar, J., Lum, L., and Blobel, C. P. (1996). Metargidin, a membrane-anchored metalloprotease-disintegrin protein with an RGD integrin binding sequence. *J. Biol. Chem.* **271**, 4593–4596.

Kurisaki, T., Masuda, A., Osumi, N., Nabeshima, Y., and Fujisawa-Sehara, A. (1998). Spatially- and temporally-restricted expression of meltrin alpha (ADAM12) and beta (ADAM19) in mouse embryo. *Mech. Dev.* **73**(2), 211–215.

Letavic, M. A., Axt, M. Z., Barberia, J. T., Carty, T. J., Danley, D. E., Geoghegan, K. F., Halim, N. S., Hoth, L. R., Kamath, A. V., Laird, E. R., Lopresti-Marrow, L. L., McClure, K. F., Mitchell, P. G.,

Natarajan, V., Noe, M. C., Pandit, J., Reeves, L., Schulte, G. K., Snow, S. L., Sweeney, F. J., Tan, D. H., and Yu, C. H. (2002). Synthesis and biological activity of selective pipecolic acid-based TNF-alpha converting enzyme (TACE) inhibitors. *Bioorg. Med. Chem. Lett.* **12**(10), 1387–1390.

Levin, J., Chen, J., Du, M., Nelson, F., Killar, L., Sakala, S., Sung, A., Jin, G., Cowling, R., Barone, D., March, C., Mohler, K., RA, B., and Skotnicki, J. (2002). Anthranilate sulfonamide hydroxamate TACE inhibitors. Part 2: SAR of the acetylenic P1 group. *Bioorg. Med. Chem. Lett.* **12**, 1199–1202.

Lieb, S., Clements, J., Lindberg, R., Heimgartner, C., Loeffler, J., Pfister, L.-A., Tauber, M., and D, L. (2001). Inhibition of matrix metalloproteinases and tumor necrosis factor alpha converting enzyme as adjuvant therapy in pneumococcal meningitis. *Brain* **124**, 1734–1742.

Lieber, T., Kidd, S., and Young, M. W. (2002). Kuzbanian-mediated cleavage of Drosophila Notch. *Genes Dev.* **16**(2), 209–221.

Loechel, F., Gilpin, B. J., Engvall, E., Albrechtsen, R., and Wewer, U. M. (1998). Human ADAM 12 (meltrin alpha) is an active metalloprotease. *J. Biol. Chem.* **273**(27), 16993–16997.

Loechel, F., Overgaard, M. T., Oxvig, C., Albrechtsen, R., and Wewer, U. M. (1999). Regulation of human ADAM 12 protease by the prodomain. Evidence for a functional cysteine switch. *J. Biol. Chem.* **274**(19), 13427–13433.

Loechel, F., Fox, J. W., Murphy, G., Albrechtsen, R., and Wewer, U. M. (2000). ADAM 12-S cleaves IGFBP-3 and IGFBP-5 and is inhibited by TIMP-3. *Biochem. Biophys. Res. Commun.* **278**(3), 511–515.

Lum, L., and Blobel, C. P. (1997). Evidence for distinct serine protease activities with a potential role in processing the sperm protein fertilin. *Dev. Biol.* **191**(1), 131–145.

Lum, L., Reid, M. S., and Blobel, C. P. (1998). Intracellular maturation of the mouse metalloprotease disintegrin MDC15. *J. Biol. Chem.* **273**, 26236–26247.

Lum, L., Wong, B. R., Josien, R., Becherer, J. D., Erdjument-Bromage, H., Schlondorff, J., Tempst, P., Choi, Y., and Blobel, C. P. (1999). Evidence for a role of a tumor necrosis factor-alpha (TNF-alpha)-converting enzyme-like protease in shedding of TRANCE, a TNF family member involved in osteoclastogenesis and dendritic cell survival. *J. Biol. Chem.* **274**(19), 13613–13618.

McCulloch, D. R., Harvey, M., and Herington, A. C. (2000). The expression of the ADAMs proteases in prostate cancer cell lines and their regulation by dihydrotestosterone. *Mol. Cell. Endocrinol.* **167**(1–2), 11–21.

McGeehan, G. M., Becherer, J. D., Bast, R. C., Jr., Boyer, C. M., Champion, B., Connolly, K. M., Conway, J. G., Furdon, P., Karp, S., Kidao, S., *et al.* (1994). Regulation of tumour necrosis factor-alpha processing by a metalloproteinase inhibitor. *Nature* **370**(6490), 558–561.

McLane, M. A., Marcinkiewicz, C., Vijay-Kumar, S., Wierzbicka-Patynowski, I., and Niewiarowski, S. (1998). Viper venom disintegrins and related molecules. *Proc. Soc. Exp. Biol. Med.* **219**(2), 109–119.

Milla, M. E., Leesnitzer, M. A., Moss, M. L., Clay, W. C., Carter, H. L., Miller, A. B., Su, J. L., Lambert, M. H., Willard, D. H., Sheeley, D. M., Kost, T. A., Burkhart, W., Moyer, M., Blackburn, R. K., Pahel, G. L., Mitchell, J. L., Hoffman, C. R., and Becherer, J. D. (1999). Specific sequence elements are required for the expression of functional tumor necrosis factor-alpha-converting enzyme (TACE). *J. Biol. Chem.* **274**(43), 30563–30570.

Mohan, M., Seaton, T., Mitchell, J., Howe, A., Blackburn, K., Burkhart, W., Moyer, M., Patel, I., Becherer, J., Moss, M., and Milla, M. (2002). The tumor necrosis factor alpha converting enzyme (TACE): a unique metalloproteinase with highly defined substrate selectivity. *Biochemistry,* **41**(30), 9462–9469.

Mohler, K. M., Sleath, P. R., Fitzner, J. N., Cerretti, D. P., Alderson, M., Kerwar, S. S., Torrance, D. S., Otten-Evans, C., Greenstreet, T., Weerawarna, K., *et al.* (1994). Protection against a lethal dose of endotoxin by an inhibitor of tumour necrosis factor processing. *Nature* **370**(6486), 218–220.

Morimoto, Y., Nishikawa, K., and Ohashi, M. (1997). KB-R7785, a novel matrix metalloproteinase inhibitor, exerts its antidiabetic effect by inhibiting tumor necrosis factor-alpha production. *Life Sci.* **61**(8), 795–803.

Moss, M. L., Jin, S.-L. C., Milla, M. E., Burkhart, W., Cartner, H. L., Chen, W.-J., Clay, W. C., Didsbury, J. R., Hassler, D., Hoffman, C. R., Kost, T. A., Lambert, M. H., Lessnitzer, M. A., McCauley, P., McGeehan, G., Mitchell, J., Moyer, M., Pahel, G., Rocque, W., Overton, L. K., Schoenen, F., Seaton, T., Su, J.-L., Warner, J., Willard, D., and Becherer, J. D. (1997). Cloning of a disintegrin metalloproteinase that processes precursor tumour-necrosis factor-α. *Nature* **385,** 733–736.

Mullberg, J., Althoff, K., Jostock, T., and Rose-John, S. (2000). The importance of shedding of membrane proteins for cytokine biology. *Eur. Cytokine Netw.* **11**(1), 27–38.

Murakami, K., Kobayashi, F., Ikegawa, R., Koyama, M., Shintani, N., Yoshida, T., Nakamura, N., and Kondo, T. (1998). Metalloproteinase inhibitor prevents hepatic injury in endotoxemic mice. *Eur. J. Pharmacol.* **341**(1), 105–110.

Myles, D. G., and Primakoff, P. (1997). Why did the sperm cross the cumulus? To get to the oocyte. Functions of the sperm surface proteins PH-20 and fertilin in arriving at, and fusing with, the egg. *Biol. Reprod.* **56**(2), 320–327.

Nath, D., Slocombe, P. M., Stephens, P. E., Warn, A., Hutchinson, G. R., Yamada, K. M., Docherty, A. J., and Murphy, G. (1999). Interaction of metargidin (ADAM-15) with $\alpha v\beta 3$ and $\alpha 5\beta 1$ integrins on different haemopoietic cells. *J. Cell. Sci.* **112**(Pt 4), 579–587.

Nath, D., Slocombe, P. M., Webster, A., Stephens, P. E., Docherty, A. J., and Murphy, G. (2000). Meltrin gamma(ADAM-9) mediates cellular adhesion through alpha(6)beta(1)integrin, leading to a marked induction of fibroblast cell motility. *J. Cell. Sci.* **113**(Pt 12), 2319–2328.

Nelson, K. K., Schlondorff, J., and Blobel, C. P. (1999). Evidence for an interaction of the metalloprotease-disintegrin tumour necrosis factor alpha convertase (TACE) with mitotic arrest deficient 2 (MAD2), and of the metalloprotease-disintegrin MDC9 with a novel MAD2-related protein, MAD2beta. *Biochem. J.* **343 Pt 3,** 673–680.

Newton, R. C., Solomon, K. A., Covington, M. B., Decicco, C. P., Haley, P. J., Friedman, S. M., and Vaddi, K. (2001). Biology of TACE inhibition. *Ann. Rheum. Dis.* **60 Suppl 3,** iii25–iii32.

Nowak, M., Gaines, G. C., Rosenberg, J., Minter, R., Bahjat, F. R., Rectenwald, J., MacKay, S. L., Edwards, C. K., 3rd, and Moldawer, L. L. (2000). LPS-induced liver injury in D-galactosamine-sensitized mice requires secreted TNF-alpha and the TNF-p55 receptor. *Am. J. Physiol. Regul. Integr. Comp. Physiol.* **278**(5), R1202–R1209.

Nye, J. S. (1997). Developmental signaling: notch signals Kuz it's cleaved. *Curr. Biol.* **7**(11), R716–R720.

Pan, D., and Rubin, J. (1997). KUZBANIAN controls proteolytic processing of NOTCH and mediates lateral inhibition during Drosophila and vertebrate neurogenesis. *Cell* **90**, 271–280.

Parkin, E. T., Trew, A., Christie, G., Faller, A., Mayer, R., Turner, A. J., and Hooper, N. M. (2002). Structure–activity relationship of hydroxamate-based inhibitors on the secretases that cleave the amyloid precursor protein, angiotensin converting enzyme, CD23, and pro-tumor necrosis factor-alpha. *Biochemistry* **41**(15), 4972–4981.

Peschon, J. J., Slack, J. L., Reddy, P., Stocking, K. L., Sunnarborg, S. W., Lee, D. C., Russel, W. E., Castner, B. J., Johnson, R. S., Fitzner, J. N., Boyce, R. W., Nelson, N., Kozlosky, C. J., Wolfson, M. F., Rauch, C. T., Cerretti, D. P., Paxton, R. J., March, C. J., and Black, R. A. (1998). An essential role for ectodomain shedding in mammalian development. *Science* **282**, 1281–1284.

Poghosyan, Z., Robbins, S. M., Houslay, M. D., Webster, A., Murphy, G., and Edwards, D. R. (2002). Phosphorylation-dependent Interactions between ADAM15 cytoplasmic domain and Src family protein-tyrosine kinases. *J. Biol. Chem.* **277**(7), 4999–5007.

Primakoff, P., and Myles, D. G. (2000). The ADAM gene family: surface proteins with an adhesion and protease activity packed into a single molecule. *Trends Genet.* **16,** 83–87.

Qi, H., Rand, M. D., Wu, X., Sestan, N., Wang, W., Rakic, P., Xu, T., and Artavanis-Tsakonas, S. (1999). Processing of the notch ligand delta by the metalloprotease Kuzbanian. *Science* **283**(5398), 91–94.

Rabinowitz, M. H., Andrews, R. C., Becherer, J. D., Bickett, D. M., Bubacz, D. G., Conway, J. G., Cowan, D. J., Gaul, M., Glennon, K., Lambert, M. H., Leesnitzer, M. A., McDougald, D. L., Moss, M. L., Musso, D. L., and Rizzolio, M. C. (2001). Design of selective and soluble inhibitors of tumor necrosis factor-alpha converting enzyme (TACE). *J. Med. Chem.* **44**(24), 4252–4267.

Reddy, P., Slack, J. L., Davis, R., Cerretti, D. P., Kozlosky, C. J., Blanton, R. A., Shows, D., Peschon, J. J., and Black, R. A. (2000). Functional analysis of the domain structure of tumor necrosis factor-alpha converting enzyme. *J. Biol. Chem.* **275**(19), 14608–14614.

Rodriguez-Manzaneque, J. C., Milchanowski, A. B., Dufour, E. K., Leduc, R., and Iruela-Arispe, M. L. (2000). Characterization of METH-1/ADAMTS1 processing reveals two distinct active forms. *J. Biol. Chem.* **275**(43), 33471–33479.

Roghani, M., Becherer, J. D., Moss, M. L., Atherton, R. E., Erdjument-Bromage, H., Arribas, J., Blackburn, R. K., Weskamp, G., Tempst, P., and Blobel, C. P. (1999). Metalloprotease-disintegrin MDC9: intracellular maturation and catalytic activity. *J. Biol. Chem.* **274**, 3531–3540.

Rooke, J., Pan, D., Xu, T., and Rubin, G. M. (1996). KUZ, a conserved metalloprotease-disintegrin protein with two roles in Drosophila neurogenesis. *Science* **273**, 1227–1230.

Rosendahl, M. S., Ko, S. C., Long, D. L., Brewer, M. T., Rosenzweig, B., Hedl, E., Anderson, L., Pyle, S. M., Moreland, J., Meyers, M. A., Kohno, T., Lyons, D., and Lichenstein, H. S. (1997). Identification and characterization of a pro-tumor necrosis factor- alpha-processing enzyme from the ADAM family of zinc metalloproteases. *J. Biol. Chem.* **272**(39), 24588–24593.

Ruuls, S. R., Hoek, R. M., Ngo, V. N., McNeil, T., Lucian, L. A., Janatpour, M. J., Korner, H., Scheerens, H., Hessel, E. M., Cyster, J. G., McEvoy, L. M., and Sedgwick, J. D. (2001). Membrane-bound TNF supports secondary lymphoid organ structure but is subservient to secreted TNF in driving autoimmune inflammation. *Immunity* **15**(4), 533–543.

Schlomann, U., Rathke-Hartlieb, S., Yamamoto, S., Jockusch, H., and Bartsch, J. W. (2000). Tumor necrosis factor alpha induces a metalloprotease-disintegrin, ADAM8 (CD 156): implications for neuron-glia interactions during neurodegeneration. *J. Neurosci.* **20**(21), 7964–7971.

Schlöndorff, J., and Blobel, C. P. (1999). Metalloprotease-disintegrins: modular proteins capable of promoting cell-cell interactions and triggering signals by protein ectodomain shedding. *J. Cell. Sci.* **112**(Pt 21), 3603–3617.

Schlöndorff, J., Becherer, J. D., and Blobel, C. P. (2000). Intracellular maturation and localization of the tumour necrosis factor alpha convertase (TACE). *Biochem. J.* **347 Pt 1**, 131–138.

Schwettmann, L., and Tschesche, H. (2001). Cloning and expression in *Pichia pastoris* of metalloprotease domain of ADAM 9 catalytically active against fibronectin. *Protein Expr. Purif.* **21**(1), 65–70.

Solorzano, C. C., Ksontini, R., Pruitt, J. H., Hess, P. J., Edwards, P. D., Kaibara, A., Abouhamze, A., Auffenberg, T., Galardy, R. E., Vauthey, J. N., Copeland, E. M., 3rd, Edwards, C. K. 3rd, Lauwers, G. Y., Clare-Salzler, M., MacKay, S. L., Moldawer, L. L., and Lazarus, D. D. (1997). Involvement of 26-kDa cell-associated TNF-alpha in experimental hepatitis and exacerbation of liver injury with a matrix metalloproteinase inhibitor. *J. Immunol.* **158**(1), 414–419.

Stocker, W., Grams, F., Baumann, U., Reinemer, P., Gomis-Ruth, F. X., McKay, D. B., and Bode, W. (1995). The metzincins—topological and sequential relations between the astacins, adamalysins, serralysins, and matrixins (collagenases) define a superfamily of zinc-peptidases. *Protein Sci.* **4**(5), 823–840.

Stone, A. L., Kroeger, M., and Sang, Q. X. (1999). Structure-function analysis of the ADAM family of disintegrin-like and metalloproteinase-containing proteins (review). *J. Protein Chem.* **18**(4), 447–465.

Suzuki, A., Kadota, N., Hara, T., Nakagami, Y., Izumi, T., Takenawa, T., Sabe, H., and Endo, T.

(2000). Meltrin alpha cytoplasmic domain interacts with SH3 domains of Src and Grb2 and is phosphorylated by v-Src. *Oncogene* **19**(51), 5842–5850.

Tang, B. L., and Hong, W. (1999). ADAMTS: a novel family of proteases with an ADAM protease domain and thrombospondin 1 repeats. *FEBS Lett.* **445**(2–3), 223–225.

Togashi, N., Ura, N., Higashiura, K., Murakami, H., and Shimamoto, K. (2002). Effect of TNF-alpha–converting enzyme inhibitor on insulin resistance in fructose-fed rats. *Hypertension* **39**(2 Pt 2), 578–580.

Trifilieff, A., Walker, C., Keller, T., Kottirsch, G., and Neumann, U. (2002). Pharmacological profile of PKF242-484 and PKF241-446, novel dual inhibitors of TNF-alpha converting enzyme and matrix metalloproteinases, in model of airway inflammation. *Br. J. Pharm.* **135**, 1655–1664.

Turner, A. J., and Hooper, N. M. (1999). Role for ADAM-family proteinases as membrane protein secretases. *Biochem. Soc. Trans.* **27**(2), 255–259.

Uysal, K. T., Wiesbrock, S. M., Marino, M. W., and Hotamisligil, G. S. (1997). Protection from obesity-induced insulin resistance in mice lacking TNF-alpha function. *Nature* **389**(6651), 610–614.

Van Wart, H. E., and Birkedal-Hansen, H. (1990). The cysteine switch: a principle of regulation of metalloproteinase activity with potential applicability to the entire matrix metalloproteinase gene family. *Proc. Natl. Acad. Sci. USA* **87**(14), 5578–5582.

Wei, P., Zhao, Y. G., Zhuang, L., Ruben, S., and Sang, Q. X. (2001). Expression and enzymatic activity of human disintegrin and metalloproteinase ADAM19/Meltrin beta. *Biochem. Biophys. Res. Commun.* **280**(3), 744–755.

Weskamp, G., Krätzschmar, J. R., Reid, M., and Blobel, C. P. (1996). MDC9, a widely expressed cellular disintegrin containing cytoplasmic SH3 ligand domains. *J. Cell Biol.* **132**, 717–726.

Weskamp, G., Cai, H., Brodie, T., Higashyama, S., Manova, K., Ludwig, T., and Blobel, C. (2002). Mice lacking the metalloprotease-disintegrin MDC9 (ADAM9) have no evident major abnormalities during development or adult life. *Mol. Cell Biol.* **22**(5), 1537–1544.

Wolfsberg, T. G., and White, J. M. (1996). ADAMs in fertilization and development. *Dev. Biol.* **180**, 389–401.

Xu, H., Uysal, K. T., Becherer, J. D., Arner, P., and Hotamisligil, G. S. (2002). Altered tumor necrosis factor-alpha (TNF-alpha) processing in adipocytes and increased expression of transmembrane TNF-alpha in obesity. *Diabetes* **51**(6), 1876–1883.

Xue, C. B., He, X., Corbett, R. L., Roderick, J., Wasserman, Z. R., Liu, R. Q., Jaffee, B. D., Covington, M. B., Qian, M., Trzaskos, J. M., Newton, R. C., Magolda, R. L., Wexler, R. R., and Decicco, C. P. (2001). Discovery of macrocyclic hydroxamic acids containing biphenylmethyl derivatives at P1', a series of selective TNF-alpha converting enzyme inhibitors with potent cellular activity in the inhibition of TNF-alpha release. *J. Med. Chem.* **44**(21), 3351–3354.

Yagami-Hiromasa, T., Sato, T., Kurisaki, T., Kamijo, K., Nabeshima, Y., and Fujisawa-Sehara, A. (1995). A metalloprotease-disintegrin participating in myoblast fusion. *Nature* **377**, 652–656.

Zhang, X.-P., Kamata, T., Yokoyama, K., Puzon-McLaughlin, W., and Takada, J. (1998). Specific interaction of the recombinant disintegrin-like domain of MDC15 (metargidin, ADAM-15) with integrin (alpha)v(beta)3. *J. Biol. Chem.* **273**, 7345–7350.

Zheng, Y., *et al.* (2002). Evidence for regulation of the tumor necrosis factor alpha-convertase (TACE) by protein tyrosine phosphatase PTPHI. *J. Biol. Chem.* **277**(45), 42463–42470.

4

Shedding of Plasma Membrane Proteins

Joaquín Arribas and Anna Merlos-Suárez
Laboratori de Recerca Oncològica
Servei d'Oncologia Mèdica
Hospital Universitari Vall d'Hebron
Barcelona 08035, Spain

I. Introduction

It has long been known that certain cell-surface proteins can release their biologically active extracellular domains to the pericellular milieu through limited proteolysis; however, the broad recognition of this type of proteolysis as a general mechanism that regulates the function of virtually all types of cell surface proteins has been relatively recent. Initially, general terms such as "cleavage and release" (Bringman *et al.,* 1987), "processing" (Esch *et al.,* 1990), or "cleavage-secretion" (Sen *et al.,* 1991) were used by different authors. Currently, the most frequently used term to designate this type of proteolysis is "ectodomain shedding."

Shedding affects type I, type II, and even GPI-bound membrane-anchored proteins that function as receptors, ligands, cell adhesion molecules, or ectoenzymes. This heterogeneity in the proteins seems to indicate that many different mechanisms are involved in ectodomain shedding. However, as the shedding of different proteins was analyzed, common features were found. The shedding of many proteins could be enhanced by phorbol esters and blocked by compounds derived from hydroxamic acid. Phorbol esters are well characterized activators of protein kinase C (PKC), while hydroxamate-based compounds are known to inhibit Zn-dependent metalloproteases of the metzincin superfamily, indicating that metzincins

Current Topics in Developmental Biology, Vol. 54

regulated by PKC were responsible for the shedding of many proteins. In this chapter, we will describe the shedding characteristics of many different proteins, trying to emphasize recent findings that support the view that, despite the variety of proteins shed, there are common components necessary for protein ectodomain shedding.

II. Functional Consequences of Ectodomain Shedding

Obviously, the functional consequences of shedding are as heterogeneous as the proteins that can be shed. Therefore, in order to keep the length of this section within reasonable limits, we will focus on representative examples of each functional type of transmembrane protein.

A. EGF-Like Transmembrane Growth Factors

The epidermal growth factor receptor (EGFR) family of tyrosine kinases, also known as HER, or ErbB, has been the prototype to study receptor tyrosine kinases. In fact, EGFR was the first signaling protein and protooncogene product to be characterized in depth (Ullrich et al., 1984). EGFR knockout mice show numerous developmental abnormalities in the gastric tract, lung, skin, and brain (Miettinen et al., 1995; Threadgill et al., 1995). The ligands of the EGFR family form a large group of growth factors that contain at least one module, known as the EGF domain, consisting in six conserved cysteine residues characteristically distributed to form three intramolecular disulfide bonds (Massagué and Pandiella, 1993).

EGF-like ligands were initially characterized as soluble proteins (Todaro et al., 1980); however, they are synthesized as type I transmembrane glycoproteins that give rise to the diffusible growth factor through ectodomain shedding (Massagué and Pandiella, 1993) (Fig. 1). Experiments performed by coculturing cells that overexpress EGFR with cells transfected with a nonsheddable form of pro-transforming growth factor-α (proTGFα), a prototypical EGF-like ligand, indicated that the transmembrane form of this growth factor are not mere precursors, since they are capable of activating EGFR (Brachmann et al., 1989; Wong et al., 1989). In support of this type of EGFR activation, known as juxtacrine stimulation (Fig. 1), it has been shown that overexpression of the transmembrane forms of the EGF-like ligands, pro-heparin binding-EGF-like growth factor (proHB-EGF) and proamphiregulin, in the presence of the transmembrane protein CD9, also induced the stimulation of EGFR (Higashiyama et al., 1991; Inui et al., 1997). Also, it has been found that CD9 strengthens the activation of EGFR by proTGFα in juxtacrine and autocrine assays by inhibiting the shedding of the growth factor and, thus, increasing the expression of its transmembrane forms (Shi et al., 2000).

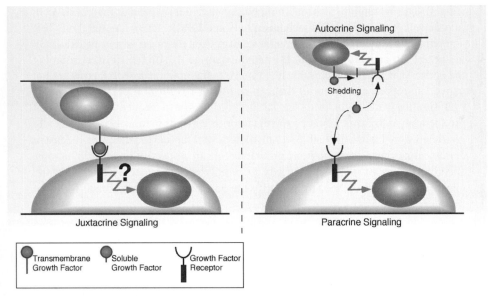

Figure 1 Mechanisms of action of EGF-like transmembrane growth factors. The ability of transmembrane growth factors of the epidermal growth factor (EGF) family to activate EGF receptors in adjacent cells (juxtacrine signaling) remains controversial (see text for details). The soluble form of the growth factor produced by ectodomain shedding is also able to bind and stimulate receptors expressed either in the same cell that produces the ligand (autocrine signaling) or in cells located at a distance from the signal-producing cell (paracrine signaling).

Finally, certain malignant cells show a defective cleavage of proTGFα that has been linked to enhanced activation of EGFR (Yang *et al.,* 2000).

Contrary to these findings, recent results cast doubts on the physiological and pathological role of transmembrane EGF-like growth factors. Inhibition of proTGFα shedding with hydroxamate compounds in human EGF-responsive mammary epithelial cells that express EGFR reduces growth and migration, as well as autocrine activation of EGFR (Dong *et al.,* 1999). These effects can be prevented by the addition of soluble EGF (Dong *et al.,* 1999), indicating that shedding is required for proTGFα-mediated signaling. In addition, the knockout of the gene that codes for the protease responsible for the shedding of proTGFα results in a phenotype very similar to that of proTGF-α knockout mice (Peschon *et al.,* 1998) indicating that transmembrane proTGFα cannot fulfill the same functions of its soluble counterpart.

Reports have indicated that cells use the shedding of transmembrane growth factors as a means to integrate physiological signals that stimulate diverse signaling pathways. Stimulation of G-protein-coupled receptors (GPCRs) causes the activation of a metalloprotease inhibitable by hydroxamates that cleaves proHB-EGF producing soluble HB-EGF that, in turn, stimulates EGFR (Prenzel *et al.,* 1999)

(Fig. 3). This result indicates that the shedding machinery that acts on proHB-EGF plays a role in the crosstalk mechanism between GPCR activation and the EGFR signaling pathway. In the experimental settings used for these assays, proHB-EGF is not able to activate EGFR to the same extent as its soluble form, reinforcing the notion that, at least in certain situations, transmembrane growth factors are not active, but represent a reservoir that can be used for very short periods of time.

In summary, although during the past decade it has been frequently proposed that the shedding of transmembrane growth factors is a means to transform a physically restricted signal into a diffusible signal, more recent reports indicate that, at least in certain situations, transmembrane growth factors are not active. In these situations, shedding could be considered a way to rapidly produce active growth factors. Clearly, more work is needed to clarify the role of the transmembrane forms of these growth factors.

B. Receptors

The immediate outcome of shedding the ectodomain receptors on hormones, cytokines, and growth factors is the release of soluble ligand-binding proteins and the production of a cell-associated fragment that usually contains domains capable of transducing intracellular signals. The soluble ectodomain of certain receptors can also arise by expression of an alternatively spliced form that lacks the transmembrane/cytoplasmic domains. Whereas the cell where shedding takes place remains desensitized to the ligand until newly synthesized receptors are delivered to the cell surface, the soluble ectodomain can modulate the ability of the ligand to activate receptors in other cells, acting as an antagonist or as an agonist by preventing or favoring the formation of signaling-competent complexes, respectively. On the other hand, it is conceivable that the cell-associated domain left behind after shedding can activate signaling pathways and, although this hypothesis has not been proven, several lines of evidence indicate that it is highly feasible (Diaz-Rodriguez *et al.,* 1999; Esparis-Ogando A and Pandiella, 1999).

1. Agonistic Action of Soluble Receptors: IL-6Rα (gp80)

IL-6 is the major growth and survival factor of several cell types. IL-6 transduces its biological activity via a receptor complex consisting of homodimers of gp130, the signal-transducing subunit, and IL-6Rα, which specifically binds and presents IL-6. The soluble form of IL-6Rα is an agonist that enhances the activity of IL-6 (Peters *et al.,* 1997), since the complex soluble IL-6R α/IL-6 recruits and activates homodimers of gp130 (Fig. 2). Thus, cells that do not express specific receptors for IL-6 become responsive to this cytokine in the presence of soluble receptors, a process that has been termed transsignaling (Peters *et al.,* 1998). Other cytokines related to IL-6, such as IL-II or CNTF (ciliary neurotrophic factor) also signal

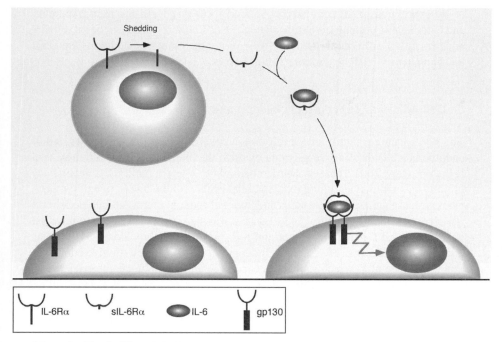

Figure 2 The shedding of IL-6Rα potentiates IL-6 signaling. The shedding of IL-6Rα releases a soluble form of the receptor (sIL-6R) which binds to IL-6R, forming a complex with the ability to recruit and activate homodimers of gp130, the signaling receptor.

through gp130 and form biologically active complexes with their soluble receptors (Peters *et al.,* 1998). Thus, for this type of receptor, shedding is a way to produce a potent agonist of the ligand that can present it even to cells that are not ordinarily responsive to such ligand. The levels of soluble IL-6R α increase under a variety of pathologies (Honda *et al.,* 1992; Thabard *et al.,* 1999) making the shedding of this receptor a promising target for therapies against different diseases.

2. Antagonistic Action of Soluble Receptors
Growth Hormone Receptor (GHR)

The circulating levels of the ectodomain of the GH Receptor (GHR), originally designated GH-Binding Protein (GHBP), caught the attention of many clinicians because of its potential value to predict GHR expression and GH responsiveness (Amit *et al.,* 2000). A substantial fraction of GHBP complexes with circulating GH and antagonizes its action by competing with the GHR. GHBP is also considered a reservoir that compensates for the oscillations of circulating GH, thus prolonging its half-life (Amit *et al.,* 2000). GHBP can be originated either by GHR shedding or by expression of an alternatively spliced form that lacks the transmembrane and cytoplasmic domains (Baumbach *et al.,* 1989). Interestingly, in the rat, GHBP arises mainly by alternative splicing, whereas in the human, shedding is the main mechanism of GHBP generation (Baumbach *et al.,* 1989).

C. Cell Adhesion Molecules: L-Selectin and Syndecans

The selectins are a family of adhesion molecules expressed in leukocytes and in endothelial cells of blood vessels that mediate the interactions between these two cell types (Ley, 2001). One of the members of the selectin family, L-selectin, was initially described as a homing receptor, responsible for the interaction of lymphocytes with high endothelial venules and their subsequent migration into peripheral lymph nodes (Gallatin *et al.,* 1983). Later, it was shown that L-selectin expression is also required for the attachment and rolling of neutrophils along the vascular endothelium (Mulligan *et al.,* 1994; Tedder *et al.,* 1995). The definitive confirmation of the role of L-selectin in these processes came with the characterization of L-selectin knockout mice, which present significant defects in leukocyte rolling and migration (Arbones *et al.,* 1994).

Shortly after hydroxamates were found to inhibit the shedding of L-selectin (Arribas *et al.,* 1996; Feehan *et al.,* 1996), the effect of these compounds on the function of the protein was tested. The rolling velocity of neutrophils on L-selectin was found to be reduced by inhibitors of shedding in both *in vitro* and *in vivo* assays (Hafezi-Moghadam and Ley, 1999; Walcheck *et al.,* 1996), indicating that L-selectin cleavage from the cell surface regulates leukocyte rolling velocity. On the other hand, L-selectin shedding seems to be also required for lymphocyte transendothelial migration; Faveeuw *et al.* (2001) have reported that L-selectin expression decreases lymphocyte migration to the lymph nodes and that inhibition of shedding correlates with arrest of lymphocytes in the endothelial cell lining of high endothelial venules. Therefore, evidence at hand support a critical role of shedding in allowing the correct function of L-selectin as a leukocyte homing receptor, since attachment to endothelial cells, rolling and transendothelial migration, the fundamental steps in the homing process, are altered in the presence of ectodomain cleavage inhibitors (Faveeuw *et al.,* 2001).

Syndecans, another type of cell adhesion molecule, seem to regulate cell–extracellular matrix interactions rather than cell–cell contact. The heparan sulfate chains of syndecans bind to cell adhesion molecules, growth factors, cytokines, proteases, and protease inhibitors (Bernfield *et al.,* 1999). The ectodomains of syndecans accumulate in bodily fluids following injury and inflammation (Subramanian *et al.,* 1997). Several functions have been attributed to soluble syndecan ectodomains including inhibition of fibroblast growth factor-2 (FGF-2)-dependent mitogenesis. Importantly, partial degradation of soluble syndecans by platelet heparanase in wound fluids produces fragments that, in contrast to the full-length soluble syndecan, markedly activate FGF-2 mitogenicity (Kato *et al.*, 1998). Soluble syndecans also reduce the affinity of cathepsin G and elastase for their physiological inhibitors, α1-antichymotrypsin and squamous cell carcinoma antigen 2, and α1-proteinase inhibitor, respectively (Kainulainen *et al.*, 1998).

D. Ectoenzymes: Angiotensin-Converting Enzyme (ACE)

Angiotensin converting enzyme (ACE) is a Zn-dependent metalloprotease that plays a key role in blood pressure homeostasis and male fertility (Krege *et al.*, 1995). ACE generates angiotensin II, a potent vasopressor peptide, by cleavage of angiotensin I and inactivates bradykinin, a vasodilator peptide (Corvol *et al.*, 1995). Inhibitors of ACE are ordinarily used to treat hypertension (Hooper, 1991). Two ACE isoforms, testis and somatic ACE, are transcribed from a single gene by two alternative promoters (Hubert *et al.*, 1991). Somatic and testis ACE consist of two and one catalytic domains, respectively (Howard *et al.*, 1990; Soubrier *et al.*, 1988). The soluble form of ACE is generated by shedding and is readily detected in blood and in cerebrospinal and seminal fluid; however, the function of soluble ACE is unknown, because most known physiological roles are performed by the transmembrane form (Esther *et al.*, 1997). It has been suggested that transmembrane and soluble ACE have different physiological functions; however, further studies of ACE need to be performed to substantiate this hypothesis.

III. Proteases Involved in Ectodomain Shedding

In 1994, different groups showed that compounds derived from hydroxamic acid block the shedding of proTNFα (Mohler *et al.*, 1994). Since then, these types of inhibitors have been shown to also block the vast majority of shedding events examined. *In vitro* studies showed that hydroxamate-derived compounds blocked the activity of matrix metalloproteases (Gearing *et al.*, 1994), indicating that this or a similar type of proteolytic activity was responsible for the ectodomain shedding. In 1997, a metalloprotease, inhibitable by hydroxamate, which cleaved proTNFα *in vitro* at the same site used to shed the ectodomain of proTNFα *in vivo* was biochemically purified and subsequently cloned simultaneously by two groups (Black *et al.*, 1997; Moss *et al.*, 1997). TACE (proTNFα converting enzyme) was convincingly shown to be responsible for the shedding of proTNFα *in vivo* because cell lines genetically deficient in TACE activity have a dramatically reduced ability to shed proTNFα (Black *et al.*, 1997). Subsequently, TACE knockout fibroblasts have been used to show that TACE is necessary for the shedding of an unexpectedly large number of proteins unrelated to proTNFα including: proTGFα, L-selectin, p75 TNFα receptor (Peschon *et al.*, 1998), βAPP (Buxbaum *et al.*, 1998; Merlos-Suárez *et al.*, 1998), the EGF-like receptor HER-4 (Rio *et al.*, 2000), IL1 receptor II, p55 TNFα receptor (Reddy *et al.*, 2000), the Notch receptor (Brou *et al.*, 2000), IL-6 receptor (Althoff *et al.*, 2000), macrophage colony-stimulating factor receptor (M-CSFR) (Rovida *et al.*, 2001), proNRGα 2c (Montero *et al.*, 2000), proHB-EGF (Merlos-Suárez *et al.*, 2001), GHR (Zhang *et al.*, 2001), and Fractalkine (Garton *et al.*, 2001). The general lack of ectodomain shedding in cells that do not express functional TACE, but do express other metalloprotease-disintegrins, such as ADAM10 (also known as kuzbanian) (Reddy *et al.*, 2000), shows a surprising lack of overlapping specificities between metalloprotease-disintegrins. In contrast, few examples of proteins correctly processed in TACE knockout fibroblasts, such as ACE (Sadhukhan *et al.*, 1999), TRANCE (Schlöndorff *et al.*, 2001), or Met (Nath *et al.*, 2001), have been reported to date. (See Chapter 3 by Becherer and Blobel for details of the ADAM family of metalloproteinases.)

In addition to TACE, other ADAMs have been suggested to shed certain cell surface proteins. In ADAM10 knockout *Drosophila* embryos the processing of Delta, one of the ligands of the Notch receptor, is severely impaired, indicating that ADAM10 is responsible for the shedding of Delta (Qi *et al.*, 1999). In mammalians, overexpression of wild-type and a mutant form of ADAM10 enhances and reduces, respectively, the shedding of βAPP (Lammich *et al.*, 1999) indicating that, in addition to TACE, ADAM10 may also play a role in the shedding of βAPP. Several metalloproteases have also been implicated in the shedding of proHB-EGF; overexpression of wild-type or dominant-negative forms of ADAM9 and 12 leads to enhanced or reduced shedding of *pro*HB-EGF, respectively, indicating that ADAM9 and 12 may be responsible for the shedding of this protein (Asakura *et al.*, 2002; Izumi *et al.*, 1998). However, more recent results show that fibroblasts from ADAM9 knockout mice do not show apparent defects in the shedding of proHB-EGF, thereby demonstrating that, at least in this cell type, ADAM9 does not play a relevant role in the shedding of the growth factor (Weskamp *et al.*, 2002). Therefore, although studies based on the overexpression of certain metalloproteases and their dominant-negative forms support that the shedding of proHB-EGF can be carried out by more than one protease, genetic evidence casts doubt on this notion and, thus, further studies are needed to confirm this point.

Additional results indicate that other proteases could be involved in the shedding of certain substrates of TACE. For example, although convincing results indicate that TACE is responsible for the activated shedding of L-selectin (Peschon *et al.*, 1998), proline substitution mutations around the cleavage site of L-selectin completely block the shedding induced by phorbol esters without affecting the basal shedding observed in unstimulated cells (Zhao *et al.*, 2001). Hence, it has been proposed that a protease different from TACE is responsible for the basal shedding of L-selectin (Zhao *et al.*, 2001). Similarly, it has been shown that while TACE is responsible for the activated shedding of IL6R, one or more alternative proteases may be involved in its basal cleavage, since it is not completely blocked in TACE knockout cells (Althoff *et al.*, 2000). Also, our group has reported the existence of a metalloprotease independent from TACE with the ability to shed the ectodomain of proTGFα in mouse fibroblasts (Merlos-Suárez *et al.*, 2001). The alternative proteases that shed these substrates and their contribution to ectodomain shedding *in vivo* remain to be determined.

Several reports have suggested that Zn-dependent metalloproteases of the matrix metalloprotease (MMP) family may also be involved in ectodomain shedding. It has been shown that MMP-3, but not MMP-2 or MMP-9, can release active, soluble HB-EGF *in vitro* (Suzuki *et al.*, 1997). Based on overexpression experiments, it has been suggested that membrane type 1 matrix metalloprotease (MT1-MMP; see Chapter 1 by Zucker, Pei, Cao, and Lopez-Otin for details of the MT-MMP family) may be responsible for the shedding of TRANCE (Schlöndorff *et al.*, 2001). The involvement of MT-MMPs in the shedding of TRANCE is further supported by the use of specific inhibitors: one of the two metalloprotease activities that seem to shed TRANCE is sensitive to TIMP-2 but not to TIMP-1 (Schlöndorff *et al.*, 2001). These details are consistent with the properties of MT1-MMP rather than TACE or other metalloprotease-disintegrins putatively involved in ectodomain shedding. Overexpression of MT1- and MT3-MMP (but not MT2-, MT4-, or MT5-MMP) promotes the shedding of CD44 (Kajita *et al.*, 2001). As is the case with TRANCE, the shedding of CD44 is sensitive to TIMP-2 but not TIMP-1 in PaCa-2 cells, further supporting the role of MMPs in this shedding event. However, different inhibitors block the shedding of CD44 in different cell lines. In PANC-1 cells and in U251MG glioma cells, TIMP-1 inhibits the cleavage while TIMP-2 has no effect, indicating that different metalloproteinases can act on CD44 in different cell lines (Okamoto *et al.*, 1999). Also, *in vitro* assays indicate that MMP-9 can specifically shed the ectodomain of IL-2Rα (Sheu *et al.*, 2001). In addition, the shedding of this receptor has been shown to be sensitive to TIMP-2 in a coculture model of these cell and tumor cells (Sheu *et al.*, 2001).

Since overexpression experiments are frequently used to assign one or more metalloproteases to a particular shedding event, it should be noted that this technique may lead to a nonphysiological situation where certain proteins, normally shed by a given metalloprotease, are shed by a different overexpressed metalloproteinase. The same rationale can be applied to experiments that overexpress a given dominant negative construct; it may sequester a common factor necessary for the activity of several metalloproteases, including the physiologically relevant one. This kind of criticism has been applied to TACE knockout cell lines. It has been shown that these cells express a form of TACE that could act potentially as a dominant negative (Montero *et al.*, 2000). However, because these cells express physiological levels of the inactive form of TACE, in order for it to act as a dominant negative, it has to be assumed that this inactive form of TACE is extremely potent and has the ability to sequester a factor needed for the activity of other metalloproteases at physiological levels. The lack of phenotype of heterozygous mice seems to indicate that this is not a likely scenario and that the defect in multiple shedding events reflects the fact that TACE acts on many different substrates. Future analysis using knockout cells for other metalloprotease-disintegrins, especially those already shown to potentially participate in ectodomain shedding, such as ADAM10 and ADAM9, will clarify the possible role of these proteases in ectodomain shedding.

IV. How Are Substrates Selected?

The activation of the shedding machinery by phorbol esters causes the shedding of hundreds of proteins with disparate structures and cleavage sites; however, the majority of cell-surface proteins are not shed (Arribas *et al.,* 1996; Arribas and Massagué, 1995). Thus, a recurrent question has been how are "sheddable" proteins selected. To answer this question, several authors have used two complementary approaches. The first evaluates the importance of different residues at the cleavage site of a given protein by analyzing the effect of deletions and point mutations. The second uses domain swapping experiments aimed to identify regions of the shedding substrates that confer "sheddability" on nonsheddable proteins. The majority of these studies were performed before the identity of the involved proteases was known. Now that TACE has been shown to be responsible for many of the proteins shed, it is possible to reevaluate old data as a result of the recently resolved structure of the catalytic domain of TACE.

For the substrates of TACE examined, the sequence surrounding the cleavage site seems to be of minor importance. Most mutations affecting the cleavage site and neighboring residues of proTGFα (Wong *et al.,* 1989), proTNFα (Tang *et al.,* 1996), βAPP (Sisodia, 1992), L-selectin (Migaki *et al.,* 1995), p55 TNFR (Brakebusch *et al.,* 1994), or IL-6R (Müllberg *et al.,* 1995) have no effect on the shedding of these molecules. The only mutations that seem to affect shedding are those which disrupt the secondary structure of the juxtamembrane region such as proline or glycine residues introduced in L-selectin (Zhao *et al.,* 2001), proTGFα (Brachmann *et al.,* 1989; Wong *et al.,* 1989), βAPP (Sisodia, 1992), p55 TNFR (Brakebusch *et al.,* 1994), or p75TNFR (Herman and Chernajovsky, 1998). In contrast, with the modest effect of point mutations, small deletions of the extracellular juxtamembrane domain prevent the shedding of all TACE substrates tested to date, indicating that the length of the juxtamembrane region that links the transmembrane domain with the cleavage site is critical. In support of this conclusion, the 10–20 amino acids of the juxtamembrane domain of several substrates of TACE, such as proTGFα, βAPP (Arribas *et al.,* 1997), and IL-6R (Althoff *et al.,* 2001), introduced into reporter molecules confer phorbolester-dependent cleavability, indicating that the juxtamembrane region of certain proteins is necessary and sufficient for the recognition by TACE. Furthermore, the importance of the juxtamembrane region was shown by naturally occurring variants of HER-4 that differ precisely in such regions (Elenius *et al.,* 1997). In the JMa and JMb isoforms, the membrane-proximal cysteine-rich region of HER-4 is linked to the transmembrane domain by different stalk regions of 23 and 13 amino acids, respectively (Elenius *et al.,* 1997). The only functional difference found between these isoforms is that while JMa is readily shed from the cell surface upon phorbol ester treatment, JMb is not (Elenius *et al.,* 1997). Since the JMa isoform is shed by TACE (Rio *et al.,* 2000), one can conclude that "naturally occurring chimeras" of TACE substrates behave just like the artificial ones.

In vitro, the catalytic domain of TACE shows preference for certain substrates; the cleavage of a peptide mimicking the proTNFα cleavage site is 9 and 2250 times more efficient than that of peptides corresponding to the cleavage sites of proTGFα and L-selectin, respectively (Peschon *et al.,* 1998). The preferential cleavage at Ala-76/Val-77 of proTNFα is partially explained by the 3D structure of TACE. The side chains of Val-77, Arg-78, and Ser-79 would appear to interact with specific pockets present in the catalytic domain of TACE. However, in cells, TACE cleaves its substrates in a similar time course ($t_{1/2} \sim 5$–10 min) and to a similar extent (cells become virtually devoid of ectodomains; Arribas *et al.,* 1996; Merlos-Suárez *et al.,* 2001), indicating that additional features facilitate the shedding mediated by TACE *in vivo.* Docking between the upper surface of the catalytic domain of TACE and the lower surface of the proTNFα complex has been suggested (Maskos *et al.,* 1998). This interaction would lock the Ala-76/Val-77 cleavage site into the active site of TACE (Maskos *et al.,* 1998). It has been shown that direct binding between TACE and proTNFα occurs (Itai *et al.,* 2001). A direct interaction between the catalytic domain of TACE and its substrates is also supported by studies showing that the membrane-tethered catalytic domain of TACE is sufficient to reconstitute the phorbol ester-activated shedding of proTNFα and p75TNFR in TACE−/− cells (Reddy *et al.,* 2000). Other domains of TACE also seem to play a role in the selection of certain substrates. For example, the cysteine-rich domain is necessary to reconstitute the PMA-activated shedding of IL-1R-II in TACE−/− cells (Reddy *et al.,* 2000). Also, a catalytic domain-lacking TACE construct interacts with proTNFα, indicating that substrate binding sites outside the catalytic domain may exist (Itai *et al.,* 2001). Since the EGF domain has been found to be necessary for the activated shedding of L-selectin, it has been suggested that EGF domains could also be docking sites for TACE (Zhao *et al.,* 2001). Several substrates of TACE such as proTGFα, proHB-EGF, or proamphiregulin also contain EGF domains; thus the possible interaction of TACE with such domains should be evaluated. Collectively, the data available indicate that although the catalytic domain of TACE shows a certain degree of specificity against peptides *in vitro, in vivo* TACE is able to cleave a great variety of sequences, providing that they are located distal to a stalk of 10–20 amino acids from the plasma membrane. This juxtamembrane region, which probably forms an alpha helix, is likely devoid of tertiary structure. The outer surface of the catalytic domain of TACE may interact with certain regions such as the TNFα domain of proTNFα or the EGF domain of L-selectin and other proteins facilitating the interaction of TACE and its substrates.

The protease(s) responsible for the shedding of ACE has not been identified yet, but studies have shown that it is different from TACE (Sadhukhan *et al.,* 1999). Several studies have investigated the structural requirements for the shedding of ACE.

Initial reports showed that, in contrast with substrates of TACE, the juxtamembrane domain of ACE does not seem to confer sheddability by itself because when introduced into a reporter molecule (CD4), even in the context of the transmembrane and cytoplasmic domains of ACE, it has no effect (Sadhukhan *et al.*, 1998). In contrast, the rest of the extracellular domain confers cleavability on a CD4 portion that conserves its own juxtamembrane domain, uncleavable in the naive reporter (Sadhukhan *et al.*, 1998). More recent studies contradict this view, because similar experiments using a different reporter, membrane dipeptidase, indicate that the juxtamembrane region of ACE is required to confer cleavability on the reporter (Pang *et al.*, 2001). It should be noted that these studies were performed using different cell lines that may express different metalloproteases with diverse requirements to select substrates. In addition, different reporters may yield chimeras with different overall structures that may expose or hide the cleavage site. Therefore, when interpreting these types of experiments the structure of the reporter molecule should be considered.

One report also emphasizes the importance of the juxtamembrane domain of ACE. Sequencing of ACE cDNAs from three apparently healthy unrelated individuals with very high plasma ACE levels revealed a single mutation in the juxtamembrane region coding sequence of ACE: Pro-1199 to Leu (Eyries *et al.*, 2001). *In vitro* analysis showed that the shedding of mutant ACE was elevated as compared to wild-type ACE. The shedding of mutant ACE occurs at the same cleavage site as in wild-type, is stimulated by phorbol esters, and is inhibited by the same concentrations of hydroxamate-based inhibitors, indicating that a very similar, if not identical, metalloprotease cleaves both versions of the enzyme (Eyries *et al.*, 2001). These results suggest that local conformational change caused by the Pro-1199 to Leu mutation leads to more accessibility at the stalk region of ACE.

To date, the studies performed indicate that TACE, and probably other metalloproteases involved in ectodomain shedding, are nonspecific proteases that cleave virtually any peptide bond, provided that it is located at a given distance from the plasma membrane and is exposed in juxtamembrane sequences devoid of tertiary structure. Nonconservative mutations that cause dramatic changes to the secondary structure of the juxtamembrane stalk may either prevent or enhance shedding, but the distance from the plasma membrane to the cleavage site remains constant. In certain instances, interactions between the exposed surface of the catalytic domain and/or additional extracellular domains of TACE and the ectodomain of the substrates may facilitate the interaction.

V. Regulation

Phorbol esters are the most widely used compounds to activate ectodomain shedding (Fig. 3; see color insert); however, the PKC isoforms that regulate ectodomain shedding have only been identified in a few cases. It has been shown that phorbol ester-induced shedding of proHB-EGF is dependent on PKCδ in Vero cells

(Izumi *et al.*, 1998). In myeloma cells PKCδ and γ control the shedding of the IL6Rα (Thabard *et al.*, 2001). PKCα is required for PMA-activated cleavage of the receptor-like protein tyrosine phosphatase LAR in 293 cells (Aicher *et al.*, 1997).

Several physiological compounds, such as chemotactic peptides, cytokines, and growth factors, have also been described as activators of ectodomain shedding (Jones *et al.*, 1999; Subramanian *et al.*, 1997; Yabkowitz *et al.*, 1999) (Fig. 3). Current efforts are focused on understanding the mechanism of activation mediated by these physiological stimuli. The activation of tyrosine kinase receptors augments the shedding of several substrates of TACE such as proTGFα (Baselga *et al.*, 1996), proTNF-α and L-selectin (Fan and Derynck, 1999). This pathway of activation involves the Erk subfamily of MAP kinases as shown by experiments with specific inhibitors and dominant negative mutants of certain MAP kinases (Fan and Derynck, 1999). Erk MAP kinases have also been involved in the activation of the shedding of proHB-EGF (Gechtman *et al.*, 1999), syndecans (Fitzgerald *et al.*, 2000), the hepatocyte growth factor receptor (Met) (Nath *et al.*, 2001), and the GHR (Guan *et al.*, 2001). Kinetic and biochemical evidences indicate that the activation of ectodomain shedding does not require synthesis of proteins *de novo*, suggesting that the metalloproteases involved could be regulated by phosphorylation/dephosphorylation (Fan and Derynck, 1999; Gechtman *et al.*, 1999). Previous reports showed that activation of different PKC isoforms by phorbol esters stimulated MAPK leading to cell proliferation (Schönwasser *et al.*, 1998). More recent experiments with inhibitors indicate that Erk kinases mediate the effects of phorbol esters on the shedding of βAPP (Desdouits-Magnen *et al.*, 1998), proHB-EGF (Gechtman *et al.*, 1999), and proTGFα (Fan and Derynck, 1999) (Fig. 3). PKC does not seem to be required, however, for the stimulation of ectodomain shedding by growth factors, because PKC inhibitors do not alter their effect (Fan and Derynck, 1999).

Agonists of G protein coupled receptors are also activators of ectodomain shedding (Fig. 3). The shedding of proHB-EGF induced by the activation of GPCRs is the cellular mechanism used to establish the crosstalk between GPCRs and receptor tyrosine kinases in certain instances (Prenzel *et al.*, 1999). Further studies have recently shown that activators of GPCRs can promote the shedding of Met by first inducing the cleavage of proHB-EGF which activates the EGFR that, in turn, induces the shedding of Met via the Erk MAP kinase pathway (Nath *et al.*, 2001).

In most cases tested to date, a basal level of shedding is present in the absence of any exogenous stimulus. In the case of proTGFα, proTNFα, and L-selectin, basal shedding seems to be mediated by the p38 MAP kinase (Fan and Derynck, 1999) (Fig. 3). Since the p38 kinase pathway can be stimulated by stress and inflammatory signals, Fan and Derynck (1999) hypothesized that basal shedding may reflect the "artifactual" conditions of cell culture rather than physiological conditions.

The shedding of syndecan-1 and -4 seems to be subjected to a complex regulation. It can be accelerated by EGF or thrombin receptors that activate the ERK

MAP kinase pathway in a process that does not seem to involve PKC. In contrast, the shedding of syndecans induced by cellular stress requires PKC activation that, in turn, activates the JNK/SAPK MAP kinase pathway. Inhibition of p38 MAP kinase activity has minimal effect on the shedding activated by any of these agonists (Fitzgerald *et al.*, 2000). Park *et al.* (2000) showed that Syndecan-1 shedding is enhanced by LasA, a Pseudomonas aeruginosa virulence factor. Importantly, it has been shown that *in vivo,* LasA enhances the virulence of gram-negative bacteria in newborn mice by generating soluble Syndecan-1 through the activation of the shedding of Syndecan-1 (Park *et al.*, 2001).

The removal of metalloprotease disintegrin prodomains is also proposed as a means of inducing shedding (Zhang *et al.*, 2000a,b). Reactive oxygen species are capable of inducing the TACE-mediated ectodomain cleavage of cell surface proteins such as TNFα and the p75TNF receptor. This presumably occurs through an oxidative attack on the cysteine residue in the prodomain that results in the disruption of the cysteine switch, which keeps the TACE inactive. (See Chapter 1 by Zucker, Pei, Cao, and Lopez-Otin for discussion of the cysteine switch mechanism in metalloproteinases.) Given the wide repertoire of TACE substrates, it is expected that these results will soon be tested on other proteins.

In summary, although several known components transduce signals that result in the activation of ectodomain shedding, the mechanisms which cause the activation of ectodomain shedding remain unclear. Elucidation of the mechanism responsible for the activation of TACE and other metalloproteases involved in ectodomain shedding will represent a milestone in this field.

Acknowledgments

We thank the members of the Arribas laboratory for critical reading of the manuscript. This work was supported by grants from the Fundació La Marató de TV3 and Fundació "la Caixa" (98/056-01), EMBO Young Investigator Progamme (YIP), and the Spanish Comisión Interministerial de Ciencia y Tecnología (SAF2000-0203) to J. A.

References

Aicher, B., Lerch, M., Muller, T., Schilling, J., and Ullrich, A. (1997). Cellular redistribution of protein tyrosine phosphatases LAR and PTPsigma by inducible proteolytic processing. *J. Cell Biol.* **138,** 681–696.

Althoff, K., Müllberg, J., Aasland, D., Voltz, N., Kallen, K., Grotzinger, J., and Rose-John, S. (2001). Recognition sequences and structural elements contribute to shedding susceptibility of membrane proteins. *Biochem J.* **353,** 663–672.

Althoff, K., Reddy, P., Voltz, N., Rose-John, S., and Müllberg, J. (2000). Shedding of interleukin-6 receptor and tumor necrosis factor alpha. Contribution of the stalk sequence to the cleavage pattern of transmembrane proteins. *Eur. J. Biochem.* **267,** 2624–2631.

Amit, T., Youdim, M. B., and Hochberg, Z. (2000). Clinical review 112: Does serum growth hormone (GH) binding protein reflect human GH receptor function? *J. Clin. Endocrinol. Metab.* **85,** 927–932.

Arbones, M. L., Ord, D. C., Ley, K., Ratech, H., Maynard-Curry, C., Otten, G., Capon, D. J., and Tedder, T. F. (1994). Lymphocyte homing and leukocyte rolling and migration are impaired in L-selectin-deficient mice. *Immunity* **1,** 247–260.

Arribas, J., Coodly, L., Vollmer, P., Kishimoto, T. K., Rosejohn, S., and Massagué, J. (1996). Diverse cell surface protein ectodomains are shed by a system sensitive to metalloprotease inhibitors. *J. Biol. Chem.* **271,** 11376–11382.

Arribas, J., López-Casillas, F., and Massagué, J. (1997). Role of the juxtamembrane domains of the transforming growth factor-α precursor and the β-amyloid precursor protein in regulated ectodomain shedding. *J. Biol. Chem.* **272,** 17160–17165.

Arribas, J., and Massagué, J. (1995). Transforming growth factor alpha and beta amyloid precursor protein share a secretory mechanism. *J. Cell Biol.* **128,** 433–441.

Asakura, M., Kitakaze, M., Takashima, S., Liao, Y., Ishikura, F., Yoshinaka, T., Ohmoto, H., Node, K., Yoshino, K., Ishiguro, H., Asanuma, H., Sanada, S., Matsumura, Y., Takeda, H., Beppu, S., Tada, M., Hori, M., and S. H. (2002). Cardiac hypertrophy is inhibited by antagonism of ADAM12 processing of HB-EGF: metalloproteinase inhibitors as a new therapy. *Nat. Med.* **8,** 35–40.

Baselga, J., Mendelsohn, J., Kim, Y.-M., and Pandiella, A. (1996). Autocrine regulation of membrane transforming growth factor-alpha cleavage. *J. Biol. Chem.* **271,** 3279–3284.

Baumbach, W. R., Horner, D. L., and Logan, J. S. (1989). The growth hormone-binding protein in rat serum is an alternatively spliced form of the rat growth hormone receptor. *Genes Dev.* **3,** 1199–1205.

Bernfield, M., Götte, M., Park, P. W., Reizes, O., Fitzgerald, M. L., Lincecum, J., and Zako, M. (1999). Functions of cell surface heparan sulfate proteoglycans. *Annu. Rev. Biochem.* **68,** 729–777.

Black, R. A., Rauch, C. T., Kozlosky, C. J., Peschon, J. J., Slack, J. L., Wolfson, M. F., Castner, B. J., Stocking, K. L., Reddy, P., Srjnivasan, S., Nelson, N., Bolani, N., Schooley, K. A., Gerhart, M., Davis, R., Fitzner, J. N., Johnson, R. S., Paxton, R. J., March, C. J., and Carreti, D. P. (1997). A metalloprotease disintegrin that releases tumour-necrosis factor-α from cells. *Nature* **385,** 729–733.

Brachmann, R., Lindquist, P. B., Nagashima, M., Kohr, W., Lipari, T., Napier, N., and Derynck, R. (1989). Transmembrane TGF-α precursors activate EGF/TGF-α receptors. *Cell* **56,** 691–700.

Brakebusch, C., Varfolomeev, E. E., Batkin, M., and Wallach, D. (1994). Structural requirements for inducible shedding of the p55 tumor necrosis factor receptor. *J. Biol. Chem.* **269,** 32488–32496.

Bringman, T. S., Lindquist, P. B., and Derynck, R. (1987). Different transforming growth factor-α species are derived from a glycosylated and palmitoylated transmembrane precursor. *Cell* **48,** 429–440.

Brou, C., Logeat, F., Gupta, N., Bessia, C., LeBail, O., Doedens, J. R., Cumano, A., Roux, P., Black, R. A., and Israel, A. (2000). A novel proteolytic cleavage involved in Notch signaling: the role of the disintegrin-metalloprotease TACE. *Mol. Cell* **5,** 207–216.

Buxbaum, J. D., Liu, K. N., Luo, Y., Slack, J. L., Stocking, K. L., Peschon, J. J., Johnson, R. S., Castner, B. J., Cerretti, D. P., and Black, R. A. (1998). Evidence that tumor necrosis factor alpha converting enzyme is involved in regulated alpha-secretase cleavage of the Alzheimer amyloid protein precursor. *J. Biol. Chem.* **273,** 27765–27767.

Corvol, P., Williams, T. A., and Soubrier, F. (1995). Peptidyl dipeptidase A: angiotensin I-converting enzyme. *Methods Enzymol.* **248,** 283–305.

Desdouits-Magnen, J., Desdouits, F., Takeda, S., Syu, L. J., Saltiel, A. R., Buxbaum, J. D., Czernik, A. J., Nairn, A. C., and Greengard, P. (1998). Regulation of secretion of Alzheimer amyloid precursor protein by the mitogen-activated protein kinase cascade. *J. Neurochem.* **70,** 524–530.

Diaz-Rodriguez, E., Cabrera, N., Esparis-Ogando, A., Montero, J. C., and Pandiella, A. (1999). Cleavage of the TrkA neurotrophin receptor by multiple metalloproteases generates signalling-competent truncated forms. *Eur. J. Neurosci.* **11,** 1421–1430.

Dong, J., Opresko, L. K., Dempsey, P. J., Lauffenburger, D. A., Coffey, R. J., and Wiley, H. S. (1999). Metalloprotease-mediated ligand release regulates autocrine signaling through the epidermal growth factor receptor. *Proc. Natl. Acad. Sci. USA* **96,** 6235–6240.

Elenius, K., Corfas, G., Paul, S., Choi, C. J., Rio, C., Plowman, G. D., and Klagsbrun, M. (1997). A novel juxtamembrane domain isoform of HER4/ErbB4. Isoform-specific tissue distribution and differential processing in response to phorbol ester. *J. Biol. Chem.* **272,** 26761–26768.

Esch, F. S., Keim, P. S., Beattie, E. C., Blacher, R. W., Culwell, A. R., Oltersdorf, T., McClure, D., and Ward, P. J. (1990). Cleavage of amyloid β peptide during constitutive processing of its precursor. *Science* **248,** 1122–1124.

Esparis-Ogando A, D.-R. E., and Pandiella, A. (1999). Signalling-competent truncated forms of ErbB2 in breast cancer cells: differential regulation by protein kinase C and phosphatidylinositol 3-kinase. *Biochem. J.* **344,** 339–348.

Esther, C. R., Marino, E. M., Howard, T. E., Machaud, A., Corvol, P., Capecchi, M. R., and Bernstein, K. E. (1997). The critical role of tissue angiotensin-converting enzyme as revealed by gene targeting in mice. *J. Clin. Invest.* **99,** 2375–2385.

Eyries, M., Michaud, A., Deinum, J., Agrapart, M., Chomilier, J., Kramers, C., and Soubrier, F. (2001). Increased shedding of angiotensin-converting enzyme by a mutation identified in the stalk region. *J. Biol. Chem.* **276,** 5525–5532.

Fan, H., and Derynck, R. (1999). Ectodomain shedding of TGF-alpha and other transmembrane proteins is induced by receptor tyrosine kinase activation and MAP kinase signaling cascades. *EMBO J.* **18,** 6962–6972.

Faveeuw, C., Preece, G., and Ager, A. (2001). Transendothelial migration of lymphocytes across high endothelial venules into lymph nodes is affected by metalloproteinases. *Blood* **98,** 688–695.

Feehan, C., Darlak, K., Kahn, J., Walcheck, B., Spatola, A., and Kishimoto, K. (1996). Shedding of the lymphocyte L-selectin adhesion molecule is inhibited by a hyroxamic acid-based protease inhibitor. *J. Biol. Chem.* **271,** 7019–7024.

Fitzgerald, M. L., Wang, Z., Park, P. W., Murphy, G., and Bernfield, M. (2000). Shedding of syndecan-1 and -4 ectodomains is regulated by multiple signaling pathways and mediated by a TIMP-3-sensitive metalloproteinase. *J. Cell Biol.* **148,** 811–824.

Gallatin, W. M., Weissman, I. L., and Butcher, E. C. (1983). A cell-surface molecule involved in organ-specific homing of lymphocytes. *Nature* **304,** 30–34.

Garton, K. J., Gough, P. J., P., B. C., Murphy, G., Greaves, D. R., Dempsey, P. J., and Raines, E. W. (2001). Tumor necrosis factor-alpha-converting enzyme (ADAM17) mediates the cleavage and shedding of fractalkine (CX3CL1). *J. Biol. Chem.* **276,** 37993–38001.

Gearing, A. J. H., Beckett, P., Christodoulou, M., Churchill, M., Clements, J., Davidson, A. H., Drummond, A. H., Galloway, W. A., Gilbert, R., Gordon, J. L., Leber, T. M., Mangan, M., Miller, K., Nayee, P., Owen, K., Patel, S., Thomas, W., Wells, G., Wood, L. M., and Wolley, K. (1994). Processing of tumour necrosis factor-α precursor by metalloproteinases. *Nature* **370,** 555–557.

Gechtman, Z., Alonso, J. L., Raab, G., Ingber, D. E., and Klagsbrun, M. (1999). The shedding of membrane-anchored heparin-binding epidermal-like growth factor is regulated by the Raf/mitogen-activated protein kinase cascade and by cell adhesion and spreading. *J. Biol. Chem.* **274,** 28828–28835.

Guan, R., Zhang, Y., Jiang, J., Baumann, C. A., Black, R. A., Baumann, G., and Frank, S. J. (2001). Phorbol ester- and growth factor-induced growth hormone (GH) receptor proteolysis and GH-binding protein shedding: relationship to GH receptor down-regulation. *Endocrinology* **142,** 1137–1147.

Hafezi-Moghadam, A., and Ley, K. (1999). Relevance of L-selectin shedding for leukocyte rolling in vivo. *J. Exp. Med.* **189,** 939–948.

Herman, C., and Chernajovsky, Y. (1998). Mutation of proline 211 reduces shedding of the human p75 TNF receptor. *J. Immunol.* **160,** 2478–2487.

Higashiyama, S., Abraham, J. A., Miller, J., Fiddes, J. C., and Klagsbrun, M. (1991). A heparin-binding growth factor secreted by macrophage-like cells that is related to EGF. *Science* **251,** 936–939.

Honda, M., Yamamoto, S., Cheng, M., Yasukawa, K., Suzuki, H., Saito, T., Osugi, Y., Tokunaga, T., and Kishimoto, T. (1992). Human soluble IL-6 receptor: its detection and enhanced release by HIV infection. *J. Immunol.* **148,** 2175–2180.

Hooper, N. M. (1991). Angiotensin converting enzyme: implications from molecular biology for its physiological functions. *Int. J. Biochem.* **23,** 641–647.

Howard, T. E., Shai, S. Y., Langford, K. G., Martin, B. M., and Bernstein, K. E. (1990). Transcription of testicular angiotensin-converting enzyme (ACE) is initiated within the 12th intron of the somatic ACE gene. *Mol. Cell Biol.* **10,** 4294–4302.

Hubert, C., Houot, A. M., Corvol, P., and Soubrier, F. (1991). Structure of the angiotensin I-converting enzyme gene. Two alternate promoters correspond to evolutionary steps of a duplicated gene. *J. Biol. Chem.* **266,** 15377–15383.

Inui, S., Higashiyama, S., Hashimoto, K., Higashiyama, M., Yoshikawa, K., and Taniguchi, N. (1997). Possible role of coexpression of CD9 with membrane-anchored heparin-binding EGF-like growth factor and amphiregulin in cultured human keratinocyte growth. *J. Cell Physiol.* **171,** 291–298.

Itai, T., Tanaka, M., and Nagata, S. (2001). Processing of tumor necrosis factor by the membrane-bound TNF-alpha-converting enzyme, but not its truncated soluble form. *Eur. J. Biochem.* **268,** 2074–2082.

Izumi, Y., Hirata, M., Hasuwa, H., Iwamoto, R., Umata, T., Miyado, K., Tamai, Y., Kurisaki, T., Sehara-Fujisawa, A., Ohno, S., and Mekada, E. (1998). A metalloprotease-disintegrin, MDC9/meltrin-gamma/ADAM9 and PKC delta are involved in TPA-induced ectodomain shedding of membrane-anchored heparin-binding EGF-like growth factor. *EMBO J.* **17,** 7260–7272.

Jones, S., Novick, D., Horiuchi, S., Yamamoto, N., Szalai, A., and Fuller, G. (1999). C-reactive protein: a physiological activator of interleukin 6 receptor shedding. *J. Exp. Med.* **189,** 599–604.

Kainulainen, V., Wang, H., Schick, C., and Bernfield, M. (1998). Syndecans, heparan sulfate proteoglycans, maintain the proteolytic balance of acute wound fluids. *J. Biol. Chem.* **273,** 11563–11569.

Kajita, M., Itoh, Y., Chiba, T., Mori, H., Okada, A., Kinoh, H., and Seiki, M. (2001). Membrane-type 1 matrix metalloproteinase cleaves CD44 and promotes cell migration. *J. Cell Biol.* **153,** 893–904.

Kato, M., Wang, H., Kainulainen, V., Fitzgerald, M. L., Ledbetter, S., Ornitz, D. M., and Bernfield, M. (1998). Physiological degradation converts the soluble syndecan-1 ectodomain from an inhibitor to a potent activator of FGF-2. *Nat. Med.* **4,** 691–697.

Krege, J. H., John, S. W., Langenbach, L. L., Hodgin, J. B., Hagaman, J. R., Bachman, E. S., Jennette, J. C., O'Brien, D. A., and Smithies, O. (1995). Male–female differences in fertility and blood pressure in ACE-deficient mice. *Nature* **375,** 146–148.

Lammich, S., Kojro, E., Postina, R., Gilbert, S., Pfeiffer, R., Jasionowski, M., Haass, C., and Fahrenholz, F. (1999). Constitutive and regulated alpha-secretase cleavage of Alzheimer's amyloid precursor protein by a disintegrin metalloprotease. *Proc. Natl. Acad. Sci. USA* **96,** 3922–3927.

Ley, K. (2001). Functions of selectins. *Results Probl. Cell Differ.* **33,** 177–200.

Maskos, K., Fernandez-Catalan, C., Huber, R., Bourenkov, G. P., Bartunik, H., Ellestad, G. A., Reddy, P., Wolfson, M. F., Rauch, C. T., Castner, B. J., Davis, R., Clarke, H. R., Petersen, M., Fitzner, J. N., Cerretti, D. P., March, C. J., Paxton, R. J., Black, R. A., and Bode, W. (1998). Crystal structure of the catalytic domain of human tumor necrosis factor-alpha-converting enzyme. *Proc. Natl. Acad. Sci. USA* **95,** 3408–3412.

Massagué, J., and Pandiella, A. (1993). Membrane-anchored growth factors. *Ann. Rev. Biochem.* **62,** 515–541.

Merlos-Suárez, A., Fernández-Larrea, J., Reddy, P., Baselga, J., and Arribas, J. (1998). proTNF-α processing activity is tightly controlled by a component that does not affect Notch processing. *J. Biol. Chem.* **273,** 24955–24962.

Merlos-Suárez, A., Ruiz-Paz, S., Baselga, J., and Arribas, J. (2001). Metalloprotease-dependent proTGF-a ectodomain shedding in the absence of TACE. *J. Biol. Chem.* **276**, 48510–48517.

Miettinen, P. J., Berger, J. E., Meneses, J., Phung, Y., Pedersen, R. A., Werb, Z., and Derynck, R. (1995). Epithelial immaturity and multiorgan failure in mice lacking epidermal growth factor receptor. *Nature* **376**, 337–341.

Migaki, G. I., Kahn, J., and Kishimoto, T. K. (1995). Mutational analysis of the membrane-proximal cleavage site of L-selectin: relaxed sequence specificity surrounding the cleavage site. *J. Exp. Med.* **182**, 549–557.

Mohler, K., Sleath, P. R., Fitzner, J. N., Cerretti, D. P., Alderson, M., Kerwar, S. S., Torrance, D. S., Otten-Evans, C., Greenstreet, T., Weerawarma, K., Kronhein, S. R., Petersen, M., Gerhart, M., Kozlosky, C. J., March, C. J., and Black, R. A. (1994). Protection against a lethal dose of endotoxin by an inhibitor of tumour necrosis factor processing. *Nature* **370**, 218–220.

Montero, J. C., Yuste, L., Diaz-Rodriguez, E., Esparis-Ogando, A., and Pandiella, A. (2000). Differential shedding of transmembrane neuregulin isoforms by the tumor necrosis factor-alpha-converting enzyme. *Mol. Cell. Neurosci.* **16**, 631–648.

Moss, M. L., Jin, C. S.-L., Milla, M. E., Burkhart, W., Carter, H. L., Chen, W.-J., Clay, W.-C., Didsbury, J. R., Hassler, D., Hoffman, C. R., Kost, T. A., Lambert, M. H., Laesnitzer, M. A., McCauley, P., McGeehan, G., Mitchell, J., Moyer, M., Pahel, G., Rocque, W., Overton, L. K., Schoenen, F., Seaton, T., Su, J.-L., Warner, J., Willard, D., and Bacherer, J. D. (1997). Cloning of a disintegrin metalloproteinase that processes precursor tumour-necrosis factor-α. *Nature* **385**, 733–736.

Müllberg, J., Durie, F., Otten-Evans, C., Alderson, M. R., Rose-John, S., Cosman, D., Black, R. A., and Mohler, K. M. (1995). A metalloprotease inhibitor blocks shedding of the IL-6 receptor and the p60 TNF receptor. *J. Immunol.* **155**, 5198–5205.

Mulligan, M. S., Miyasaka, M., Tamatani, T., Jones, M. L., and Ward, P. A. (1994). Requirements for L-selectin in neutrophil-mediated lung injury in rats. *J. Immunol.* **152**, 832–840.

Nath, D., Williamson, N. J., Jarvis, R., and Murphy, G. (2001). Shedding of c-Met is regulated by crosstalk between a G-protein coupled receptor and the EGF receptor and is mediated by a TIMP-3 sensitive metalloproteinase. *J. Cell Sci.* **114**, 1213–1220.

Okamoto, I., Kawano, Y., Tsuiki, H., Sasaki, J., Nakao, M., Matsumoto, M., Suga, M., Ando, M., Nakajima, M., and Saya, H. (1999). CD44 cleavage induced by a membrane-associated metalloprotease plays a critical role in tumor cell migration. *Oncogene* **18**, 1435–1446.

Pang, S., Chubb, A. J., Schwager, S. L., Ehlers, M. R., Sturrock, E. D., and Hooper, N. M. (2001). Roles of the juxtamembrane and extracellular domains of angiotensin-converting enzyme in ectodomain shedding. *Biochem. J.* **358**, 185–192.

Park, P., Pier, G., Hinkes, M., and Bernfield, M. (2001). Exploitation of syndecan-1 shedding by Pseudomonas aeruginosa enhances virulence. *Nature* **411**, 98–102.

Park, P., Pier, G., Preston, M., Goldberger, O., Fitzgerald, M., and Bernfield, M. (2000). Syndecan-1 shedding is enhanced by LasA, a secreted virulence factor of *Pseudomonas aeruginosa. J. Biol. Chem.* **275**, 3057–3064.

Peschon, J., Slack, J., Reddy, P., Stocking, K., Sunnarborg, S., Lee, D., Rusell, W., Castner, R., Johnson, R., Fitzner, J., Boyce, N., Nelson, C., Kozlosky, M., Wolfson, M., Rauch, C., Cerretti, D., Paxton, R., March, C., and Black, R. (1998). An essential role for ectodomains shedding in mammalian development. *Science* **282**, 1281–1284.

Peters, M., Muller, A. M., and Rose-John, S. (1998). Interleukin-6 and soluble interleukin-6 receptor: direct stimulation of gp130 and hematopoiesis. *Blood* **92**, 3495–3504.

Peters, M. F., Adams, M. E., and Froehner, S. C. (1997). Differential association of syntrophin pairs with the dystrophin complex. *J. Cell Biol.* **138**, 81–93.

Prenzel, N., Zwick, E., Daub, H., Leserer, M., Abraham, R., Wallasch, C., and Ullrich, A. (1999). EGF receptor transactivation by G-protein-coupled receptors requires metalloproteinase cleavage of proHB-EGF. *Nature* **402**, 23–30.

Qi, H., Rand, M. D., Wu, X., Sestan, N., Wang, W., Rakic, P., Xu, T., and Artavanis-Tsakonas, S.

(1999). Processing of the Notch ligand Delta by the metalloprotease Kuzbanian. *Science.* **283,** 91–94.

Reddy, P., Slack, J. L., Davis, R., Cerretti, D. P., Kozlosky, C. J., Blanton, R. A., Shows, D., Peschon, J. J., and Black, R. A. (2000). Functional analysis of the domain structure of tumor necrosis factor-α converting enzyme. *J. Biol. Chem.* **275,** 14608–14614.

Rio, C., Buxbaum, J. D., Peschon, J. J., and Corfas, G. (2000). Tumor necrosis factor–converting enzyme is required for cleavage of erbB4/HER4. *J. Biol. Chem.* **275,** 10379–10387.

Rovida, E., Paccagnini, A., Del Rosso, M., Peschon, J., and Dello Sbarba, P. (2001). TNF-alpha-converting enzyme cleaves the macrophage colony-stimulating factor receptor in macrophages undergoing activation. *J. Immunol.* **166,** 1583–1589.

Sadhukhan, R., Santhamma, K. R., Reddy, P., Peschon, J. J., Black, R. A., and Sen, I. (1999). Unaltered cleavage and secretion of angiotensin-converting enzyme in tumor necrosis factor–converting enzyme-deficient mice. *J. Biol. Chem.* **274,** 10511–10516.

Sadhukhan, R., Sen, G. C., Ramchandran, R., and Sen, I. (1998). The distal ectodomain of angiotensin-converting enzyme regulates its cleavage-secretion from the cell surface. *Proc. Natl. Acad. Sci. USA* **95,** 138–143.

Schlöndorff, J., Lum, L., and Blobel, C. P. (2001). Biochemical and pharmacological criteria define two shedding activities for TRANCE/OPGL that are distinct from the tumor necrosis factor alpha convertase. *J. Biol. Chem.* **276,** 14665–14674.

Schönwasser, D. C., Marais, R. M., Marshall, C. J., and Parker, P. J. (1998). Activation of the mitogen-activated protein kinase/extracellular signal-regulated kinase pathway by conventional, novel, and atypical protein kinase C isotypes. *Mol. Cell Biol.* **18,** 790–798.

Sen, I., Samanta, H., Livingston, W., and Sen, G. C. (1991). Establishment of transfected cell lines producing testicular angiotensin-converting enzyme. Structural relationship between its secreted and cellular forms. *J. Biol. Chem.* **266,** 21985–21990.

Sheu, B. C., Hsu, S. M., Ho, H. N., Lien, H. C., Huang, S. C., and Lin, R. H. (2001). A novel role of metalloproteinase in cancer-mediated immunosuppression. *Cancer Res.* **61,** 237–242.

Shi, W., Fan, H., Shum, L., and Derynck, R. (2000). The tetraspanin CD9 associates with transmembrane TGF-alpha and regulates TGF-alpha-induced EGF receptor activation and cell proliferation. *J. Cell Biol.* **148,** 591–602.

Sisodia, S. S. (1992). β-Amyloid precursor protein cleavage by a membrane-bound protease. *Proc. Natl. Acad. Sci. USA* **89,** 6075–6079.

Soubrier, F., Alhenc-Gelas, F., Hubert, C., Allegrini, J., John, M., Tregear, G., and Corvol, P. (1988). Two putative active centers in human angiotensin I-converting enzyme revealed by molecular cloning. *Proc. Natl. Acad. Sci. USA* **85,** 9386–9390.

Subramanian, S. V., Fitzgerald, M. L., and Bernfield, M. (1997). Regulated shedding of syndecan-1 and -4 ectodomains by thrombin and growth factor receptor activation. *J. Biol. Chem.* **272,** 14713–14720.

Suzuki, M., Raab, G., Moses, M., Fernandez, C., and Klagsbrun, M. (1997). Matrix metalloproteinase-3 releases active heparin-binding EGF-like growth factor by cleavage at a specific juxtamembrane site. *J. Biol. Chem.* **272,** 31730–31737.

Tang, P., Hung, M. C., and Klostergaard, J. (1996). Length of the linking domain of human pro-tumor necrosis factor determines the cleavage processing. *Biochemistry* **35,** 8226–8233.

Tedder, T. F., Steeber, D. A., and Pizcueta, P. (1995). L-selectin-deficient mice have impaired leukocyte recruitment into inflammatory sites. *J. Exp. Med.* **181,** 2259–2264.

Thabard, W., Barille, S., Collette, M., Harousseau, J. L., Rapp, M. J., Bataille, R., and Amiot, M. (1999). Myeloma cells release soluble interleukin-6Ra in relation to disease progression by two distinct mechanisms : alternative splicing and proteolytic cleavage. *Clin. Cancer Res.* **5,** 2693–2697.

Thabard, W., Collette, M., Bataille, R., and Amiot, M. (2001). Protein kinase C delta and eta isoenzymes control the shedding of the interleukin 6 receptor alpha in myeloma cells. *Biochem. J.* **358,** 193–200.

Threadgill, D. W., Dlugosz, A. A., Hansen, L. A., Tennenbaum, T., Lichti, U., Yee, D., LaMantia, C., Mourton, T., Herrup, K., Harris, R. C., Raymond, C., Barnard, J. A., Yuspa, S. H., Stuart, H., Coffey, R. J., and Magnuson, T. (1995). Targeted disruption of mouse EGF receptor: effect of genetic background on mutant phenotype. *Science* **269**, 230–234.

Todaro, G. J., Fryling, C., and Delarco, J. E. (1980). Transforming growth factors produced by certain human tumor cells: polypeptides that interact with epidermal growth factor receptors. *Proc. Natl. Acad. Sci. USA* **77**, 5258–5262.

Ullrich, A., Coussens, L., Hayflick, J. S., Dull, T. J., Gray, A., Tam, A. W., Lee, J., Yarden, Y., Libermann, T. A., and Schlessinger, J. (1984). Human epidermal growth factor receptor cDNA sequence and aberrant expression of the amplified gene in A431 epidermoid carcinoma cells. *Nature* **309**, 418–425.

Walcheck, B., Kahn, J., Fisher, J. M., Wang, B. B., Fisk, R. S., Payan, D. G., Feehan, C., Betageri, R., Darlak, K., Spatola, A. F., and Kishimoto, T. K. (1996). Neutrophil rolling altered by inhibition of L-selectin shedding in vitro. *Nature* **380**, 720–723.

Weskamp, G., Cai, H., Brodie, T. A., Higashyama, S., Manova, K., Ludwig, T., and Blobel, C. P. (2002). Mice lacking the metalloprotease-disintegrin MDC9 (ADAM9) have no evident major abnormalities during development or adult life. *Mol. Cell Biol.* **22**, 1537–1544.

Wong, S. T., Winchell, L. F., McCune, B. K., Earp, H. S., Teixidó, J., Massagué, J., Herman, B., and Lee, D. C. (1989). The TGF-α precursor expressed on the cell surface binds to the EGF receptor on adjacent cells, leading to signal transduction. *Cell* **56**, 495–506.

Yabkowitz, R., Meyer, S., Black, T., Elliott, G., Merewether, L. A., and Yamane, H. K. (1999). Inflammatory cytokines and vascular endothelial growth factor stimulate the release of soluble tie receptor from human endothelial cells via metalloprotease activation. *Blood* **93**, 1969–1979.

Yang, H., Jiang, D., Li, W., Liang, J., Gentry, L. E., and Brattain, M. G. (2000). Defective cleavage of membrane bound TGFalpha leads to enhanced activation of the EGF receptor in malignant cells. *Oncogene* **19**, 1901–1914.

Zhang, Y., Jiang, J., Black, R. A., Baumann, G., and Frank, S. J. (2001). Tumor necrosis factor-converting enzyme (TACE) is a growth hormone binding protein (GHBP) sheddase: the metalloprotease TACE/ADAM-17 is critical for (PMA-induced) GH receptor proteolysis and GHBP generation. *Endocrinology* **141**, 4342–4348.

Zhang, Z., Cork, J., Ye, P., D., L., Schwarzenberger, P., Summer, W., Shellito, J., Nelson, S., and Kolls, J. (2000a). Inhibition of TNF-alpha processing and TACE-mediated ectodomain shedding by ethanol. *J. Leukoc. Biol.* **67**, 856–862.

Zhang, Z., Kolls, J. K., Oliver, P., Good, D., Schwarzenberger, P. O., Joshi, M. S., Ponthier, J. L., and Lancaster, J. R., Jr. (2000b). Activation of tumor necrosis factor-alpha-converting enzyme-mediated ectodomain shedding by nitric oxide. *J. Biol. Chem.* **275**, 15839–15844.

Zhao, L.-C., Shey, M., Farnsworth, M., and Dailey, M. (2001). Regulation of membrane metalloproteolytic cleavage of L-selectin (cd62l) by the epidermal growth factor domain. *J. Biol. Chem.* **276**, 30631–30640.

5

Expression of Meprins in Health and Disease

Lourdes P. Norman, Gail L. Matters, Jacqueline M. Crisman,
and *Judith S. Bond*
Department of Biochemistry and Molecular Biology
The Pennsylvania State University College of Medicine
Hershey, Pennsylvania 17033-0850

I. Introduction

A. Overview of Meprins: Discovery, Genes, and Localization

Meprins are complex, multidomain, oligomeric metalloproteinases that are highly regulated at the transcriptional and posttranslational levels. They are tissue-specific enzymes that are normally targeted to apical membranes of polarized epithelial cells (Bond and Beynon, 1995; Eldering *et al.,* 1997; Kadowaki *et al.,* 2000). The complex regulation of meprin activity and expression implies specific functions for these proteases and emphasizes the necessity for controlling the localization, concentration, and enzymatic activity of these proteases lest they become destructive.

Current Topics in Developmental Biology, Vol. 54

Meprins were discovered in the 1980s and purified as large metalloproteinases from the kidneys of BALB/c mice (Beynon *et al.*, 1981). Meprins have been identified in leukocytes, certain cancer cells, and mouse, rat, and human intestine and kidney. Other names for meprin are kinase-splitting membranal proteinase (KSMP), neutral endopeptidase 2 (NEP 2), and PABA-peptide hydrolase (PPH). They are particularly highly expressed in rodent brush border membranes of the kidney and intestine (Fig. 1; Barnes *et al.*, 1989; Craig *et al.*, 1987). In 1983, it

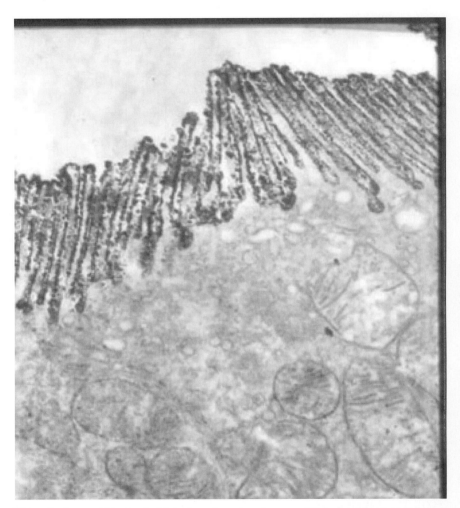

Figure 1 Immunohistochemical detection of meprin α in mouse kidney brush border membranes. Electron micrograph of proximal tubule cells from juxtamedullary region of kidney cortex of C57BL/6 mice. Kidney sections were incubated with anti-meprin α serum and peroxidase-conjugted goat anti-rabbit IgG. The pattern of immunoperoxidase staining shows that meprin is evenly distributed on the luminal surface of the microvilli and in apical invaginations.

was discovered that different strains of inbred mice had markedly different levels of kidney meprin activity (Beynon and Bond, 1983). More specifically, C Stock inbred mouse strains (e.g., C3H, CBA) had approximately 100-fold lower kidney meprin activity than random-bred mouse strains or other inbred strains (e.g., ICR, C57BL/6 mice). The description of the variable levels of meprin activity in kidneys of a large number of mouse strains led to the localization of one of the genes for meprins on mouse chromosome 17 near the histocompatibility complex (Bond *et al.,* 1984; Jiang *et al.,* 1992a; Reckelhoff *et al.,* 1985).

It is now clear that meprins consist of two types of subunits, α and β, that are coded for by two distinct genes: *Mep1a* on mouse chromosome 17 near the histocompatibility complex (*MEP1A* on human chromosome 6p21.2-p21.1), and *Mep1b* in the proximal region of mouse chromosome 18 (*MEP1B* on human chromosome 18q12.2-q12.3) (Bond *et al.,* 1995; Gorbea *et al.,* 1993; Jiang *et al.,* 1993, 1995). The structures of the genes have been described and promoter elements for tissues-specific expression have been identified (Hahn *et al.,* 2000; Matters and Bond, 1999a). Studies of the gene structure have been particularly important for defining mechanisms of transcriptional regulation (see Section III).

Meprins have been implicated in kidney fibrosis, injury, and end-stage kidney disease (Ricardo *et al.,* 1996; Sampson *et al.,* 2001; Trachtman *et al.,* 1993, 1995). In addition, linkage studies with the Pima Indian population of Phoenix, Arizona, have identified meprin β as a candidate gene for diabetic nephropathy (Imperatore *et al.,* 2001). Studies are now being conducted to determine whether there are polymorphisms in the gene or upstream regions that correlate with the susceptibility to nephropathy in this population.

B. Relationship to Other Membrane-Associated Metalloproteinases

Meprins are members of the "astacin family" of metalloproteinases (Section II) (Bond and Beynon, 1985, 1995). This family has a characteristic signature sequence in the protease domain that includes the zinc-binding motif (HEXXH); the 19 amino acid sequence is HEXXHXXGFXHEXXRXDRD. Astacin family protease domains also have a "Met-turn" with a conserved sequence, SXMHY, at the base of the active site in the three-dimensional structure (Dumermuth *et al.,* 1991; Stocker *et al.,* 1995). There are four evolutionarily defined families of zinc metalloendopeptidases that have this Met-turn (astacins, adamalysins, serrilysins, matrixins) (Hooper, 1994; Stocker *et al.,* 1995). The primary sequence similarity among the four families is low (except for the zinc binding motif and a third histidine which is 11 amino acids from the beginning of the motif); however, there is considerable three-dimensional similarity in the protease domains of these families (Stocker *et al.,* 1993, 1995). The superfamily has been called "metzincins" or Clan MA(M) (Fig. 2). Three types of membrane-bound metalloproteinases that

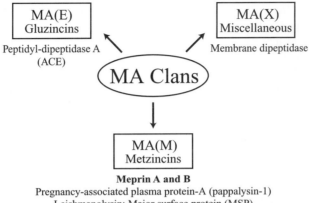

Figure 2 Metalloproteinases of the MA Clan. Metalloproteinases are a diverse and complex class of proteases—more than 40 distinct evolutionary families have been identified in this class. Several metalloprotease clans (consisting of evolutionarily related families) have been identified (see MEROPS classification: www.merops.co.uk). The MA Clan contains gluzincins (MA(E)), metzincins (MA(M)), and miscellaneous enzymes (MA(X)). Examples of enzymes in each of these superfamilies are shown in this diagram. Membrane-associated enzymes in the metzincin superfamily that are highlighted in chapters of this book are shown in bold.

are highlighted in this and other chapters of this book are indicated in bold print in the figure [meprins; ADAMS such as tumor necrosis factor α (TACE); and membrane-type matrix metalloproteinases (MT-MMPs)].

C. Structural Characteristics

The domain and oligomeric structure of meprins have been studied in detail. The amino acid sequences of the meprin α and β subunits are 42% identical, and the domain structures are similar (Fig. 3; Jiang *et al.,* 1992b). One marked difference between the subunits is the absence of the I (inserted) domain in the meprin β subunit. The I domain enables proteolytic cleavage of the meprin α subunit during biosynthesis, which releases this subunit from the membrane. The functional consequence of this difference is that the meprin β subunit remains a type I integral membrane protein at the plasma membrane, whereas meprin α is a secreted protein unless it is associated with a meprin dimer that contains the anchored meprin β

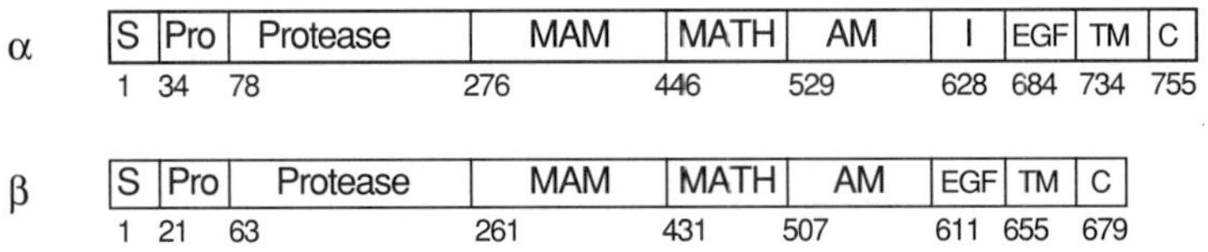

α | S | Pro | Protease | MAM | MATH | AM | I | EGF | TM | C |
1 34 78   276   446   529   628 684 734 755

β | S | Pro | Protease | MAM | MATH | AM | EGF | TM | C |
1 21 63   261   431   507   611 655 679

Figure 3 Domain structure from the cDNA deduced amino acid sequences of the mouse meprin α and β subunits. Domains are S, signal sequence; Pro, prosequence; Protease, catalytic domain; MAM, meprin, A5 protein, protein tyrosine phosphatase μ; MATH, meprin and TRAF homology; AM, after MATH; I, inserted domain; EGF, epidermal growth factor-like; TM, transmembrane-spanning; C, cytoplasmic domain. Numbers are amino acid number at the beginning of the domain.

subunit (Hahn *et al.*, 1997; Johnson and Hersh, 1992; Marchand *et al.*, 1994, 1995). The subunits differ in latency (removal of prosequence), in substrate specificity, and in their ability to form large homooligomers (Bertenshaw *et al.*, 2001, 2002; Ishmael *et al.*, 2001; Kounnas *et al.*, 1991).

All isoforms of meprins contain disulfide-linked dimers that are highly glycosylated (Fig. 4, see color insert; Butler and Bond, 1988; Gorbea *et al.*, 1991; Marchand *et al.*, 1994). Homooligomers of meprin β (meprin B; EC 3.4.24.63) are primarily membrane-bound dimers; meprin B has been isolated from mouse kidney membranes and produced by recombinant methods. Meprin A (EC 3.4.24.18) is any isoform containing the α subunit. Meprin α/β heterooligomers are membrane-bound in most cases (by virtue of containing the β subunit); however, there is evidence that a small amount of heterooligomeric meprin A can be released from human intestinal membranes and as a consequence is soluble (Eldering *et al.*, 1997; Johnson and Hersh, 1994; Pischitzis *et al.*, 1999). Homooligomers of mouse meprin α are large secreted multimers (primarily dodecamers) (Fig. 4; Ishmael *et al.*, 2001). The function of large secreted meprin α multimers may be to concentrate proteolytic activity in the extracellular milieu and enhance the ability of the complex to degrade large proteins or complexes.

D. Functional Characteristics

Meprin α and β have distinct peptide bond specificities. Meprin β prefers peptides that contain acidic amino acids (Asp, Glu) in or near the scissile bond; meprin α prefers small or hydrophobic amino acids at the cleavage site, and proline residues in the P2' position (Bertenshaw *et al.*, 2001). There are some overlapping substrate specificities; two of the best substrates hydrolyzed by both subunits are gastrin-releasing peptide and cholecystokinin (CCK). There are other substrates that are specific to one or the other subunits [e.g., mouse meprin β cleaves gastrin and osteopontin (Bertenshaw *et al.*, 2001); meprin α does not hydrolyze these peptides, but cleaves the cytokine monocyte chemoattractant protein-1 (MCP-1), unlike meprin β (see section IV)]. Several peptides of the gastrointestinal tract are among the best substrates for meprins; this is of interest because meprins are highly expressed in the intestine. The ability of meprins to hydrolyze chemokines is also of interest because of the identification of meprins in leukocytes (see Section IV).

II. Expression and Proposed Functions of Astacin Family Members

A. Localization and Proposed Functions of Astacins

The astacin family of proteases contains unique metalloproteases, important in a wide range of processes from development to adulthood. They exhibit a multifaceted nature from involvement in food digestion and hatching, to growth factor processing and extracellular matrix turnover (Bond and Beynon, 1995; Stocker and Zwilling, 1995; Yasumasu *et al.*, 1992). Astacins have been identified in a diverse set of organisms ranging from flavobacteria to mammals, and have amino acid similarity ranging from 30 to 95% (Table I; Jiang and Bond, 1992). Members of the astacin family are all secreted proteins that work at or near the cell surface, with the notable exception of the integral membrane protein, meprin β.

Many astacin family proteases are capable of proteolytically cleaving extracellular matrix components such as collagens, laminin, and fibronectin (Bertenshaw *et al.*, 2001; Kaushal *et al.*, 1994). Bone morphogenetic protein 1(BMP-1), tolloid (TLD), and tolloid-like (TLL) proteases are important in organizing zones of embryonic differentiation and shaping the extracellular milieu through processing fibrillar procollagens, prolysl oxidase, and probiglycan (Imamura *et al.*, 1998; Scott *et al.*, 2000; Uzel *et al.*, 2001). Meprins are capable of degrading a wide panel of substrates from extracellular matrix (ECM) proteins to cytokines and thus they likely mediate the actions of numerous bioactive proteins *in vivo* (Bertenshaw *et al.*, 2001; Kaushal *et al.*, 1994).

B. Meprins in the Embryo and Postnatally

Meprins α and β are both expressed during rat and mouse embryonic (E) and postnatal (PN) development. Microarray analyses to determine global renal gene expression in rats from gestation to adulthood revealed that meprin β was expressed in the kidney at the earliest gestation point tested (day E13). Furthermore,

Table I Members of the Astacin Family of Metalloproteases

Name (selected references)	Organism	Location/Proposed Function
Alveolin (Shibata et al., 2000)	Fish (medaka)	Induces chorion hardening
Astacin (Stocker and Zwilling, 1995)	Crayfish	Hepato-pancreas
Astacus embryonic astacin (AEA) (Geier and Zwilling, 1998)	Crayfish	Embryo
Blastula protease-10 (BP-10) (Lhomond et al., 1996)	Sea urchin	Blastula (16–32 cell stage)
Bone morphogenetic protein-1 (BMP-1, procollagen C-proteinase, PCP) (Reynolds et al., 2000)	Human, rodent, other mammals, chicken, Xenopus, Drosophila	Embryo and various adult tissues; removes C-terminal propeptide from procollagen
Chorioallantoic membrane protein 1 (CAM-1) (Elaroussi and DeLuca, 1994)	Quail	Utilization of egg shell calcium by avian embryo
Flavastacin (Tarentino et al., 1995)	Flavobacterium	Extracellular
Foot activator responsive matrix metalloprotease (Farm-1) (Kumpfmuller et al., 1999)	Cnideria (hydra)	Gastric epithelium
High and Low choriolytic enzymes (HCE and LCE) (Yasumasu et al., 1989, 1992)	Fish (medaka)	Hatching enzyme
Hydra metalloproteases 1 and 2 (HMP-1 and HMP-2) (Yan et al., 2000, 1995)	Cnideria (hydra)	Head regeneration and foot morphogenesis
Meprins α and β (Beynon et al., 1981; Sterchi et al., 1988)	Human, rodents, other mammals	Kidney, intestine, colon, appendix, leukocytes, in certain cancers and in various embryonic tissues (see text)
Myosinase III (Tamori et al., 1999)	Spear squid	Hydrolyzes myosine heavy chain, isolated from liver
Nephrosin (Hung et al., 1997)	Carp	Lymphohematopoietic tissue such as head kidney, kidney, and spleen
Podocornyne metalloprotease 1 (PMP-1) (Pan et al., 1998)	Cnideria (jellyfish)	Medusa bud development and adult manubrium (feeding organ)
SpAN (Wardle et al., 1999)	Sea urchin	Embryo
Tolloid, Tolkin and Tolloid-like proteins (TLD and TLL) (Nguyen et al., 1994)	Human, rodents, other mammals, Xenopus, Drosophila, zebrafish	Embryo and various adult tissues; dorsal/ventral patterning
Xenopus hatching enzyme (UVS.2) (Katagiri et al., 1997)	Xenopus	Hatching gland cells and neurula stage embryos
More than 30 ESTs cataloged as astacin family members	Caenorhabditis elegans	Unknown

the level of renal meprin β expression continued to increase until adulthood (Stuart et al., 2001). In rats, meprin β expression was highest at PN day 4 in the jejunum; expression then dropped dramatically at PN days 7–19. By weaning (PN day 22) expression returned to levels seen at PN day 4. In contrast, meprin α mRNA levels remained relatively constant in rat jejunum throughout the suckling period, and then declined at weaning. Additionally, it was demonstrated that administering dexamethasone during the suckling period dramatically induced meprin β but not meprin α expression. Thus, it was proposed that the pattern of meprin β expression in the jejunum of suckling rats could be related to changes in circulating gluco-corticoid levels, which are known to increase at weaning (Henning *et al.*, 1999).

In mouse embryos, meprin α is expressed by day E11, which coincides with the development of the metanephric kidney (Horster *et al.*, 1999; Kumar and Bond, 2001). *In situ* hybridization analyses of ICR and C3H/He embryos at different stages revealed that although these strains exhibit very different expression of meprin α as adults, the embryonic expression of both meprins α and β was similar. By day E16.5 both ICR and C3H/He mice express meprins α and β in the kidney and intestine (Fig. 5; Kumar and Bond, 2001). The expression of meprins at E16.5 in the intestine coincides with the development of the intestinal epithelium into a single cell layer with defined villi (Angres *et al.*, 2001). For both mouse strains meprin expression in the intestine during the suckling/weaning period was similar.

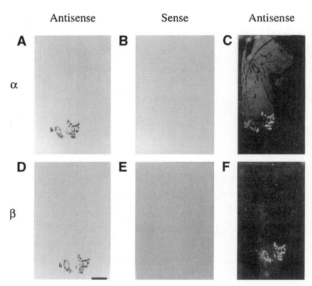

Figure 5 *In situ* hybridization of ICR mouse embryos with mouse meprin α and β riboprobes. Serial sagittal sections, 8 μm thick, of E16.5 embryos were probed with meprin α antisense (A, C), meprin α sense (B), meprin β antisense (D, F), or meprin β sense (E) ^{35}S-radiolabeled cRNAs. After hybridization, sections were exposed to X-ray film for 1 week and then to photographic emulsion for 3 weeks. The mRNAs in A, B, D, E were visualized after autoradiography; C and F are dark-field images of A and D, respectively. The mRNA localized to the kidney (k) and intestine (i) for both subunits. Bar = 1 mm. This figure is reprinted from *Biochem. Biophys. Acta*, Vol. 1518, Kumar, J. M., and Bond, J. S. Developmental expression of meprin metalloproteases subunits in ICR and C3H/He mouse kidney and intestine in the embryo, postnatally and after weaning, pp.106–114. ©2001, with permission from Elsevier Science.

Meprin α tended to rise during suckling and then decreased at weaning, whereas meprin β increased through suckling and then leveled off by adulthood (Fig. 6). These data indicate that meprin α plays a role in the digestion of milk during suckling. This point is further substantiated by the observation that casein, a major milk protein, is an excellent *in vitro* meprin substrate (Kumar and Bond, 2001).

Meprins have also been detected in development and in adult mammals at sites other than the kidney and intestine (Spencer-Dene *et al.*, 1994). In rats, meprins

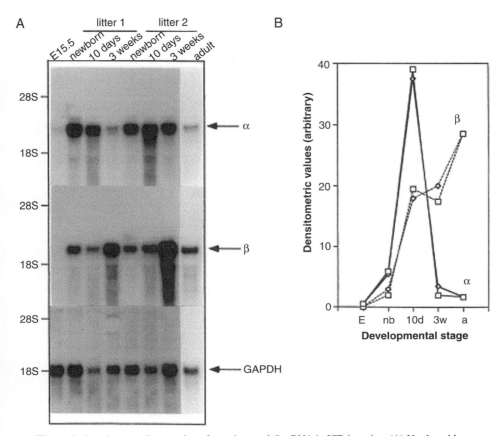

Figure 6 Developmental expression of meprin α and β mRNA in ICR intestine. (A) Northern blots of total intestinal RNA. Total RNA (15 μg) isolated from E15.5, newborn, 10-day- and 3-week-old pups from two different litters, and adult was loaded on a formaldehyde/agarose gel as indicated, electrophoresed, transferred, and probed with meprin and GAPDH cDNAs. The mRNAs specific for meprin α and β and for GAPDH are indicated by arrows. (B) Values obtained from densitometry were analyzed and graphed. This figure is reprinted from *Biochem. Biophys. Acta*, Vol. 1518, Kumar, J. M., and Bond, J. S. Developmental expression of meprin metalloproteases subunits in ICR and C3H/He mouse kidney and intestine in the embryo, postnatally and after weaning, pp. 106–114. © 2001, with permission from Elsevier Science.

have been detected on neuroepithelial cells at E12. By E14 and E16 meprins are also detected in the craniofacial region in tissues such as the choroid plexus, the ependymal layer of brain ventricles, the nasal conchae, the developing lens, the pigmented layer of the retina, the inner ear and in postcranial sites such as the gut, bladder, and ureter. In adult mice meprin was immunohistochemically detected in the salivary glands in both striated and intercalated ducts (Craig *et al.*, 1991). In mouse embryos an expressed sequence tag corresponding to meprin α was detected in the heart by day E13 while a meprin β expressed sequence tag was detected as early as the first blastocyst stage (B1) (WashU-HHMI Mouse EST Project, dbEST Id:1317387 @ www.ncbi.nlm.nih.gov). Meprins are also expressed by certain leukocytes found in the lamina propria of the intestine (Lottaz *et al.*, 1999a) (see Section IV). Adult humans express meprin α in the kidney, small intestine, colon, and appendix (Jiang and Le, 2000).

C. Developmental Functions of Bone Morphogenetic Protein 1 (BMP-1), Tolloid (TLD), and Hydra Metalloproteases (HMPs)

BMP-1 (procollagen C-proteinase) and TLD are products of alternatively spliced mRNAs encoded by a single gene (Takahara *et al.*, 1994). TLD possesses an extra EGF-like repeat and two additional CUB (complement components C1r/C1s, sea urchin protein UeGF, and BMP-1) domains at its C terminus (Stocker *et al.*, 1993). A third protein, TLL-1 (tolloid-like), was cloned from a cDNA library from BMP1/TLD knockout mice (Scott *et al.*, 1999b). Yet another protease, TLL-2, was cloned using degenerate PCR primers which corresponded to conserved regions of BMP-1/TLD and TLL-1(Scott *et al.*, 1999a, 1999b). All four of these secreted proteases are important in ECM and cellular maturation in early development and their expression overlaps in numerous tissues. Each is produced as a latent proenzyme, which requires removal of a prosequence for activity, and each has protein interaction domains such as CUB and epidermal growth factor domains. Although the domain and sequence structures of these four proteases are highly similar, they differ in their proteolytic activities and as such are likely to differentially affect development (Scott *et al.*, 1999a,b).

In development, chordin can associate with certain BMPs and consequently prevent them for associating with their respective receptors. *In vivo,* BMP-1 and TLL-1 have been shown to cleave chordin–BMP complexes, and in so doing they can modulate the amount of these BMPs available for interaction with their receptors. This action is an example of proteolysis in the extracellular matrix affecting the activity of a growth factor at the cell surface. BMP-1 and TLL-1 generate growth factor gradients, which direct axis formation in embryogenesis. Interestingly, TLL-2 and mammalian TLD do not cleave chordin. However, all four of these proteases (BMP-1, TLD, TLL-1, and TLL-2) can cleave pro-lysyl oxidase, the enzyme that catalyzes the final step of elastin–collagen cross-linking (Uzel *et al.*, 2001). Furthermore, of the four proteases only TLL-2 lacks procollagen C-proteinase activity (Scott *et al.*, 1999a). Thus these structurally similar enzymes have marked differences in proteolytic activity.

Two hydra proteases, HMP-1 and -2, have been characterized as key modulators of head regeneration and foot morphogenesis, respectively. Hydras are organized as a gastric tube with basal and apical poles. The mouth and tentacles are at the apical pole while the foot process is at the basal pole. HMP-1 is expressed in the apical pole and HMP-2 is predominantly expressed in the basal pole. By exploiting the highly regenerative properties of hydra, it has been shown that HMP-1 expression is highly induced during head regeneration. Using an antisense construct approach it was shown that HMP-2 is necessary for foot morphogenesis (Yan *et al.*, 1995, 2000). Interestingly, HMP-2 but not HMP-1 contains a MAM domain (meprin, A-5 protein, and receptor tyrosine phosphatase μ) which is only found in two other astacin family proteins, meprins α and β.

III. Meprin Expression in Cancer

Human carcinomas and cultured human cancer cell lines from numerous tissues secrete a variety of proteases, including serine, cysteine, aspartic, and metallo-proteases. Proteases are expressed at every stage of tumor progression, and their targets can include growth factors and their receptors, chemokines, adhesion proteins, and structural matrix components. A unique isoform of meprin β mRNA has been detected in several cultured human cancer cell lines, and meprin α is expressed in cancer cell lines with a more metastatic phenotype.

A. Intestinal Carcinomas and Invasive Cell Lines

Expression of meprin α in human tumor cell lines was first demonstrated in the differentiated small intestine–like Caco-2 cell line (Lottaz *et al.*, 1999b). Caco-2 cells secrete meprin α from both the apical and basolateral surfaces, as compared to the strictly apical secretion of meprin α in enterocytes and proximal tubule cells (Bond and Beynon, 1995). Several optimal binding sites for the GATA class of transcription factors, especially GATA-6, were identified in the human MEP1A promoter (Hahn *et al.*, 2000). Caco-2 cells have previously been shown to produce these GATA class factors. Functional promoter assays, using 3 kb of genomic DNA upstream of the MEP1A transcription start site, showed that this region is capable of directing reporter gene expression in transiently transfected Caco-2 cells.

In human colorectal adenocarcinomas, active meprin α was detected in about half of the tumors tested (Lottaz *et al.,* 1999b). The secreted meprin α accumulated in the tumor stroma; thus the location of meprin α in the tumor stroma could facilitate matrix degradation and increase the invasive and metastatic potential of tumor cells. *In vitro,* meprins can cleave laminins, fibronectin, nidogen, and gelatin (Bertenshaw *et al.,* 2001; Kohler *et al.,* 2000). Data from our laboratory support a role for meprin α in creating an invasive phenotype in cancer cells. Meprin α mRNA and protein are produced by the invasive colorectal carcinoma cell line SW620, but are not found in a noninvasive colorectal carcinoma cell line isolated from the same individual, SW480 (Matters, G. L., Norman, L. P., and Bond, J. S. unpublished). In addition, we have confirmed the expression of meprin α protein in human colorectal tumors (Fig. 7).

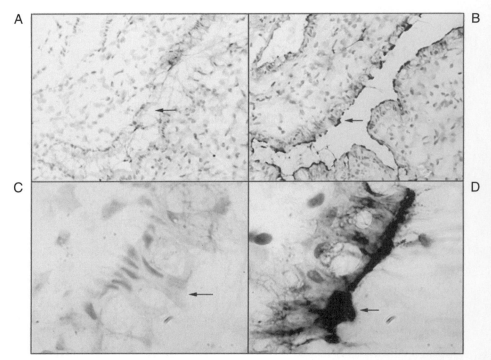

Figure 7 Immunohistochemical detection of meprin α in human colorectal tumors. (A) Preimmune serum negative control (original magnification 40×) (B) section stained using polyclonal anti-human meprin α antibody (40×), (C) preimmune serum negative control (100×), (D) section stained using polyclonal anti-human meprin α antibody (100×). The arrows indicates staining for meprin α. Note that staining in the indicated regions is specific to the anti-human meprin antibody and is not detected in the negative controls.

B. Breast Cancer Cell Lines

A novel form of the meprin β subunit mRNA, meprin β', has been detected in the cultured human breast cancer cell lines MCF-7, SK-BR-3, MDA-MB-231, and MDA-MB-435; none of these cells express meprin α. To test the role of meprin α in cancer cell invasion, MDA-MB-231 cells were stably transfected with the meprin α gene. When the meprin α cDNA was constitutively expressed in 231 cells, the cells were significantly more invasive *in vitro* than nontransfected 231 cells, meprin β transfected 231 cells, or vector transfected 231 cells (Matters, G. L., Norman, L. P., and Bond, J. S. unpublished).

C. Molecular Mechanisms That Regulate Meprin Transcripts in Transformed Cells

Adenocarcinoma and osteosarcoma cell lines from a variety of tissues express a unique isoform of meprin β mRNA, meprin β prime (β') (Dietrich *et al.*, 1996; Matters and Bond, 1999a,b). 5′ RACE analysis of meprin β transcripts from normal intestine and from cultured colon cancer cells (HT29-18C1) indicated that the human MEP1B gene is expressed from two independent transcription start sites: one is utilized in normal epithelium and the other, about 70 bp upstream, is used in cancer cells (Matters and Bond, 1999b). Use of the upstream start site results in a slightly longer 5′ untranslated region of the mRNA, but does not change the protein coding regions. The size of this mRNA is consistent with data from Northern blots, which have shown that the HT29-18C1 colorectal carcinoma cell line produces a larger meprin β mRNA (Dietrich *et al.*, 1996). A potential function for the longer 5′ UTR in the cancer cell mRNA may be to interact with mRNA binding proteins. Transcript binding proteins could affect the translation and stability of the meprin β' mRNA.

Meprin β' mRNA expression is increased in cancer cells by tumor promoters such as the phorbol ester PMA, as is the case for many cancer-related gene products. The PMA-enhanced expression is consistent with the presence of multiple AP-1 and PEA3 elements (Fos/Jun and Ets binding sites) in the MEP1B promoter, whereas these elements are rarely found in the MEP1A promoter (Hahn *et al.*, 2000; Matters and Bond, 1999b).

The mouse meprin β gene, *Mep1b*, also produces a longer, β', meprin mRNA in teratocarcinoma cells (Dietrich *et al.*, 1996). However the mechanism by which the mouse meprin β' mRNA is generated is different from that for the human meprin β' mRNA. The mouse β' mRNA is a consequence of alternative splicing that removes the first two β exons, and incorporates three new upstream exons (exons β' 1–3) (Fig. 8; Jiang and Le, 2000). The β' exons are located 1.6 kb upstream of the β exons, and transcription from the β' start site utilizes independent

Mouse *Mep1b* Gene

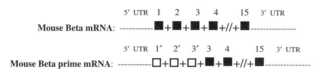

Human MEP1B Gene

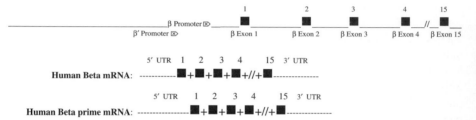

Figure 8 Schematic of mouse and human meprin β and meprin β′ mRNAs. In mouse, three additional exons can be transcribed to generate a meprin β′ mRNA. White boxes represent the extra exons incorporated into the meprin β′ message. Black boxes represent exons present in both normal mouse meprin β and mouse meprin β′ mRNAs. A human meprin β′ message is generated when an alternative promoter is utilized in the human MEP1B gene; additional exons are not incorporated into the human meprin β′ mRNA.

promoter elements. The protein encoded by the β′ mRNA contains a unique 5′ untranslated region, signal peptide, and propeptide. The sequence of the fully processed, active protease encoded by the β′ mRNA should be identical to the mouse β mRNA (Dietrich *et al.*, 1996). Comparison of the human MEP1B promoter and the mouse *Mep1b* promoter revealed a few areas of sequence similarity, but there are no identifiable transcription factor binding sites in common (Jiang and Le, 2000). Thus, the regulation of meprin β in human adenocarcinomas and in mouse teratocarcinomas differs significantly.

IV. Meprins in Inflammatory Diseases

A. The Role of Chemokines and Proteases in the Immune Response

Leukocyte extravasation into tissue is driven by the interplay of three molecular processes: adhesion molecule/receptor interactions, cytokine/receptor interactions, and the digestion of extracellular matrix and basement membrane proteins (Ebnet and Vestweber, 1999). Factors that affect the expression, activation, and half-life of these three classes of molecules have the potential to dramatically affect leukocyte compartmentalization. Extracellular proteases have the potential to hydrolyze all the molecules involved in leukocyte movement and biology, and thereby have an impact on the course of leukocyte biology and inflammation.

Chemokines (chemotactic cytokines) represent a separate supergene family of cytokines which induce the chemotaxis of specific types of leukocytes at very low concentrations (Mackay, 2001). As a result, chemokines are important mediators of leukocyte emigration into tissue. Chemokines comprise a free-rotating N terminus protruding from a relatively large globular domain (comprising a β-pleated sheet and α-helix) (Ward and Westwick, 1998). Mutagenesis studies indicate that the N termini of chemokines are essential for receptor binding and activation (Ward and Westwick, 1998). Removal of as few as two amino acids from the N terminus of chemokines can alter receptor binding preferences, or ablate receptor binding and activation (Gong and Clark-Lewis, 1995). The chemokine globular domain is believed to protect chemokines from degradation by many proteases (Van Damme *et al.,* 1999). However, the chemokine N terminus, which protrudes from the globular domain, is much more exposed than the rest of the molecule, making it particularly susceptible to proteolysis. Several proteases are known to cleave chemokines *in vitro;* for example, CD 26/DPP IV hydrolyzes the N termini of chemokines (Van Damme *et al.,* 1999). Interestingly, CD 26/DPP IV cleaves the N terminus of RANTES and ablates its ability to bind to CCR5, but not CCR1. A truncated RANTES analog was synthesized that mimicked the CD 26/DPP IV-truncated form of RANTES (Van Damme *et al.,* 1999). This analog induced T cell chemotaxis, but not macrophage chemotaxis. These findings are likely to have functional significance, since CCR5 is expressed at high levels on T cells, and may explain the shift in the chemotactic potential of RANTES induced by CD26/DPP IV. Thus, the coexpression of chemokines and proteases can have important implications for the resident chemokine biology in a tissue.

B. Meprins in the Immune System and in Intestinal Inflammation

Meprins are abundantly expressed on the brush border membrane of the intestine, particularly in the ileum (Bankus and Bond, 1996). The apical distribution of meprins in normal intestine indicates that meprins may be involved in the degradation of antigens in the lumen, which is necessary for the development of mucosal immune responses. Lottaz *et al.* (1999a) have shown by *in situ* hybridization that meprin α and β are expressed in leukocytes isolated from the lamina propria of the human inflamed bowel. These data indicate that meprins have the potential to regulate leukocyte activities *in vivo*. As detailed above, proteases such as meprin could target leukocyte emigration and activity by two primary mechanisms: the modulation of cytokine activity and the degradation of extracellular matrix components. There is evidence that meprins in fact do both.

Meprin A degrades the cytokine and transitional extracellular matrix protein, osteopontin (Bertenshaw *et al.*, 2001). Recent results indicate that meprin α, and not meprin β, selectively digests the N termini of some chemokines including MIP-1 β, MIP-1α, RANTES, and MCP-1 (Fig. 9, see color insert). Meprin A hydrolysis of MCP-1 yields a molecule that is truncated between amino acids 6 and 7. The truncated (7-76) MCP-1 molecule is known to have diminished ability to activate a major receptor (CCR2) for MCP-1 (Gong and Clark-Lewis, 1995).

Another important process is the ability of chemokines to guide immune responses. MCP-1 is involved in the maintenance of oral tolerance in the intestine. Karpus *et al.* (1998) demonstrated that the oral administration of proteolipid protein induces oral tolerance in mice, as well as the expression of MCP-1 in the intestinal mucosa, Peyer's patches, and mesenteric lymph nodes. The increase in MCP-1 expression resulted in a decrease in mucosal interleukin-12 (associated with Th1 immune responses), and a concomitant increase in mucosal interleukin-4 expression (associated with Th2 immune responses). Neutralizing antibodies to MCP-1 inhibited tolerance induction and resulted in the restoration of mucosal interleukin-12 expression, as well as peripheral antigen-specific T helper cell 1 responses.

The tissue localization of meprins is a key issue, since meprin β is apically distributed on the intestinal epithelium. Meprin β spans the plasma membrane and localizes meprin α subunits to the apical surface by heterooligomerization. Meprin α homooligomers can also be secreted from cells. During inflammation, the distribution of meprin is likely to localize to apical and basolateral surfaces of the intestinal epithelium. Thus, MCP-1 would be exposed to meprin A and ultimately result in the ablation of MCP-1-mediated tolerance mechanisms. This would have important implications for diseases whose etiologies lie in the disregulation of oral tolerance to food antigens, such as inflammatory bowel diseases.

Another chemokine, TECK, is completely digested by meprin β. TECK is specifically expressed in the crypts of the small intestine, particularly in the ileum and jejunum (Kunkel *et al.*, 2000). CCR9 (the receptor for TECK) is expressed by CD4+ and CD8+ T cells in circulation. These T cells also express the $\alpha_4\beta_7$ integrin and thus are targeted to the intestine. Immunohistochemistry has localized these CCR9 (+) cells to pockets of leukocytes found beneath the crypts. The significance of these observations is underscored by the role of the crypt in intestinal immune responses (Ganz, 2000). Crypts contain a source of pluripotent stem cells needed to regenerate the epithelial surface. Beneath the stem cells are Paneth cells, which protect the stem cells from infection by the release of granules after the acetylcholine stimulation induced by bacterial or food antigens. The granules contain a group of proteins known as defensins that have antimicrobial activity. Defensins are remarkably similar to chemokines structurally. In fact, some chemokines (ELR (−) CXC chemokines) have antimicrobial activity (Cole *et al.*, 2001). Interestingly, meprins localize to epithelial cells found in the crypts. Our data suggest that because meprins cleave chemokines they may also cleave defensins. Furthermore, the meprin-induced cleavage products of some chemokines may also have antimicrobial activity. The role of meprins in this process has yet to be elucidated, but our observations open intriguing possibilities.

Overall, meprins are likely to play an integral role in regulating leukocyte emigration and other chemokine-mediated events in the intestine, including oral tolerance and resistance to bacterial infection.

V. Summary, Concluding Remarks, Future Directions

The expression patterns of meprin subunits observed thus far implicate meprins in embryogenesis, tissue-specific functions (especially for kidney, intestine, and leucocytes), and disease processes of cancer and inflammation. The abundance of meprins in brush border membranes and the multimeric nature of secreted isoforms indicate that localized high concentrations of latent and activated proteases are fundamental to the action of these metalloproteinases. The enzymes are highly regulated transcriptionally and posttranslationally, indicating that they have specific functions and are not general "housekeeping" proteases, and are targeted to specific sites of action. There is now a base of knowledge about the structure/function of the meprins, and about potential physiological substrates for the enzymes. With the information provided by genomics, the mouse and human genome projects, microarray analysis, molecular modeling, and high-throughput technologies, we are learning of new sites of expression of the meprins (e.g., prostate and pancreatic cells), about the enzymatic activities, and the potential contributions of these enzymes to the degradation and activation of peptides and proteins at the cell surface and in extracellular compartments. The creation of specific inhibitors of meprins and of genetically engineered mice and the identification of polymorphisms in the meprin genes and regulatory elements for the genes will help decipher functions of the enzyme *in vivo* and further define the role of these enzymes in heath and disease.

References

Angres, B., Kim, L., Jung, R., Gessner, R., and Tauber, R. (2001). LI-cadherin gene expression during mouse intestinal development. *Dev. Dynam.* **221,** 182–193.

Bankus, J. M., and Bond, J. S. (1996). Expression and distribution of meprin protease subunits in mouse intestine. *Arch. Biochem. Biophys.* **331,** 87–94.

Barnes, K., Ingram, J., and Kenny, A. J. (1989). Proteins of the kidney microvillar membrane. Structural and immunochemical properties of rat endopeptidase-2 and its immunohistochemical localization in tissues of rat and mouse. *Biochem. J.* **264,** 335–346.

Bertenshaw, G. P., Turk, B. E., Hubbard, S. J., Matters, G. L., Bylander, J. E., Crisman, J. M., Cantley, L. C., and Bond, J. S. (2001). Marked differences between metalloproteases meprin A and B in substrate and peptide bond specificity. *J. Biol. Chem.* **276,** 13248–13255.

Bertenshaw, G. P., Norcum, M .T., and Bond, J. S. (2002). Structure of homo- and hetero-oligomeric meprin metalloproteases: Dimers, tetramers, and high molecular mass multimers. *J. Biol. Chem.* Papers in Press, DOI 10.1074;bc. M2088200.

Beynon, R. J., and Bond, J. S. (1983). Deficiency of a kidney metalloproteinase activity in inbred mouse strains. *Science* **219,** 1351–1353.

Beynon, R. J., Shannon, J. D., and Bond, J. S. (1981). Purification and characterization of a metallo-endoproteinase from mouse kidney. *Biochem. J.* **199,** 591–598.

Bond, J. S., and Beynon, R. J. (1985). Mammalian metalloendopeptidases. *Int. J. Biochem.* **17,** 565–574.

Bond, J. S., and Beynon, R. J. (1995). The astacin family of metalloendopeptidases. *Protein Sci.* **4,** 1247–1261.

Bond, J. S., Beynon, R. J., Reckelhoff, J. F., and David, C. S. (1984). Mep-1 gene controlling a kidney metalloendopeptidase is linked to the major histocompatibility complex in mice. *Proc. Natl. Acad. Sci. USA* **81,** 5542–5545.

Bond, J. S., Rojas, K., Overhauser, J., Zoghbi, H. Y., and Jiang, W. (1995). The structural genes, MEP1A and MEP1B, for the alpha and beta subunits of the metalloendopeptidase meprin map to human chromosomes 6p and 18q, respectively. *Genomics* **25,** 300–303.

Butler, P. E., and Bond, J. S. (1988). A latent proteinase in mouse kidney membranes. Characterization and relationship to meprin. *J. Biol. Chem.* **263,** 13419–13426.

Cole, A. M., Ganz, T., Liese, A. M., Burdick, M. D., Liu, L., and Strieter, R. M. (2001). Cutting edge: IFN-inducible ELR- CXC chemokines display defensin-like antimicrobial activity. *J. Immunol.* **167,** 623–627.

Craig, S. S., Reckelhoff, J. F., and Bond, J. S. (1987). Distribution of meprin in kidneys from mice with high- and low-meprin activity. *Am. J. Physiol.* **253,** C535–540.

Craig, S. S., Mader, C., and Bond, J. S. (1991). Immunohistochemical localization of the metalloproteinase meprin in salivary glands of male and female mice. *J. Histochem. Cytochem.* **39,** 123–129.

Dietrich, J. M., Jiang, W., and Bond, J. S. (1996). A novel meprin beta′ mRNA in mouse embryonal and human colon carcinoma cells. *J. Biol. Chem.* **271,** 2271–2278.

Dumermuth, E., Sterchi, E. E., Jiang, W. P., Wolz, R. L., Bond, J. S., Flannery, A. V., and Beynon, R. J. (1991). The astacin family of metalloendopeptidases. *J. Biol. Chem.* **266,** 21381–21385.

Ebnet, K., and Vestweber, D. (1999). Molecular mechanisms that control leukocyte extravasation: the selectins and the chemokines. *Histochem. Cell Biol.* **112,** 1–23.

Elaroussi, M. A., and DeLuca, H. F. (1994). A new member to the astacin family of metalloendopeptidases: a novel 1,25-dihydroxyvitamin D-3-stimulated mRNA from chorioallantoic membrane of quail. *Biochim. Biophys. Acta* **1217,** 1–8.

Eldering, J. A., Grunberg, J., Hahn, D., Croes, H. J., Fransen, J. A., and Sterchi, E. E. (1997). Polarised expression of human intestinal *N*-benzoyl-L-tyrosyl-*p*-aminobenzoic acid hydrolase (human meprin) alpha and beta subunits in Madin–Darby canine kidney cells. *Eur. J. Biochem.* **247,** 920–932.

Ganz, T. (2000). Paneth cells—guardians of the gut cell hatchery. *Nat. Immunol.* **1,** 99–100.

Geier, G., and Zwilling, R. (1998). Cloning and characterization of a cDNA coding for *Astacus* embryonic astacin, a member of the astacin family of metalloproteases from the crayfish *Astacus astacus. Eur. J. Biochem.* **253,** 796–803.

Gong, J. H., and Clark-Lewis, I. (1995). Antagonists of monocyte chemoattractant protein 1 identified by modification of functionally critical NH_2-terminal residues. *J. Exp. Med.* **181,** 631–640.

Gorbea, C. M., Flannery, A. V., and Bond, J. S. (1991). Homo- and heterotetrameric forms of the membrane-bound metalloendopeptidases meprin A and B. *Arch. Biochem. Biophys.* **290,** 549–553.

Gorbea, C. M., Marchand, P., Jiang, W., Copeland, N. G., Gilbert, D. J., Jenkins, N. A., and Bond, J. S. (1993). Cloning, expression, and chromosomal localization of the mouse meprin beta subunit. *J. Biol. Chem.* **268,** 21035–21043.

Hahn, D., Lottaz, D., and Sterchi, E. E. (1997). C-cytosolic and transmembrane domains of the *N*-benzoyl-L-tyrosyl-*p*-aminobenzoic acid hydrolase alpha subunit (human meprin alpha) are essential for its retention in the endoplasmic reticulum and C-terminal processing. *Eur. J. Biochem.* **247,** 933–941.

Hahn, D., Illisson, R., Metspalu, A., and Sterchi, E. E. (2000). Human *N*-benzoyl-L-tyrosyl-*p*-aminobenzoic acid hydrolase (human meprin): genomic structure of the alpha and beta subunits. *Biochem. J.* **346,** 83–91.

Henning, S. J., Oesterreicher, T. J., Osterholm, D. E., Lottaz, D., Hahn, D., and Sterchi, E. E. (1999). Meprin mRNA in rat intestine during normal and glucocortocoid-induced maturation: divergent patterns of expression of alpha and beta subunits. *FEBS Lett.* **462,** 368–372.

Hooper, N. M. (1994). Families of zinc metalloproteases. *FEBS Lett.* **354,** 1–6.

Horster, M. F., Braun, G. S., and Huber, S. M. (1999). Embryonic renal epithelia: induction, nephrogenesis, and cell differentiation. *Physiol. Rev.* **79,** 1157–1191.

Hung, C. H., Huang, H. R., Huang, C. J., Huang, F. L., and Chang, G. D. (1997). Purification and cloning of carp nephrosin, a secreted zinc endopeptidase of the astacin family. *J. Biol. Chem.* **272,** 13772–13778.

Imamura, Y., Steiglitz, B. M., and Greenspan, D. S. (1998). Bone morphogenetic protein-1 processes the NH2-terminal propeptide, and a furin-like proprotein convertase processes the COOH-terminal propeptide of pro-alpha1(V) collagen. *J. Biol. Chem.* **273,** 27511–27517.

Imperatore, G., Knowler, W. C., Nelson, R. G., and Hanson, R. L. (2001). Genetics of diabetic nephropathy in the Pima Indians. *Curr. Diabetes Rep.* **1:3,** 275–281.

Ishmael, F. T., Norcum, M. T., Benkovic, S. J., and Bond, J. S. (2001). Multimeric structure of the secreted meprin A metalloproteinase and characterization of the functional protomer. *J. Biol. Chem.* **276,** 23207–23211.

Jiang, W., and Bond, J. S. (1992). Families of metalloendopeptidases and their relationships. *FEBS Lett.* **312,** 110–114.

Jiang, W., and Le, B. (2000). Structure and expression of the human MEP1A gene encoding the alpha subunit of metalloendopeptidase meprin A. *Arch. Biochem. Biophys.* **379,** 183–187.

Jiang, W., Gorbea, C. M., Flannery, A. V., Beynon, R. J., Grant, G. A., and Bond, J. S. (1992a). The alpha subunit of meprin A. Molecular cloning and sequencing, differential expression in inbred mouse strains, and evidence for divergent evolution of the alpha and beta subunits. *J. Biol. Chem.* **267,** 9185–9193.

Jiang, W., Gorbea, C. M., Flannery, A. V., Beynon, R. J., Grant, G. A., and Bond, J. S. (1992b). The alpha subunit of meprin A. Molecular cloning and sequencing, differential expression in inbred mouse strains, and evidence for divergent evolution of the alpha and beta subunits [published erratum appears in *J. Biol. Chem.* **267,** 13779]. *J. Biol. Chem.* **267,** 9185–9193.

Jiang, W., Sadler, P. M., Jenkins, N. A., Gilbert, D. J., Copeland, N. G., and Bond, J. S. (1993). Tissue-specific expression and chromosomal localization of the alpha subunit of mouse meprin A. *J. Biol. Chem.* **268,** 10380–10385.

Jiang, W., Dewald, G., Brundage, E., M, Cher, G. M., Schildhaus, H. U., Zerres, K., and Bond, J. S. (1995). Fine mapping of MEP1A, the gene encoding the alpha subunit of the metalloendopeptidase meprin, to human chromosome 6P21. *Biochem. Biophys. Res. Commun.* **216,** 630–635.

Johnson, G. D., and Hersh, L. B. (1992). Cloning a rat meprin cDNA reveals the enzyme is a heterodimer [published erratum appears in *J. Biol. Chem.* **268,** 17647]. *J. Biol. Chem.* **267,** 13505–13512.

Johnson, G. D., and Hersh, L. B. (1994). Expression of meprin subunit precursors. Membrane anchoring through the beta subunit and mechanism of zymogen activation. *J. Biol. Chem.* **269,** 7682–7688.

Kadowaki, T., Tsukuba, T., Bertenshaw, G.P., and Bond, J.S. (2000). N-Linked oligosaccharides on the meprin A metalloprotease are important for secretion and enzymatic activity, but not for apical targeting. *J. Biol. Chem.* **275**, 25577–25584.

Karpus, W. J., Kennedy, K. J., Kunkel, S. L., and Lukacs, N. W. (1998). Monocyte chemotactic protein 1 regulates oral tolerance induction by inhibition of T helper cell 1-related cytokines. *J. Exp. Med.* **187**, 733–741.

Katagiri, C., Maeda, R., Yamashika, C., Mita, K., Sargent, T. D., and Yasumasu, S. (1997). Molecular cloning of *Xenopus* hatching enzyme and its specific expression in hatching gland cells. *Internatl. J. Develop. Biol.* **41**, 19–25.

Kaushal, G. P., Walker, P. D., and Shah, S. V. (1994). An old enzyme with a new function: purification and characterization of a distinct matrix-degrading metalloproteinase in rat kidney cortex and its identification as meprin. *J. Cell Biol.* **126**, 1319–1327.

Kohler, D., Kruse, M., Stocker, W., and Sterchi, E. E. (2000). Heterologously overexpressed, affinity-purified human meprin alpha is functionally active and cleaves components of the basement membrane in vitro. *FEBS Lett.* **465**, 2–7.

Kounnas, M. Z., Wolz, R. L., Gorbea, C. M., and Bond, J. S. (1991). Meprin-A and -B. Cell surface endopeptidases of the mouse kidney. *J. Biol. Chem.* **266**, 17350–17357.

Kumar, J. M., and Bond, J. S. (2001). Developmental expression of meprin metalloprotease subunits in ICR and C3H/He mouse kidney and intestine in the embryo, postnatally and after weaning. *Biochim. Biophys. Acta* **1518**, 106–114.

Kumpfmuller, G., Rybakine, V., Takahashi, T., Fujisawa, T., and Bosch, T. C. (1999). Identification of an astacin matrix metalloprotease as target gene for Hydra foot activator peptides. *Dev. Genes Evol.* **209**, 601–607.

Kunkel, E. J., Campbell, J. J., Haraldsen, G., Pan, J., Boisvert, J., Roberts, A. I., Ebert, E. C., Vierra, M. A., Goodman, S. B., Genovese, M. C., Wardlaw, A. J., Greenberg, H. B., Parker, C. M., Butcher, E. C., Andrew, D. P., and Agace, W. W. (2000). Lymphocyte CC chemokine receptor 9 and epithelial thymus-expressed chemokine (TECK) expression distinguish the small intestinal immune compartment: Epithelial expression of tissue-specific chemokines as an organizing principle in regional immunity. *J. Exp. Med.* **192**, 761–768.

Lhomond, G., Ghiglione, C., Lepage, T., and Gache, C. (1996). Structure of the gene encoding the sea urchin blastula protease 10 (BP10), a member of the astacin family of Zn^{2+}-metalloproteases. *Eur. J. Biochem.* **238**, 744–751.

Lottaz, D., Hahn, D., Muller, S., Muller, C., and Sterchi, E. E. (1999a). Secretion of human meprin from intestinal epithelial cells depends on differential expression of the alpha and beta subunits. *Eur. J. Biochem.* **259**, 496–504.

Lottaz, D., Maurer, C. A., Hahn, D., Buchler, M. W., and Sterchi, E. E. (1999b). Nonpolarized secretion of human meprin alpha in colorectal cancer generates an increased proteolytic potential in the stroma. *Cancer Res.* **59**, 1127–1133.

Mackay, C. R. (2001). Chemokines: immunology's high impact factors. *Nature Immunol.* **2**, 95–101.

Marchand, P., Tang, J., and Bond, J. S. (1994). Membrane association and oligomeric organization of the alpha and beta subunits of mouse meprin A. *J. Biol. Chem.* **269**, 15388–15393.

Marchand, P., Tang, J., Johnson, G. D., and Bond, J. S. (1995). COOH-terminal proteolytic processing of secreted and membrane forms of the alpha subunit of the metalloprotease meprin A. Requirement of the I domain for processing in the endoplasmic reticulum. *J. Biol. Chem.* **270**, 5449–5456.

Matters, G. L., and Bond, J. S. (1999a). Expression and regulation of the meprin beta gene in human cancer cells. *Mol. Carcinog.* **25**, 169–178.

Matters, G. L., and Bond, J. S. (1999b). Meprin B: transcriptional and posttranscriptional regulation of the meprin beta metalloproteinase subunit in human and mouse cancer cells. *APMIS* **107**, 19–27.

Nguyen, T., Jamal, J., Shimell, M. J., Arora, K., and O'Connor, M. B. (1994). Characterization of tolloid-related-1: a BMP-1-like product that is required during larval and pupal stages of *Drosophila* development. *Dev. Biol.* **166**, 569–586.

Pan, T., Groger, H., Schmid, V., and Spring, J. (1998). A toxin homology domain in an astacin-like metalloproteinase of the jellyfish *Podocoryne carnea* with a dual role in digestion and development. *Dev. Genes Evol.* **208**, 259–266.

Pischitzis, A., Hahn, D., Leuenberger, B., and Sterchi, E. E. (1999). *N*-Benzoyl-L-tyrosyl-*p*-aminobenzoic acid hydrolase beta (human meprin beta). A 13-amino-acid sequence is required for proteolytic processing and subsequent secretion. *Eur. J. Biochem.* **261**, 421–429.

Reckelhoff, J. F., Bond, J. S., Beynon, R. J., Savarirayan, S., and David, C. S. (1985). Proximity of the Mep-1 gene to H-2D on chromosome 17 in mice. *Immunogenetics* **22**, 617–623.

Reynolds, S. D., Zhang, D., Puzas, J. E., O'Keefe, R. J., Rosier, R. N., and Reynolds, P. R. (2000). Cloning of the chick BMP1/Tolloid cDNA and expression in skeletal tissues. *Gene* **248**, 233–243.

Ricardo, S. D., Bond, J. S., Johnson, G. D., Kaspar, J., and Diamond, J. R. (1996). Expression of subunits of the metalloendopeptidase meprin in renal cortex in experimental hydronephrosis. *Am. J. Physiol.* **270**, F669–676.

Sampson, N. S., Ryan, S. T., Enke, D. A., Cosgrove, D., Koteliansky, V., and Gotwals, P. (2001). Global gene expression analysis reveals a role for the alpha 1 integrin in renal pathogenesis. *J. Biol. Chem.* **276**, 34182–34188.

Scott, I. C., Blitz, I. L., Pappano, W. N., Imamura, Y., Clark, T. G., Steiglitz, B. M., Thomas, C. L., Maas, S. A., Takahara, K., Cho, K. W., and Greenspan, D. S. (1999a). Mammalian BMP-1/Tolloid-related metalloproteinases, including novel family member mammalian Tolloid-like 2, have differential enzymatic activities and distributions of expression relevant to patterning and skeletogenesis. *Dev. Biol.* **213**, 283–300.

Scott, I. C., Clark, T. G., Takahara, K., Hoffman, G. G., Eddy, R. L., Haley, L. L., Shows, T. B., and Greenspan, D. S. (1999b). Assignment of TLL1 and TLL2, which encode human BMP-1/Tolloid-related metalloproteases, to chromosomes 4q32→q33 and 10q23→q24 and assignment of murine Tll2 to chromosome 19. *Cytogenet. Cell Genet.* **86**, 64–65.

Scott, I. C., Imamura, Y., Pappano, W. N., Troedel, J. M., Recklies, A. D., Roughley, P. J., and Greenspan, D. S. (2000). Bone morphogenetic protein-1 processes probiglycan. *J. Biol. Chem.* **275**, 30504–30511.

Shibata, Y., Iwamatsu, T., Oba, Y., Kobayashi, D., Tanaka, M., Nagahama, Y., Suzuki, N., and Yoshikuni, M. (2000). Identification and cDNA cloning of alveolin, an extracellular metalloproteinase, which induces chorion hardening of medaka (*Oryzias latipes*) eggs upon fertilization. *J. Biol. Chem.* **275**, 8349–8354.

Spencer-Dene, B., Thorogood, P., Nair, S., Kenny, A. J., Harris, M., and Henderson, B. (1994). Distribution of, and a putative role for, the cell-surface neutral metallo-endopeptidases during mammalian craniofacial development. *Development* **120**, 3213–3226.

Sterchi, E. E., Naim, H. Y., Lentze, M. J., Hauri, H. P., and Fransen, J. A. (1988). *N*-benzoyl-L-tyrosyl-*p*-aminobenzoic acid hydrolase: a metalloendopeptidase of the human intestinal microvillus membrane which degrades biologically active peptides. *Arch. Biochem. Biophys.* **265**, 105–118.

Stocker, W., and Zwilling, R. (1995). Astacin. *Methods Enzymol.* **248**, 305–325.

Stocker, W., Gomis-Ruth, F. X., Bode, W., and Zwilling, R. (1993). Implications of the three-dimensional structure of astacin for the structure and function of the astacin family of zinc-endopeptidases. *Eur. J. Biochem.* **214**, 215–231.

Stocker, W., Grams, F., Baumann, U., Reinemer, P., Gomis-Ruth, F. X., McKay, D. B., and Bode, W. (1995). The metzincins—topological and sequential relations between the astacins, adamalysins, serralysins, and matrixins (collagenases) define a superfamily of zinc-peptidases. *Protein Sci.* **4**, 823–840.

Stuart, R.O., Bush, K.T., and Nigam, S.K. (2001). Changes in global gene expression patterns during development and maturation of the rat kidney. *Proc. Natl. Acad. Sci. USA* **98**, 5649–5654.

Takahara, K., Lyons, G. E., and Greenspan, D. S. (1994). Bone morphogenetic protein-1 and a mammalian tolloid homolog (mTld) are encoded by alternatively spliced transcripts which are differentially expressed in some tissues. *J. Biol. Chem.* **269**, 32572–32578.

Tamori, J., Kanzawa, N., Tajima, T., Tamiya, T., and Tsuchiya, T. (1999). Purification and characterization of a novel isoform of myosinase from spear squid liver. *J. Biochem.* **126,** 969–974.

Tarentino, A. L., Quinones, G., Grimwood, B. G., Hauer, C. R., and Plummer, T. H. (1995). Molecular cloning and sequence analysis of flavastacin: an O-glycosylated prokaryotic zinc metalloendopeptidase. *Arch. Biochem. Biophys.* **319,** 281–285.

Trachtman, H., Greenwald, R., Moak, S., Tang, J., and Bond, J. S. (1993). Meprin activity in rats with experimental renal disease. *Life Sci.* **53,** 1339–1344.

Trachtman, H., Valderrama, E., Dietrich, J. M., and Bond, J. S. (1995). The role of meprin A in the pathogenesis of acute renal failure. *Biochem. Biophys. Res. Commun.* **208,** 498–505.

Uzel, M. I., Scott, I. C., Babakhanlou-Chase, H., Palamakumbura, A. H., Pappano, W. N., Hong, H. H., Greenspan, D. S., and Trackman, P. C. (2001). Multiple bone morphogenetic protein 1-related mammalian metalloproteinases process pro-lysyl oxidase at the correct physiological site and control lysyl oxidase activation in mouse embryo fibroblast cultures. *J. Biol. Chem.* **276,** 22537–22543.

Van Damme, J., Struyf, S., Wuyts, A., Van Coillie, E., Menten, P., Schols, D., Sozzani, S., De Meester, I., and Proost, P. (1999). The role of CD26/DPP IV in chemokine processing. *Chem. Immunol.* **72,** 42–56.

Ward, S. G., and Westwick, J. (1998). Chemokines: understanding their role in T-lymphocyte biology. *Biochem. J.* **333,** 457–470.

Wardle, F. C., Angerer, L. M., Angerer, R. C., and Dale, L. (1999). Regulation of BMP signaling by the BMP1/TLD-related metalloprotease, SpAN. *Dev. Biol.* **206,** 63–72.

Yan, L., Pollock, G. H., Nagase, H., and Sarras, M. P. (1995). A 25.7×10^3 M_r hydra metalloproteinase (HMP1), a member of the astacin family, localizes to the extracellular matrix of *Hydra vulgaris* in a head-specific manner and has a developmental function. *Development* **121,** 1591–1602.

Yan, L., Leontovich, A., Fei, K., and Sarras, M. P., Jr. (2000). Hydra metalloproteinase 1: a secreted astacin metalloproteinase whose apical axis expression is differentially regulated during head regeneration. *Dev. Biol.* **219,** 115–128.

Yasumasu, S., Iuchi, I., and Yamagami, K. (1989). Purification and partial characterization of high choriolytic enzyme (HCE), a component of the hatching enzyme of the teleost, *Oryzias latipes*. *J. Biochem.* **105,** 204–211.

Yasumasu, S., Yamada, K., Akasaka, K., Mitsunaga, K., Iuchi, I., Shimada, H., and Yamagami, K. (1992). Isolation of cDNAs for LCE and HCE, two constituent proteases of the hatching enzyme of *Oryzias latipes,* and concurrent expression of their mRNAs during development. *Dev. Biol.* **153,** 250–258.

6

Type II Transmembrane Serine Proteases

Qingyu Wu
Department of Cardiovascular Research
Berlex Biosciences
Richmond, California 94806

Current Topics in Developmental Biology, Vol. 54

I. Introduction

Serine proteases are essential for a variety of biological processes including food digestion, hormone processing, blood coagulation, complement activation, wound healing, and embryonic development (Davie *et al.*, 1991; Kraut, 1977; Neurath, 1986; Stroud, 1974). Most trypsin-like serine proteases are secreted proteins. More recently, several new serine proteases such as corin (Yan *et al.*, 1999) and matriptase (Lin *et al.*, 1999c; Takeuchi *et al.*, 1999) have been identified that are transmembrane proteins containing a trypsin-like protease domain at their extracellular C terminus. Together with enteropeptidase and hepsin, these proteins represent a distinct subclass of type II transmembrane proteases within the trypsin superfamily (Hooper *et al.*, 2001).

The biological importance of membrane-bound serine proteases was recognized more than 100 years ago. Ivan Pavlov, who won the Nobel Prize in Physiology or Medicine in 1904, was the first to discover that the walls of the upper section of the small intestine contained a special enzyme necessary for the activation of pancreatic digestive enzymes (Pavlov, 1904). Using canine models, Schepovalnikov in Pavlov's laboratory established in 1899 that this special enzyme was abundant in the duodenal mucosal fluid. The enzyme was named enterokinase and later renamed enteropeptidase to emphasize its proteolytic function (Light and Janska, 1989; Lu and Sadler, 1998; Mann and Mann, 1994). In the following decades after Pavlov's discovery, enteropeptidase was purified and studied extensively. The cloning of enteropeptidase cDNA sequences from several species in the 1990s revealed that the enzyme is a type II transmembrane serine protease (Kitamoto *et al.*, 1994, 1995).

Cell surface serine proteases have also been shown to play a major role in tumor invasion and metastasis (Mignatti and Rifkin, 1993; Mullins and Rohrlich, 1983). Tumor-associated proteolytic activities were first observed at the beginning of the 20th century when tumor cells were found to have enhanced plasma clot-degrading activities in tissue culture. The significance of this observation was the implication that the proteolytic activities of tumor cells could be a molecular mechanism in cancer progression and metastasis. Further studies in the 1970s led to the identification of plasminogen activators as proteolytic enzymes associated with oncogenic transformation (Ossowski *et al.*, 1973a,b; Quigley *et al.*, 1974; Unkeless *et al.*, 1973, 1974). In the early 1980s, Zucker and colleagues identified a trypsin-like serine protease on the surface of rat Walker 256 carcino-sarcoma cells that was responsible for cancer cell-mediated cytolysis (DiStefano *et al.*, 1982; LaBombardi *et al.*, 1983). In polyacrylamide gel electrophoresis–based assays, the protein had a molecular mass of ~120 kDa but the sequence of this protease was not determined (DiStefano *et al.*, 1983). This enzyme may well be a hepsin-like type II transmembrane serine protease abundantly expressed on the surface of tumor cells.

Trypsin-like serine proteases also play an essential role in embryonic development in *Drosophila melanogaster* (Belvin and Anderson, 1996). Serine proteases encoded by the *nudel* (Hong and Hashimoto, 1995), *gastrulation defective* (Han *et al.*, 2000), *easter* (Jin and Anderson, 1990), and *snake* (DeLotto and Spierer, 1986) genes are key components of a signaling pathway important for the dorsal–ventral polarity in a developing embryo. The *Stubble-stubbloid* gene encodes a type II transmembrane serine protease with an overall domain structure similar to that of hepsin (Appel *et al.*, 1993). Genetic studies have demonstrated that *Stubble-stubbloid* is critical for epithelial morphogenesis and development. Defects in *Stubble-stubbloid* cause malformation of legs, wings, and bristles of the fruitfly. In a recent genome-wide analysis of *Drosophila* genes in response to bacterial and fungal infection, several type II transmembrane serine proteases have been identified that share structural and sequence similarities with corin and hepsin. These *Drosophila* transmembrane proteases may play a role in innate immunity of the fruitfly (Irving *et al.*, 2001).

This chapter will review the molecular biology and biochemistry of mammalian type II transmembrane serine proteases of the trypsin superfamily and discuss their potential roles in physiological and pathological conditions.

II. General Features of Type II Transmembrane Serine Proteases

The common structural features that define the type II transmembrane serine proteases include a transmembrane domain near the N terminus and a trypsin-like protease domain at the C terminus. Table I summarizes the properties of the genes, mRNAs, and proteins of the published human type II transmembrane serine proteases. The total number of these proteases is expected to increase as more and more genomic sequences become publicly available. The general domain structures of these proteases are shown in Fig. 1 and briefly described in the following sections.

A. Cytoplasmic and Transmembrane Domains

The length of the cytoplasmic domains of the type II transmembrane serine proteases varies significantly. The functional significance of the cytoplasmic domains is completely unknown. Although putative phosphorylation recognition sites have been found in several proteins such as corin (Hooper *et al.*, 2000; Yan *et al.*, 1999),

Table I Properties of the Genes, mRNAs, and Proteins of Human Type II Transmembrane Serine Proteases

Name	Other Name	Chromosome	Gene (kb)	Exons	mRNA (kb)	Amino Acids (#)	M_r	Reference
Corin	LDL-RP4	4p12-13	150	22	5	1042	150,000	Yan et al. (1999)
Enteropeptidase	Enterokinase	21q21	90	25	4.4	1019	150,000	Kitamoto et al. (1995)
HAT		4q13.2	62	10	1.9	418	48,000	Yamaoka et al. (1998)
Hepsin	TMPRSS1	19q11-13.2	26	13	1.8	417	51,000	Leytus et al. (1988)
Matriptase	MT-SP1	11q24.3	50	21	3.3	855	87,000	Lin et al. (1999); Takeuchi et al. (1999)
MSPL		11q23.2	29	12	2.4	581	60,000	Kim et al. (2001)
Spinesin	TMPRSS5	11q23.2	19	13	2.3	457	52,000	Yamaguchi et al. (2002)
TMPRSS2	Epitheliasin	21q22.2-3	44	14	2.5	492	70,000	Paoloni-Giacobino et al. (1997)
TMPRSS3	TADG-12	21q22.3	24	13	2.4	454		Scott et al. (2001)
TMPRSS4	TMPRSS3	11q23.2	41	13	2.3	437	64,000	Wallrapp et al. (2000)

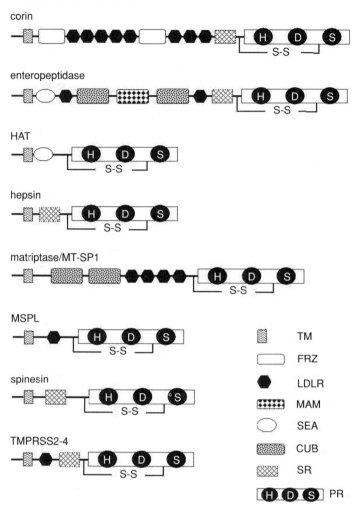

Figure 1 Schematic structure of human type II transmembrane serine proteases. A transmembrane domain (TM) at the N terminus and a protease catalytic domain (PR) with active site residues His (H), Asp (D), and Ser (S) at the C terminus are common features in these proteases. In the extracellular region, a variety of protein domain structures are present, which include frizzled-like cysteine-rich repeats (FRZ), LDL receptor repeats (LDLR), meprin-like domain (MAM), scavenger receptor cysteine-rich repeat (SR), CUB repeats found in complement, urchin embryonic growth factor, and bone morphogeic protein (CUB), and SEA repeats found in sea urchin sperm protein, enteropeptidase, and agrin (SEA). A disulfide bond (S–S) that connects the propeptide and the protease domain is indicated.

TMPRSS2 (Paoloni-Giacobino *et al.,* 1997), and matriptase (Lin *et al.,* 1999c; Takeuchi *et al.,* 1999), their signaling activities have never been demonstrated experimentally in these proteins. Potential functions of the cytoplasmic domains in membrane recycling, cytoskeletal protein binding, and signal transduction cannot be excluded.

The transmembrane domains in these proteases are approximately 20 amino acids in length. There are, in general, more positively charged amino acid residues immediately preceding the transmembrane segments, as predicted for the type II transmembrane proteins (Hartmann *et al.,* 1989). Most of the proteases also contain putative N-linked glycosylation sites in their extracellular regions. For example, human enteropeptidase (Kitamoto *et al.,* 1995) and corin (Yan *et al.,* 1999) proteins contain 17 and 19 potential N-linked glycosylation sites, respectively. The predicted protein topology on the cell surface has been confirmed by immuno-histochemistry, membrane fractionation, cell surface labeling, tryptic digestion, and Western analysis in several proteases including enteropeptidase (Zheng *et al.,* 1999), hepsin (Tsuji *et al.,* 1991), and matriptase (Lin *et al.,* 1997; Takeuchi *et al.,* 2000).

Under physiological conditions, proteases are highly regulated so that their enzymatic activities are limited to specific sites and for specific substrates, preventing potential proteolytic damage to normal tissues. This regulation is usually achieved by controlled zymogen activation of the protease and interactions with cofactors and inhibitors. For the transmembrane proteases, the transmembrane domain may provide an additional mechanism to restrict the proteases at desired locations. The transmembrane domain, however, is not required for the enzymatic activity of these proteases because recombinant soluble enteropeptidase (Lu *et al.,* 1997), matriptase (Lin *et al.,* 1997; Takeuchi *et al.,* 2000), hepsin (Kurachi *et al.,* 1994), spinesin (Yamaguchi *et al.,* 2002) and human airway trypsin-like protease (HAT) (Yamaoka *et al.,* 1998) were shown to be catalytically active.

B. Motifs in the Propeptide Region

A variety of protein motifs have been identified in the extracellular space regions between the transmembrane domain and the protease domain. These protein motifs include frizzled-like cysteine-rich motifs, low-density lipoprotein (LDL) receptor repeats, scavenger receptor cysteine-rich domains, CUB repeats found in complement C1s/C1r, the urchin embryonic growth factor and bone morphogenic protein, and protein motifs found in meprin, mucin, and sea urchin sperm protein (Fig. 1). The presence of the diverse domain structures may reflect long evolutionary histories of these proteases. Interestingly, kringle-, epidermal growth factor (EGF)-, fibronectin finger-, and apple-like domains that are common in the serine proteases involved in blood coagulation and fibrinolysis are absent in the type II transmembrane serine proteases (Furie and Furie, 1988). The protein motifs in the propeptide region are most likely to have a role in the regulation of the proteases. In fact, the protein motifs in the enteropeptidase propeptide have been shown to be important for its cell surface expression and substrate–inhibitor interactions (Lu *et al.,* 1997; Zheng and Sadler, 2002).

C. Protease Domains

The protease domains at the C termini of these proteins are highly homologous to the catalytic domain of other trypsin-like proteases. All the essential features of serine protease sequences, including the catalytic triad (His, Asp, and Ser) and the residues forming the substrate specificity pocket (Asp, Gly, and Gly), are well conserved. There is a conserved activation cleavage sequence, Arg/Lys-Ile/Val-Val/Leu-Gly-Gly, between the propeptide domain and the protease domain, whose cleavage is required to activate the zymogen. There are five pairs of conserved cysteine residues that are predicted to form disulfide bonds, one of which connects the protease domain to the propeptide after the activation cleavage. Thus, the activated enzymes are expected to remain on the cell surface. In corin, there is an additional pair of cysteine residues, Cys-817 and Cys-830, in the protease domain that are not found in any other serine proteases in vertebrates (Yan et al., 1999). Computer modeling indicates that these two cysteine residues are likely to form a disulfide bond connecting two β-sheets in the core of the protease domain. The crystal structures of the protease domains from enteropeptidase (Lu et al., 1999) and matriptase (Friedrich et al., 2002) have been determined. As expected, the overall fold of these proteases is similar to that of chymotrypsin.

III. Corin

A. Atrial Natriuretic Peptide

Atrial natriuretic peptide (ANP), also called atrial natriuretic factor (ANF), and related natriuretic peptides were discovered in the early 1980s. To examine potential renal effects of myocardium, de Bold and colleagues homogenized the atria and ventricles of rats and injected the homogenates intravenously into anesthetized Sprague-Dawley rats (de Bold et al., 1981). Muscle extracts from the atrium, but not ventricle, caused a rapid increase of sodium and chloride excretions and urine volume, suggesting the presence of a biological substance in atrial myocardium that could regulate body fluid and electrolyte homeostasis. The results implied for the first time that the heart acted not only as a mechanical pump in the circulation system but also as an endocrine organ by secreting a hormone(s) that altered kidney function.

This seminal finding quickly led to the isolation of ANP from human, porcine, and rat hearts (de Bold, 1985). Biochemical and pharmacological studies have established ANP as a cardiac hormone involved in the regulation of body fluid and electrolyte homeostasis. In response to high blood pressure, ANP is secreted from cardiomyocytes into the circulation. In target organs such as kidney and peripheral blood vessels, ANP binds to its receptor and stimulates the intrinsic guanylyl cyclase activity of the receptor, leading to accumulation of intracellular cGMP. The biological effects of ANP are to promote salt excretion, reduce blood volume, and relax vessel tension, thereby reducing blood pressure (Inagami, 1989; Koller and Goeddel, 1992; Levin *et al.*, 1998; Wilkins *et al.*, 1997).

In cardiomyocytes, ANP is synthesized as a prepropeptide of 126 amino acids (Bloch *et al.*, 1985; Schwartz *et al.*, 1985). After removal of the signal peptide, pro-ANP is stored in the dense granules of the cells. Upon secretion, pro-ANP is activated on the cell surface by proteolytic cleavage at residue Arg-98, generating an N-terminal propeptide and a biologically active peptide of 26 amino acids (Ito *et al.*, 1988; Lang *et al.*, 1985; Manning *et al.*, 1985; Shields and Glembotski, 1988). Several studies showed that a high-molecular-weight trypsin-like enzyme associated with the membrane of cardiomyocytes was responsible for the activation cleavage of pro-ANP but the identity of the pro-ANP convertase remained elusive (Imada *et al.*, 1987, 1988; Sei *et al.*, 1992; Seidah *et al.*, 1986).

B. Discovery of Corin

1. Corin cDNA and Gene Structure

Corin was first identified as an expressed sequence tag (EST) clone from a human heart cDNA library that shared sequence homology with trypsin-like serine proteases. In Northern analysis, the EST-derived probes detected an ~5-kb mRNA in samples from human heart but not other tissues such as placenta, liver, lung, skeletal muscle, spleen, and brain. A full-length cDNA was isolated that encodes a novel serine protease, designated corin for its abundant cardiac expression (Yan *et al.*, 1999). The human *corin* gene, which spans ~150 kb and contains 22 exons, has been mapped to chromosome 4 (4p12-13) by fluorescence *in situ* hybridization (FISH) analysis (Yan *et al.*, 1999). In an independent study, mouse corin cDNA was isolated from a heart library in a search for LDL receptor binding proteins (Tomita *et al.*, 1998). The protein was named LDLR-related protein 4. In cell-based binding assays, however, no binding of recombinant mouse corin to very low density lipoprotein (VLDL) was detected (Tomita *et al.*, 1998). The mouse *corin* gene is located on chromosome 5D, a syntenic locus to human chromosome 4p12-13.

2. Corin Protein Domain Structures

The open reading frame of human corin cDNA encodes a polypeptide of 1042 amino acids with a calculated mass of 116 kDa (Yan *et al.*, 1999). As shown in Fig. 1, corin is a mosaic protein with a unique combination of domain structures. In particular, corin is the only trypsin-like serine protease that contains frizzled-like cysteine-rich domains. The frizzled-like cysteine-rich repeat, which is about 120 amino acids long and contains 10 conserved cysteine residues, has been found in a number of proteins from different species (Cadigan and Nusse, 1997; Shulman *et al.*, 1998). Many of the frizzled domain–containing proteins are involved in the Wnt signaling pathways by interacting with Wnt proteins. It is possible that the frizzled-like cysteine-rich domains in corin also are involved in protein–protein interactions.

The propeptide region of corin contains eight LDL receptor class A repeats and a macrophage scavenger receptor-like cysteine-rich repeat. These protein motifs are also found in other type II transmembrane serine proteases such as enteropeptidase, matriptase, and TMPRSS2-4 (Fig. 1), as well as other proteins such as LDL receptor-related proteins (Krieger and Herz, 1994), megalin (Kounnas *et al.*, 1993), complement proteins (Catterall *et al.*, 1987), and sea urchin spermatozoa speract receptor (Dangott *et al.*, 1989; Thorpe and Garbers, 1989). In the LDL receptor, the cysteine-rich repeats have been shown to contain calcium ions and play a role in endocytosis of the extracellular ligands (Brown *et al.*, 1997). The potential function of the LDL repeats in corin remains to be defined.

The protease domain of corin is homologous to those of trypsin-like proteases. All essential features of trypsin-like serine protease are conserved, which include the residues of the catalytic triad at His-843, Asp-892, and Ser-985 and the residues forming the substrate specificity pocket at Asp-979, Gly-1007, and Gly-1018 (Yan *et al.*, 1999). The substrate specificity of corin is predicted to favor basic residues in the P1 position. The sequence of R-ILGG between residues 801 and 805 represents a conserved activation cleavage site, indicating that proteolytic cleavage of the peptide bond between Arg801 and Ile802 is required to generate a catalytically active enzyme.

3. Tissue Distribution of Corin mRNA

Corin mRNA expression is most abundant in the human heart (Yan *et al.*, 1999). By *in situ* hybridization, corin mRNA was detected in cardiomyocytes from both the atrium and ventricle and the level of expression appeared to be higher in the atrium than in the ventricle. In other tissues such as uterus, small intestine, bladder, stomach, and prostate, no corin mRNA was detected, suggesting that smooth muscle cells do not express corin. The cardiac expression of corin was confirmed by immunohistochemical analysis using polyclonal antibodies against human corin (Hooper *et al.*, 2000). During embryonic development, corin mRNA was first detected at E9.5 in both atrium and ventricle of a mouse developing heart. By E15.5, corin mRNA expression in the heart was very abundant (Yan *et al.*, 1999).

Corin mRNA was also detected in several other mouse tissues by *in situ* hybridization (Yan *et al.*, 1999). For example, corin mRNA was found in the uterus from pregnant mice and in the developing kidneys. In mouse embryos, corin mRNA was also identified in cartilage-derived structures, such as the vertebrae in the tail, the turbinate in the head, and the long bones in the limbs. The expression in developing bones appeared to be limited in a specific cell type, the prehypertrophic chrondrocytes, suggesting that corin may play a role in chondrocyte differentiation.

4. Recombinant Corin Protein

In sodium dodecyl sulfate–polyacrylamide gel electrophoresis (SDS–PAGE) and Western analysis, recombinant human corin with a C-terminal epitope tag expressed in human embryonic kidney 293 cells appeared as a single band with a molecular mass of ~150 kDa, indicating that corin is synthesized as a zymogen (Yan *et al.,* 2000). The protein was detected in total cell lysate and membrane fractions but not in cell culture medium, consistent with the prediction that corin is a transmembrane protein. Native corin from the human heart appeared as a major band of ~130 kDa in SDS–PAGE and Western analysis (Hooper *et al.,* 2000). These results are consistent with the calculated mass of 116 kDa for human corin protein that also contains 19 potential N-linked glycosylation sites in its extracellular domains (Yan *et al.,* 1999).

C. Identification of Corin as the Pro-ANP Convertase

As in many genomics studies, determination of the function of a novel gene identified by database mining is a major challenge. Initial characterization of corin, however, suggested that it might be the long-missing pro-ANP convertase. The predicted substrate specificity, the apparent molecular mass, and tissue distribution patterns of corin all match the characteristics of the pro-ANP convertase suggested from previous studies. To determine if corin is indeed the pro-ANP convertase, a series of functional studies were performed. In 293 cells, cotransfection of plasmids expressing corin and pro-ANP led to cleavage of pro-ANP, producing a small peptide fragment that was indistinguishable from ANP on Western blots and active in cGMP assays (Wu *et al.,* 2002; Yan *et al.,* 2000). The corin-mediated cleavage in pro-ANP is highly sequence-specific because a single mutation at pro-ANP Arg-98, the predicted cleavage site, completely prevented the processing by corin whereas mutations at nearby residues such as Arg-101 and Arg-102 did not. The activity of corin in pro-ANP processing is inhibited by benzamidine, leupeptin, and aprotinin but not soybean trypsin inhibitor (Wu *et al.,* 2002; Yan *et al.,* 2000).

To prove that corin is indeed the endogenous pro-ANP convertase, the corin-mediated processing of pro-ANP was studied in a murine cardiomyocyte cell line, HL-5 (Wu *et al.,* 2002). The processing of pro-ANP in HL-5 cells was indistinguishable from recombinant corin-mediated processing of pro-ANP in 293 cells. Overexpression of an active site mutant of corin, S985A, inhibited the processing of pro-ANP in HL-5 cells. Furthermore, in RNA interference (RNAi)-based gene silencing experiments, small interfering RNA duplexes directed against the *corin* gene completely prevented the processing of pro-ANP in the cardiomyocytes.

These results strongly support corin being the pro-ANP convertase that acts at the top of the ANP-mediated pathway.

D. Corin in Human Diseases

1. Hypertension and Congestive Heart Failure

The biological importance of the ANP-mediated pathway in maintaining normal blood pressure has been well established. In knockout mice, for example, deficiency in either ANP or its receptor leads to spontaneous hypertension (John et al., 1995; Lopez et al., 1995). High plasma concentrations of ANP and brain-type natriuretic peptide (BNP) are consistently found in patients with congestive heart failure. The levels of these natriuretic peptides often correlate with the extent of ventricular dysfunction and subsequent development of cardiac arrhythmias (Burnett et al., 1986; Gottlieb et al., 1989). Clinically, ANP and BNP have been used as therapeutic agents in patients with decompensated congestive heart failure and acute myocardial infarction to improve cardiac function. Administration of ANP has also been used as a therapy in patients with renal failure (Allgren et al., 1997; Colucci et al., 2000).

Identification of corin as the pro-ANP convertase suggests a potential new regulatory mechanism for the ANP-mediated pathway. In principle, deficiency in corin could reduce ANP production, leading to pathological conditions such as hypertension. In addition to the heart, corin mRNA expression was found in pregnant, but not normal, uterus in mice (Yan et al., 1999). The biological significance of the upregulation of corin in the uterus during pregnancy is not clear. ANP expression in the uterus has been reported. Plasma levels of ANP are increased significantly in pregnant women, especially in women with preeclampsia. It is possible that corin is induced in the uterus during pregnancy to increase ANP production and prevent preeclampsia.

2. Total Anomalous Pulmonary Venous Return

The human corin gene is located on chromosome 4p12-13 (Yan et al., 1999). Interestingly, a disease locus, TAPVR, has also been mapped to this region. Total anomalous pulmonary venous return (TAPVR) is a rare cyanotic form of congenital heart defects in which the pulmonary vein connected abnormally to the right atrium or one of the venous tributaries instead of the left atrium. The molecular mechanism responsible for this developmental defect in the heart is completely unknown. A genetic study of a large Utah–Idaho family with 14 affected individuals localized the TAPVR locus to a 30-centimorgan interval on 4p13-q12 (Bleyl et al., 1995). The close proximity of the corin gene and TAPVR locus suggests an intriguing possibility that corin deficiency might contribute to the TAPVR phenotype. Hooper *et al.* (2000) examined by immunohistochemistry corin protein expression in heart tissue sections derived from a TAPVR patient and found no obvious reduction of corin protein in these samples. To date, no sequence studies have been performed to examine if there are point mutations present in the *corin* gene in TAPVR patients.

IV. Enteropeptidase

A. cDNA and Gene Structure

Although enteropeptidase (EC 3.4.21.9) was discovered more than 100 years ago and its physiological importance in the digestive system had been well recognized, the structure of this remarkable enzyme remained poorly defined until enteropeptidase cDNA clones were isolated in the 1990s. Based on the partial amino acid sequence (Light and Janska, 1991), bovine enteropeptidase cDNA clones were first isolated from small intestine libraries (Kitamoto *et al.*, 1994; LaVallie *et al.*, 1993). Subsequently, human (Kitamoto *et al.*, 1995), porcine (Matsushima *et al.*, 1994), mouse (Yuan *et al.*, 1998), and rat (Yahagi *et al.*, 1996) enteropeptidase cDNA clones were also reported. By Northern analysis, enteropeptidase mRNA was most abundantly expressed in the duodenum. Lower levels of the expression were also detected in the proximal segment of jejunum. In the small intestinal tissues, the enteropeptidase mRNA expression was found in the enterocytes throughout the villus (Yuan *et al.*, 1998). These results are consistent with the previous findings that enteropeptidase activity and protein are associated with the brush border of enterocytes in the duodenum and proximal jejunum (Hermon-Taylor *et al.*, 1977; Lojda and Gossrau, 1983). The human *enteropeptidase* gene is mapped to chromosome 21q21 by FISH analysis (Kitamoto *et al.*, 1995). The gene spans ~90 kb and contains 25 exons (Holzinger *et al.*, 2002).

B. Protein Domain Structures

The open reading frame of human enteropeptidase cDNA encodes a mosaic protein of 1019 amino acids with a calculated mass of 113 kDa (Kitamoto *et al.*, 1995) (Fig. 1). The presence of a transmembrane domain explains the cell surface association of this enzyme. The combination of domain structures in the extracellular region of enteropeptidase is unique among trypsin family members and suggests that these protein domains play a role in its regulation. Indeed, the SEA domain has been shown to be critical for targeting enteropeptidase to the apical cell surface (Zheng *et al.*, 1999; Zheng and Sadler, 2002). The other protein motifs in the propeptide region have also been found to be important for efficient activation of trypsinogen. Recombinant bovine enteropeptidase that contained only the protease domain retained its catalytic activity toward small peptide substrates but was ~500-fold less active in processing trypsinogen (LaVallie *et al.*, 1993; Lu *et al.*, 1997).

C. Biochemical Properties

Porcine (Maroux *et al.*, 1971), bovine (Anderson *et al.*, 1977; Liepnieks and Light, 1979), and human (Grant *et al.*, 1978) enteropeptidase was purified from duodenal juices. The duodenum-derived protein consists of a heavy chain of 100–140 kDa and a light chain of 35–62 kDa that are linked by a disulfide bond. The enteropeptidase sequence contains a well-conserved activation site (K-IVGG) between the propeptide and the protease domain, indicating that proteolytic cleavage is required to convert the zymogen to an active two-chain enzyme (Kitamoto *et al.*, 1995). The sequence also predicts that the activated two-chain molecule is linked by a disulfide bond between Cys-788 and Cys-912. Consistent with these predictions, expression of recombinant bovine enteropeptidase in baby hamster kidney (BHK) cells led to a single-chain species of ~150 kDa which had no detectable activity toward peptide substrates. Incubation with trypsin converted the zymogen to an active enzyme containing two fragments of 133 and 43 kDa connected by a disulfide bond (Lu *et al.*, 1997). The catalytic activity of bovine enteropeptidase is inhibited by bovine pancreatic trypsin inhibitor and small-molecule serine protease inhibitors such as diisopropyl fluorophosphate (DFP), *p*-aminobenzamidine, and benzamidine (Lu and Sadler, 1998).

In the 1930s, Kunitz was first to partially purify enteropeptidase from swine duodenal fluid by ammonium sulfate precipitation and identify the enzyme as the physiological activator of trypsinogen (Kunitz, 1939a,b). Enteropeptidase activates trypsinogen by cleaving the VDDDDK-I peptide bond. This cleavage sequence with a basic residue at the P1 and acidic residues at the P2–P5 sites is well conserved in trypsinogen molecules from various species. Based on the enteropeptidase amino acid residues, Asp-981, Gly-1008, and Gly-1018, which form the S1 subsite, the basic residue at the P1 position is expected. However, the sequence of tandem Asp residues at P2–P5 is unusual. Studies using protein substrates and synthetic peptides have shown that a basic residue at P1 and acidic residues at P2 and P3 are required for the recognition by the enzyme. Acidic residues at P4 and P5 are also preferred but less critical (Light and Janska, 1989; Lu and Sadler, 1998). Such remarkable sequence specificity may provide an important mechanism to ensure that trypsinogen is activated only by enteropeptidase after it has been secreted into the small intestine but not by other proteases in pancreas. Excessive activation of trypsinogen in pancreas leads to devastating conditions such as acute pancreatitis. Despite the stringent sequence specificity for P1–P5 residues, enteropeptidase is much less specific for the amino acid residues C-terminal to the scissile bond. This property of enteropeptidase has been exploited for the *in vitro* cleavage of fusion proteins by designing recombinant fusion proteins with the enteropeptidase recognition sequence to achieve cleavage at a specific site (Collins-Racie *et al.*, 1995).

The structural determinants of the substrate specificity of enteropeptidase have been localized within the protease domain. In addition to the common features such as the catalytic triad and the substrate specificity pocket, the enteropeptidase light chain from most species contains a conserved sequence of four basic residues, R/KRRK, at positions 96–99. Molecular modeling based on three-dimensional structures from other serine proteases suggested that these basic residues may interact with the acidic P2–P5 residues within the trypsinogen activation sequence (Kitamoto et al., 1994; Matsushima et al., 1994). This hypothesis has been confirmed in a crystal structure of the enteropeptidase light chain complexed with an analog of the trypsinogen activation peptide (Lu et al., 1999). As one of the major determinants in the substrate recognition, the Lys-99 residue in the enteropeptidase light chain makes extensive contacts with P2 and P4 Asp residues in the trypsinogen activation peptide. Substitution of Lys-99 with Ala by site-directed mutagenesis virtually abolished the trypsinogen activation activity of enteropeptidase. In contrast, mutations at Lys-96, Arg-97, and Arg-98 in bovine enteropeptidase had much less dramatic effects (Lu et al., 1999).

D. Zymogen Activation

As the physiological trypsinogen activator, enteropeptidase appears to act at the top of the protease cascade in the digestive system. However, the presence of the conserved activation cleavage site in the proenteropeptidase sequence and the lack of apparent autoactivation of the zymogen (Lu et al., 1997) raise the question of how proenteropeptidase is activated. In assays using purified proteins, trypsin was shown to efficiently convert recombinant proenteropeptidase to an active enzyme (Lu et al., 1997), suggesting the possible reciprocal activation of proenteropeptidase and trypsinogen, but not answering the question of how the "first cut" takes place. The identification of a new serine protease, duodenase, from duodenal mucosa appears to solve this long-standing puzzle.

In 1995, Zamolodchikova et al. purified a novel serine protease of ∼30 kDa from bovine duodenal mucosa and named it duodenase (Zamolodchikova et al., 1995a). Protein sequence analysis indicated that the enzyme is a chymase-like enzyme composed of 226 amino acids (Zamolodchikova et al., 1995b). By immunofluorescence and immunoelectron microscopy, duodenase was found to be highly expressed in the secretory epithelial cells of Brunner's glands that are abundant in the proximal segment of duodenum (Zamolodchikova et al., 1997). Studies using protein and peptide substrates indicated that the preferred cleavage sequences of duodenase contain Lys at P1 and Pro at P2, matching the activation cleavage sequence of proenteropeptidase (Zamolodchikova et al., 1997). These findings suggested that duodenase could be the activator for proenteropeptidase. Further experiments have shown that duodenase indeed activates recombinant proenteropeptidase although it is ∼100 times slower compared with trypsin under similar experimental conditions (Zamolodchikova et al., 2000). It is not known whether a cofactor is still missing that may enhance the duodenase-mediated activation of proenteropeptidase. Nevertheless, these results support a possible protease cascade in which duodenase converts proenteropeptidase to enteropeptidase which in turn activates trypsinogen to trypsin (Zamolodchikova *et al.*, 2000). In addition to activating the downstream proteolytic enzymes such as chymotrypsinogen and proelastase, trypsin further amplifies the enzymatic reactions by cleaving proenteropeptidase. Such a positive feedback mechanism is well known in the serine protease-mediated cascades in blood coagulation and fibrinolysis.

E. Congenital Enteropeptidase Deficiency

The first case of congenital enteropeptidase deficiency was described in 1969 in a 3-week-old female infant who suffered from diarrhea and failed to gain body weight (Hadorn *et al.*, 1969). Since then, additional cases of congenital enteropeptidase deficiency have been reported (Ghishan *et al.*, 1983; Hadorn *et al.*, 1975). In these patients, diarrhea, vomiting, edema, anemia, hypoproteinemia, and failure to gain weight are common symptoms in early infancy. Very low or no enteropeptidase activity can be detected in intestinal biopsies and duodenal fluid samples. Treatment with pancreatic extracts usually is effective in these patients. Interestingly, the pancreatic replacement therapy can be discontinued in later life and patients live with essentially normal body weights and free of gastrointestinal symptoms (Hadorn *et al.*, 1975). This implies that, once initiated, the enzymatic reactions of the digestive proteases can be sustained, possibly by a trypsin(ogen)-mediated autoactivation mechanism.

The *enteropeptidase* gene in enteropeptidase-deficient patients from two unrelated families have been sequenced (Holzinger *et al.*, 2002). In one family with two affected siblings, the patients carried compound heterozygous nonsense mutations, one in exon 18 (2135 C > G) and another in exon 22 (2569 C > T), resulting in premature stop codons at residues Ser-712 and Arg-857, respectively. In another family with one patient, a heterozygous nonsense mutation was found in exon 8 (781 C > T) and another heterozygous deletion mutation (2707-8 del GT) in exon 23, resulting in a premature stop codon at residue Gln261 and a frameshift at residue Gln-902, respectively.

V. Hepsin

A. cDNA and Gene Structure

Hepsin was discovered in Earl Davie's laboratory in 1988 by screening a human liver cDNA library with degenerate oligonucleotides derived from a serine protease consensus sequence (Leytus *et al.*, 1988). A novel cDNA was identified that was ~1.8 kb in length and encoded a trypsin-like serine protease. The protease was designated hepsin for its expression in hepatocytes. In Northern analysis, hepsin mRNA was detected in several human tissues including liver, kidney, pancreas, prostate, and thyroid. Rat and mouse hepsin cDNA clones were also isolated that share high sequence homology (~88%) with the human sequence (Farley *et al.,* 1993; Kawamura *et al.,* 1999; Vu *et al.,* 1997). In mice, two alternatively spliced forms of hepsin mRNA were reported. On Northern blots, these two alternatively spliced forms appeared as 1.8- and 1.9-kb bands, respectively. Characterization of PCR products and genomic sequences showed that the 1.9-kb species was generated by an in-frame insertion of a 60-bp sequence near the 5'-end of the open reading frame, resulting in an addition of 20 amino acids within the cytoplasmic domain of mouse hepsin (Vu *et al.,* 1997). The functional significance of the alternatively spliced forms of hepsin is not known at this time. The human *hepsin* gene is mapped to chromosome 19q11-13.2.

B. Protein Domain Structures

The full-length human hepsin cDNA encodes a polypeptide of 417 amino acids with a calculated molecular mass of 45 kDa (Leytus *et al.,* 1988). A highly hydrophobic segment between residues 18 and 44 represents a putative transmembrane domain. In the extracellular region, there is a macrophage scavenger receptor class A cysteine-rich domain and a trypsin-like protease domain at the C terminus (Fig. 1). A conserved activation cleavage sequence R-IVGG is present between residues 162 and 166, suggesting that hepsin is synthesized as a zymogen. After the activation cleavage at Arg-162, the catalytic domain of hepsin is expected to be attached to the propeptide fragment on the cell surface by a disulfide bond between Cys-153 and Cys-277.

C. Biochemical Properties

Human and rat hepsin was purified from the membrane fractions of hepatocytes and liver tissues (Tsuji *et al.,* 1991; Zhukov *et al.,* 1997). In SDS–PAGE and Western analysis, hepsin appeared as a major band at ~51 kDa, indicating that hepsin is synthesized as a zymogen. The apparent molecular mass is consistent with predictions from the hepsin protein sequence that also contains a putative N-linked glycosylation site at residue 112 (Leytus *et al.,* 1988). A smaller minor band of 28 kDa was also detected by Western analysis, which represented the protease domain of hepsin generated by cleavage at the activation site during purification. The predicted topology for type II transmembrane protein was confirmed by immunohistochemistry and proteolytic digestion followed by SDS–PAGE and Western analysis (Kurachi *et al.,* 1994). The substrate specificity of the purified hepsin toward small synthetic peptides has been examined. The basic residues lysine and arginine, but not aromatic or aliphatic amino acids, are preferred at the P1 position. The catalytic activity of hepsin is inhibited by protein and small molecule inhibitors such as leupeptin, antipain, N^α-tosyl-L-lysine chloromethyl ketone, 4-amidiophenylmethylsulfonyl fluoride, and antithrombin III (Kurachi et al., 1994; Zhukov et al., 1997).

D. Functions of Hepsin

Several potential functions for hepsin have been suggested in the past. Kazama et al. (1995) first reported that hepsin might have a function in the activation of blood clotting factors. Recombinant human hepsin expressed on the surface of BHK cells was shown to convert factor VII (FVII) to active enzyme, FVIIa. The reaction appeared to be specific because hepsin did not activate other structurally related proteases such as factors IX and X, prothrombin, and protein C under similar experimental conditions. Because FVIIa is an important enzyme in the tissue factor-dependent coagulation pathway, this finding implied that hepsin may be involved in the initiation of blood coagulation.

In 1993, Torres-Rosado et al. reported that inhibition of hepsin by either antibodies or antisense oligonucleotides altered the morphology of hepatoma cells and impaired their growth in cell culture, suggesting that hepsin may act as a growth factor for hepatocytes. Several other serine proteases including thrombin and factor Xa are known to have signaling functions and, for example, thrombin is a potent mitogen for vascular fibroblasts and smooth muscle cells. To date, however, the molecular mechanism by which hepsin stimulates hepatocyte growth has not been defined.

In 1997, Vu et al. suggested another potential function of hepsin in embryonic development. In a search for membrane-associated serine proteases involved in fertilization, hepsin mRNA was found in mouse embryos as early as the two-cell stage. Furthermore, hepsin mRNA expression became most abundant in blastocysts prior to implantation, suggesting that hepsin may be upregulated to digest the zona pellucida, the extracellular coat of blastocysts, before the hatching of fertilized embryos. This proteolytic process is critical for a fertilized embryo to be implanted in the uterus. It was proposed that hepsin might be the enzyme responsible for blastocyst hatching in mammalian development.

E. Hepsin-Deficient Mice

To assess the biological importance of hepsin, hepsin-deficient mice were generated by homologous recombination techniques (Wu, 2001; Wu et al., 1998; Yu et al., 2000). Surprisingly, hepsin-deficient mice were viable and developed normally. Both male and female hepsin-deficient mice were fertile and produced viable offspring, indicating that hepsin is not essential for embryonic development. Necropsies and histologic examinations failed to identify any gross abnormalities in major organs including liver, kidney, brain, lung, heart, pancreas, spleen, and bone. No inflammatory cell infiltration and hepatocyte degeneration, necrosis, or apoptosis were detected in liver samples from hepsin-deficient mice.

In hematological examinations, values for platelet, red and white blood cell counts, hematocrit, and hemoglobin were similar in hepsin-deficient and wild-type mice. The tail bleeding time and plasma clotting times (aPTT and PT) were also normal in hepsin-deficient mice, indicating that hepsin deficiency has a minimal effect on normal hemostasis in mice. These data are consistent with the observation that no spontaneous bleedings occur in hepsin-deficient mice. To examine the potential function of hepsin under pathological conditions, hepsin-deficient mice were challenged in thromboplastin-induced disseminated intravascular coagulation (DIC) and lipopolysaccharide (LPS)-induced septic shock models (Wu *et al.*, 1998). The results were very similar in both hepsin-deficient mice and wild-type controls, indicating that the tissue-factor-mediated coagulation pathway is not significantly affected in hepsin-deficient mice under these experimental conditions.

To examine the effects of hepsin deficiency on liver function, blood samples were analyzed for liver-derived proteins. Serum concentrations of alanine aminotransferase (ALT), aspartate aminotransferase (AST), albumin, globulin, γ-glutamyltranspeptidase (GGT), amylase, creatine kinase, and lactate dehydrogenase were all similar in hepsin-deficient and wild-type control mice, indicating that the liver function is not significantly affected by hepsin deficiency. Hepsin-deficient mice were also challenged in an acute liver regeneration model (Yu *et al.*, 2000). In this model, a partial hepatectomy was performed to remove two-thirds of the liver and the liver mass restoration was examined at 24, 48, and 96 h after the surgery. At each time point, the rate of liver mass restoration was similar in hepsin-deficient and control mice, indicating that hepsin is not essential for hepatocyte proliferation during liver regeneration in mice. Together, studies with hepsin knockout mice demonstrate that hepsin is not an essential enzyme for normal hemostasis, embryogenesis, and maintenance of normal liver function.

Unexpectedly, serum alkaline phosphatase levels are $\sim$2-fold higher in hepsin-deficient male and female mice compared with wild-type controls (Wu *et al.*, 1998; Yu *et al.*, 2000). Analysis of alkaline phosphatase isotype showed that the elevated serum alkaline phosphatase from hepsin-deficient mice is mostly of bone origin. X-ray examinations performed at the age of 5 months did not detect structural abnormalities in long bones, pelvis, and vertebrae in hepsin-deficient mice. The molecular mechanism responsible for the mild elevation of serum alkaline phosphatase and the significance of this phenotype in hepsin-deficient mice remain unknown at present time.

F. Hepsin in Cancer

High levels of hepsin mRNA expression were first reported in advanced ovarian cancer and kidney carcinoma (Tanimoto *et al.*, 1997; Zacharski *et al.*, 1998). In ovarian cancer, hepsin mRNA overexpression correlated with the malignancy of the tumor, suggesting that hepsin may play a role in cancer progression and metastasis.

More recently, five laboratories independently reported that the hepsin gene was highly upregulated in advanced prostate cancers (Dhanasekaran *et al.*, 2001; Luo *et al.*, 2001; Magee *et al.*, 2001; Stamey *et al.*, 2001; Welsh *et al.*, 2001). In these studies, gene chips covering ~6500 to 9900 human genes were used to survey expression profiles in tissue samples from prostate cancer patients. *Hepsin* appeared to be the most upregulated gene in the cancer tissues and the mRNA expression was ~10 to 40-fold higher in advanced/metastatic prostate cancer than in normal prostate or benign prostatic hyperplasia (BPH) controls. By immunohistochemical staining, hepsin protein expression appeared to be higher in median-size (~1.3 cm) prostate cancer than large-size (~1.5 cm) or hormone-refractory prostate cancers (Dhanasekaran *et al.*, 2001). Interestingly, no differential expression of prostate specific antigen (PSA) mRNA was found in prostate cancer and BPH tissue samples (Stamey *et al.*, 2001), despite plasma PSA being widely used as a molecular marker for progressive prostate cancer in the clinic (Stamey *et al.*, 1999).

The finding of hepsin overexpression in prostate cancers is intriguing but its biological significance has not been determined. Hepsin could contribute to prostate cancer progression and metastasis by several molecular mechanisms. For example, hepsin may act as a growth factor to stimulate the proliferation of cancer cells. Such an activity of hepsin has already been reported for cultured hepatocytes (Torres-Rosado *et al.*, 1993). Alternatively, hepsin may function as a processing enzyme that activates polypeptide growth hormones required for cancer progression. Hepsin may also degrade extracellular matrix proteins either directly or indirectly by activating matrix metalloproteinases (MMPs). Proteolytic digestion of extracellular matrix proteins is well known to play a critical role in tumor invasion and metastasis. It is equally possible, however, that hepsin overexpression in prostate cancer may simply reflect the state of poorly differentiated prostate cells but has no significant impact on cancer growth. This question can be addressed experimentally using hepsin-deficient mice to determine if the lack of hepsin expression prevents or impairs cancer development. Testing prostate cancer models in these knockout animals should provide insights into the biological importance of hepsin in cancer.

VI. Matriptase

A. Molecular Cloning

Matriptase, also called membrane type-serine protease 1 (MT-SP1), is an epithelial-derived trypsin-like serine protease that was identified independently using different approaches. In 1993, Shi *et al.* discovered a protease of 80 kDa in the conditioned medium derived from a human breast cancer cell line, T-47D (Shi *et al.*, 1993). By zymography on gelatin-containing gels, the enzyme was proteolytically active and it was named matriptase. The protein was partially purified and sequenced. Based on the protein sequence, a 2830-bp partial cDNA sequence was found in an EST database and subsequently cloned from T-47D human breast cancer cells by RT-PCR-based methods (Lin et al., 1999c).

Independently, Craik and colleagues identified the same protease from prostate cancer cells (Takeuchi et al., 1999). In a mouse model of cancer using human prostate PC-3 cells, a bacterial polypeptide serine protease inhibitor, ecotin (McGrath et al., 1991), was shown to inhibit primary tumor growth and metastasis in nude mice. The results implied that serine protease activities derived from PC-3 cells might contribute to tumor progression in vivo. Using RT-PCR-based methods, Takeuchi et al. (1999) cloned a full-length cDNA from PC-3 cells that encoded a novel type II transmembrane protease, named MT-SP1 (Fig. 1). It is clear that matriptase and MT-SP1 are the same protein, but its official name still remains to be resolved. By Northern analysis, matriptase mRNA was found in a variety of human tissues including stomach, lung, placenta, prostate, small intestine, and colon (Takeuchi et al., 1999). Lower levels of matriptase mRNA were also detected in samples from the spleen and thymus. The human matriptase gene is located on chromosome 11q24.3.

The mouse homolog of matriptase, called epithin, was identified in fetal thymic stromal cells by screening genes differentially expressed in specific cell types (Kim et al., 1999). The full-length epithin cDNA was isolated from a thymus library from SCID mice. The predicted epithin protein shares 81% sequence identity with human matriptase. The epithin gene was mapped at 17 centimorgan from the centromere on mouse chromosome 9. More recently, rat matriptase cDNA was also isolated by two independent groups (Kishi et al., 2001; Satomi et al., 2001).

B. Biochemical Properties

1. Matriptase Protein and Its Inhibitors

Matriptase protein contains 855 amino acids with a calculated molecular mass of 95 kDa. In Western analysis under reducing conditions, native matriptase from PC-3 cell lysate appeared as a doublet at ∼87 kDa. A smaller band at ∼30 kDa was also detected that might represent the cleaved protease domain (Takeuchi et al., 2000). Treatment with PNGase F reduced the 87-kDa doublet to a single band at 85 kDa, consistent with the presence of four putative N-glycosylation sites in matriptase (Takeuchi et al., 1999). The topology of matriptase on the cell surface has been confirmed by immunofluorescent staining, biotinylation of cell-surface proteins, and membrane fractionation (Lin et al., 1997; Takeuchi et al., 2000). Soluble forms of matriptase were also found in cell culture medium or human milk either in a free form or in a complex with another protein (Lin et al., 1997, 1999b). Apparently, soluble matriptase is generated by ectodomain shedding. Cho et al. (2001) showed that proteolytic cleavage of Gly-149 in mouse matriptase was critical for generating the soluble protein in cultured cells.

The protein forming the complex with soluble matriptase has been isolated from human milk and identified as hepatocyte growth factor activator inhibitor-1 (HAI-1) (Lin *et al.*, 1999b). HAI-1 is a Kunitz-type serine protease inhibitor that was first isolated from the conditioned medium of a human stomach carcinoma cell line and shown to inhibit hepaptocyte growth factor activator (Shimomura *et al.*, 1997). HAI-1 binds to active matriptase, but not the zymogen, and inhibits the activity of matriptase in gelatin zymography and peptide substrate-based assays, suggesting that HAI-1 is a physiological inhibitor of matriptase (Benaud *et al.*, 2001). In *in vitro* assays, the catalytic activity of matriptase is also inhibited by other serine protease inhibitors including ecotin, leupeptin, DFP, and phenylmethylsulfonyl fluoride (PMSF) (Cho *et al.*, 2001; Shi *et al.*, 1993; Takeuchi *et al.*, 1999).

2. Activation Mechanism

Human matriptase contains a conserved activation cleavage sequence, R-VVGG, between residues 614 and 618, indicating that proteolytic cleavage at Arg-614 is required to activate the enzyme. Takeuchi *et al.* reported that soluble matriptase containing only the protease domain autoactivated when expressed and purified from *Escherichia coli* (Takeuchi *et al.*, 1999). Native matriptase from PC-3 or 184A1N4 cancer cells, however, appeared as a major band of ~87 kDa in Western blots under reducing conditions, suggesting that the majority of the protein remained as a zymogen (Benaud *et al.*, 2001; Takeuchi *et al.*, 1999). Matriptase on 184A1N4 cell surface was rapidly activated and cleaved from the cell membrane when exposed to sera from various species, suggesting that another protease or cofactor in the serum may contribute to the activation of matriptase (Benaud *et al.*, 2001). Most recently, a lipid component, sphingosine 1-phosphate (S1P), has been identified in human serum as a matriptase activator (Benaud *et al.*, 2002). In 184A1N4 cell culture, S1P rapidly induced matriptase activation on the cell surface in a calcium-dependent manner via an unknown molecular mechanism. It is unlikely that S1P activates matriptase directly because it is not an enzyme and does not bind to matriptase. In addition, S1P did not activate matriptase stably expressed on the surface of a different cell line. It is possible, however, that in 184A1N4 cells S1P elicits a cellular response leading to exposure or activation of an unknown protease that in turn activates matriptase.

3. Substrate Specificity and Protein Structure

The substrate specificity of recombinant matriptase toward small peptides has been examined using a positional scanning synthetic combinatorial library and substrate phage display (Takeuchi *et al.*, 2000). The preferred cleavage sequences are Arg/Lys at P4, basic residues or Gln at P3, Ser/Phe/Gly at P2, Arg/Lys at P1, and Ala at P1'. Similar results were also found using soluble matriptase purified from human milk (Lee *et al.*, 2000). A crystal structure of the matriptase catalytic domain has been reported (Friedrich *et al.,* 2002). The overall structure of the matriptase catalytic domain is typically trypsin/chymotrypsin-like. It has a trypsin-like S1 pocket, a small hydrophobic S2 subsite, and a negatively charged S4 cavity that favors the binding of basic P3/P4 residues. The structure also contains unique surface loops surrounding the active-site cleft that may contribute to the narrow substrate specificity of the enzyme. These results are consistent with the preferred substrate cleavage sequences determined experimentally.

C. Functions of Matriptase

Matriptase was initially identified as a gelatinase-like matrix-degrading enzyme from breast cancer cells. The physiological function and macromolecular substrate(s) of the enzyme have not been unambiguously defined. Takeuchi *et al.* (2000) searched potential matriptase substrates based on the preferred substrate cleavage sequences. Among potential candidates, protease-activated receptor-2 (PAR2) and urokinase (uPA) contain cleavage sequences matching the profile of P4(Arg/Lys)-P3(basic/Gln)-P2(Ser/Phe/Gly)-P1(Arg/Lys). In addition, the tissue expression patterns of PAR2 and uPA are similar to that of matriptase. When tested *in vitro,* soluble matriptase was shown to activate both PAR2 and uPA at low nanomolar concentrations (Takeuchi *et al.,* 2000).

In a similar search for potential matriptase substrates, Lee *et al.* (2000) looked for molecules that may interact with epithelial cells but require proteolytic activation cleavage at Arg or Lys residues. Hepatocyte growth factor (HGF), also called scatter factor, appears to be a good candidate. HGF is a potent growth factor secreted by stromal cells (Stoker *et al.,* 1987) and its growth-stimulating function is mediated by the cell surface receptor c-Met (Stewart, 1996). Previous studies have shown that proteolytic cleavage at Arg-495 is required to convert the single-chain form of HGF to an active two-chain molecule (Naka *et al.,* 1992). Using purified protein, Lee *et al.* (2000) showed that matriptase converted single-chain HGF to an active two-chain molecule as measured by c-Met tyrosine phosphorylation and Madin–Darby canine kidney (MDCK) cell scattering assays. In addition, matriptase also activated uPA, confirming the finding by Takeuchi *et al.* In contrast, matriptase did not cleave plasminogen that shares structural and sequence homology with HGF.

By activating HGF, uPA, and PAR2, matriptase may play a role in epithelial cell differentiation, extracellular matrix degradation, and tissue remodeling. HGF is essential for epithelial cell growth and differentiation. HGF deficiency impaired the development of epithelial organs including liver and placenta, leading to lethality in knockout mice (Schmidt *et al.,* 1995; Uehara *et al.,* 1995). Proteases of the fibrinolytic system such as uPA and plasmin are also known to play an important role in cell adhesion, tissue remodeling, and wound healing (Collen and Lijnen, 1991). It is not known, however, whether the matriptase-mediated activation of HGF, uPA, and PAR2 is physiologically important. Other proteases such as the

HGF activator and tryptase have also been reported to activate HGF and PAR2, respectively (Miyazawa *et al.*, 1993; Molino *et al.*, 1997). Additional biochemical and genetic experiments are needed to define the physiological importance of matriptase *in vivo*.

D. Matriptase in Cancer

In addition to PC-3 and T47D cancer cells where matriptase was originally identified, matriptase mRNA expression has also been found in many other cancers. By Northern analysis, *in situ* hybridization, and immunohistochemistry, matriptase mRNA expression was abundant in the primary breast carcinomas but barely detectable in surrounding noncancerous tissues (Oberst *et al.*, 2001). In ovarian tumors, the expression was positive in tumors of epithelial origin but negative in stromal cell-derived tumors. Matriptase mRNA was also found in most colon tumors examined. In contrast, matriptase mRNA was not detected in osteosarcomas, liposarcomas, histocytomas, and leiomyosarcoma. Dhanasekaran *et al.* (2001) also reported matriptase mRNA overexpression in prostate cancer as compared with normal prostate or BPH tissues. These studies indicate that matriptase is selectively expressed in epithelium-derived tumors and may contribute to tumor progression, invasion, and metastasis in cancer patients.

The role of matriptase in cancer is supported by studies with animal cancer models. Craik and colleagues have shown that ecotin and its derivatives are highly potent matriptase inhibitors. In both mouse and rat models of prostate cancer, subcutaneous injection of these potent matriptase inhibitors led to significant suppression of primary tumor growth and metastasis (Takeuchi *et al.*, 1999). Ihara *et al.* (2002) have reported an unexpected finding that further links matriptase to cancer. It is known that overexpression of GlcNAc-branched *N*-glycan correlates with poor prognosis in cancer patients. Increased activity of beta 1-6 *N*-acetylglucosaminyltransferase V (GnT-V), a Golgi N-linked glycan processing enzyme, is associated with malignant transformation of fibroblasts and epithelial cells (Dennis *et al.*, 1999). Mice with GnT-V deficiency are more resistant to tumor growth and metastasis (Granovsky *et al.*, 2000). To study the pro-metastatic effect of GnT-V, Ihara *et al.* established tumor cell lines overexpressing GnT-V and tested them in nude mice. The results showed that the transfected cells had a higher incidence of tumor metastasis compared with control cells (Ihara *et al.*, 2002). Analysis of the conditioned medium from the GnT-V transfected cells found significant increase in both matriptase protein and activity. Apparently, overexpression of GnT-V increased the attachment of GlcNAc-branched *N*-glycans to matriptase, thereby making the enzyme more resistant to degradation. To further confirm the role of matriptase in tumor metastasis, Ihara *et al.* established tumor cell lines overexpressing matriptase. When injected in nude mice, these cells also had a high incidence of metastasis to the lymph node.

If matriptase is indeed critical in cancer cell migration and metastasis in human patients, it will be possible to develop potent and selective matriptase inhibitors as an anticancer therapy. In fact, Enyedy *et al.* (2001) have reported a structure-based approach to design small-molecule matriptase inhibitors. By computer modeling and structure-based compound screening, a series of bis-benzamidines were identified as potential matriptase inhibitors. In fluorescent substrate-based assays, a lead compound inhibited the activity of human soluble matriptase with a K_i value of $\sim$200 nM. This compound had 9- and 13-fold selectivity toward uPA and thrombin, respectively. The results demonstrate the feasibility of developing potent and selective small molecule matriptase inhibitors. Additional studies are needed to test these matriptase inhibitors in animal cancer models and demonstrate their antitumor efficacy.

VII. Transmembrane Protease, Serine 2 (TMPRSS2)

A. Cloning of TMPRSS2 cDNA

Chromosome 21 is the smallest human autosome and known to be associated with a number of genetic diseases. For example, patients with an extra copy of chromosome 21 suffer from a series of developmental abnormalities, clinically known as Down syndrome. There has been strong interest in determining the structure of human chromosome 21 and identifying genes associated with human diseases. Using an exon-trapping method to search for genes on human chromosome 21, Paoloni-Giacobino *et al.* (1997) cloned a transmembrane serine protease gene, called *TMPRSS2*. The *TMPRSS2* gene, which spans 44 kb and contains 14 exons (Jacquinet *et al.,* 2001), is mapped to chromosome 21q22.3 (Paoloni-Giacobino *et al.,* 1997).

The full-length TMPRSS2 cDNA is $\sim$2.5 kb in length and encodes a polypeptide of 492 amino acids. The overall domain structure is similar to that of TMPRSS3 and TMPRSS4 (Fig. 1). The mouse TMPRSS2 homolog, also called epitheliasin, was first identified by searching EST databases for potential serine protease genes and subsequently cloned from a kidney cDNA library (Jacquinet *et al.,* 2000). Epitheliasin shares $\sim$80% sequence identity with human TMPRSS2. By FISH analysis, the mouse *epitheliasin* gene has been mapped to the telomeric region in the long arm of chromosome 16 (16C2), which is syntenic to human chromosome 21q22.2-3.

B. Tissue Expression of TMPRSS2

By Northern hybridization, human TMPRSS2 mRNA was detected in many epithelial cell-rich tissues such as prostate, kidney, colon, small intestine, pancreas, and lung (Jacquinet *et al.*, 2001). TMPRSS2 mRNA expression in the human heart was reported initially (Paoloni-Giacobino *et al.*, 1997), but not confirmed by other investigators (Afar *et al.*, 2001; Jacquinet *et al.*, 2001; Lin *et al.*, 1999a). By immunohistochemistry and *in situ* hybridization, the expression of TMPRSS2 protein and mRNA in kidney and prostate was restricted to epithelial cells (Jacquinet *et al.*, 2000; Lin *et al.*, 1999a). In these cells, TMPRSS2 protein is present mainly on the apical side of the cell (Afar *et al.*, 2001).

C. TMPRSS2 Protein

By SDS–PAGE analysis, *in vitro*-translated TMPRSS2 protein appeared as a major band at 54 kDa, consistent with the calculated mass based on the protein sequence (Afar *et al.*, 2001). Recombinant and native TMPRSS2 proteins had an apparent molecular mass of ~70 kDa on Western blots. The apparent difference in molecular mass between the *in vitro*-translated and native proteins is most likely due to different glycosylation patterns in the different production systems. TMPRSS2 contains three putative N-linked glycosylation sites at residues 128, 213, and 249. In human prostate tissue and cell lines, TMPRSS2 appears to be activated because Western analysis of tissue or cell lysate detected a band at ~32 kDa that may represent the proteolytically cleaved protease domain. Mutagenesis studies confirmed that the protein was cleaved at the predicted activation cleavage site. Afar *et al.* (2001) showed that the proteolytic cleavage was mediated by an autoactivation mechanism because mutation at the active site Ser-441 prevented the cleavage of recombinant TMPRSS2 in transfected 293T cells.

D. TMPRSS2 Function

The function of TMPRSS2 has not been well defined. Although the gene is mapped to human chromosome 21, TMPRSS2 is unlikely to be involved in Down syndrome because the *TMPRSS2* locus lies outside the Down syndrome critical region (Delabar *et al.*, 1993). Consistent with this notion, no TMPRSS2 mRNA expression was detected in the human brain. In a recent report, TMPRSS2 was shown to reduce epithelial sodium channel activity in a *Xenopus* oocyte expression system, suggesting a potential role of TMPRSS2 in the regulation of sodium transport in epithelial cells (Donaldson *et al.*, 2002). Several other reports, however, suggest that TMPRSS2 may be involved in prostate cancer progression. In a microarray-based study, Lin *et al.* (1999a) detected a six-fold increase of TM-PRSS2 mRNA in androgen-stimulated prostate cancer (LNCaP) cells compared with untreated controls. The upregulation of TMPRSS2 mRNA by androgen is apparently mediated by the androgen receptor. In PC-3 cells, which do not express the androgen receptor, TMPRSS2 mRNA was barely detectable. In contrast, in a PC-3 cell-derived cell line (PC-3 AR) that expresses recombinant wild-type androgen receptor, TMPRSS2 mRNA was significantly increased after androgen treatment (Afar *et al.*, 2001). In addition, androgen treatment also enhanced the activation cleavage of TMPRSS2, generating a soluble form of the protein by LNCaP cells in both cell culture and a mouse xenograft model (Afar *et al.*, 2001). Sequence analysis of the human *TMPRSS2* gene promoter revealed a 15-bp androgen response element at position −148 relative to the putative transcription start site (Lin *et al.*, 1999a). Since androgen plays an important role during the early stage of prostate cancer development, the discovery of the androgen-mediated upregulation of TMPRSS2 suggests an intriguing possibility that TMPRSS2 may contribute to prostate cancer growth and invasion. Identification of the physiological substrate(s) of TMPRSS2 will help to understand the role of this epithelium-derived transmembrane protease in prostate cancer.

VIII. Transmembrane Protease, Serine 3 (TMPRSS3)

TMPRSS3 was identified by positional cloning in a search for genes responsible for congenital deafness. In 1996, several independent genetic linkage studies using DNA samples from families with autosomal recessive deafness mapped a disease locus (*DFNB10*) to chromosome 21q22 (Bonne-Tamir *et al.*, 1996; Veske *et al.*, 1996). Cloning and characterization of the *DFNB10* critical region identified at least nine genes in this locus. These genes, however, were eliminated later as disease gene candidates because direct DNA sequencing failed to identify any mutations in these genes (Bartoloni *et al.*, 2000; Berry *et al.*, 2000; Wattenhofer *et al.*, 2001).

In 2001, Scott *et al.* analyzed the DFNB10 locus by searching the newly completed genomic sequences from human chromosome 21 and discovered four additional genes in this locus. One of the genes, *TMPRSS3*, contains 13 exons and spans 24 kb. Northern analysis showed that the *TMPRSS3* gene had two transcripts of 2.4 and 1.4 kb in length, respectively. The mRNA was detected in a variety of tissues including kidney, lung, colon, thymus, and fetal cochlea (Scott *et al.*, 2001). The full-length TMPRSS3 cDNA encodes a polypeptide of 454 amino acids (Fig. 1). TMPRSS3 and TMPRSS2 share 63% sequence similarities.

In a Palestinian family with a congenital form of deafness, a mutation in exon 11 of the *TMPRSS3* gene was identified that consisted of an 8-bp deletion and insertion of 18 β-satellite tandem units (Scott *et al.*, 2001). The β-satellite tandem units are repetitive sequences of ∼68 bp in length and normally present on the short arms of acrocentric chromosomes. Apparently, these repetitive sequences were inserted into the *TMPRSS3* gene through a recombination event. In another Pakistani family with a childhood-onset form of deafness, a splice site mutation in exon 4 was identified in the *TMPRSS3* gene, resulting in a frameshift (Scott *et al.*, 2001). In a separate study, Masmoudi *et al.* (2001) screened 39 Tunisian families affected by autosomal recessive deafness and identified two additional *TMPRSS3* missense mutations in two unrelated consanguineous families. In one family, a homozygous G to C mutation was found in exon 8 (753 G > C), resulting in a Trp to Cys substitution at residue 251 (W251C). In another family, a homozygous C to T mutation occurred in exon 12 (1221 C > T), resulting in a Pro to Leu substitution at residue 404 (P404L). These two mutations are close to active site residues, His-257 and Ser-401, and would be expected to alter the conformation of the active center, thereby affecting the proteolytic activity of the enzyme. In another study, Ben-Yosef et al. (2001) identified additional *TMPRSS3* mutations in four Pakistani families with congenital autosomal recessive deafness. These mutations include 325 C > T in exon 5, resulting in an Arg to Trp (R109W) substitution, 581 C > T in exon 7, resulting in a Cys to Phe (C194F) substitution, and 1219 T > C in exon 12, resulting in a Cys to Arg (C407R) substitution. Both R109W and C194F mutations are located in the scavenger receptor cysteine-rich domain whereas C407R mutation is near the active site Ser-401 in the protease domain.

These genetic studies indicate that the transmembrane serine protease TMPRSS3 is important for normal hearing. At this time, the biochemical function of TMPRSS3 has not been characterized. None of the mutant TMPRSS3 proteins has been expressed and characterized. Since TMPRSS3 is a transmembrane protease and its mRNA was detected in fetal cochlea, it is possible that TMPRSS3 plays a role in the development and maintenance of the function of the inner ear. In a separate study, TMPRSS3, also called TADG-12, was reported to be overexpressed in ovarian cancers (Underwood et al., 2000). The biological significance of this finding is not known. Further study of TMPRSS3 will help to identify the physiological substrate(s) of this enzyme and may lead to a better understanding of its potential function in the normal hearing process.

IX. Transmembrane Protease, Serine 4 (TMPRSS4)

TMPRSS4 was first identified as a cDNA clone that was overexpressed in pancreatic cancer. In a comparison of genes differentially expressed in pancreatic cancer and normal pancreatic or pancreatitis tissues, Gress et al. (1997) identified 16 distinct cancer-specific cDNA fragments. One of the cDNA clones, RDA12, shared sequence homology with other trypsin-like serine proteases. Further cloning of the full-length cDNA identified a novel transmembrane serine protease gene that was named *TMPRSS3* for transmembrane protease, serine 3 (Wallrapp et al., 2000). However, the name of TMPRSS3 had already been used for the transmembrane serine protease gene localized on human chromosome 21 (Scott et al., 2001). Subsequently, the name of TMPRSS4 was given to this pancreatic cancer-derived transmembrane serine protease (Hooper et al., 2001). The *TMPRSS4* gene is mapped to human chromosome 11q23.3 (Wallrapp et al., 2000).

The predicted TMPRSS4 protein consists of 437 amino acids with a calculated molecular mass of 48 kDa. The overall domain structure of TMPRSS4 is similar to those of TMPRSS2 and TMPRSS3 (Fig. 1). TMPRSS2, TMPRSS3, and

TMPRSS4 proteins share ~50% sequence similarities, suggesting that they may have evolved from the same ancestral gene. In TMPRSS4, there are two putative N-linked glycosylation sites at residues 130 and 178 in the extracellular region. In SDS–PAGE and Western analysis, recombinant TMPRSS4 containing an epitope tag had an apparent molecular mass of ~68 kDa (Wallrapp *et al.*, 2000).

The function of this protease has not been defined. In Northern analysis, TMPRSS4 mRNA was detected in a high percentage of primary pancreatic cancer tissues (9/13) and cell lines (10/16). In addition, strong TMPRSS4 mRNA expression was also detected in other tumor tissues such as gastric and colorectal cancers (Wallrapp *et al.*, 2000). In contrast, TMPRSS4 mRNA was not detectable by Northern blotting in normal pancreas, heart, brain, placenta, lung, liver, and uterus. Low levels of expression of TMPRSS4 were found in tissues of the normal gastrointestinal and urogenital tracts such as stomach, small intestine, colon, kidney, and bladder. Interestingly, TMPRSS4 mRNA expression in several pancreative cancer cell subclones positively correlated with their metastatic potentials, suggesting a possible role of TMPRSS4 in tumor progression and metastasis.

X. Other Transmembrane Serine Proteases

A. Human Airway Trypsin-Like Protease

Human airway trypsin-like protease (HAT) was first isolated as a ~27-kDa trypsin-like protease in the sputum from patients with chronic airway diseases (Yasuoka *et al.*, 1997). In assays with synthetic peptide substrates, HAT has a trypsin-like substrate specificity preferring Arg at the P1 position. The enzymatic activity was inhibited by DFP, leupeptin, antipain, aprotinin, and soybean trypsin inhibitor but not secretory leukocyte protease inhibitor. By immunohistochemical staining, the protein was detected mainly in cells of submucosal serous glands of the bronchi and trachea.

The full-length HAT cDNA was later cloned from the trachea that encodes a polypeptide of 418 amino acids (Yamaoka *et al.*, 1998) (Fig. 1). By Northern hybridization, HAT mRNA of ~1.9 kb in length was detected only in samples derived from human trachea but not in other tissues including heart, brain, pancreas, lung, and liver (Yamaoka *et al.*, 1998). Recombinant HAT protein expressed in insect cells appeared as a major band of ~48 kDa on SDS–PAGE and Western analysis, consistent with the calculated molecular mass of 46 kDa (Yamaoka *et al.*, 1998). The HAT protein sequence also contains two putative N-linked glycosylation sites at resides 144 and 152. A minor species of ~28 kDa was also present in the preparation of recombinant HAT, which may represent the protease domain activated during the purification. In assays using purified proteins, both recombinant HAT expressed in the insect cells and native HAT purified from the sputum of patients degraded fibrinogen α-chain but not β- and γ-chains. However, the physiological significance of the fibrinogen-degrading activity still remains unclear. Because

HAT is specifically expressed in the serous gland of the bronchi and trachea, it has been postulated that HAT may be a part of the host defense system in the human airway. The mucous membrane in the airway contains a host of secreted proteins such as lysozyme, secretory IgA, and leukocyte protease inhibitor. It will be important to determine whether HAT is involved in the processing or degradation of mucosal proteins in the airway.

B. Spinesin

Spinesin was cloned from a human spinal cord library by a PCR-base strategy (Yamaguchi *et al.*, 2002). The full-length cDNA is 2265 bp and encodes a type II transmembrane serine protease of 457 amino acids, designated spinesin for spinal cord-enriched trypsin-like protease. The protein sequence also contains five putative N-linked glycosylation sites in the extracellular region, two of which are located in the protease domain. The overall domain structure of spinesin is similar to those of hepsin, TMPRSS2, and TMPRSS3 (Fig. 1). In fact, these proteases share ∼40% sequence similarity in their protease domains. The human *spinesin* gene is located on chromosome 11q23. By Northern analysis, spinesin mRNA was found in samples derived from human brain but not kidney, liver, lung, placenta, and heart, suggesting that spinesin is a specific protease in the central nervous system. In immunohistochemical examinations using antibodies against human spinesin, positive staining was identified in tissue sections from the spinal cord. It appeared that spinesin protein was mainly present in neuronal cells and their axons at the anterior horn of the spinal cord. The positive staining was also found at the synapses of motoneurons in the spinal cord.

In SDS–PAGE and Western analysis, anti-spinesin antibodies detected a major band of ∼52 kDa in human brain homogenate, indicating that spinesin is synthesized as a zymogen (Yamaguchi *et al.*, 2002). An ∼50 kDa band was also detected in protein samples from the cerebrospinal fluid, suggesting the presence of soluble forms of spinesin in the cerebrospinal fluid, possibly generated by protease-mediated shedding. Recombinant soluble spinesin that contained only the protease domain was active toward synthetic substrates with an optimal pH of ∼10. The recombinant protein was also active in gelatin-based zymography assays. The catalytic active site of soluble spinesin was inhibited by *p*-amidinophenylmethanesulfonyl fluoride (1 μM) but not antipain and leupeptin (1 mM). The physiological substrates and potential function of spinesin in the central nervous system are not known at present time.

C. Mosaic Serine Protease L

In 2001, Kim *et al.* reported a cDNA isolated from a human lung library that encodes a novel mosaic serine protease, named MSPL. Apparently, the cDNA has another alternatively spliced form that encodes a shorter version of the protein,

MSPS. The full-length MSPL cDNA encodes a putative type II transmembrane serine protease of 581 amino acids. The predicted protein contains a cytoplasmic tail, a transmembrane domain, an LDL receptor repeat and a trypsin-like domain (Fig. 1). The protease domain shares sequence homology with those of human kallikrein (42%), hepsin (39%), and TMPRSS2 (43%). The alternatively spliced cDNA encodes a putative soluble form that lacks the transmembrane domain. By PCR analysis, *MSPL* transcripts were detected in human lung, placenta, pancreas, and prostate but not brain, liver, small intestine, kidney, thymus, muscle, or ovary. In SDS–PAGE analysis, recombinant MSPL and MSPS appeared as single bands of 60 and 57 kDa, respectively. To date, there is limited information on the activity and function of this lung-derived serine protease.

XI. Perspective

The serine proteases of the trypsin superfamily are by far the most versatile and best-characterized proteolytic enzymes. The importance of these enzymes in a variety of biological processes has been demonstrated both biochemically and genetically. Although the membrane-bound proteases have been known for many years, the significance of type II transmembrane serine proteases as a distinct subclass was not recognized until the recent discovery of several new members by functional genomics studies. It is well known that cell membrane surfaces contribute significantly to protease-mediated events. In blood coagulation, for example, assembly of clotting factors on the cell surface greatly enhances the rate of catalytic reactions. For soluble clotting enzymes, this is achieved by their interactions with phospholipids on the cell membrane and by binding to integral membrane protein cofactors such as tissue factor and thrombomodulin (Davie *et al.*, 1991; Furie and Furie, 1988; Mann *et al.*, 1990). The presence of a transmembrane domain in the type II transmembrane serine proteases clearly provides a unique regulatory mechanism to control the activity of these proteases.

Despite their structural similarities, the type II transmembrane serine proteases evidently have distinct functions. For example, enteropeptidase is a key enzyme in the protease cascade involved in food digestion whereas corin has a function in processing natriuretic peptides important in maintaining normal blood pressure. The biological functions of most type II transmembrane serine proteases have not yet been defined. The specific expression of several enzymes such as hepsin, spinesin, TMPRSS3, and HAT in different cell types and tissues suggests that these proteases are involved in diverse physiological or pathological processes. Although *Drosophila* transmembrane serine proteases such as *Stubble-stubbloid* are critical in embryonic development, defects in mammalian transmembrane serine proteases such as enteropeptidase and hepsin have less severe phenotypes. Further genetic studies are expected to assess the biological importance of other type II transmembrane serine proteases.

The observation that many type II transmembrane serine proteases such as hepsin, matriptase, TMPRSS2, TMPRSS3, and TMPRSS4 are upregulated in cancer tissues is intriguing, but its functional significance remains to be determined. If the type II transmembrane serine proteases are indeed critical in tumor invasion and metastasis, these enzymes should represent valuable molecular targets for developing new anticancer drugs that have distinct molecular mechanisms.

Acknowledgments

The author thanks Drs. Junliang Pan, Joyce Chan, John Morser, and Bill Dole for their helpful discussions and critical reading of the manuscript.

References

Afar, D. E., Vivanco, I., Hubert, R. S., Kuo, J., Chen, E., Saffran, D. C., Raitano, A. B., and Jakobovits, A. (2001). Catalytic cleavage of the androgen-regulated TMPRSS2 protease results in its secretion by prostate and prostate cancer epithelia. *Cancer Res.* **61,** 1686–1692.

Allgren, R. L., Marbury, T. C., Rahman, S. N., Weisberg, L. S., Fenves, A. Z., Lafayette, R. A., Sweet, R. M., Genter, F. C., Kurnik, B. R., Conger, J. D., and Sayegh, M. H. (1997). Anaritide in acute tubular necrosis. Auriculin Anaritide Acute Renal Failure Study Group. *N. Engl. J. Med.* **336,** 828–834.

Anderson, L. E., Walsh, K. A., and Neurath, H. (1977). Bovine enterokinase. Purification, specificity, and some molecular properties. *Biochemistry* **16,** 3354–3360.

Appel, L. F., Prout, M., Abu-Shumays, R., Hammonds, A., Garbe, J. C., Fristrom, D., and Fristrom, J (1993). The *Drosophila Stubble-stubbloid* gene encodes an apparent transmembrane serine protease required for epithelial morphogenesis. *Proc. Natl. Acad. Sci. USA* **90,** 4937–4941.

Bartoloni, L., Wattenhofer, M., Kudoh, J., Berry, A., Shibuya, K., Kawasaki, K., Wang, J., Asakawa, S., Talior, I., Bonne-Tamir, B., Rossier, C., Michaud, J., McCabe, E. R., Minoshima, S., Shimizu, N., Scott, H. S., and Antonarakis, S. E. (2000). Cloning and characterization of a putative human glycerol 3-phosphate permease gene (SLC37A1 or G3PP) on 21q22.3: mutation analysis in two candidate phenotypes, DFNB10 and a glycerol kinase deficiency. *Genomics* **70,** 190–200

Belvin, M. P., and Anderson, K. V. (1996). A conserved signaling pathway: the *Drosophila* toll-dorsal pathway. *Annu. Rev. Cell. Dev. Biol.* **12,** 393–416.

Benaud, C., Dickson, R. B., and Lin, C. Y. (2001). Regulation of the activity of matriptase on epithelial cell surfaces by a blood-derived factor. *Eur. J. Biochem.* **268,** 1439–1447.

Benaud, C., Oberst, M., Hobson, J. P., Spiegel, S., Dickson, R. B., and Lin, C. Y. (2002). Sphingosine 1-phosphate, present in serum-derived lipoproteins, activates matriptase. *J. Biol. Chem.* **277,** 10539–10546.

Ben-Yosef, T., Wattenhofer, M., Riazuddin, S., Ahmed, Z. M., Scott, H. S., Kudoh, J., Shibuya, K., Antonarakis, S. E., Bonne-Tamir, B., Radhakrishna, U., Naz, S., Ahmed, Z., Pandya, A., Nance, W. E., Wilcox, E. R., Friedman, T. B., and Morell, R. J. (2001). Novel mutations of TMPRSS3 in four DFNB8/B10 families segregating congenital autosomal recessive deafness. *J. Med. Genet.* **38,** 396–400.

Berry, A., Scott, H. S., Kudoh, J., Talior, I., Korostishevsky, M., Wattenhofer, M., Guipponi, M., Barras, C., Rossier, C., Shibuya, K., Wang, J., Kawasaki, K., Asakawa, S., Minoshima, S., Shimizu, N., Antonarakis, S., and Bonne-Tamir, B. (2000). Refined localization of autosomal recessive nonsyndromic deafness DFNB10 locus using 34 novel microsatellite markers, genomic structure, and exclusion of six known genes in the region. *Genomics* **68,** 22–29.

Bleyl, S., Nelson, L., Odelberg, S. J., Ruttenberg, H. D., Otterud, B., Leppert, M., and Ward, K. (1995). A gene for familial total anomalous pulmonary venous return maps to chromosome 4p13-q12. *Am. J. Hum. Genet.* **56,** 408–415.

Bloch, K. D., Scott, J. A., Zisfein, J. B., Fallon, J. T., Margolies, M. N., Seidman, C. E., Matsueda, G. R., Homcy, C. J., Graham, R. M., and Seidman, J. G. (1985). Biosynthesis and secretion of proatrial natriuretic factor by cultured rat cardiocytes. *Science* **230,** 1168–1171.

Bonne-Tamir, B., DeStefano, A. L., Briggs, C. E., Adair, R., Franklyn, B., Weiss, S., Korostishevsky, M., Frydman, M., Baldwin, C. T., and Farrer, L. A. (1996). Linkage of congenital recessive deafness (gene DFNB10) to chromosome 21q22.3. *Am. J. Hum. Genet.* **58,** 1254–1259.

Brown, M. S., Herz, J., and Goldstein, J. L. (1997). LDL-receptor structure. Calcium cages, acid baths and recycling receptors [news; comment]. *Nature* **388,** 629–630.

Burnett, J. C., Jr., Kao, P. C., Hu, D. C., Heser, D. W., Heublein, D., Granger, J. P., Opgenorth, T. J., and Reeder, G. S. (1986). Atrial natriuretic peptide elevation in congestive heart failure in the human. *Science* **231,** 1145–1147.

Cadigan, K. M., and Nusse, R. (1997). Wnt signaling: a common theme in animal development. *Genes Dev.* **11,** 3286–3305.

Catterall, C. F., Lyons, A., Sim, R. B., Day, A. J., and Harris, T. J. (1987). Characterization of primary amino acid sequence of human complement control protein factor I from an analysis of cDNA clones. *Biochem. J.* **242,** 849–856.

Cho, E. G., Kim, M. G., Kim, C., Kim, S. R., Seong, I. S., Chung, C., Schwartz, R. H., and Park, D. (2001). N-terminal processing is essential for release of epithin, a mouse type II membrane serine protease. *J. Biol. Chem.* **276,** 44581–44589.

Collen, D., and Lijnen, H. R. (1991). Basic and clinical aspects of fibrinolysis and thrombolysis. *Blood* **78,** 3114–3124.

Collins-Racie, L. A., McColgan, J. M., Grant, K. L., DiBlasio-Smith, E. A., McCoy, J. M., and LaVallie, E. R. (1995). Production of recombinant bovine enterokinase catalytic subunit in *Escherichia coli* using the novel secretory fusion partner DsbA. *Biotechnology (NY)* **13,** 982–987.

Colucci, W. S., Elkayam, U., Horton, D. P., Abraham, W. T., Bourge, R. C., Johnson, A. D., Wagoner, L. E., Givertz, M. M., Liang, C. S., Neibaur, M., Haught, W. H., and LeJemtel, T. H. (2000). Intravenous nesiritide, a natriuretic peptide, in the treatment of decompensated congestive heart failure. Nesiritide Study Group. *N. Engl. J. Med.* **343,** 246–253.

Dangott, L. J., Jordan, J. E., Bellet, R. A., and Garbers, D. L. (1989). Cloning of the mRNA for the protein that crosslinks to the egg peptide speract. *Proc. Natl. Acad. Sci. USA* **86,** 2128–2132.

Davie, E. W., Fujikawa, K., and Kisiel, W. (1991). The coagulation cascade: initiation, maintenance, and regulation. *Biochemistry* **30,** 10363–10370.

de Bold, A. J. (1985). Atrial natriuretic factor: a hormone produced by the heart. *Science* **230,** 767–770.

de Bold, A. J., Borenstein, H. B., Veress, A. T., and Sonnenberg, H. (1981). A rapid and potent natriuretic response to intravenous injection of atrial myocardial extract in rats. *Life Sci.* **28,** 89–94.

Delabar, J. M., Theophile, D., Rahmani, Z., Chettouh, Z., Blouin, J. L., Prieur, M., Noel, B., and Sinet, P. M. (1993). Molecular mapping of twenty-four features of Down syndrome on chromosome 21. *Eur. J. Hum. Genet.* **1,** 114–124.

DeLotto, R., and Spierer, P. (1986). A gene required for the specification of dorsal–ventral pattern in *Drosophila* appears to encode a serine protease. *Nature* **323,** 688–692.

Dennis, J. W., Granovsky, M., and Warren, C. E. (1999). Glycoprotein glycosylation and cancer progression. *Biochim. Biophys. Acta* **1473,** 21–34.

Dhanasekaran, S. M., Barrette, T. R., Ghosh, D., Shah, R., Varambally, S., Kurachi, K., Pienta, K. J., Rubin, M. A., and Chinnaiyan, A. M. (2001). Delineation of prognostic biomarkers in prostate cancer. *Nature* **412,** 822–826.

DiStefano, J. F., Beck, G., Lane, B., and Zucker, S. (1982). Role of tumor cell membrane-bound serine proteases in tumor-induced target cytolysis. *Cancer Res.* **42,** 207–218.

DiStefano, J. F., Beck, G., and Zucker, S. (1983). Characterization of tumor cell membrane serine proteases by polyacrylamide gel electrophoresis. *J. Histochem. Cytochem.* **31,** 1233–1240.

Donaldson, S. H., Hirsh, A., Li, D. C., Holloway, G., Chao, J., Boucher, R. C., and Gabriel, S. E. (2002). Regulation of the epithelial sodium channel by serine proteases in human airways. *J. Biol. Chem.* **277,** 8338–8345.

Enyedy, I. J., Lee, S. L., Kuo, A. H., Dickson, R. B., Lin, C. Y., and Wang, S. (2001). Structure-based approach for the discovery of bis-benzamidines as novel inhibitors of matriptase. *J. Med. Chem.* **44,** 1349–1355.

Farley, D., Reymond, F., and Nick, H. (1993). Cloning and sequence analysis of rat hepsin, a cell surface serine proteinase. *Biochim. Biophys. Acta* **1173,** 350–352.

Friedrich, R., Fuentes-Prior, P., Ong, E., Coombs, G., Hunter, M., Oehler, R., Pierson, D., Gonzalez, R., Huber, R., Bode, W., and Madison, E. L. (2002). Catalytic domain structures of MT-SP1/matriptase, a matrix-degrading transmembrane serine proteinase. *J. Biol. Chem.* **277,** 2160–2168.

Furie, B., and Furie, B. C. (1988). The molecular basis of blood coagulation. *Cell* **53,** 505–518.

Ghishan, F. K., Lee, P. C., Lebenthal, E., Johnson, P., Bradley, C. A., and Greene, H. L. (1983). Isolated congenital enterokinase deficiency. Recent findings and review of the literature. *Gastroenterology* **85,** 727–731.

Gottlieb, S. S., Kukin, M. L., Ahern, D., and Packer, M. (1989). Prognostic importance of atrial natriuretic peptide in patients with chronic heart failure. *J. Am. Coll. Cardiol.* **13,** 1534–1539.

Granovsky, M., Fata, J., Pawling, J., Muller, W. J., Khokha, R., and Dennis, J. W. (2000). Suppression of tumor growth and metastasis in Mgat5-deficient mice. *Nat. Med.* **6,** 306–312.

Grant, D. A., Magee, A. I., and Hermon-Taylor, J. (1978). Optimisation of conditions for the affinity chromatography of human enterokinase on immobilised *p*-aminobenzamidine. Improvement of the preparative procedure by inclusion of negative affinity chromatography with glycylglycyl-aniline. *Eur. J. Biochem.* **88,** 183–189.

Gress, T. M., Wallrapp, C., Frohme, M., Muller-Pillasch, F., Lacher, U., Friess, H., Buchler, M., Adler, G., and Hoheisel, J. D. (1997). Identification of genes with specific expression in pancreatic cancer by cDNA representational difference analysis. *Genes Chromosomes Cancer* **19,** 97–103.

Hadorn, B., Haworth, J. C., Gourley, B., Prasad, A., and Troesch, V. (1975). Intestinal enterokinase deficiency. Occurrence in two sibs and age dependency of clinical expression. *Arch. Dis. Child* **50,** 277–282.

Hadorn, B., Tarlow, M. J., Lloyd, J. K., and Wolff, O. H. (1969). Intestinal enterokinase deficiency. *Lancet* **1,** 812–813.

Han, J. H., Lee, S. H., Tan, Y. Q., LeMosy, E. K., and Hashimoto, C. (2000). Gastrulation defective is a serine protease involved in activating the receptor toll to polarize the *Drosophila* embryo. *Proc. Natl. Acad. Sci. USA* **97,** 9093–9097.

Hartmann, E., Rapoport, T. A., and Lodish, H. F. (1989). Predicting the orientation of eukaryotic membrane-spanning proteins. *Proc. Natl. Acad. Sci. USA* **86,** 5786–5790.

Hermon-Taylor, J., Perrin, J., Grant, D. A., Appleyard, A., Bubel, M., and Magee, A. I. (1977). Immunofluorescent localisation of enterokinase in human small intestine. *Gut* **18,** 259–265.

Holzinger, A., Maier, E. M., Buck, C., Mayerhofer, P. U., Kappler, M., Haworth, J. C., Moroz, S. P., Hadorn, H. B., Sadler, J. E., and Roscher, A. A. (2002). Mutations in the proenteropeptidase gene are the molecular cause of congenital enteropeptidase deficiency. *Am. J. Hum. Genet.* **70,** 20–25.

Hong, C. C., and Hashimoto, C. (1995). An unusual mosaic protein with a protease domain, encoded by the nudel gene, is involved in defining embryonic dorsoventral polarity in *Drosophila. Cell* **82,** 785–794.

Hooper, J. D., Clements, J. A., Quigley, J. P., and Antalis, T. M. (2001). Type II transmembrane serine proteases. Insights into an emerging class of cell surface proteolytic enzymes. *J. Biol. Chem.* **276,** 857–860.

Hooper, J. D., Scarman, A. L., Clarke, B. E., Normyle, J. F., and Antalis, T. M. (2000). Localization of the mosaic transmembrane serine protease corin to heart myocytes. *Eur. J. Biochem.* **267**, 6931–6937.

Ihara, S., Miyoshi, E., Ko, J. H., Murata, K., Nakahara, S., Honke, K., Dickson, R. B., Lin, C. Y., and Taniguchi, N. (2002). Pro-metastatic effect of *N*-acetylglucosaminyltransferase V is due to modification and stabilization of active matriptase by adding beta 1–6 GlcNAc-branching. *J. Biol. Chem.* **25**, 25.

Imada, T., Takayanagi, R., and Inagami, T. (1987). Identification of a peptidase which processes atrial natriuretic factor precursor to its active form with 28 amino acid residues in particulate fractions of rat atrial homogenate. *Biochem. Biophys. Res. Commun.* **143**, 587–592.

Imada, T., Takayanagi, R., and Inagami, T. (1988). Atrioactivase, a specific peptidase in bovine atria for the processing of pro-atrial natriuretic factor. Purification and characterization. *J. Biol. Chem.* **263**, 9515–9519.

Inagami, T. (1989). Atrial natriuretic factor. *J. Biol. Chem.* **264**, 3043–3046.

Irving, P., Troxler, L., Heuer, T. S., Belvin, M., Kopczynski, C., Reichhart, J. M., Hoffmann, J. A., and Hetru, C. (2001). A genome-wide analysis of immune responses in *Drosophila*. *Proc. Natl. Acad. Sci. USA* **98**, 15119–15124.

Ito, T., Toki, Y., Siegel, N., Gierse, J. K., and Needleman, P. (1988). Manipulation of stretch-induced atriopeptin prohormone release and processing in the perfused rat heart. *Proc. Natl. Acad. Sci. USA* **85**, 8365–8369.

Jacquinet, E., Rao, N. V., Rao, G. V., and Hoidal, J. R. (2000). Cloning, genomic organization, chromosomal assignment and expression of a novel mosaic serine proteinase: epitheliasin. *FEBS Lett.* **468**, 93–100.

Jacquinet, E., Rao, N. V., Rao, G. V., Zhengming, W., Albertine, K. H., and Hoidal, J. R. (2001). Cloning and characterization of the cDNA and gene for human epitheliasin. *Eur. J. Biochem.* **268**, 2687–2699.

Jin, Y. S., and Anderson, K. V. (1990). Dominant and recessive alleles of the *Drosophila easter* gene are point mutations at conserved sites in the serine protease catalytic domain. *Cell* **60**, 873–881.

John, S. W., Krege, J. H., Oliver, P. M., Hagaman, J. R., Hodgin, J. B., Pang, S. C., Flynn, T. G., and Smithies, O. (1995). Genetic decreases in atrial natriuretic peptide and salt-sensitive hypertension [published erratum appears in *Science* (1995) 267, 1753]. *Science* **267**, 679–681.

Kawamura, S., Kurachi, S., Deyashiki, Y., and Kurachi, K. (1999). Complete nucleotide sequence, origin of isoform and functional characterization of the mouse hepsin gene. *Eur. J. Biochem.* **262**, 755–764.

Kazama, Y., Hamamoto, T., Foster, D. C., and Kisiel, W. (1995). Hepsin, a putative membrane-associated serine protease, activates human factor VII and initiates a pathway of blood coagulation on the cell surface leading to thrombin formation. *J. Biol. Chem.* **270**, 66–72.

Kim, D. R., Sharmin, S., Inoue, M., and Kido, H. (2001). Cloning and expression of novel mosaic serine proteases with and without a transmembrane domain from human lung. *Biochim. Biophys. Acta* **1518**, 204–209.

Kim, M. G., Chen, C., Lyu, M. S., Cho, E. G., Park, D., Kozak, C., and Schwartz, R. H. (1999). Cloning and chromosomal mapping of a gene isolated from thymic stromal cells encoding a new mouse type II membrane serine protease, epithin, containing four LDL receptor modules and two CUB domains. *Immunogenetics* **49**, 420–428.

Kishi, K., Yamazaki, K., Yasuda, I., Yahagi, N., Ichinose, M., Tsuchiya, Y., Athauda, S. B., Inoue, H., and Takahashi, K. (2001). Characterization of a membrane-bound arginine-specific serine protease from rat intestinal mucosa. *J. Biochem. (Tokyo)* **130**, 425–430.

Kitamoto, Y., Veile, R. A., Donis-Keller, H., and Sadler, J. E. (1995). cDna sequence and chromosomal localization of human enterokinase, the proteolytic activator of trypsinogen. *Biochemistry* **34**, 4562–4568.

Kitamoto, Y., Yuan, X., Wu, Q., McCourt, D. W., and Sadler, J. E. (1994). Enterokinase, the initiator of intestinal digestion, is a mosaic protease composed of a distinctive assortment of domains. *Proc. Natl. Acad. Sci. USA* **91**, 7588–7592.

Koller, K. J., and Goeddel, D. V. (1992). Molecular biology of the natriuretic peptides and their receptors. *Circulation* **86**, 1081–1088.

Kounnas, M. Z., Chappell, D. A., Strickland, D. K., and Argraves, W. S. (1993). Glycoprotein 330, a member of the low density lipoprotein receptor family, binds lipoprotein lipase in vitro. *J. Biol. Chem.* **268**, 14176–14781.

Kraut, J. (1977). Serine proteases: structure and mechanism of catalysis. *Annu. Rev. Biochem.* **46**, 331–358.

Krieger, M., and Herz, J. (1994). Structures and functions of multiligand lipoprotein receptors: macrophage scavenger receptors and LDL receptor-related protein (LRP). *Annu. Rev. Biochem.* **63**, 601–637.

Kunitz, M. (1939a). Formation of trypsin from crystalline trypsinogen by means of enterokinase. *J. Gen. Physiol.* **22**, 429–446.

Kunitz, M. (1939b). Purification and concentration of enterokinase. *J. Gen. Physiol.* **22**, 447–450.

Kurachi, K., Torres-Rosado, A., and Tsuji, A. (1994). Hepsin. *Methods Enzymol.* **244**, 100–114.

LaBombardi, V. J., Shaw, E., DiStefano, J. F., Beck, G., Brown, F., and Zucker, S. (1983). Isolation and characterization of a trypsin-like serine proteinase from the membranes of Walker 256 carcino-sarcoma cells. *Biochem. J.* **211**, 695–700.

Lang, R. E., Tholken, H., Ganten, D., Luft, F. C., Ruskoaho, H., and Unger, T. (1985). Atrial natriuretic factor—a circulating hormone stimulated by volume loading. *Nature* **314**, 264–266.

LaVallie, E. R., Rehemtulla, A., Racie, L. A., DiBlasio, E. A., Ferenz, C., Grant, K. L, Light, A., and McCoy, J. M. (1993). Cloning and functional expression of a cDNA encoding the catalytic subunit of bovine enterokinase. *J. Biol. Chem.* **268**, 23311–23317.

Lee, S. L., Dickson, R. B., and Lin, C. Y. (2000). Activation of hepatocyte growth factor and urokinase/plasminogen activator by matriptase, an epithelial membrane serine protease. *J. Biol. Chem.* **275**, 36720–36725.

Levin, E. R., Gardner, D. G., and Samson, W. K. (1998). Natriuretic peptides. *N. Engl. J. Med.* **339**, 321–328.

Leytus, S. P., Loeb, K. R., Hagen, F. S., Kurachi, K., and Davie, E. W. (1988). A novel trypsin-like serine protease (hepsin) with a putative transmembrane domain expressed by human liver and hepatoma cells. *Biochemistry* **27**, 1067–1074.

Liepnieks, J. J., and Light, A. (1979). The preparation and properties of bovine enterokinase. *J. Biol. Chem.* **254**, 1677–1683.

Light, A., and Janska, H. (1989). Enterokinase (enteropeptidase): comparative aspects. *Trends Biochem. Sci.* **14**, 110–112.

Light, A., and Janska, H. (1991). The amino-terminal sequence of the catalytic subunit of bovine enterokinase. *J. Protein Chem.* **10**, 475–480.

Lin, B., Ferguson, C., White, J. T., Wang, S., Vessella, R., True, L. D., Hood, L., and Nelson, P. S. (1999a). Prostate-localized and androgen-regulated expression of the membrane-bound serine protease TMPRSS2. *Cancer Res.* **59**, 4180–4184.

Lin, C. Y., Anders, J., Johnson, M., and Dickson, R. B. (1999b). Purification and characterization of a complex containing matriptase and a Kunitz-type serine protease inhibitor from human milk. *J. Biol. Chem.* **274**, 18237–18242.

Lin, C. Y., Anders, J., Johnson, M., Sang, Q. A., and Dickson, R. B. (1999c). Molecular cloning of cDNA for matriptase, a matrix-degrading serine protease with trypsin-like activity. *J. Biol. Chem.* **274**, 18231–18236.

Lin, C. Y., Wang, J. K., Torri, J., Dou, L., Sang, Q. A., and Dickson, R. B. (1997). Characterization of a novel, membrane-bound, 80-kDa matrix-degrading protease from human breast cancer cells. Monoclonal antibody production, isolation, and localization. *J. Biol. Chem.* **272**, 9147–9152.

Lojda, Z., and Gossrau, R. (1983). Histochemical demonstration of enteropeptidase activity. New

202   Qingyu Wu

method with a synthetic substrate and its comparison with the trypsinogen procedure. *Histochemistry* **78,** 251–270.

Lopez, M. J., Wong, S. K., Kishimoto, I., Dubois, S., Mach, V., Friesen, J., Garbers, D. L., and Beuve, A. (1995). Salt-resistant hypertension in mice lacking the guanylyl cyclase-A receptor for atrial natriuretic peptide. *Nature* **378,** 65–68.

Lu, D., and Sadler, J. E. (1998). "Handbook of Proteolytic Enzymes." Academic Press Ltd., London.

Lu, D., Futterer, K., Korolev, S., Zheng, X., Tan, K., Waksman, G., and Sadler, J. E. (1999). Crystal structure of enteropeptidase light chain complexed with an analog of the trypsinogen activation peptide. *J. Mol. Biol.* **292,** 361–373.

Lu, D., Yuan, X., Zheng, X., and Sadler, J. E. (1997). Bovine proenteropeptidase is activated by trypsin, and the specificity of enteropeptidase depends on the heavy chain. *J. Biol. Chem.* **272,** 31293–31300.

Luo, J., Duggan, D. J., Chen, Y., Sauvageot, J., Ewing, C. M., Bittner, M. L., Trent, J. M., and Isaacs, W. B. (2001). Human prostate cancer and benign prostatic hyperplasia: molecular dissection by gene expression profiling. *Cancer Res.* **61,** 4683–4688.

Magee, J. A., Araki, T., Patil, S., Ehrig, T., True, L., Humphrey, P. A., Catalona, W. J., Watson, M. A., and Milbrandt, J. (2001). Expression profiling reveals hepsin overexpression in prostate cancer. *Cancer Res.* **61,** 5692–5696.

Mann, K. G., Nesheim, M. E., Church, W. R., Haley, P., and Krishnaswamy, S. (1990). Surface-dependent reactions of the vitamin K-dependent enzyme complexes. *Blood* **76,** 1–16.

Mann, N. S., and Mann, S. K. (1994). Enterokinase. *Proc. Soc. Exp. Biol. Med.* **206,** 114–118.

Manning, P. T., Schwartz, D., Katsube, N. C., Holmberg, S. W., and Needleman, P. (1985). Vasopressin-stimulated release of atriopeptin: endocrine antagonists in fluid homeostasis. *Science* **229,** 395–397.

Maroux, S., Baratti, J., and Desnuelle, P. (1971). Purification and specificity of porcine enterokinase. *J. Biol. Chem.* **246,** 5031–5039.

Masmoudi, S., Antonarakis, S. E., Schwede, T., Ghorbel, A. M., Gratri, M., Pappasavas, M. P., Drira, M., Elgaied-Boulila, A., Wattenhofer, M., Rossier, C., Scott, H. S., Ayadi, H., and Guipponi, M. (2001). Novel missense mutations of TMPRSS3 in two consanguineous Tunisian families with non-syndromic autosomal recessive deafness. *Hum. Mutat.* **18,** 101–108.

Matsushima, M., Ichinose, M., Yahagi, N., Kakei, N., Tsukada, S., Miki, K., Kurokawa, K., Tashiro, K., Shiokawa, K., Shinomiya, K., Umeyama, H., Inoue, H., Takahashi, T., and Takahashi, K. (1994). Structural characterization of porcine enteropeptidase. *J. Biol. Chem.* **269,** 19976–19982.

McGrath, M. E., Hines, W. M., Sakanari, J. A., Fletterick, R. J., and Craik, C. S. (1991). The sequence and reactive site of ecotin. A general inhibitor of pancreatic serine proteases from *Escherichia coli. J. Biol. Chem.* **266,** 6620–6625.

Mignatti, P., and Rifkin, D. B. (1993). Biology and biochemistry of proteinases in tumor invasion. *Physiol. Rev.* **73,** 161–195.

Miyazawa, K., Shimomura, T., Kitamura, A., Kondo, J., Morimoto, Y., and Kitamura, N. (1993). Molecular cloning and sequence analysis of the cDNA for a human serine protease reponsible for activation of hepatocyte growth factor. Structural similarity of the protease precursor to blood coagulation factor XII. *J. Biol. Chem.* **268,** 10024–10028.

Molino, M., Barnathan, E. S., Numerof, R., Clark, J., Dreyer, M., Cumashi, A., Hoxie, J. A., Schechter, N., Woolkalis, M., and Brass, L. F. (1997). Interactions of mast cell tryptase with thrombin receptors and PAR-2. *J. Biol. Chem.* **272,** 4043–4049.

Mullins, D. E., and Rohrlich, S. T. (1983). The role of proteinases in cellular invasiveness. *Biochim. Biophys. Acta* **695,** 177–214.

Naka, D., Ishii, T., Yoshiyama, Y., Miyazawa, K., Hara, H., Hishida, T., and Kidamura, N. (1992). Activation of hepatocyte growth factor by proteolytic conversion of a single chain form to a heterodimer. *J. Biol. Chem.* **267,** 20114–20119.

Neurath, H. (1986). The versatility of proteolytic enzymes. *J. Cell. Biochem.* **32**, 35–49.

Oberst, M., Anders, J., Xie, B., Singh, B., Ossandon, M., Johnson, M., Dickson, R. B., and Lin, C. Y. (2001). Matriptase and HAI-1 are expressed by normal and malignant epithelial cells in vitro and in vivo. *Am. J. Pathol.* **158**, 1301–1311.

Ossowski, L., Quigley, J. P., Kellerman, G. M., and Reich, E. (1973a). Fibrinolysis associated with oncogenic transformation. Requirement of plasminogen for correlated changes in cellular morphology, colony formation in agar, and cell migration. *J. Exp. Med.* **138**, 1056–1064.

Ossowski, L., Unkeless, J. C., Tobia, A., Quigley, J. P., Rifkin, D. B., and Reich, E. (1973b). An enzymatic function associated with transformation of fibroblasts by oncogenic viruses. II. Mammalian fibroblast cultures transformed by DNA and RNA tumor viruses. *J. Exp. Med* **137**, 112–126.

Paoloni-Giacobino, A., Chen, H., Peitsch, M. C., Rossier, C., and Antonarakis, S. E. (1997). Cloning of the TMPRSS2 gene, which encodes a novel serine protease with transmembrane, LDLRA, and SRCR domains and maps to 21q22.3. *Genomics* **44**, 309–320.

Pavlov, I. P. (1904). "Nobel Lecture, Physiology or Medicine 1901–1921." The Nobel Foundation.

Quigley, J. P., Ossowski, L., and Reich, E. (1974). Plasminogen, the serum proenzyme activated by factors from cells transformed by oncogenic viruses. *J. Biol. Chem.* **249**, 4306–4311.

Satomi, S., Yamasaki, Y., Tsuzuki, S., Hitomi, Y., Iwanaga, T., and Fushiki, T. (2001). A role for membrane-type serine protease (MT-SP1) in intestinal epithelial turnover. *Biochem. Biophys. Res. Commun.* **287**, 995–1002.

Schmidt, C., Bladt, F., Goedecke, S., Brinkmann, V., Zschiesche, W., Sharpe, M., Gherardi, E., and Birchmeier, C. (1995). Scatter factor/hepatocyte growth factor is essential for liver development. *Nature* **373**, 699–702.

Schwartz, D., Geller, D. M., Manning, P. T., Siegel, N. R., Fok, K. F., Smith, C. E., and Needleman, P. (1985). Ser-Leu-Arg-Arg-atriopeptin III: the major circulating form of atrial peptide. *Science* **229**, 397–400.

Scott, H. S., Kudoh, J., Wattenhofer, M., Shibuya, K., Berry, A., Chrast, R., Guipponi, M., Wang, J., Kawasaki, K., Asakawa, S., Minoshima, S., Younus, F., Mehdi, S. Q., Radhakrishna, U., Papasavvas, M. P., Gehrig, C., Rossier, C., Korostishevsky, M., Gal, A., Shimizu, N., Bonne-Tamir, B., and Antonarakis, S. E. (2001). Insertion of beta-satellite repeats identifies a transmembrane protease causing both congenital and childhood onset autosomal recessive deafness. *Nat. Genet.* **27**, 59–63.

Sei, C. A., Hand, G. L., Murray, S. F., and Glembotski, C. C. (1992). The cosecretional maturation of atrial natriuretic factor by primary atrial myocytes. *Mol. Endocrinol.* **6**, 309–319.

Seidah, N. G., Cromlish, J. A., Hamelin, J., Thibault, G., and Chretien, M. (1986). Homologous IRCM-serine protease 1 from pituitary, heart atrium and ventricle: a common pro-hormone maturation enzyme. *Biosci. Rep.* **6**, 835–844.

Shi, Y. E., Torri, J., Yieh, L., Wellstein, A., Lippman, M. E., and Dickson, R. B. (1993). Identification and characterization of a novel matrix-degrading protease from hormone-dependent human breast cancer cells. *Cancer Res.* **53**, 1409–1415.

Shields, P. P., and Glembotski, C. C. (1988). The post-translational processing of rat pro-atrial natriuretic factor by primary atrial myocyte cultures. *J. Biol. Chem.* **263**, 8091–8098.

Shimomura, T., Denda, K., Kitamura, A., Kawaguchi, T., Kito, M., Kondo, J., Kagaya, S., Qin, L., Takata, H., Miyazawa, K., and Kitamura, N. (1997). Hepatocyte growth factor activator inhibitor, a novel Kunitz-type serine protease inhibitor. *J. Biol. Chem.* **272**, 6370–6376.

Shulman, J. M., Perrimon, N., and Axelrod, J. D. (1998). Frizzled signaling and the developmental control of cell polarity. *Trends Genet.* **14**, 452–458.

Stamey, T. A., McNeal, J. E., Yemoto, C. M., Sigal, B. M., and Johnstone, I. M. (1999). Biological determinants of cancer progression in men with prostate cancer. *J. Am. Med. Assoc.* **281**, 1395–1400.

Stamey, T. A., Warrington, J. A., Caldwell, M. C., Chen, Z., Fan, Z., Mahadevappa, M., McNeal, J. E., Nolley, R., and Zhang, Z. (2001). Molecular genetic profiling of Gleason grade 4/5 prostate cancers compared to benign prostatic hyperplasia. *J. Urol.* **166,** 2171–2177.

Stewart, F. (1996). Roles of mesenchymal-epithelial interactions and hepatocyte growth factor-scatter factor (HGF-SF) in placental development. *Rev. Reprod.* **1,** 144–148.

Stoker, M., Gherardi, E., Perryman, M., and Gray, J. (1987). Scatter factor is a fibroblast-derived modulator of epithelial cell mobility. *Nature* **327,** 239–242.

Stroud, R. M. (1974). A family of protein-cutting proteins. *Sci. Am.* **231,** 74–88.

Takeuchi, T., Harris, J. L., Huang, W., Yan, K. W., Coughlin, S. R., and Craik, C. S. (2000). Cellular localization of membrane-type serine protease 1 and identification of protease-activated receptor-2 and single-chain urokinase-type plasminogen activator as substrates. *J. Biol. Chem.* **275,** 26333–26342.

Takeuchi, T., Shuman, M. A., and Craik, C. S. (1999). Reverse biochemistry: use of macromolecular protease inhibitors to dissect complex biological processes and identify a membrane-type serine protease in epithelial cancer and normal tissue. *Proc. Natl. Acad. Sci. USA* **96,** 11054–11061.

Tanimoto, H., Yan, Y., Clarke, J., Korourian, S., Shigemasa, K., Parmley, T. H., Parham, G. P., and TJ, O. B. (1997). Hepsin, a cell surface serine protease identified in hepatoma cells, is overexpressed in ovarian cancer. *Cancer Res.* **57,** 2884–2887.

Thorpe, D. S., and Garbers, D. L. (1989). The membrane form of guanylate cyclase. Homology with a subunit of the cytoplasmic form of the enzyme. *J. Biol. Chem.* **264,** 6545–6549.

Tomita, Y., Kim, D. H., Magoori, K., Fujino, T., and Yamamoto, T. T. (1998). A novel low-density lipoprotein receptor-related protein with type II membrane protein-like structure is abundant in heart. *J. Biochem. (Tokyo)* **124,** 784–789.

Torres-Rosado, A., KS, O. S., Tsuji, A., Chou, S. H., and Kurachi, K. (1993). Hepsin, a putative cell-surface serine protease, is required for mammalian cell growth. *Proc. Natl. Acad. Sci. USA* **90,** 7181–7185.

Tsuji, A., Torres-Rosado, A., Arai, T., Le Beau, M. M., Lemons, R. S., Chou, S. H., and Kurachi, K. (1991). Hepsin, a cell membrane-associated protease. Characterization, tissue distribution, and gene localization. *J. Biol. Chem.* **266,** 16948–16953.

Uehara, Y., Minowa, O., Mori, C., Shiota, K., Kuno, J., Noda, T., and Kitamura, N. (1995). Placental defect and embryonic lethality in mice lacking hepatocyte growth factor/scatter factor. *Nature* **373,** 702–705.

Underwood, L. J., Shigemasa, K., Tanimoto, H., Beard, J. B., Schneider, E. N., Wang, Y., Parmley, T. H., and O'Brien, T. J. (2000). Ovarian tumor cells express a novel multi-domain cell surface serine protease. *Biochim. Biophys. Acta* **1502,** 337–350.

Unkeless, J., Dano, K., Kellerman, G. M., and Reich, E. (1974). Fibrinolysis associated with oncogenic transformation. Partial purification and characterization of the cell factor, a plasminogen activator. *J. Biol. Chem.* **249,** 4295–4305.

Unkeless, J. C., Tobia, A., Ossowski, L., Quigley, J. P., Rifkin, D. B., and Reich, E. (1973). An enzymatic function associated with transformation of fibroblasts by oncogenic viruses. I. Chick embryo fibroblast cultures transformed by avian RNA tumor viruses. *J. Exp. Med.* **137,** 85–111.

Veske, A., Oehlmann, R., Younus, F., Mohyuddin, A., Muller-Myhsok, B., Mehdi, S. Q., and Gal, A. (1996). Autosomal recessive non-syndromic deafness locus (DFNB8) maps on chromosome 21q22 in a large consanguineous kindred from Pakistan. *Hum. Mol. Genet.* **5,** 165–168.

Vu, T. K. H., Liu, R. W., Haaksma, C. J., Tomasek, J. J., and Howard, E. W. (1997). Identification and cloning of the membrane-associated serine protease, hepsin, from mouse preimplantation embryos. *J. Biol. Chem.* **272,** 31315–31320.

Wallrapp, C., Hahnel, S., Muller-Pillasch, F., Burghardt, B., Iwamura, T., Ruthenburger, M., Lerch, M. M., Adler, G., and Gress, T. M. (2000). A novel transmembrane serine protease (TMPRSS3) overexpressed in pancreatic cancer. *Cancer Res.* **60,** 2602–2606.

Wattenhofer, M., Shibuya, K., Kudoh, J., Lyle, R., Michaud, J., Rossier, C., Kawasaki, K.,

Asakawa, S., Minoshima, S., Berry, A., Bonne-Tamir, B., Shimizu, N., Antonarakis, S. E., and Scott, H. S. (2001). Isolation and characterization of the UBASH3A gene on 21q22.3 encoding a potential nuclear protein with a novel combination of domains. *Hum. Genet.* **108**, 140–147.

Welsh, J. B., Sapinoso, L. M., Su, A. I., Kern, S. G., Wang-Rodriguez, J., Moskaluk, C. A., Frierson, H. F., and Hampton, G. M. (2001). Analysis of gene expression identifies candidate markers and pharmacological targets in prostate cancer. *Cancer Res.* **61**, 5974–5978.

Wilkins, M. R., Redondo, J., and Brown, L. A. (1997). The natriuretic-peptide family. *Lancet* **349**, 1307–1310.

Wu, F., Yan, W., Pan, J., Morser, J., and Wu, Q. (2002). Processing of pro-atrial natriuretic peptide by corin in cardiac myocytes. *J. Biol. Chem.* **277**, 16900–16905.

Wu, Q. (2001). Gene targeting in hemostasis. *Hepsin. Front. Biosci.* **6**, D192–D200.

Wu, Q., Yu, D., Post, J., Halks-Miller, M., Sadler, J. E., and Morser, J. (1998). Generation and characterization of mice deficient in hepsin, a hepatic transmembrane serine protease. *J. Clin. Invest.* **101**, 321–326.

Yahagi, N., Ichinose, M., Matsushima, M., Matsubara, Y., Miki, K., Kurokawa, K., Fukamachi, H., Tashiro, K., Shiokawa, K., Kageyama, T., Takahashi, T., Inoue, H., and Takahashi, K. (1996). Complementary DNA cloning and sequencing of rat enteropeptidase and tissue distribution of its mRNA. *Biochem. Biophys. Res. Commun.* **219**, 806–812.

Yamaguchi, N., Okui, A., Yamada, T., Nakazato, H., and Mitsui, S. (2002). Spinesin/TMPRSS5, a novel transmembrane serine protease, cloned from human spinal cord. *J. Biol. Chem.* **277**, 6806–6812.

Yamaoka, K., Masuda, K., Ogawa, H., Takagi, K., Umemoto, N., and Yasuoka, S. (1998). Cloning and characterization of the cDNA for human airway trypsin-like protease. *J. Biol. Chem.* **273**, 11895–11901.

Yan, W., Sheng, N., Seto, M., Morser, J., and Wu, Q. (1999). Corin, a mosaic transmembrane serine protease encoded by a novel cDNA from human heart. *J. Biol. Chem.* **274**, 14926–14935.

Yan, W., Wu, F., Morser, J., and Wu, Q. (2000). Corin, a transmembrane cardiac serine protease, acts as a pro-atrial natriuretic peptide-converting enzyme. *Proc. Natl. Acad. Sci. USA* **97**, 8525–8529.

Yasuoka, S., Ohnishi, T., Kawano, S., Tsuchihashi, S., Ogawara, M., Masuda, K., Yamaoka. K., Takahashi, M., and Sano, T. (1997). Purification, characterization, and localization of a novel trypsin-like protease found in the human airway. *Am. J. Respir. Cell. Mol. Biol.* **16**, 300–308.

Yu, I. S., Chen, H. J., Lee, Y. S., Huang, P. H., Lin, S. R., Tsai, T. W., and Lin, S. W. (2000). Mice deficient in hepsin, a serine protease, exhibit normal embryogenesis and unchanged hepatocyte regeneration. *Thromb. Haemost.* **84**, 865–870.

Yuan, X., Zheng, X., Lu, D., Rubin, D. C., Pung, C. Y., and Sadler, J. E. (1998). Structure of murine enterokinase (enteropeptidase) and expression in small intestine during development. *Am. J. Physiol.* **274**, G342–G349.

Zacharski, L. R., Ornstein, D. L., Memoli, V. A., Rousseau, S. M., and Kisiel, W. (1998). Expression of the factor VII activating protease, hepsin, in situ in renal cell carcinoma [letter] [In Process Citation]. *Thromb. Haemost.* **79**, 876–877.

Zamolodchikova, T. S., Sokolova, E. A., Alexandrov, S. L., Mikhaleva, II, Prudchenko, I. A., Morozov, I. A., Kononenko, N. V., Mirgorodskaya, O. A., Da, U., Larionova, N. I., Pozdnev, V. F., Ghosh, D., Duax, W. L., and Vorotyntseva, T. I. (1997). Subcellular localization, substrate specificity and crystallization of duodenase, a potential activator of enteropeptidase. *Eur. J. Biochem.* **249**, 612–621.

Zamolodchikova, T. S., Sokolova, E. A., Lu, D., and Sadler, J. E. (2000). Activation of recombinant proenteropeptidase by duodenase. *FEBS Lett.* **466**, 295–299.

Zamolodchikova, T. S., Vorotyntseva, T. I., and Antonov, V. K. (1995a). Duodenase, a new serine protease of unusual specificity from bovine duodenal mucosa. Purification and properties. *Eur. J. Biochem.* **227**, 866–872.

Zamolodchikova, T. S., Vorotyntseva, T. I., Nazimov, I. V., and Grishina, G. A. (1995b). Duodenase, a new serine protease of unusual specificity from bovine duodenal mucosa. Primary structure of the enzyme. *Eur. J. Biochem.* **227,** 873–879.

Zheng, X., Lu, D., and Sadler, J. E. (1999). Apical sorting of bovine enteropeptidase does not involve detergent-resistant association with sphingolipid-cholesterol rafts. *J. Biol. Chem.* **274,** 1596–1605.

Zheng, X., and Sadler, J. E. (2002). Mucin-like domain of enteropeptidase directs apical targeting in Madin–Darby canine kidney cells. *J. Biol. Chem.* **277,** 6858–6863.

Zhukov, A., Hellman, U., and Ingelman-Sundberg, M. (1997). Purification and characterization of hepsin from rat liver microsomes. *Biochim. Biophys. Acta* **1337,** 85–95.

7

DPPIV, Seprase, and Related Serine Peptidases in Multiple Cellular Functions

Wen-Tien Chen,[1] Thomas Kelly,[2] and Giulio Ghersi[3]
[1]Department of Medicine/Division of Neoplastic Diseases
State University of New York at Stony Brook, Stony Brook
New York, 11794-8154

[2]Department of Pathology
Arkansas Cancer Research Center
Little Rock, Arkansas, 72205-7199

[3]Department of Cellular and Developmental Biology
University of Palermo, Viale delle Scienze
90128 Palermo, Italy

I. Introduction

Several families of membrane proteases are distinguishable on the basis of their proteolytic activities, biologic functions, and structural organization. There are the membrane-type matrix metalloproteinases (MT-MMPs), the ADAM

Current Topics in Developmental Biology, Vol. 54

(a disintegrin and metalloprotease) family, the meprins, the secretases (also termed sheddases or convertases), and the metallo- and serine peptidases. Localization of proteases is critical for their function in cellular activities. Increasing evidence indicates that the serine peptidases and MT-MMPs accumulate at cell surface protrusions, termed invadopodia, that may have a prominent role in processing soluble factors (including growth factors, hormones, chemokines, and other bioactive peptides) in addition to the well-established role of invadopodia in degrading the components of the ECM. Moreover, these membrane proteases may direct activation of either themselves or other workhorse soluble enzymes such as the 72-kDa matrix metalloprotease (MMP), MMP-2, and plasmin (Sato *et al.*, 1994). There is an extensive literature addressing metalloproteases and peptidases. It is generally agreed that a given membrane protease may have several functions (diversity) and that more than one membrane protease or one protease family may mediate the same function (redundancy) (see Bauvois, 2001, for a review). Different membrane proteases form complexes at invadopodia or other specialized locations that could provide distinct and overlapping actions. This may be necessary for regulation of complex regulatory processes where modulation is achieved by proteolysis of several different molecules: for example, processing of various chemokines, activation of associated proteases, and ECM degradation.

Although extensive laboratory research has supported that the MMP family plays a major role in tumor angiogenesis and metastasis, clinical trials of MMP inhibitors (including Marimastat, BAY12-9566, AG3340, and Ro32-3555) in patients with cancer have not produced obvious evidence of antiangiogenic and antimetastatic effects (Zucker *et al.*, 2000). These findings suggest that MMPs are not critical to these disease processes, but there could be alternative explanations for the apparent failure of the MMP inhibitors. For example, the concentration of the inhibitors at the local sites may not have been high enough, or the stage of the cancer was too late to be treated effectively. On the other hand, other protease systems may be required to permit cell migration/invasion activity. These proteases are likely to be membrane-bound, transiently expressed, and closely linked to the cell invasion or activation process.

This article intends to focus on a small group of membrane serine peptidases, the serine-integral membrane peptidases (SIMP), that are inducible, specific for proline-containing peptides and macromolecules, and active on the cell surface. Prototypes of SIMP members are DPPIV and seprase. Other SIMP-related peptidases, including quiescent cell proline aminodipeptidase (QPP), prolyl carboxypeptidase (PCP), prolyl endopeptidase (PEP), dipeptidyl peptidase 6 (DPP6), dipeptidyl peptidase 8 (DPP8), dipeptidyl peptidase 9 (DDP9), attractin, dipeptidyl peptidase II (DPPII), and dipeptidyl peptidase IV-β (DPPIV-β), have subtle structural and functional differences from DPPIV and seprase, and they are included in the discussion.

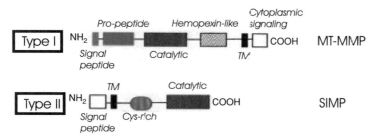

Figure 1 Diagram of the structures of two distinct families of membrane proteases: MT-MMP and SIMP. TM, transmembrane domain; Cys-rich, cysteine-rich.

SIMP members are classified on the basis of their proteolytic activities, biologic functions, structural organization, and chromosomal localization. SIMP members are type II transmembrane proteins (Fig. 1), with cytoplasmic tails that contain six amino acids (a.a.) followed by a 20 a.a. (seprase) or 22 a.a. (DPPIV) transmembrane domain at the N terminus, an N-glycosylation- and cysteine-rich substrate-binding domain, and a stretch of 200 a.a. at the C terminus that constitutes a catalytic region with the catalytic serine in a nonclassical orientation (Goldstein *et al.,* 1997; Pineiro-Sanchez *et al.,* 1997). Figure 1 shows a comparison of the structural organization of SIMP with that of MT-MMP.

MT-MMP has the classic type I transmembrane topology where its TM locates at the carboxyl end with a cytoplasmic signaling domain. For example, an earlier study demonstrated that MT1-MMP overexpression resulted in the localization of this protease at invadopodia and initiated a proteolytic cascade for cell invasion. Furthermore, the TM/cytoplasmic domain of MT1-MMP ($TM_{MT-MMP,542}$VVLPVL LLLLVLAVGLAVFFFRRHGTPRRLLYCQRSLLDKV$_{582}$) was required for invadopodial localization of the enzyme and therefore for directing cellular invasion (Nakahara *et al.,* 1997). In keeping with the dynamic action of invadopodia, MT1-MMP is internalized from the cell surface in a process that is dependent upon the cytoplasmic tail and internalization is required for MT1-MMP to promote invasion (Uekita *et al.,* 2001).

On the other hand, SIMP have the type II transmembrane topology retaining its signal peptide that becomes the TM domain of the amino terminus and its membrane anchorage site. The cytoplasmic tails of SIMP members are short; thus, it is doubtful whether they may be involved directly in the transduction of extracellular signals. Since the orientation of residues of the catalytic triad (Ser, Asp, His) of the SIMP is inverted respect to chymotrypsin-like serine proteases (His, Asp, Ser), SIMP members are classified as nonclassical serine proteases. Both human DPPIV and seprase have over 68% identity at the catalytic region and a conserved serine protease motif G-X-S-X-G. Studies have also showed that, like DPPIV, all SIMP peptidases cleave prolyl peptide bonds (Pro-Xaa). Protease activation of the MT-MMP is known to involve the classic mechanism by cleavage of the signal and propeptide domains. These domains have not been identified in SIMP, and it is unclear how SIMP becomes active. However, dimerization of some SIMP members is required for their prolyl peptidase and gelatinase activities. The glycosylation- and cysteine-rich domains of SIMP and the hemopexin-like domain of MT-MMP may function in recognition and binding to their substrates, including peptides, macromolecules, other enzymes, or themselves.

II. Structural Chemistry and Biochemistry of SIMP

Table I summarizes the characteristics of SIMP and related peptidases, for which proteins or cDNA clones have been obtained, expressed and/or characterized. Figure 2 illustrates simplified structural features of SIMP and related peptidases. So far, more than 10 SIMP and related peptidases have been identified. No endogenous inhibitor of these peptidases has been identified. Most of these peptidases are distributed widely in different cell types with a noticeable induction of their expression in activated cells, i.e., CD26 in activated T-cells and FAPα in activated fibroblasts. As both DPPIV and seprase are found in various cell types in their activated states, they should not be treated as markers for specific cell types. They are regulated tightly during fetal/postnatal development, tissue repair, inflammation, and disease. They are not found in most adult tissue cells, suggesting that they are not essential enzymes involving in housekeeping activities. Instead, they are inducible enzymes that are expressed in response to wounding, inflammation, and cancer invasion, and various growth factors and cytokines can control their expression by inducing or repressing their levels of gene transcription and/or protein synthesis. Finally, soluble isoforms of membrane peptidases with undefined functions are beginning to be identified in biological fluids.

A. Domains and Gene Structure

1. Dipeptidyl Peptidase IV (DPPIV; EC 3.4.14.5)

DPPIV was first identified in 1966 as a glycylproline naphthylamidase, because of its ability to liberate naphthylamine from gly-pro-2-naphthylamide (Hopsu-Havu and Glenner, 1966). Relatively few enzymes are able to cleave the prolyl peptide bond (Pro-Xaa); and DPPIV is the most widely studied of these enzymes. Accordingly, DPPIV was classified in the peptidase family S9b, which along with prolyl endopeptidase (PEP, S9a) and acylaminoacyl peptidase (S9c) forms the prolyl oligopeptidase family (see Barrett and Rawlings, 1992, for a review). A DPPIV-like gene family that corresponds to peptidase family S9b has also been proposed (Abbott *et al.*, 1999a).

Table I Main Features of SIMP and Related Peptidases (References Cited in the Text)

Protease	Actve Form	Proteolysis	Major Localization; Soluble Form	Inducible Expression	Chromosome
DPPIV/CD26	Dimer	Peptides, ECM, membrane proteins	Cell surface; shed or AS in plasma?	All	2q24.3
Seprase/FAPα	Dimer	Peptides, ECM, membrane proteins	Cell surface; AS in cytoplasm	Fibroblasts; Endothelia; Smooth muscle; tumor	2q23
QPP	Dimer	Peptides	Post-Golgi membrane	Lymphocytes	?
PCP/angiotensinase	Dimer	Peptides	Lysosome membrane	Kidney cells	?
DPPII	Dimer	Peptides	Lysosome membrane	Brain cells	?
PEP	Monomer	Peptides	Cytoplasm	?	6q22
DPP6	Monomer	Peptides	?	?	7
DPP8	Monomer	Peptides	Cytoplasm	Lymphocytes	15q22
DPP9	Monomer	Peptides	Cytoplasm	Lymphocytes	19p13.3
Attractin	Monomer	Peptides	Plasma	Lymphocytes	20p13
DPPIV-β	Monomer	Peptides	Cell surface	Lymphocytes	?
NAALADase	Monomer	Peptides	?	Ovary, testis, brain cells	11

All, all *in vivo* cell types that are activated by physiological stimulants; cells including hematopoietic cells (T - or B-cells), stromal fibroblasts, angiogenic endothelia or smooth muscle cells, and invasive tumor cells; AS, alternatively spliced form; ECM, extracellular matrix; ?, not known.

A. Dimer: DPPIV/Seprase

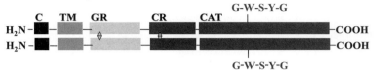

B. Dimer: QPP

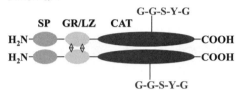

C. Monomer diverse: DPP8

Figure 2 General structures of SIMP-related post prolyl peptidases. (A) SIMP prototypes such as DPPIV and seprase are proteolytically active as dimers and have a cytoplasmic domain (C); transmembrane domain (TM); glycosylation-rich region (GR); cysteine-rich domain (CR); and catalytic domain (CAT). (B) QPP exemplifies SIMP-related postprolyl peptidases that are proteolytically active as dimers but do not have sequence homology to DPPIV. QPP consists of a signal peptide (SP), a glycosylation-rich region (GR), a leucine zipper (LZ), and a catalytic domain (CAT). (C) DPP8 is an example of a SIMP-related postprolyl peptiodase that is active as a monomer. DPP8 is not glycosylated but the catalytic domain (CAT) exhibits homology to DPPIV.

Several groups have cloned cDNAs encoding human, mouse and rat DPPIV. Table II summarizes some of characteristics of DPPIV and seprase genes. The DPPIV sequences show a surprising conservation between different species (up to 98% between mammalian *DPPIV* cDNAs and 36% with yeast *DPPIV*). The human *DPPIV* cDNA is 2.9 kb with an open reading frame of 2298 bp; the deduced DPPIV

Table II CD26/DPPIV and Seprase/FAPα Genes[a]

	DPPIV	Seprase
Size	90 kb	60 kb
Exons	26 exons, 45 b to 1.4 kb	26 exons, 46 b to 195 b
Exon–intron organization	Same in mice and humans	Same in mice and human
5′ untranslated region	No TATA box or CAAT box	?
Binding site for transcription factors	GC-rich region	?
Transcriptional activation	Hepatocyte nuclear factor 1	TGF-β, 12-*O*-tetradecanoyl phorbol-13-acetate, retinoids
Chromosome location	2q24	2q23

[a] Abbott *et al.* (1994); Mathew *et al.* (1995); Niedermeyer *et al.* (1998); Rettig *et al.* (1994).

structure is 766 amino acid residues in length and produces a protein of molecular weight 88.3 kDa. The predicted DPPIV polypeptide shows six highly conserved cytoplasmic amino acids, a 22 amino acid hydrophobic transmembrane region, and a 738 amino acid extracellular domain. The extracellular domain can be divided into three regions: an N-terminal glycosylated region containing seven glycosylation sites and starting with a 20 amino acid flexible "stalk" region; a cysteine-rich region; and a 260 amino acid C-terminal region containing the catalytic sequence. The formation of complex N-glycans plays an important role in the stability and function of DPPIV (Fan et $al.$, 1997). A cysteine-rich region containing 10 of the 12 cysteine residues is potential substrate binding site that has been referred as the "adhesion domain." The remaining sequence of 738 residues at the C terminus of the molecule contain the catalytic triad constituted of Ser-624 in mouse (David et $al.$, 1993) or Ser-631 in human (Ogata et $al.$, 1989), Asp-702/Asp-709 and His-734/His-741.

2. Seprase

The human 170-kDa membrane serine gelatinase, termed $seprase$ (surface expressed protease), was originally identified as a cell surface glycoprotein peptidase selectively expressed on cell surface invadopodia of melanoma and breast carcinoma cells for its potential role in cell invasion (Aoyama and Chen, 1990; Monsky et $al.$, 1994; Chen et $al.$, 1994). Seprase was isolated from a human malignant melanoma cell line LOX that exhibits extremely aggressive behavior in experimental metastasis (Fodstad et $al.$, 1988). However, the protease was not found by gelatin zymography in 32 other tumor cell lines that did not show extracellular gelatin degradation. Active seprase was then isolated from membranes of LOX cells, which were lacking DPPIV (Pineiro-Sanchez et $al.$, 1997). The active seprase is a homodimer that contains two N-glycosylated 97-kDa subunits. Inhibitor profiling of seprase was found to be identical with that of DPPIV. The gelatinase activity of seprase was completely blocked by serine-protease inhibitors, including diisopropyl fluorophosphate (DFP). Seprase could be affinity-labeled by [^{3}H]DFP, but the proteolytically inactive 97-kDa subunit could not, confirming the existence of a serine protease active site on the dimeric form, which was further demonstrated by loss of proteolytic activity upon dissociation into its 97-kDa subunit following treatment with acid, heat, or cysteine and histidine modifying agents. In addition, analysis of the deduced amino acid sequence from a cDNA that encodes the 97-kDa subunit (Goldstein et $al.$, 1997) revealed that it is homologous to DPPIV and is essentially identical to the human fibroblast activation protein α (FAPα) (Scanlan et $al.$, 1994; Levy et $al.$, 1999; Park et $al.$, 1999).

Analysis of the deduced amino acid sequence from a complete cDNA sequence of the seprase subunit (Goldstein et $al.$, 1997) revealed that seprase monomer is homologous to DPPIV and identical to FAPα after correction of the original report regarding the Gly-626 to Lys-670 region and the consensus sequence motif Gly-X-Ser-X-Gly of seprase (Scanlan et $al.$, 1994). The predicted seprase polypeptide shows six highly conserved cytoplasmic amino acids; a 20 amino acid hydrophobic transmembrane region, a 20 amino acid 27–48 flexible "stalk" region, a glycosylated region 49–314 containing seven glycosylation sites; a cysteine-rich region 321–466; and a 267 amino acid 570–766 C-terminal region containing the catalytic sequence. An alternatively spliced seprase messenger has been identified from the human melanoma cell line LOX that encodes a novel truncated 27-kDa isoform, which precisely overlaps the carboxyl-terminal catalytic region of the 97-kDa seprase subunit (Goldstein and Chen, 2000). The splice variant mRNA is generated by an out-of-frame deletion of a 1223-base pair exonic region that encodes part of the cytoplasmic tail, transmembrane, and the membrane proximal-central regions of the extracellular domain (Val-5 through Ser-412) of the seprase 97-kDa subunit. The 27-kDa isoform is located in the cytosol of melanoma cells but its proteolytic activity is currently unknown.

Three other peptidases have been reported to have SIMP-like activities in that they require dimerization for hydrolyzing Proline-Xaa peptides, but are structurally unrelated to DPPIV: the intracellular serine protease, quiescent cell proline amino-dipeptidase (QPP) (Underwood *et al.*, 1999; Chiravuri *et al.*, 2000a); the prolyl-carboxypeptidase (PCP, angiotensinase), a postproline cleaving carboxypeptidase that cleaves amino acids off the C terminus of proteins (Tan *et al.*, 1993); and DPPII (Mentlein and Struckhoff, 1989). DPPII and PCP are both found in lyso-somes and liberate amino acids from the carboxyl terminus of proteins at acidic pH. DPPIV, seprase, QPP, and PCP have the same sequential ordering of the active site residues, Ser-Asp-His; share dipeptidase substrate specificity; are functionally ac-tive as homodimers. However, there is no significant sequence homology between DPPIV-seprase and QPP-PCP. QPP does, however, share significant sequence ho-mology with PCP (42% sequence identity). In addition, DPPIV and seprase are located on the cell surface, while QPP is found to be a post-Golgi membrane en-zyme and PCP and DPPII are lysosomal membrane proteases. Unlike the role of CD26 in activated lymphocytes, QPP plays an important role in resting cells, as its aminodipeptidase inhibitors cause cell death in quiescent lymphocytes.

A novel human DPPIV-related postproline peptidase, named DPP8, has been described (Abbott *et al.*, 2000). DPP8 exhibited dipeptidyl aminopeptidase activ-ity and showed significant mRNA expression in activated T cells. Sequence and localization data indicate that DPP8 and PEP are monomeric, nonglycosylated, soluble, cytoplasmic proteins. DPP8 and PEP are catalytically active as monomers and cleave Pro-Xaa bonds.

DPPIV and FAPα/seprase have both been localized to the long arm of chromo-some 2, 2q24.3 and 2q23, respectively (Abbott *et al.*, 1994). DPP8 was localized to 15q22 and DPP9 to 19p13.3 (Abbott *et al.*, 2000). The related genes DPP6 and PEP have been localized to chromosome 7 (Yokotani *et al.*, 1993) and 6q22 (Goossens *et al.*, 1996), respectively.

In addition, a report has described the cloning of N-acetylated α-linked acidic dipeptidase (NAALADase), which codes for a type II integral membrane protein, is localized to chromosome 11, exhibits both NAALADase and prolyl peptidase activities, and is highly expressed in ovary and testis as well as within discrete brain areas (Pangalos *et al.*, 1999). Hydrolysis of the neuropeptide *N*-acetyl-L-aspartyl-L-glutamate (NAAG) by NAALADase II to release glutamate may be important in a number of neurodegenerative disorders in which excitotoxic mechanisms are implicated. The gene coding for human prostate-specific membrane antigen, a marker of prostatic carcinomas, and its rat homolog glutamate carboxypeptidase II has been shown to possess such NAALADase activity.

Based on assays for DPPIV-like activities, attractin (DPPT-L), a 175-kDa soluble glycoprotein reported to hydrolyze Pro-Xaa bonds, has been isolated from the peripheral blood (Duke-Cohan *et al.*, 1995; 1998). Like CD26 and DDP8, attractin associates with T-cell activation. Unlike CD26, attractin does not bind to adenosine deaminase (ADA; EC 3.5.4.4), but it induces the spreading of monocytes and clustering of T cells, contains a kelch repeat domain, and shares no significant sequence homology with DPPIV or any other peptidase (Duke-Cohan *et al.*, 1998). The genomic structure of human attractin reveals that soluble attractin arises from transcription of 25 sequential exons on human chromosome 20p13, where the 3' terminal exon contains sequence from a long interspersed nuclear element-1 (LINE-1) retrotransposon element that includes a stop codon and a polyadenylation signal (Tang *et al.*, 2000). Both membrane-bound and secreted forms of human attractin arise as a result of alternate splicing of the same gene. However, using ADA-affinity chromatography, it is shown that 95% of the serum dipeptidyl peptidase activity is associated with a protein with ADA-binding properties, i.e., CD26 rather than the defined attractin (Durinx *et al.*, 2000). Finally, DPPIV-β, a cell surface glycoprotein of 75–80 kDa, has been identified that binds DPPIV inhibitors but with less affinity than DPPIV, and no sequence data for this enzyme are available (Blanco *et al.*, 1998). Thus, new enzymes of this family or enzymes of similar function but diverse structures are emerging.

B. Catalytic Features: Nonclassical Serine Active Site, Glycosylation, and Dimerization

The catalytic domain at the C terminus of DPPIV probably forms an α/β hydrolase fold, having a structure similar to those of proline imidopeptidase (EC 3.4.11.5) and prolyl oligopeptidase (POP; EC 3.4.21.26) (Medrano *et al.*, 1998; Fulop *et al.*, 1998). DPPIV and seprase have the serine protease Gly-X-Ser-X-Gly consensus motif in its extracellular domain and share a conserved structural domain of approximately 200 amino acids with several nonclassical serine hydrolases. These amino acids (Ser-624, Asp-702, and His-734) are arranged in a novel sequential order as compared with that of archetypal serine proteases [e.g., nucleophile (Ser)-acid-His versus His-acid-nucleophile (Ser), respectively] (David *et al.*, 1993). The Ser-624 of seprase is important in both the dipeptidyl peptidase and gelatinolytic activities of the enzyme, suggesting a single active site for various proteolytic activities of SIMP members (Park *et al.*, 1999). In addition, the proteolytic activities of QPP, PCP, seprase, and DPPIV are dependent upon dimerization (Pineiro-Sanchez *et al.*, 1997; Underwood *et al.*, 1999; Tan *et al.*, 1993).

Interestingly, DPPIV, seprase, QPP, and PCP include common structural features that may be reflective of their convergent evolution to form efficient post-proline cleaving enzymes: (1) they have the same ordering of the catalytic triad: Ser, Asp, His; (2) they are glycoproteins, and glycosylation is essential for the enzymatic activity; and (3) they form homodimers to activate enzymatic activity (Chiravuri *et al.*, 2000b). The leucine zipper QPP mutants showed a complete lack of enzymatic activity, but an enzyme active site mutant retained its ability to homodimerize. It appears that QPP homodimerization is mediated through a leucine zipper and that this homodimerization is required for QPP enzymatic activity, to assume its correct structural conformation and/or to recognize and cleave its substrate. Disulfide bonds do not appear to play a role in QPP homodimerization as QPP migrates as a 58-kDa species under both reducing and nonreducing conditions.

The cysteine-rich region of rat DPPIV harbors 10 of the 12 highly conserved cysteine residues that may be highly involved in the formation of the native DPPIV structure and therefore essential for interactions with other proteins. Titration experiments have shown the presence of six free cysteines and three disulfide bridges in native rat DPPIV. Site-directed mutagenesis studies demonstrate that cysteine residues 326, 337, 445, 448, 455, 473, and 552 in rat DPPIV are involved in cell-surface expression, dimerization, peptidase activity, and protein stability (Dobers *et al.*, 2000). It is believed that six of the 12 cysteine residues are essential for the correct folding and intracellular trafficking of this glycoprotein, and therefore for its normal biological properties.

C. Protein Folding; Substrate Binding; Interactions with Other Proteins That Affect Function of the Protease

DPPIV is a multifunctional molecule, and various biological activities of DPPIV appear to be due to its interactions with other membrane-associated molecules. The many synonyms that have been assigned to DPPIV (e.g., CD26, rat liver biomatrix glycoprotein, Ta1, THAM, Tp103 and ADA-binding protein, collagen and fibronectin receptors) reflect the many functions of DPPIV and the multiple interactions in which it participates.

Human CD26/DPPIV binds adenosine deaminase (ADA) to the T-cell surface, thus protecting the T-cell from adenosine-mediated inhibition of proliferation. Physicochemical studies suggest a two-domain structure for the substrate-binding region of DPPIV: a heavily glycosylated membrane proximal region (residues 47–323), and a cysteine-rich region which contains 10 of the 12 cysteines (residues 290–552). The structures of the glycosylation-rich and cysteine-rich regions of DPPIV are unknown, but could include an atypical β propeller domain of repeated four-stranded antiparallel β sheets as occurs in POP (Fulop *et al.*, 1998). Mutation analysis and monoclonal antibody binding studies showed that the binding sites on DPPIV of ADA and antibodies that inhibit ADA binding are discontinuous and overlapping, which concords with their location on a β propeller fold consisting of repeated β sheets of about 50 amino acids (Abbott *et al.*, 1999b). Consistently, the ADA-binding domain is located within the cysteine-rich domain (Dong *et al.*, 1998). Using truncated, human-rat CD26 swap mutants and antibody cross-blocking studies, epitopes in the 248–358th and 359–449th amino acid regions were shown to associate with inducing modulation of CD26 and T-cell costimulation through the CD3 pathway; the 359–449th amino acid region appeared to encompass the ADA binding domain (Dong *et al.*, 1998).

The cysteine-rich domain is also believed to be responsible for DPPIV-binding to collagen I and fibronectin (Loster *et al.*, 1995). By *in vitro* binding assays, DPPIV binds to collagens; preferentially to collagens I and III, which are both characterized by the formation of large triple-helical domains. Within collagen I, the alpha 1(I) chain was found to be the most prominent binding ligand of DPPIV. A monoclonal anti-DPPIV antibody (13.4) specifically inhibited the interaction of DPPIV with collagen I. Peptide mapping and N-terminal sequencing revealed that the corresponding epitope of mAb 13.4 is located in the cysteine-rich domain of DPPIV (Loster *et al.*, 1995).

The possible association of DPPIV or seprase with other proteases on the cell surface is not known. However, DPPIV was often coexpressed with FAPα in activated fibroblasts and it was considered as the second fibroblast activation protein, FAPβ (Rettig *et al.*, 1993).

D. Substrate Specificity

The prolyl peptidases cleave the Pro-Xaa bond located at the amino terminus (prolyl aminopeptidases such as DPPIV), at the carboxyl terminus (prolyl carboxypeptidases such as PCP), and in the mid region (prolyl endopeptidases such as seprase) of polypeptides or proteins. DPPIV is a prolyl exopeptidase with potential natural substrates including neuropeptide Y, substance P, and β chemokines such as eotaxin, SDF-1 (stromal derived factor), and RANTES (regulated on activation normal T-cell expressed and secreted) with L-proline, L-hydroxyproline, or L-alanine at the penultimate position. In addition to its well-known exopeptidase activity, DPPIV also exhibits endopeptidase activity toward denatured collagen (Bermpohl *et al.*, 1998). Protease inhibitor profile analysis showed that the endo- and exopeptidase activities of DPPIV share a common active site. This substrate specificity was detected for denatured collagen types I, II, III, and V, suggesting that DPPIV might contribute to collagen trimming and metabolism.

Early studies have demonstrated by gelatin zymography that seprase is a neutral, serine type 170-kDa membrane gelatinase. Recent studies have also showed that, using a sensitive fluorogenic assay to measure prolyl peptidase activity of FAPα, it can cleave a number of dipeptidyl peptidase substrates, including Ala-Pro-AFC; similar dipeptidase activity is found in the purified soluble FAPα fusion protein (Park *et al.*, 1999). Several other studies from our laboratories have confirmed this result, suggesting that seprase exhibits both prolyl peptidase and gelatinase activities.

Assessment of proteolytic activities of seprase and DPPIV by zymography and substrate gel overlay assays indicates that purified seprase and DPPIV exhibit individual gelatinase and prolyl dipeptidase activities, respectively (Pineiro-Sanchez *et al.*, 1997). Zymography and substrate overlay assays are not quantitative and have low sensitivity for detecting proteolytic activities. Soluble reaction assays, i.e., using colorimetric, fluorogenic, or ^{3}H-labeled substrates, have been established for determining seprase activity. The degradation of a ^{3}H-gelatin substrate by seprase has been measured in the presence of 5 mM EDTA, which inhibits matrix metalloproteinases but not seprase (Kelly, 1999). Interestingly, this assay showed: (1) that exposure to 60°C abolished seprase activity as judged by zymography (Pineiro-Sanchez *et al.*, 1997; Kelly, 1999), but about 50% gelatinase activity was observed using the ^{3}H-gelatin substrate (Kelly, 1999); and (2) that the seprase purified from human breast cancer tissue exhibited a specific activity (cpm gelatin fragments released/mg protein × h) five times greater than that of seprase purified from chicken embryos (Kelly, 1999). This approach contributes significantly to understanding of enzymatic properties of SIMP members, and important knowledge in this area is forthcoming. In the same line, using a fluorogenic assay membrane extracts prepared from 293FAP^{wt} cells that express recombinant FAPα cleave a number of dipeptidyl peptidase substrates, including Ala-Pro-AFC; similar prolyl peptidase activity is found in the purified soluble FAPα fusion protein (Park *et al.*, 1999).

E. Processing of Enzyme, Trafficking, and Localization

DPPIV is expressed as a noncovalently linked 210-kDa homodimer at the cell surface. Dimerization occurs after hybrid and complex glycosylation in the late Golgi apparatus (Danielsen, 1994). A highly conserved residue of mouse DPPIV, Asp-599 upstream within the catalytic domain, has been shown to be important in enzyme processing. Substitution of Asp-599 for Ala (D599A) specifically decreases the cell-surface expression of DPPIV in stably transfected mouse fibroblasts, suggesting that conserved residue Asp-599 is important for the proper folding, dimerization, and transport of mouse DPPIV (David *et al.,* 1996).

The DPPIV-like enzyme, QPP, undergoes N-glycosylation and is targeted to intracellular vesicles that are distinct from lysosomes (Chiravuri *et al.,* 2000a). Interestingly, this glycosylation is required for QPP enzymatic activity (possibly through dimerization), but not for its localization. Proteinase K treatment of intact vesicles indicates that QPP is located within the vesicles. These vesicles appear to have a secretory component, as QPP is secreted in a functionally active form in response to calcium release. The presence of QPP in the vesicular compartment suggests that molecules bearing the N-terminal X-Pro motif can be cleaved at multiple sites within and outside the cell. These results expand the potential site(s) and scope of a process that appears to be an important mechanism of posttranslational regulation (Chiravuri *et al.,* 2000a).

III. Biologic and Pathologic Roles

A. Natural Substrates

Although physiological targets of SIMP are still not established, several peptides *in vitro* are found to be substrates of these peptidases. Several cytokines, hematopoietic growth factors, neuropeptides, hormones, and chemokines share the X-Pro or X-Ala motif at their N termini. The presence of a proline near the N terminus of a protein could form N-terminal cyclization and protect against nonspecific proteolytic degradation (De Meester *et al.,* 1999). Truncation of bioactive peptides may result in inactivation of peptides or alteration of its receptor specificity. Two approaches for determining DPPIV natural substrates have been applied simultaneously: the *in vitro* ability of DPPIV to truncate the substrate and the effect of truncation on the substrate's biological activities. In some studies on chemokines, complete removal of the N-terminal dipeptide required a large amount of enzyme and inordinately long incubation times (Lambeir *et al.,* 2001). This raises the questions of whether chemokines may serve as substrates *in vivo* or whether DPPIV has a preference for some chemokines over others.

Another aspect of SIMP function relates with the proteolytic degradation of extracellular matrix (ECM) components during tissue remodeling that plays a pivotal role in normal and pathological processes including wound healing, inflammation, angiogenesis, cancer invasion, and metastasis. Proteolytic enzymes in tumors may activate or release growth factors from the ECM or act directly on the ECM itself, thereby facilitating angiogenesis or cancer cell invasion. As described above, DPPIV and seprase exhibit both dipeptidyl peptidase activity and a gelatinase activity capable of degrading denature collagens or other ECM components that belong to the new ECM-degrading serine protease family rather than the matrix metalloprotease family.

B. Role of DPPIV in Blood Glucose Regulation

Among the substrates of DPPIV enzyme activity are two hormones important for glucose regulation, glucagon-like peptide 1 (GLP-1) and glucose-dependent insulinotropic polypeptide (GIP). N-terminally truncated GLP-1 and GIP can be detected in the circulation, suggesting such cleavage may be a physiologically relevant mechanism of regulating their activity (Deacon *et al.*, 1998). In addition, administration of inhibitors of DPPIV leads to increased levels of GLP-1 in the circulation and causes enhanced insulin secretion and improved glucose tolerance in normal and diabetic animals. However, the precise molecular target of these pharmacological inhibitors has not been identified, in part because of the multiplicity of enzymes exhibiting DPPIV-like activity.

An animal DPPIV knockout study has determined the contribution of DPPIV to blood glucose control, showing enhanced insulin secretion and improved glucose tolerance in mice lacking DPPIV (Marguet *et al.*, 2000). Targeted inactivation of the *DPPIV* gene yielded healthy mice that have normal blood glucose levels in the fasted state, but reduced glycemic excursion after a glucose challenge. Levels of glucose-stimulated circulating insulin and the intact insulinotropic form of GLP-1 are increased in DPPIV$^{-/-}$ mice. A pharmacological inhibitor (valine-pyrrolidide) of DPPIV enzymatic activity improved glucose tolerance in wild-type, but not in DPPIV$^{-/-}$, mice. This inhibitor also improved glucose tolerance in GLP-1 receptor$^{-/-}$ mice, indicating that DPPIV contributes to blood glucose regulation by controlling the activity of GLP-1 as well as additional substrates. It is believed that DPPIV may become a potential target for therapy in type II diabetes (Marguet *et al.*, 2000).

C. Truncation of Chemokines by DPPIV: Role in Leukocyte Migration

Cell migration is essential for development, inflammation, and tissue repair, but it also allows malignant cells to exert their lethal ability to invade tissues and metastasize. Cell migration is linked intrinsically to concomitant induction of cell mobility and localized degradation of the ECM, together called cell adhesiveness. Inductive cells, including local inflammatory keratinocytes and cancer cells, express soluble factors, chemokines that interact with chemokine receptors on responsive cells such as leukocytes to promote the directional cell migration. Chemokines, a family of relatively small cytokines that induce leukocyte migration (chemotaxis), might be substrates for DPPIV. They have been classified into four groups, C-, CC-, CXC-, and CXXXC-chemokines (X being any amino acid), based on the characteristic position of the N-terminal cysteine residue(s).

Chemokines act by binding to seven transmembrane G-protein coupled receptors, resulting in chemotaxis, mobilization of intracellular calcium or activation of other signaling pathways. Several chemokines bind to multiple receptors, and the receptors often respond to more than one chemokine. Chemokine receptors are exploited by HIV for cell entry and disease transmission. Many chemokines provide a certain degree of protection against HIV infection (Lambeir *et al.,* 2001).

There are supporting data for the role of DPPIV in N-terminal truncation of chemokines to modulate chemokine action. The biological activity of several hormones and chemokines can be abolished by DPPIV *in vitro* (Mentlein, 1999). The localization of DPPIV on cell surfaces and in biological fluids, its primary specificity, and the type of naturally occurring truncated chemokines are consistent with such a function (Lambeir *et al.,* 2001). The steady-state catalytic parameters for a relevant selection of chemokines (CCL3b, CCL5, CCL11, CCL22, CXCL9, CXCL10, CXCL11, and CXCL12) have been investigated, and, among these, stromal cell-derived factor-1 (CXCL12) and macrophage-derived chemokine (CCL22) appear to be processed by DPPIV (Lambeir *et al.,* 2001). Removal by DPPIV of the two N-terminal amino acids from CXCL12 resulted in significantly reduced chemotactic and calcium-signaling activity due to a decreased affinity for CXCR4 (Proost *et al.,* 1998). Accordingly, CXCL12(3-68) processed by DPPIV lacks antiviral activity against T-tropic human immunodeficiency virus type 1 (HIV-1) strains (Proost *et al.,* 1998).

An amino-terminal truncation of RANTES by DPPIV may introduce receptor selectivity and specificity. DPPIV-processed RANTES(3-68) has a more than 10 times lower chemotactic potency for monocytes and eosinophils (Proost *et al.,* 1998). RANTES(3-68) also has impaired binding and signaling properties through CXCR1 and CXCR3, but remains fully active on CXCR5.

The chemokine receptor CXCR3 is specific for the chemokines CXCL9 and 10 (Mig and IP10) and the recently identified CXCL11 (I-TAC/IP-9). This receptor is primarily expressed on activated T-lymphocytes and the expression of its ligands has been implicated in several pathological processes including several skin diseases. Processing of CXCL10 and CXCL11 by DPPIV resulted in reduced CXCR3-binding properties, loss of calcium-signaling capacity through CXCR3, and more than 10-fold reduced chemotactic potency. Expression of CXCR3 on microvascular endothelial cells has been reported to be dependent on the cell cycle. Proliferating microvascular endothelial cells in the S/G_2-M phase of their cell cycle express CXCR3. Moreover, CXCL9 and CXCL10 inhibited endothelial cell proliferation *in vitro,* suggesting a possible downregulatory role for CXCR3 in angiogenic processes. A study has shown that CXCL10 and CXCL11 cleaved by CD26/DPPIV acted as chemotaxis antagonists and DPPIV-truncated CXCL10 and CXCL9 retained their ability to inhibit the angiogenic activity of interleukin-8 in the rabbit cornea micropocket model. These data demonstrated a negative feedback regulation by DPPIV in CXCR3-mediated chemotaxis without affecting the angiostatic potential of the CXCR3 ligands CXCL10 and CXCL9 (Proost *et al.,* 2001). In addition, expression of CXCR3-targeting chemokines by keratinocytes is thought to direct the influx of T-lymphocytes seen in inflammatory skin diseases and cutaneous T-cell lymphoma. Truncation of both CXCL10 and CXCL11 by DPPIV was found in human keratinocytes that showed impaired CXCR3 activating capacity (Hensbergen *et al.*, 2001).

D. Truncation of Neuropeptide Y by DPPIV: Role in Angiogenesis

Members of the pancreatic polypeptide family, including neuropeptide Y and peptide YY, form another important and well-conserved group of DPPIV substrates. Truncation of neuropeptide Y (NPY), a sympathetic cotransmitter that is released during nerve activation and ischemia and causes vasoconstriction and smooth muscle cell proliferation, by DPPIV generates a shift in their receptor selectivity and alters their biological specificity from vasoconstriction to growth factor activity and proangiogenesis (Zukowska-Grojec *et al.*, 1998). The NPY action is mediated by Y1 and Y2 receptors, and Y2 appears to be the main NPY angiogenic receptor. Its upregulation parallels the NPY-induced capillary tube formation on Matrigel; the Y2 agonist mimics the tube-forming activity of NPY, whereas the Y2 antagonist blocks it. Endothelial cells *in vitro* were shown to express NPY, their receptors, and DPPIV, which terminates the Y1 activity of NPY and cleaves the Tyr1-Pro2 from NPY to form an angiogenic Y2 agonist, NPY3-36. Furthermore, NPY(3-36) could activate Y2, Y3, and/or Y5 to mediate angiogenic effects of NPY on migratory endothelial cells (Ghersi *et al.*, 2001).

Another example is angiotensin, which undergoes a loss of vasoactive function following postproline cleavage of the C-terminal amino acid by prolylcarboxypeptidase (PCP; angiotensinase C), suggesting that PCP is a candidate gene for mediating essential hypertension (Tan *et al.*, 1993).

E. CD26 and T-Cell Activation

The expression of CD26/DPPIV is regulated by the differentiation and activation status of immune cells (see Morimoto and Schlossman, 1998; De Meester *et al.*, 1999, for detailed reviews). CD26 interacts, presumably via its extracellular domain, with CD45, a protein tyrosine-phosphatase. CD26 and CD45 localize at lipid rafts on the plasma membrane, as aggregation of CD26 by anti-CD26 mAb crosslinking also causes coaggregation of CD45 into rafts (Ishii *et al.*, 2001), thereby enhancing protein tyrosine phosphorylation of various signaling molecules, e.g., $p56^{Lck}$, ZAP-70, and TCRζ, and subsequent interleukin-2 production. The extracellular domain of human CD26 on T cells also forms a complex with ADA, which reduces the immunosuppressive activity of local adenosine by its catalytic removal (Dong *et al.*, 1997). The most striking evidence for the importance of ADA for immune function is that a defect in ADA activity results in severe combined immunodeficiency disease in humans.

The internalization of CD26 has been suggested to play a role in CD26-mediated T cell costimulation (Ikushima *et al.,* 2000). Early works show that crosslinking of CD26 and CD3 with immobilized mAbs can induce T cell activation and IL-2 production. Moreover, anti-CD26 antibody treatment of T cells leads to a decrease in the surface expression of CD26 via its internalization, and such modulation results in an enhanced proliferative response to anti-CD3 or anti-CD2 stimulation, as well as enhanced tyrosine phosphorylation of signaling molecules such as CD3ζ and p56^{lck} (Hegen *et al.,* 1997). In a more recent study (Ikushima *et al.,* 2000), CD26 was found to bind the mannose 6-phosphate/insulin-like growth factor II receptor (M6P/IGFIIR) via M6P residues in the CD26 carbohydrate moiety of CD26. T cell activation resulted in enhanced mannose-6-phosphorylation of CD26, whereas crosslinking of CD26 with antibody induced not only internalization of CD26 but also the colocalization of CD26 with the M6P/IGFIIR. The interaction of CD26 and M6P/IGFIIR is inhibited by the addition of M6P.

F. Fibroblast Activation and Tissue Remodeling

FAPα, originally defined by a mouse mAb F19, is a 95-kDa cell surface glycoprotein antigen selectively expressed in cultured fibroblasts, fetal mesenchymal tissues, reactive stromal fibroblasts of epithelial cancers, granulation tissue of healing wounds, and malignant cells of bone and soft tissue sarcomas, but not in carcinoma cells themselves and in cells of all other adult tissues including hematopoietic and blood vessel cells (Rettig *et al.,* 1988, 1993; Garin-Chesa *et al.,* 1990). In these original works, Lloyd J. Old and colleagues demonstrated the existence of fibroblast activation antigens that were present on the cell surface of mesenchymal cells at developing stages and of fibroblasts reactive to wound healing and carcinoma. Stromal expression of FAPα has a significant impact on survival of patients with invasive ductal carcinoma (Ariga *et al.,* 2001). Ohtani and colleagues, thus, concluded that invasive ductal carcinomas with fewer stromal reactions expressing FAPα might be more aggressive.

Based upon the F19 epitope, a FAP protein complex is induced in cultured fibroblasts and, in these cells, the FAP complex consists of a 95-kDa subunit (FAPα) and a CD26/DPPIV subunit (FAPβ) (Rettig *et al.,* 1994). Interestingly, studies with FAP^{low} leptomeningeal fibroblasts revealed that transforming growth factor-β (TGF-β), 12-*O*-tetradecanoyl phorbol-13-acetate, and retinoids can upregulate FAP expression, whereas serum and several other factors had no or little effect on FAP (Rettig *et al.,* 1994). The finding that TGF-β upregulates FAPα expression is important. The TGF-β family member, *lefty,* decreased collagen type I mRNA expression and simultaneously increased ECM-degrading activities of fibroblasts (Mason *et al.,* 2001). These works point to the possibility that FAPα/seprase might be a key cell surface protease involved in promoting ECM degradation, tissue remodeling, and fibrosis. Accordingly, FAPα was shown to be expressed at sites of liver tissue remodeling, e.g., stellate cells in cirrhotic human liver (Levy *et al.*, 1999). FAPα immunoreactivity was most intense on perisinusoidal cells of the periseptal regions within regenerative nodules (15 of 15 cases); this pattern coincides with the tissue remodeling interface.

Consistently, a *Xenopus laevis* FAPα homolog has been described, and its expression is upregulated during hormone-induced tail resorption, indicating a possible role in ECM degradation and in promoting tissue remodeling.

Furthermore, the enzyme may not be essential or involving in housekeeping function, as a report shows that FAPα/seprase $(-/-)$ mice are fertile, show no overt developmental defects, and have no general change in cancer susceptibility (Niedermeyer *et al.*, 2000). Further studies with Fap$-/-$ lacZ showed that mice express β-galactosidase at regions of active tissue remodeling during embryogenesis including somites and perichondrial mesenchyme from cartilage primordial (Niedermeyer *et al.*, 2001).

G. Localized ECM Degradation

Membrane proteases may contribute to ECM degradation by virtue of their localization in membrane structures called invadopodia that contact the ECM. There are secreted proteases, including MMPs and cathepsins, elastase, and plasminogen activators, and membrane proteases, including ADAMs $(-10, -12, -17)$ and MT-MMPs $(-1, -2, -3, -5)$, meprin A, seprase, and DPPIV. They are involved directly in the degradation of collagenous matrix components, which are abundant in inflamed tissues and sites of cancer invasion.

Recent studies suggested that DPPIV may also play important roles in ECM degradation; hence tissue remodeling such as in fibrosis and tumor invasion. The pattern of DPPIV expression is altered in cirrhotic human liver, with normal liver showing DPPIV expression in the bile canalicular domain of hepatocytes, whereas cirrhotic liver shows a loss of zonal expression with DPPIV is reorganized on proliferating bile ductules, leukocytes, and the basolateral domain of hepatocytes (Levy *et al.*, 1999). A retrospective study on follicular thyroid carcinoma supports this proposal that DPPIV could be a cell invasiveness marker (Hirai *et al.*, 1999). DPPIV immunoreactivity on paraffin sections of follicular thyroid carcinoma group with distant metastasis (7/10 cases) was much higher than that of the control group (1/29), which consisted of 15 cases of follicular thyroid adenoma and 14 cases of nodular hyperplasia, suggesting its value in predicting metastatic potential of "benign" thyroid tumors (Hirai *et al.*, 1999). On the other hand, loss of DPPIV expression occurs during melanoma progression at a stage where transformed melanocytes become independent of exogenous growth factors for survival. Reexpressing DPPIV in melanoma cells induced a profound change in phenotype that was characteristic of normal melanocytes, including a loss of tumorigenicity, anchorage-independent growth, a reversal in block in differentiation, and an acquired dependence on exogenous growth factors for cell survival (Wesley *et al.*, 1999). In addition, reexpression of DPPIV rescued expression of FAPα. These results support the view that downregulation of DPPIV is an important early event in the pathogenesis of melanoma.

Seprase was originally identified as a 170-kDa membrane serine gelatinase selectively expressed on cell surface invadopodia of melanoma and breast carcinoma cells for its potential role in cell invasion (Aoyama and Chen, 1990; Monsky *et al.*, 1994; Chen *et al.*, 1994). In contrast to the use of mAb F19 for immunohistochemistry staining of cancer tissues (Rettig *et al.*, 1988, 1993; Garin-Chesa *et al.*, 1990), a detailed analysis using polyclonal antibodies directed against seprase was performed on malignant, premalignant, benign, and normal breast tissues (Kelly *et al.*, 1998). Both 170-kDa gelatinase activity and immunoreactivity of seprase were identified in tumor cells but not the stromal cells or morphologically normal epithelium of infiltrating ductal carcinomas. In addition, lymph-node metastases of infiltrating ductal carcinomas were also strongly positive, but the lymphoid tissue in affected nodes, neoplastic cells in ductal carcinoma *in situ,* epithelial cells of benign fibroadenoma, and benign proliferative breast disease exhibited little or no immunoreactivity. Kelly and colleagues, thus, concluded that the overexpression of seprase by carcinoma cells is consistent with seprase having a role in facilitating invasion and metastasis of infiltrating ductal carcinomas of the breast (Kelly *et al.*, 1998). This cell localization result has been confirmed by two other studies from our laboratories using a panel of mAbs directed against seprase on breast cancer and gastric cancer. The apparent difference in cellular localization of FAPα and seprase depicted by immunohistochemistry could be partially due to the use of antibodies that recognize with different affinity to common and different epitopes exhibited by FAPα (derived from activated fibroblasts) and seprase (derived from invasive cancer cells).

H. Cell Adhesion: Substrate-Binding Sites of Membrane Proteases

Consistent with the potential prolyl peptidase activity toward macromolecules, it is conceivable that seprase or DPPIV may form transient, adhesive bonds with collagen and other ECM components. For examples, outside-out luminal membrane vesicles isolated from rat lung microvascular endothelia by *in situ* perfusion with a low-strength paraformaldehyde solution were shown to bind in significantly larger numbers to lung-metastatic than to nonmetastatic rat breast carcinoma cells (Johnson *et al.*, 1993; Cheng *et al.*, 1998). The mAb 6A3 generated against lung-derived endothelial cell membrane vesicles was shown to be reactive to DPPIV and to inhibit specific adhesion of lung endothelial vesicles to lung-metastatic breast cancer cells. The DPPIV ligand was identified as tumor cell surface-associated fibronectin, and concomitantly, a correlation between the level of fibronectin expression and the tumor cells' ability to bind to DPPIV on endothelial cells and metastasize to the lungs was demonstrated (Cheng *et al.,* 1998). DPPIV- and fibronectin-mediated adhesion and metastasis are blocked when tumor cells are incubated with soluble DPPIV, anti-DPPIV mAb 6A3, and anti-fibronectin anti-serum prior to conducting adhesion and lung colony assays, but is unaffected by soluble plasma fibronectin. Further supporting this, Fischer 344/CRJ rats, which harbor a G633R substitution in DPPIV leading to retention and degradation of the mutant protein in the endoplasmic reticulum, reduce lung colonization of the highly metastatic MTF7 rat breast cancer cell line by 33% relative to normal Fischer 344 rats (Cheng *et al.,* 1999). These results suggested that a single site mutation of DPPIV could drastically change the role of DPPIV in metastasis. Two different cell surface proteases (membrane dipeptidase and amino peptidase N) have been identified by phage display as adhesion molecules important in homing of tumor cells to specific tissues (Rajotte and Ruoslahti, 1999; Ruoslahti and Rajotte, 2000). These findings support the potential importance of SIMP adhesion in the metastatic cascade.

IV. Conclusions and Future Directions

Serine-type, integral membrane peptidases (SIMP), including DPPIV, seprase, and related prolyl peptidases, that are both Pro-Xaa cleaving enzymes and adhesion molecules, are likely to emerge as an important protease family. The main functions of SIMPs reside in their proteolytic and adhesive capacities, thus influencing cellular activities, migration, and invasion. These membrane proteases may form a physically and functionally linked complex at invadopodia during cellular invasion. The cysteine-rich domain of some peptidases exhibits the capability of binding to multiple molecules. This could allow not only activation of the peptidases themselves but also association with other membrane proteases to participate in cooperative ECM protein degradation at invadopodia during cancer invasion.

DPPIV has a cytoplasmic region consisting of only six amino acids, which may be too short to explain signal-transducing activity of DPPIV. Signaling via DPPIV is likely to occur by the interaction of DPPIV with other molecules on the cell surface. This structural complexity of DPPIV could contribute to its multifunctional activities. DPPIV may associate with other integral membrane proteases to form heteromeric complexes with multiple enzymatic activities. DPPIV is now known to form heterooligomeric complexes with seprase. Extensive biological activities of DPPIV that are complementary to DPPIV active sites may be mediated through this complex formation. Other membrane proteases may interact at invadopodia in which proteases of different classes activate other proteases to amplify the proteolytic process. A challenging future direction is to determine the role of protease complexes in protease regulation of invasion.

It is also not understood about mechanisms of activation of membrane proteases. Some proteases (meprin, DPPIV, seprase, QPP, and PCP) must form an oligomeric structure for expression of proteolytic activity rather than proteolytic activation of a zymogen form. It is possible that membrane proteases have a lot in common with other well-characterized integral membrane glycoproteins such as integrins where heterodimeric interactions among subunits govern their functions. Much still needs to be learned about the structure, activation, synthesis, turnover, processing, and regulation of membrane proteases.

Overexpression and inappropriate regulation of proteolytic activity occur often in diseases. There are many studies of the dysregulated expression of DPPIV in leukocyte malignancies (leukemias, lymphomas, autoimmune diseases, HIV) as well as in solid tumor malignancies. Although FAPα and seprase are products of identical genes, extensive model studies have suggested an overlapping role in tissue remodeling and cell invasion, respectively. It is important to determine whether the dysfunction in cancer reflects different isoforms expressed on fibroblasts and invasive cancer cells. Furthermore, recent advances in DNA microarray, SAGE, and real-time PCR technologies provide a "nonbiased" means of analyzing expression profiling at the transcriptional levels of thousands of genes in cancers and other diseases. These technologies can reproducibly generate molecular data from specimens taken directly from patients. The data will generate a database of the molecular "signature" of different cancers and other diseases. Thus, multiplex analysis tools are forthcoming and they will become most valuable as an index of prognosis and as a guide in the design of antiproteolytic strategies aimed at controlling progression of the disease.

Finally, soluble forms of these membrane proteases are found in biological fluids. Although their roles are still poorly understood, it may be expected that the released proteases retain their proteolytic activities and therefore add to the battery of secreted enzymes such as MMPs, elastase, cathepsins, the plasminogen activators, and the related ADAM-TSs (ADAMs with thrombospondin-type 1 motifs), all proteolytic enzymes involved in ECM degradation. By addressing these issues in future studies, we should gain insight into the relationships between membrane proteases and the roles that they play in physiological and pathophysiological processes.

Acknowledgments

We are most grateful to Stan Zucker, Donghai Chen, Huan Dong Wei Zeng, and Jaclyn Freudenberg for critical reviews of this paper. The work was supported by USPHS grant number CA-39077 and HL33711.

References

Abbott, C. A., Baker, E., Sutherland, G. R., and McCaughan, G. W. (1994). Genomic organization, exact localization, and tissue expression of the human CD26 (dipeptidyl peptidase IV) gene [published erratum appears in *Immunogenetics* (1995) **42**, 76]. *Immunogenetics* **40**, 331–338.

228   Chen *et al.*

Abbott, C. A., McCaughan, G. W., and Gorrell, M. D. (1999a). Two highly conserved glutamic acid residues in the predicted beta propeller domain of dipeptidyl peptidase IV are required for its enzyme activity. *FEBS Letters* **458,** 278–284.

Abbott, C. A., McCaughan, G. W., Levy, M. T., Church, W. B., and Gorrell, M. D. (1999b). Binding to human dipeptidyl peptidase IV by adenosine deaminase and antibodies that inhibit ligand binding involves overlapping, discontinuous sites on a predicted beta propeller domain. *Eur. J. Biochem.* **266,** 798–810.

Abbott, C. A., Yu, D. M., Woollatt, E., Sutherland, G. R., McCaughan, G. W., and Gorrell, M. D. (2000). Cloning, expression and chromosomal localization of a novel human dipeptidyl peptidase (DPP) IV homolog, DPP8. *Eur. J. Biochem.* **267,** 6140–6150.

Aoyama, A., and Chen, W.-T. (1990). A 170-kDa membrane-bound protease is associated with the expression of invasiveness by human malignant melanoma cells. *Proc. Natl. Acad. Sci. USA* **87,** 8296–8300.

Ariga, N., Sato, E., Ohuchi, N., Nagura, H., and Ohtani, H. (2001). Stromal expression of fibroblast activation protein/seprase, a cell membrane serine proteinase and gelatinase, is associated with longer survival in patients with invasive ductal carcinoma of breast. *Int. J. Cancer* **95,** 67–72.

Barrett, A. J., and Rawlings, N. D. (1992). Oligopeptidases, and the emergence of the prolyl oligopeptidase family. *Biol. Chem. Hoppe Seyler* **373,** 353–360.

Bauvois, B. (2001). Transmembrane proteases in focus: diversity and redundancy. *J. Leukoc. Biol.* **70,** 11–17.

Bermpohl, F., Löster, K., Reutter, W., and Baum, O. (1998). Rat dipeptidyl peptidase IV (DPP IV) exhibits endopeptidase activity with specificity for denatured fibrillar collagens. *FEBS Lett.* **428,** 152–156.

Blanco, J., Nguyen, C., Callebaut, C., Jacotot, E., Krust, B., Mazaleyrat, J. P., Wakselman, M., and Hovanessian, A. G. (1998). Dipeptidyl-peptidase IV-beta—further characterization and comparison to dipeptidyl-peptidase IV activity of CD26. *Eur. J. Biochem.* **256,** 369–378.

Chen, W.-T., Lee, C. C., Goldstein, L., Bernier, S., Liu, C. H., Lin, C. Y., Yeh, Y., Monsky, W. L., Kelly, T., Dai, M., and Mueller, S. C. (1994). Membrane proteases as potential diagnostic and therapeutic targets for breast malignancy. *Breast Cancer Res. Treat.* **31,** 217–226.

Cheng, H. C., Abdel-Ghany, M., Elble, R. C., and Pauli, B. U. (1998). Lung endothelial dipeptidyl peptidase IV promotes adhesion and metastasis of rat breast cancer cells via tumor cell surface-associated fibronectin. *J. Biol. Chem.* **273,** 24207–24215.

Cheng, H. C., Abdel-Ghany, M., Zhang, S., and Pauli, B. U. (1999). Is the Fischer 344/CRJ rat a protein-knock-out model for dipeptidyl peptidase IV-mediated lung metastasis of breast cancer. *Clin. Exp. Metastasis* **17,** 609–615.

Chiravuri, M., Lee, H., Mathieu, S. L., and Huber, B. T. (2000b). Homodimerization via a leucine zipper motif is required for enzymatic activity of quiescent cell proline dipeptidase. *J. Biol. Chem.* **275,** 26994–26999.

Chiravuri, M., Agarraberes, F., Mathieu, S. L., Lee, H., and Huber, B. T. (2000a). Vesicular localization and characterization of a novel post-proline-cleaving aminodipeptidase, quiescent cell proline dipeptidase. *J. Immunol.* **165,** 5695–5702.

Danielsen, E. M. (1994). Dimeric assembly of enterocyte brush border enzymes. *Biochemistry* **33,** 1599–1605.

David, F., Baricault, L., Sapin, C., Gallet, X., Marguet, D., Thomas-Soumarmon, A., and Trugnan, G. (1996). Reduced cell surface expression of a mutated dipeptidyl peptidase iv (dpp iv/cd26) correlates with the generation of a beta strand in its c-terminal domain. *Biochem. Biophys. Res. Commun.* **222,** 833–838.

David, F., Bernard, A. M., Pierres, M., and Marguet, D. (1993). Identification of serine 624, aspartic acid 702, and histidine 734 as the catalytic triad residues of mouse dipeptidyl-peptidase IV (CD26). A member of a novel family of nonclassical serine hydrolases. *J. Biol. Chem.* **268,** 17247–17252.

De Meester, I., Korom, S., Van Damme, J., and Scharpe, S. (1999). CD26, let it cut or cut it down [Review]. *Immunol. Today* **20**, 367–375.

Deacon, C. F., Hughes, T. E., and Holst, J. J. (1998). Dipeptidyl peptidase IV inhibition potentiates the insulinotropic effect of glucagon-like peptide 1 in the anesthetized pig. *Diabetes* **47**, 764–769.

Dobers, J., Grams, S., Reutter, W., and Fan, H. (2000). Roles of cysteines in rat dipeptidyl peptidase IV/CD26 in processing and proteolytic activity. *Eur. J. Biochem.* **267**, 5093–5100.

Dong, R. P., Tachibana, K., Hegen, M., Munakata, Y., Cho, D., Schlossman, S. F., and Morimoto, C. (1997). Determination of adenosine deaminase binding domain on CD26 and its immunoregulatory effect on T cell activation. *J. Immunol.* **159**, 6070–6076.

Dong, R. P., Tachibana, K., Hegen, M., Scharpe, S., Cho, D., Schlossman, S. F., and Morimoto, C. (1998). Correlation of the epitopes defined by anti-CD26 mAbs and CD26 function. *Mol. Immunol.* **35**, 13–21.

Duke-Cohan, J. S., Morimoto, C., Rocker, J. A., and Schlossman, S. F. (1995). A novel form of dipeptidylpeptidase IV found in human serum. Isolation, characterization, and comparison with T lymphocyte membrane dipeptidylpeptidase IV (CD26). *J. Biol. Chem.* **270**, 14107–14114.

Duke-Cohan, J. S., Gu, J., McLaughlin, D. F., Xu, Y., Freeman, G. J., and Schlossman, S. F. (1998). Attractin (DPPT-L), a member of the CUB family of cell adhesion and guidance proteins, is secreted by activated human T lymphocytes and modulates immune cell interactions. *Proc. Natl. Acad. Sci. USA* **95**, 11336–11341.

Durinx, C., Lambeir, A. M., Bosmans, E., Falmagne, J. B., Berghmans, R., Haemers, A., Scharpe, S., and De, M. I. (2000). Molecular characterization of dipeptidyl peptidase activity in serum: soluble CD26/dipeptidyl peptidase IV is responsible for the release of X-Pro dipeptides. *Eur. J. Biochem.* **267**, 5608–5613.

Fan, H., Meng, W. M., Kilian, C., Grams, S., and Reutter, W. (1997). Domain-specific N-glycosylation of the membrane glycoprotein dipeptidylpeptidase IV (CD26) influences its subcellular trafficking, biological stability, enzyme activity and protein folding. *Eur. J. Biochem.* **246**, 243–251.

Fodstad, O., Aamdal, S., McMenamin, M., Nesland, J. M., and Pihl, A. (1988). A new experimental metastasis model in athymic nude mice, the human malignant melanoma LOX. *Int. J. Cancer* **41**, 442–449.

Fulop, V., Bocskei, Z., and Polgar, L. (1998). Prolyl oligopeptidase: an unusual beta-propeller domain regulates proteolysis. *Cell* **94**, 161–170.

Garin-Chesa, P., Old, L. J., and Rettig, W. J. (1990). Cell surface glycoprotein of reactive stromal fibroblasts as a potential antibody target in human epithelial cancers. *Proc. Natl. Acad. Sci. USA* **87**, 7235–7239.

Ghersi, G., Chen, W.-T., Lee, E. W., and Zukowska, Z. (2001). Critical role of dipeptidyl peptidase IV in neuropeptide Y-mediated endothelial cell migration in response to wounding. *Peptides* **22**, 453–458.

Goldstein, L. A., and Chen, W.-T. (2000). Identification of an alternatively spliced seprase mRNA that encodes a novel intracellular isoform. *J. Biol. Chem.* **275**, 2554–2559.

Goldstein, L. A., Ghersi, G., Piñeiro-Sánchez, M. L., Salamone, M., Yeh, Y. Y., Flessate, D., and Chen, W.-T. (1997). Molecular cloning of seprase: A serine integral membrane protease from human melanoma. *Biochim. Biophys. Acta* **1361**, 11–19.

Goossens, F. J., Wauters, J. G., Vanhoof, G. C., Bossuyt, P. J., Schatteman, K. A., Loens, K., and Scharpe, S. L. (1996). Subregional mapping of the human lymphocyte prolyl oligopeptidase gene (PREP) to human chromosome 6q22. *Cytogenet. Cell Genet.* **74**, 99–101.

Hegen, M., Kameoka, J., Dong, R. P., Schlossman, S. F., and Morimoto, C. (1997). Cross-linking of CD26 by antibody induces tyrosine phosphorylation and activation of mitogen-activated protein kinase. *Immunology* **90**, 257–264.

Hensbergen, P. J., Raaij-Helmer, E. M., Dijkman, R., van der Schors, R. C., Werner-Felmayer, G., Boorsma, D. M., Scheper, R. J., Willemze, R., and Tensen, C. P. (2001). Processing of natural and recombinant CXCR3-targeting chemokines and implications for biological activity. *Eur. J. Biochem.* **268,** 4992–4999.

Hirai, K., Kotani, T., Aratake, Y., Ohtaki, S., and Kuma, K. (1999). Dipeptidyl peptidase IV (DPP IV/CD26) staining predicts distant metastasis of "benign" thyroid tumor [letter]. *Pathol. Int.* **49,** 264–265.

Hopsu-Havu, V. K., and Glenner, G. G. (1966). A new dipeptide naphthylamidase hydrolyzing glycyl-prolyl-beta-naphthylamide. *Histochemie* **7,** 197–201.

Ikushima, H., Munakata, Y., Ishii, T., Iwata, S., Terashima, M., Tanaka, H., Schlossman, S. F., and Morimoto, C. (2000). Internalization of CD26 by mannose 6-phosphate/insulin-like growth factor II receptor contributes to T cell activation. *Proc. Natl. Acad. Sci. USA* **97,** 8439–8444.

Ishii, T., Ohnuma, K., Murakami, A., Takasawa, N., Kobayashi, S., Dang, N. H., Schlossman, S. F., and Morimoto, C. (2001). CD26-mediated signaling for T cell activation occurs in lipid rafts through its association with CD45RO. *Proc. Natl. Acad. Sci. USA* **98,** 12138–12143.

Johnson, R. C., Zhu, D., Augustin-Voss, H. G., and Pauli, B. U. (1993). Lung endothelial dipeptidyl peptidase IV is an adhesion molecule for lung-metastatic rat breast and prostate carcinoma cells. *J. Cell Biol.* **121,** 1423–1432.

Kelly, T. (1999). Evaluation of seprase activity. *Clin. Exp. Metastasis* **17,** 57–62.

Kelly, T., Kechelava, S., Rozypal, T. L., West, K. W., and Korourian, S. (1998). Seprase, a membrane-bound protease, is overexpressed by invasive ductal carcinoma cells of human breast cancers. *Mod. Pathol.* **11,** 855–863.

Lambeir, A. M., Proost, P., Durinx, C., Bal, G., Senten, K., Augustyns, K., Scharpe, S., Van Damme, J., and De Meester, I. (2001). Kinetic investigation of chemokine truncation by CD26/dipeptidyl peptidase IV reveals a striking selectivity within the chemokine family. *J. Biol. Chem.* **276,** 29839–29845.

Levy, M. T., McCaughan, G. W., Abbott, C. A., Park, J. E., Cunningham, A. M., Muller, E., Rettig, W. J., and Gorrell, M. D. (1999). Fibroblast activation protein: a cell surface dipeptidyl peptidase and gelatinase expressed by stellate cells at the tissue remodelling interface in human cirrhosis. *Hepatology* **29,** 1768–1778.

Loster, K., Zeilinger, K., Schuppan, D., and Reutter, W. (1995). The cysteine-rich region of dipeptidyl peptidase IV (CD 26) is the collagen-binding site. *Biochem. Biophys. Res. Commun.* **217,** 341–348.

Marguet, D., Baggio, L., Kobayashi, T., Bernard, A. M., Pierres, M., Nielsen, P. F., Ribel, U., Watanabe, T., Drucker, D. J., and Wagtmann, N. (2000). Enhanced insulin secretion and improved glucose tolerance in mice lacking CD26. *Proc. Natl. Acad. Sci. USA* **97,** 6874–6879.

Mason, J. M., Xu, H. P., Rao, S. K., Leask, A., Barcia, M., Shan, J., Stephenson, R., and Tabibzadeh, S. (2001). Lefty contributes to the remodeling of extracellular matrix by inhibition of connective tissue growth factor and collagen mRNA expression and increased proteolytic activity in a fibrosarcoma model. *J. Biol. Chem.* **277,** 407–415.

Medrano, F. J., Alonso, J., Garcia, J. L., Romero, A., Bode, W., and Gomis-Ruth, F. X. (1998). Structure of proline iminopeptidase from *Xanthomonas campestris* pv. *citri:* a prototype for the prolyl oligopeptidase family. *EMBO J.* **17,** 1–9.

Mentlein, R. (1999). Dipeptidyl-peptidase IV (CD26)—role in the inactivation of regulatory peptides. *Regul. Pept.* **85,** 9–24.

Mentlein, R., and Struckhoff, G. (1989). Purification of two dipeptidyl aminopeptidases II from rat brain and their action on proline-containing neuropeptides. *J. Neurochem.* **52,** 1284–1293.

Monsky, W. L., Lin, C.-Y., Aoyama, A., Kelly, T., Mueller, S. C., Akiyama, S. K., and Chen, W.-T. (1994). A potential marker protease of invasiveness, seprase, is localized on invadopodia of human malignant melanoma cells. *Cancer Res.* **54,** 5702–5710.

Morimoto, C., and Schlossman, S. F. (1998). The structure and function of CD26 in the T-cell immune response. [Review]. *Immunol. Rev.* **161,** 55–70.

Nakahara, H., Howard, L., Thompson, E. W., Sato, H., Seiki, M., Yeh, Y., and Chen, W.-T. (1997). Transmembrane/cytoplasmic domain-mediated membrane type 1-matrix metalloprotease docking to invadopodia is required for cell invasion. *Proc. Natl. Acad. Sci. USA* **94,** 7959–7964.

Niedermeyer, J., Garin-Chesa, P., Kriz, M., Hilberg, F., Mueller, E., Bamberger, U., Rettig, W. J., and Schnapp, A. (2001). Expression of the fibroblast activation protein during mouse embryo development. *Int. J. Dev. Biol.* **45,** 445–447.

Niedermeyer, J., Kriz, M., Hilberg, F., Garin-Chesa, P., Bamberger, U., Lenter, M. C., Park, J., Viertel, B., Puschner, H., Mauz, M., Rettig, W. J., and Schnapp, A. (2000). Targeted disruption of mouse fibroblast activation protein. *Mol. Cell. Biol.* **20,** 1089–1094.

Niedermeyer, J., Enenkel, B., Park, J. E., Lenter, M., Rettig, W. J., Damm, K., and Schnapp, A. (1998). Mouse fibroblast-activation protein–conserved Fap gene organization and biochemical function as a serine protease. *Eur. J. Biochem.* **254,** 650–654.

Ogata, S., Misumi, Y., and Ikehara, Y. (1989). Primary structure of rat liver dipeptidyl peptidase IV deduced from its cDNA and identification of the NH_2-terminal signal sequence as the membrane-anchoring domain. *J. Biol. Chem.* **264,** 3596–3601.

Pangalos, M. N., Neefs, J. M., Somers, M., Verhasselt, P., Bekkers, M., van der Helm, L., Fraiponts, E., Ashton, D., and Gordon, R. D. (1999). Isolation and expression of novel human glutamate carboxypeptidases with N-acetylated alpha-linked acidic dipeptidase and dipeptidyl peptidase IV activity. *J. Biol. Chem.* **274,** 8470–8483.

Park, J. E., Lenter, M. C., Zimmermann, R. N., Garin-Chesa, P., Old, L. J., and Rettig, W. J. (1999). Fibroblast activation protein, a dual specificity serine protease expressed in reactive human tumor stromal fibroblasts. *J. Biol. Chem.* **274,** 36505–36512.

Pineiro-Sanchez, M. L., Goldstein, L. A., Dodt, J., Howard, L., Yeh, Y., Tran, H., Argraves, W. S., and Chen, W.-T. (1997). Identification of the 170-kDa melanoma membrane-bound gelatinase (seprase) as a serine integral membrane protease. *J. Biol. Chem.* **272,** 7595–7601 [Correction (1998). *J. Biol. Chem.* **273,** 13366].

Proost, P., De Meester, I., Schols, D., Struyf, S., Lambeir, A. M., Wuyts, A., Opdenakker, G., De Clercq, E., Scharpe, S., and Van Damme, J. (1998). Amino-terminal truncation of chemokines by CD26/dipeptidylpeptidase IV—Conversion of RANTES into a potent inhibitor of monocyte chemotaxis and HIV-1-infection. *J. Biol. Chem.* **273,** 7222–7227.

Proost, P., Schutyser, E., Menten, P., Struyf, S., Wuyts, A., Opdenakker, G., Detheux, M., Parmentier, M., Durinx, C., Lambeir, A. M., Neyts, J., Liekens, S., Maudgal, P. C., Billiau, A., and Van Damme, J. (2001). Amino-terminal truncation of CXCR3 agonists impairs receptor signaling and lymphocyte chemotaxis, while preserving antiangiogenic properties. *Blood* **98,** 3554–3561.

Rajotte, D., and Ruoslahti, E. (1999). Membrane dipeptidase is the receptor for a lung-targeting peptide identified by in vivo phage display. *J. Biol. Chem.* **274,** 11593–11598.

Rettig, W. J., Garin-Chesa, P., Beresford, H. R., Oettgen, H. F., Melamed, M. R., and Old, L. J. (1988). Cell-surface glycoproteins of human sarcomas: differential expression in normal and malignant tissues and cultured cells. *Proc. Natl. Acad. Sci. USA* **85,** 3110–3114.

Rettig, W. J., Garin-Chesa, P., Healey, J. H., Su, S. L., Ozer, H. L., Schwab, M., Albino, A. P., and Old, L. J. (1993). Regulation and heteromeric structure of the fibroblast activation protein in normal and transformed cells of mesenchymal and neuroectodermal origin. *Cancer Res.* **53,** 3327–3335.

Rettig, W. J., Su, S. L., Fortunato, S. R., Scanlan, M. J., Raj, B. K., Garin-Chesa, P., Healey, J. H., and Old, L. J. (1994). Fibroblast activation protein: purification, epitope mapping and induction by growth factors. *Int. J. Cancer* **58,** 385–392.

Ruoslahti, E., and Rajotte, D. (2000). An address system in the vasculature of normal tissues and tumors. *Annu. Rev. Immunol.* **18,** 813–827.

Sato, H., Takino, T., Okada, Y., Cao, J., Shinagawa, A., Yamamoto, E., and Seiki, M. (1994). A matrix metalloproteinase expressed on the surface of invasive tumour cells. *Nature* **370,** 61–65.

Scanlan, M. J., Raj, B. K., Calvo, B., Garin-Chesa, P., Sanz-Moncasi, M. P., Healey, J. H., Old, L. J., and Rettig, W. J. (1994). Molecular cloning of fibroblast activation protein alpha, a member of the serine protease family selectively expressed in stromal fibroblasts of epithelial cancers. *Proc. Natl. Acad. Sci. USA* **91,** 5657–5661.

Tan, F., Morris, P. W., Skidgel, R. A., and Erdos, E. G. (1993). Sequencing and cloning of human prolylcarboxypeptidase (angiotensinase C). Similarity to both serine carboxypeptidase and prolylendopeptidase families [published erratum appears in *J. Biol. Chem.* (1993). **268,** 26032]. *J. Biol. Chem.* **268,** 16631–16638.

Tang, W., Gunn, T. M., McLaughlin, D. F., Barsh, G. S., Schlossman, S. F., and Duke-Cohan, J. S. (2000). Secreted and membrane attractin result from alternative splicing of the human ATRN gene. *Proc. Natl. Acad. Sci. USA* **97,** 6025–6030.

Uekita, T., Itoh, Y., Yana, I., Ohno, H., and Seiki, M. (2001). Cytoplasmic tail-dependent internalization of membrane-type 1 matrix metalloproteinase is important for its invasion-promoting activity. *J. Cell Biol.* **155,** 1345–1356.

Underwood, R., Chiravuri, M., Lee, H., Schmitz, T., Kabcenell, A. K., Yardley, K., and Huber, B. T. (1999). Sequence, purification, and cloning of an intracellular serine protease, quiescent cell proline dipeptidase. *J. Biol. Chem.* **274,** 34053–34058.

Wesley, U. V., Albino, A. P., Tiwari, S., and Houghton, A. N. (1999). A role for dipeptidyl peptidase IV in suppressing the malignant phenotype of melanocytic cells. *J. Exp. Med.* **190,** 311–322.

Yokotani, N., Doi, K., Wenthold, R. J., and Wada, K. (1993). Non-conservation of a catalytic residue in a dipeptidyl aminopeptidase IV-related protein encoded by a gene on human chromosome 7. *Hum. Mol. Genet.* **2,** 1037–1039.

Zucker, S., Cao, J., and Chen, W.-T. (2000). Critical appraisal of the use of matrix metalloproteinase inhibitors in cancer treatment. *Oncogene* **19,** 6642–6650.

Zukowska-Grojec, Z., Karwatowska, P., Rose, W., Rone, J., Movafagh, S., Ji, H., Yeh, Y., Chen, W.-T., Kleinman, H. K., Grouzmann, E., and Grant, D. S. (1998). Neuropeptide Y: a novel angiogenic factor from the sympathetic nerves and endothelium. *Circ. Res.* **83,** 187–195.

8

The Secretases of Alzheimer's Disease

Michael S. Wolfe
Center for Neurologic Diseases
Brigham and Women's Hospital and Harvard Medical School
Boston, Massachusetts 02115

I. The Amyloid Hypothesis of Alzheimer's Disease
    A. Amyloid Plaques and Neurofibrillary Tangles
    B. Proteolytic Processing of the Amyloid-β Precursor Protein
    C. Genetic Evidence: Mutations in APP and the Presenilins
II. β-Secretase
    A. BACE: A Membrane-Tethered Aspartyl Protease
    B. Specificity and Structure of BACE and BACE2
    C. Protein Maturation
    D. BACE Knockout and Amyloid Production
    E. Biological Roles of β-Secretases
III. γ-Secretase
    A. Substrate Specificity
    B. Pharmacological Profiling as an Aspartyl Protease
    C. The Role of Presenilins: Alzheimer-Causing Mutations and Knockout Mice
    D. Membrane Topology and Protein Maturation
    E. Presenilins as Aspartyl Proteases
    F. Biological Roles of γ-Secretases
    G. Identification and Role of Nicastrin
    H. An Emerging Family of Polytopic Membrane Proteases
IV. α-Secretase
    A. Substrate Specificity
    B. Pharmacological Profiling and Tentative Identification as TACE and Kuzbanian
    C. Regulation of α-Secretase Processing of APP
V. Concluding Remarks
    References

I. The Amyloid Hypothesis of Alzheimer's Disease

Alzheimer's disease (AD) is a progressive neurodegenerative disorder that leads to loss of memory and other cognitive functions and eventually death. The leading hypothesis is that aggregation of the 4-kDa amyloid-β peptide (Aβ) is the initial molecular event that causes this devastation. It may come as a surprise that one small protein can be so critical to a disease as complex as AD. Nevertheless, a substantial body of evidence now strongly supports the amyloid hypothesis of AD

Current Topics in Developmental Biology, Vol. 54

pathogenesis (Selkoe, 1999). Because the levels of $A\beta$ are apparently so important to pathogenesis, modulation of the proteases that process the amyloid-β precursor protein (APP) is considered a reasonable strategy for the prevention and treatment of AD (Wolfe, 2001). The biochemistry of these proteases and their role in normal and pathological events are the subject of this chapter. To put things into proper context, it is reasonable to first provide some of the evidence for the amyloid hypothesis, introducing the APP-cleaving proteases in the process.

A. Amyloid Plaques and Neurofibrillary Tangles

Postmortem analysis of the AD brain reveals the extraneuronal plaques and intraneuronal tangles first described by Alois Alzheimer nearly a century ago. The major protein component of the plaques is the amyloid-β protein ($A\beta$), and the tangles are composed of filaments of the microtubule-associated protein tau. Dense deposits containing fibrillar forms of $A\beta$ are intimately associated with dystrophic axons and dendrites, and these neuritic plaques are found in cerebral and midbrain regions associated with cognition and memory (Blessed et al., 1968; Perry et al., 1978; Cummings and Cotman, 1995). Within and surrounding the neuritic plaques are activated microglia and astrocytes (Meda et al., 1995; El Khoury et al., 1996). $A\beta$-specific antibodies reveal less dense, "diffuse" plaques that are not associated with dystrophic neurites, activated microglia, or reactive astrocytes (Stalder et al., 1999; Sasaki et al., 1997; Funato et al., 1998). These plaques contain amorphous, nonfibrillary $A\beta$ and are found in areas of the brain generally not implicated in clinical AD. The diffuse plaques are often found in abundance in elderly, cognitively normal people, leading to the suggestion that these diffuse plaques may be the precursors to pathogenic dense plaques. Neurofibrillary tangles are also found in the brain regions critical to higher brain function. Biochemical analysis reveals that the filamentous form of the tau protein found in these tangles is hyperphosphorylated. Tau hyperphosphorylation renders insoluble this otherwise highly soluble cytosolic protein, and this modified form of tau is also found in many plaque-associated dystrophic neurites. Interestingly, tau-containing neurofibrillary tangles occur in a number of other, uncommon neurodegenerative diseases. In contrast, the amyloid-containing neuritic plaques are unique to AD and occur prior to tangles.

B. Proteolytic Processing of the Amyloid-β Precursor Protein

APP is an integral membrane protein proteolyzed by several different secretases (Fig. 1). β-Secretase generates the N terminus of $A\beta$, cleaving APP to produce the soluble β-APP$_s$ and a 99-residue C-terminal fragment (C99) that remains membrane bound. Alternatively, α-secretase cuts within the $A\beta$ region to produce

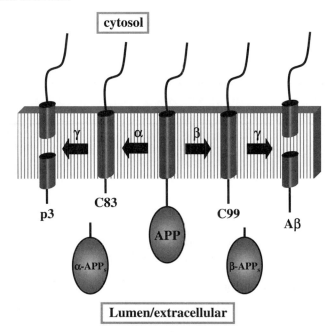

Figure 1 Topology and proteolytic processing of APP. Cleavage by α-secretase forms C83, while processing by β-secretase produces C99. In either event, the APP ectodomain is shed (α- and β-APP$_s$). C83 and C99 are further processed by γ-secretase within the transmembrane domain to produce p3 and Aβ, respectively.

α-APP$_s$ and an 83-residue C-terminal fragment (C83). Both C99 and C83 are substrates for γ-secretase, which performs an unusual proteolysis in the middle of the transmembrane domain to produce the 4-kDa Aβ from C99 and a 3-kDa peptide called p3 from C83. Proteolysis by γ-secretase is heterogeneous: most of the full-length Aβ species produced is a 40-residue peptide (Aβ_{40}), while a small proportion is a 42-residue C-terminal variant (Aβ_{42}). The longer and more hydrophobic Aβ_{42} is much more prone to fibril formation than is Aβ_{40} (Jarrett *et al.*, 1993), and even though Aβ_{42} is a minor form of Aβ, it is the major Aβ species found in cerebral plaques (Iwatsubo *et al.*, 1994).

C. Genetic Evidence: Mutations in APP and the Presenilins

Major clues to the etiology of AD came from families suffering from autosomal dominant, early-onset forms of the disease (familial AD or FAD). Other than the fact that FAD is clearly hereditary and manifests itself at earlier ages (<60 years), it is indistinguishable from the sporadic form of AD. The first identified FAD-causing mutations were found in the gene encoding APP on chromosome 21

(Chartier-Harlin *et al.*, 1991; Goate *et al.*, 1991; Murrell *et al.*, 1991). At least five such mutations have been identified. These are found near the proteolytic cleavage sites in APP that result in Aβ, and all lead to increased production of Aβ in general or specifically a 42-residue form of Aβ (Aβ_{42}) in transfected cells (Citron *et al.*, 1992, 1994; Cai *et al.*, 1993; Suzuki *et al.*, 1994; Haass *et al.*, 1994), in transgenic mice (Johnson-Wood *et al.*, 1997; Citron *et al.*, 1997), and in the plasma of mutant carriers (Scheuner *et al.*, 1996).

A related genetic clue that Aβ is involved in the early molecular events leading to AD is the fact that all Down syndrome (trisomy 21) patients invariably develop AD by age 50. Down syndrome patients carry an extra copy of *APP*, located on chromosome 21, and they produce more Aβ from birth and develop amyloid plaques as early as age 12 (Teller *et al.*, 1996; Lemere *et al.*, 1999). These early plaques are of the diffuse kind and contain Aβ_{42} almost exclusively. Because victims of Down syndrome are fated to develop AD at early ages, the observance of Aβ_{42}-specific diffuse plaques suggests that these plaques could be the precursors of the dense, neuritic plaques found in the AD brain.

APP mutations, however, account for only about 10% of FAD cases and only 2% of all incidences of AD. Most FAD is caused by missense mutations in the genes for presenilin-1 (PS1) and presenilin-2 (PS2), located on chromosomes 14 and 1, respectively (Sherrington *et al.*, 1995; Levy-Lahad *et al.*, 1995). These genes encode multitransmembrane proteins, the normal biological roles of which were completely unknown upon discovery. More than 70 different mutations in the presenilins have been identified that lead to FAD. Virtually all are missense mutations, and these are located in various regions of the primary sequence (see later discussion). Remarkably, all such mutations analyzed to date lead to specific increases in Aβ_{42} production, implicating this particular Aβ species in the etiology of AD (Scheuner *et al.*, 1996; Citron *et al.*, 1997; Tomita *et al.*, 1997; Duff *et al.*, 1996).

The three principal proteolytic activities that process APP are described next. β-Secretase and the presenilin-associated γ-secretase are discussed first because of their primary role in the etiology of AD and the fact that they are considered major therapeutic targets. α-Secretase is discussed last and in considerably less detail because the enzymes putatively responsible for this activity are members of the ADAM family of metalloproteases that are more fully discussed in Chapter 4 on membrane sheddases by Arribas and Merlos-Suarez.

II. β-Secretase

A. BACE: A Membrane-Tethered Aspartyl Protease

β-Secretase generates the N terminus of Aβ, cleaving APP on the lumenal/extracellular side at approximately 30 residues from the transmembrane domain (Selkoe, 1994). Most cell types produce Aβ, indicating broad expression of

A

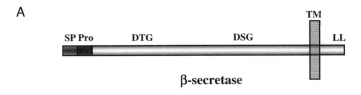

β-secretase

B

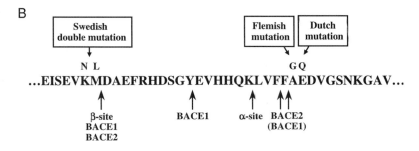

β-secretase cleavage sites in APP

Figure 2 (A) Structure of β-secretase. SP, signal peptide; Pro, pro region; TM, transmembrane region. DTG and DSG regions are conserved sequences containing the two catalytic aspartates. LL is a dileucine motif necessary for recycling β-secretase into endosomes. (B) Sequence of APP in and around the α- and β-secretase cleavage sites and disease-causing missense mutations. See text for details.

β-secretase. However, considerably more Aβ is generated in primary brain cultures than in peripheral cells (Seubert *et al.,* 1993), and neurons display more β-secretase activity than astrocytes (Zhao *et al.,* 1996) (perhaps partly explaining why Aβ selectively aggregates in the brain). β-Secretase cuts at the sequence EVKM*DAEF, where the asterisk denotes the cleavage site (see Fig. 2). However, a KM-to-NL double mutation immediately adjacent to the β-secretase cleavage site causes FAD in a Swedish family and leads to increased Aβ production by enhancing proteolysis at the β-site (Citron *et al.,* 1992). Whether the same β-secretase processed both wild-type and Swedish mutant APP was not clear, especially because β-secretase otherwise appears to have a fairly restricted sequence specificity (Citron *et al.,* 1995).

This enzyme long eluded vigorous attempts at identification, but in 1999 five laboratories independently reported the same protein as β-secretase (Vassar *et al.,* 1999; Sinha *et al.,* 1999; Yan *et al.,* 1999; Hussain *et al.,* 1999; Lin *et al.,* 2000). In all five reports, β-secretase (also named beta-site APP-cleaving enzyme or BACE, BACE1, asp2, or memapsin 2 by the various research groups) was identified as a novel aspartyl protease of 501 amino acids containing a single transmembrane domain near the C terminus as well as a signal sequence and propeptide region at the N terminus (Fig. 2). Consistent with observations from cell-based APP mutagenesis studies, the enzyme processes Swedish mutant APP or short peptides based on this mutant much better than the corresponding wild-type protein or peptide (Vassar *et al.,* 1999; Yan *et al.,* 1999; Lin *et al.,* 2000), but it does not cleave a peptide containing a Met-to-Val change at P1 (Vassar *et al.,* 1999). Two aspartates, D93 and D289, are required for activity. Mutation of either aspartate, however, does not affect removal of the propeptide region in transfected cells, indicating that β-secretase does not autoproteolyze (McConlogue *et al.,* 2000). Indeed, installation of an ER retention sequence demonstrated that pro-peptide cleavage occurs after exit from this organelle (Huse *et al.,* 2000). The responsible β-secretase-activating protease appears to be a furin-like convertase (Bennett *et al.,* 2000). Interestingly, the homologous BACE2 apparently does undergo autoproteolysis (Hussain *et al.,* 2001). β-Secretase shows a pH optimum of 4-4.5 for the cleavage of both wild-type and Swedish mutant peptides (Vassar *et al.,* 1999; Sinha *et al.,* 1999; Lin *et al.,* 2000). Although clearly an aspartyl protease, β-secretase is not blocked by the broad-spectrum aspartyl protease inhibitor pepstatin A at 30-50 μM (Vassar *et al.,* 1999; Sinha *et al.,* 1999). However, incorporation of a statine moiety into peptide substrates provided potent inhibitors (Sinha *et al.,* 1999).

B. Specificity and Structure of BACE and BACE2

Sequencing of APP fragment C99 generated from overexpressed β-secretase (Vassar *et al.,* 1999) confirmed that the major cleavage site is at the expected Asp1 (Aβ numbering); however, β-secretase expression also induced the formation of another protein, slightly smaller and beginning at Glu-11 (Fig. 2). Glu11-Aβ is also formed naturally from APP, and apparently β-secretase initiates the formation of this alternative truncated Aβ species as well. Use of decameric FRET substrates showed that BACE has relatively poor kinetic constants for its known substrates (the wild-type and Swedish mutant sequences) (Gruninger-Leitch *et al.,* 2002). In common with several other aspartyl proteases (e.g., cathepsin D), BACE prefers a leucine residue at position P1. However, unlike other aspartyl proteases, BACE accepts polar or acidic residues at position P2' and P1 but prefers bulky hydrophobic residues at position P3 (Gruninger-Leitch *et al.,* 2002).

The gene for β-secretase (BACE1) has been localized to chromosome 11, but no FAD-causing mutation in this gene has yet been identified (Saunders *et al.,* 1999). However, a β-secretase homolog, BACE2, maps to the Down syndrome chromosome 21, raising the possibility that this protease contributes to the development of AD in Down patients, because they carry an extra copy of chromosome 21 (Saunders *et al.,* 1999; Acquati *et al.,* 2000). BACE1 and BACE2 exhibit 52% amino acid sequence identity and 68% similarity, and BACE2 cleaves APP and short peptides in a β-secretase-like manner: activity is similarly increased at Asp-1 by the Swedish double mutation and prevented by the P1 Met-to-Val mutation (Farzan *et al.,* 2000). However, BACE2 is not expressed well in the brain, suggesting that it may play little, if any, role in AD neuritic plaque formation. Moreover,

BACE2 generates little Glu11-Aβ but efficiently performs an additional proteolysis in the middle of the Aβ region, resulting in the production of Phe20-Aβ and Ala21-Aβ (Fig. 2), with the implication that BACE2 might limit the production of pathogenic forms of Aβ.

A crystal structure of β-secretase bound to a hydroxyethyl-containing transition-state analog inhibitor at 1.9 Å resolution has been determined (Hong *et al.*, 2000). The bilobal structure of β-secretase has the conserved general folding of aspartyl proteases. Indeed, the backbone structure could be overlapped quite well onto that of pepsin. The six cysteine residues in the ectodomain form three disulfide bonds. The inhibitor is located in the substrate binding cleft between the amino- and carboxy-terminal lobes, and as expected the transition-state mimicking hydroxyethyl moiety is coordinated with the two active-site aspartates (Asp-32 and Asp-228). As with other aspartyl proteases, β-secretase possesses a "flap" that partially covers the cleft, and the backbone of the inhibitor is mostly in an extended conformation. However, β-secretase does display some structural differences, at least compared with pepsin. Four insertions are located on the molecular surface near the amino terminus of the inhibitor, and together these insertions significantly enlarge the molecular boundary of β-secretase compared with pepsin. Two other insertions are located at the surface near the inhibitor C terminus. In general, the β-secretase active site is more open and accessible than that of pepsin, and most of the hydrogen bond interactions between the enzyme and the backbone of the inhibitor are highly conserved among eukaryotic and HIV aspartyl proteases. However, the S2 and S4 subsites are relatively hydrophilic and open to solvent, and the hydrophilic character of these subsites is not conserved in the corresponding subsites of other human aspartyl proteases, such as pepsin, gastricsin, and cathepsins D and E, suggesting that these differences could be exploited in the design of selective inhibitors. In contrast, the P3' and P4' inhibitor side chains of the bound inhibitor point toward the molecular surface and have little interaction with the protease, and the backbone of residues P2'–P4' deviates from the regular extended conformation, with a kink at P2'.

C. Protein Maturation

The β-secretase protein is apparently N-glycosylated (Haniu *et al.*, 2000; Huse *et al.*, 2000) and is expressed primarily in the Golgi and in endosomes, although the enzyme could be detected at the plasma membrane as well. A cytoplasmic dileucine motif is apparently necessary for the recycling of β-secretase into endosomes (Huse *et al.*, 2000). The k_{cat}/K_m of proBACE when assayed with a polypeptide substrate was only 2.3-fold less than that of mature BACE (Shi *et al.*, 2001). Moreover, a statine-based inhibitor was equally potent toward blocking proBACE and mature BACE activities. Thus, the prodomain has little effect on the enzyme active site, and proBACE is apparently not a true zymogen. However, the prodomain does facilitate proper folding of the active protease (Shi *et al.*, 2001).

D. BACE Knockout and Amyloid Production

Several reports on BACE knockout mice have concluded that this enzyme is the major β-secretase in the brain (Cai *et al.*, 2001; Luo *et al.*, 2001; Roberds *et al.*, 2001). In two of these studies, knockout of BACE was shown to abolish murine $A\beta$1-40/42 and $A\beta$11-40/42 in cultured primary cortical neurons. The other study crossed BACE knockout mice with mice transgenic for human APP. Brain extracts from the APP+/BACE−/− progeny of this cross displayed a decrease in $A\beta$x-40 to 5–7% compared to APP+/BACE+/+ or APP+/BACE+/− mice. $A\beta$x-42 levels were undetectable in the APP+/BACE−/− mouse brain extracts. The BACE knockout mice were healthy, viable, and appeared normal in gross anatomy, tissue histology, hematology, and clinical chemistry. These findings indicate that inhibition of BACE should lead to dramatic decreases in brain $A\beta$ levels and that such inhibition may not lead to mechanism-based toxicity. Concerns about such toxicity have been raised about γ-secretase (see later discussion).

E. Biological Roles of β-Secretases

Despite the promising results from the BACE knockout studies, it remains to be seen whether long-term BACE inhibition is tolerable in aging adult humans. BACE1 apparently cleaves at least one other endogenous substrate besides APP: a sialyltransferase called ST6Gal I (Kitazume *et al.*, 2001). This finding is unexpected because, unlike APP, ST6Gal I is a type II integral membrane protein. If true, these observations raise the question of mechanism-based toxicity as a result of BACE inhibition. ST6Gal I is important for B cell function, and mice deficient in ST6Gal I show reduced levels of serum IgM. It will be interesting to learn whether BACE-deficient mice display abnormalities as they age.

III. γ-Secretase

A. Substrate Specificity

This enzyme has been considered central to understanding the etiology of AD, because it determines the proportion of the highly fibrillogenic $A\beta_{42}$ peptide. After either α- or β-secretase release the APP ectodomain, the resulting C83 and C99 APP C-terminal fragments are clipped in the middle of their transmembrane regions by γ-secretase (Fig. 3) (Selkoe, 1994). Normally, about 90% of the proteolysis occurs between Val-40 and Ile-41 ($A\beta$ numbering) to provide $A\beta_{40}$; roughly 10% takes place between Ala-42 and Thr-43 to produce $A\beta_{42}$. Minor proportions of other C-terminal variants, such as $A\beta_{39}$ and $A\beta_{43}$, are formed as well. γ-Secretase

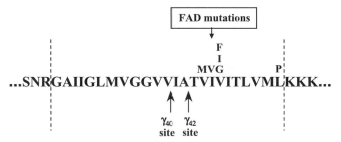

Figure 3 Sequence of APP in and around the γ-secretase cleavage site and disease-causing missense mutations. See text for details.

has been of interest not only because of its key role in AD pathogenesis, but also because it presents an intriguing biochemical problem: how does this enzyme catalyze a hydrolysis at a site located within a membrane?

Does γ-secretase actually catalyze a hydrolysis within the boundaries of the lipid bilayer? Molecular modeling of the γ-secretase cleavage site in APP as an α-helix, a conformation typical of transmembrane domains, showed that the two major cleavage sites, one leading to Aβ_{40} and the other to Aβ_{42}, are on opposite faces of the helix (Wolfe *et al.*, 1999b). Moreover, in this model FAD-causing APP mutations are immediately adjacent to the scissile amide bond that leads to Aβ_{42}. Thus, the helical model provides a simple biochemical explanation for why these FAD mutations result in selective increases in Aβ_{42} production. Lichtenthaler *et al.* (1999) provided important experimental support for this model via a phenylalanine scanning mutagenesis study. Systematic replacement of each APP transmembrane residue after the γ_{42} cleavage site with phenylalanine resulted in periodic changes in effects on Aβ_{40} and Aβ_{42} production that were consistent with a helical conformation of substrate upon initial interaction with the protease. In a second mutagenesis study, Lichtenthaler *et al.* (2002) further showed that the site of γ-secretase cleavage is changed if the length of the transmembrane domain of APP is altered (e.g., by insertion or deletion of two hydrophobic residues on either end of the transmembrane domain). Thus, the length of the whole transmembrane domain is apparently a major determinant for the cleavage site of γ-secretase. These findings are consistent with γ-secretase being a novel intramembrane-cleaving protease (I-CLiP), which hydrolyzes its substrates within the confines of the membrane (Wolfe *et al.*, 1999a).

B. Pharmacological Profiling as an Aspartyl Protease

Important mechanistic evidence that γ-secretase is an aspartyl protease came from observations that hydroxyl-containing peptidomimetics inhibit γ-secretase activity (Wolfe *et al.*, 1999b; Moore *et al.*, 2000; Shearman *et al.*, 2000). Such analogs mimic the *gem*-diol transition state of aspartyl protease catalysis, and a variety of related compounds have been shown to inhibit other aspartyl proteases. For example, the general aspartyl protease inhibitor pepstatin A contains two hydroxyethyl moieties (Marciniszyn *et al.*, 1976), and the difluoro alcohol group is found in peptidomimetic inhibitors of renin (Thaisrivongs *et al.*, 1986) and penicillopepsin (James *et al.*, 1992). Potent substrate-based inhibitors of β-secretase contain the hydroxyethyl moiety (Ghosh *et al.*, 2000). Moreover, all of the clinically marketed inhibitors of HIV protease contain a hydroxyethyl group (Huff, 1991). In many cases, cocrystal structures of these compounds and their cognate proteases have been determined (e.g., James *et al.*, 1992; Silva *et al.*, 1996; Hong *et al.*, 2000), demonstrating that the hydroxyl group of the inhibitor indeed coordinates with the two active-site aspartates in the protease.

C. The Role of Presenilins: Alzheimer-Causing Mutations and Knockout Mice

During the search for the major FAD-causing genes on chromosomes 14 and 1, many thought the encoded proteins would reveal at least one, if not both, of the proteases involved in Aβ production. When the search identified the presenilins in 1995 (Sherrington *et al.*, 1995; Levy-Lahad *et al.*, 1995), it was far from clear what the normal function of these proteins might be and why mutant forms might lead to AD. The two proteins, presenilin-1 (PS1) and presenilin-2 (PS2), are 65% identical, but the only homology they had to anything else known at the time was to a set of otherwise obscure proteins in worms involved in egg-laying and spermatogenesis (L'Hernault and Arduengo, 1992; Levitan and Greenwald, 1995). Remarkably, these proteins were the sites of dozens of FAD-causing missense mutations (Hardy, 1997). More than 70 such mutations have now been identified, with all but six occurring in PS1. These mutations are found in many regions of the linear sequence of presenilins, although they tend to cluster in certain areas (Fig. 4).

How could all these different PS mutations cause AD? Some mutations are extremely deleterious, a single copy of the mutant gene resulting in the onset of dementia as early as 25 years old (Campion *et al.*, 1999). Because of the amyloid plaque pathology of AD and because FAD-causing APP mutations are near β- and γ-secretase cleavage sites and affect Aβ production, investigating the effects of FAD-causing PS mutations on Aβ production, deposition, or toxicity seemed reasonable. Indeed, all PS mutations examined to date have been found to cause specific increases in Aβ_{42} production, and this effect can be observed in transfected cells, in transgenic mice, and in blood plasma or media from cultured fibroblasts of FAD patients (Scheuner *et al.*, 1996; Citron *et al.*, 1997; Tomita *et al.*, 1997; Duff *et al.*, 1996). Thus, presenilins could somehow modulate γ-secretase activity to enhance cleavage of the Ala42-Thr43 amide bond. A major clue to the function

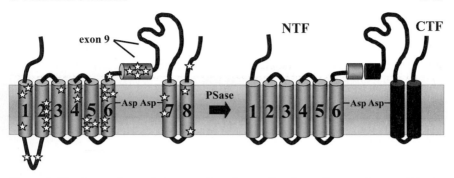

Figure 4 Topology and proteolytic processing of presenilins. Presenilins are cleaved within the hydrophobic region of the large cytosolic loop between TM6 and TM7, resulting in the formation of a heterodimeric complex composed of the N-terminal fragment (NTF) and the C-terminal fragment (CTF). Stars represent sites of missense mutation that cause familial Alzheimer's disease. The two conserved aspartates required for presenilin endoproteolysis and γ-secretase processing of APP and Notch are predicted to be within TM6 and TM7. The region of the protein encoded by exon 9 is denoted; natural deletion of this exon leads to a noncleavable but functional presenilin.

of the presenilins came via PS1 knockout mice: deletion of PS1 in these mice was lethal *in utero,* indicating the clear requirement of this gene for proper development of the organism (Wong *et al.,* 1997; Shen *et al.,* 1997).

Unfortunately, the embryonic lethality resulting from deleting PS1 did not suggest reasons why PS mutations might cause AD relatively late in life. Nevertheless, neurons from PS1-deficient embryos could be cultured, and transfection of these cells with APP revealed that γ-secretase activity was markedly reduced (De Strooper *et al.,* 1998). The maturation and distribution of APP was not affected by the deletion of the PS1 gene, nor was the release of α- or β-APP$_s$ altered. However, γ-secretase substrates C83 and C99 were dramatically elevated, and Aβ production was substantially lowered. Formation of total Aβ and Aβ_{42} was reduced to similar degrees (to roughly 20% of levels seen in neurons from PS1 +/+ littermates), indicating that PS1 plays a role in the production of both Aβ_{40} (which makes up 90% of all Aβ) and Aβ_{42}. The remaining γ-secretase activity was thought to be due to PS2. Indeed, the development of PS1/PS2 double knockout mice (Donoviel *et al.,* 1999; Herreman *et al.,* 1999) allowed the culturing of embryonic stem cells, and transfection of APP demonstrated that complete absence of γ-secretase activity (Herreman *et al.,* 2000; Zhang *et al.,* 2000). Thus, presenilins are absolutely required for the γ-secretase cleavage of APP.

D. Membrane Topology and Protein Maturation

Presenilins themselves undergo proteolytic processing within the hydrophobic region of the large cytosolic loop between transmembrane domain (TM) 6 and TM 7

(Fig. 4) to form stable heterodimeric complexes composed of the N- and C-terminal fragments (Thinakaran *et al.*, 1996; Capell *et al.*, 1998). These heterodimers are only produced to limited levels even upon overexpression of the holoprotein and may be found at the cell surface (Thinakaran *et al.*, 1996; Ratovitski *et al.*, 1997; Podlisny *et al.*, 1997; Steiner *et al.*, 1998). Expression of exogenous presenilins leads to replacement of endogenous presenilin heterodimers with the corresponding exogenous heterodimers, indicating competition for limiting cellular factors needed for stabilization and endoproteolysis (Thinakaran *et al.*, 1997).

FAD-causing presenilin mutants are likewise processed to stable heterodimers with one exception, a missense mutation in PS1 that leads to the aberrant splicing out of exon 9, a region that encodes the endoproteolytic cleavage site (Thinakaran *et al.*, 1996; Ratovitski *et al.*, 1997). This PS1 ΔE9 variant is an active presenilin, able to partially rescue a loss of function presenilin mutation in the worm *Caenorhabditis elegans* (Levitan *et al.*, 1996; Steiner *et al.*, 1999b), and like other FAD-causing presenilin mutants, causes increased Aβ_{42} production (Borchelt *et al.*, 1996; Steiner *et al.*, 1999b). Upon overexpression, most PS1 ΔE9 is rapidly degraded similar to unprocessed wild-type presenilins; however, a small portion of this PS1 variant is stabilized in cells (Ratovitski *et al.*, 1997; Zhang *et al.*, 1998) and forms a high molecular weight complex like the N- and C-terminal fragments (Capell *et al.*, 1998; Yu *et al.*, 2000), suggesting that it can interact with the same limiting cellular factors as wild-type presenilins. These observations are consistent with the idea that the bioactive form of presenilin is the heterodimer and that the hydrophobic region is an inhibitory domain.

E. Presenilins as Aspartyl Proteases

Presenilins contain two transmembrane aspartates (Fig. 4), one found in TM6 and one in TM7, predicted to lie the same distance within the membrane (i.e., they could interact with each other) and roughly aligned with the γ-secretase cleavage site in APP (i.e., they might work together to cut C99 and C83). These two aspartates are completely conserved from worms to humans and are even found in a plant presenilin (Lin *et al.*, 1999b). Mutation of either TM aspartate to alanine did not affect the expression or subcellular location of APP, and the subcellular distribution of the mutant presenilins was also similar to the wild type (Wolfe *et al.*, 1999c). However, the mutant presenilins were completely incapable of undergoing endoproteolysis and acted in a dominant-negative manner with respect to γ-secretase processing of APP. Similar effects on APP processing were observed even when conservative mutations to glutamate (Wolfe *et al.*, 1999c) or asparagine (Steiner *et al.*, 1999a; Leimer *et al.*, 1999) were made, indicating the crucial identity of these two key residues as aspartates and suggesting that the effects are not likely due to misfolding. These effects have been corroborated by several different laboratories and have been seen for both PS1 and PS2 (Kimberly *et al.*, 2000; Leimer *et al.*, 1999; Steiner *et al.*, 1999a; Yu *et al.*, 2000).

The aspartates are critical for γ-secretase activity independent of their role in presenilin endoproteolysis: aspartate mutation in the PS1 ΔE9 variant still blocked γ-secretase activity, even though endoproteolysis is not required of this presenilin variant (Wolfe *et al.*, 1999c). Together these results suggest that presenilins might be the catalytic component of γ-secretase: upon interaction with as yet unidentified limiting cellular factors, presenilin undergoes autoproteolysis via the two aspartates, and the two presenilin subunits remain together, each contributing one aspartate to the active site of γ-secretase. The issues of PS autoproteolysis and the role of PS endoproteolysis is controversial, especially in light of the identification of certain uncleavable artificial missense PS1 mutants that are still functional with respect to γ-secretase activity (Steiner *et al.*, 1999c). On the other hand, these mutations may disrupt the putative prodomain so that it no longer blocks the active site.

Advancing the understanding of γ-secretase and the role of presenilins in this activity had been hampered by the lack of an isolated enzyme assay. Li *et al.* (2000) reported a solubilized γ-secretase assay that faithfully reproduces the properties of the protease activity observed in whole cells. Isolated microsomes were solubilized with detergent, and γ-secretase activity was determined by measuring Aβ production from a C-terminally modified version of C99. Aβ_{40} and Aβ_{42} were produced in the same ratio as seen in living cells ($\sim$9:1), and peptidomimetics that blocked Aβ_{40} and Aβ_{42} formation in cells likewise inhibited production of these Aβ species in the solubilized protease assay. Choice of detergent was critical for Aβ production in the assay: CHAPSO was optimal, although CHAPS, a detergent known to keep presenilin subunits together (Capell *et al.*, 1998), was also compatible with activity, and Triton X-100 did not allow any Aβ formation. After separation of the detergent-solubilized material by size-exclusion chromatography, γ-secretase activity coeluted with the two subunits of PS1. Remarkably, immunoprecipitated PS1 heterodimers also produced Aβ from the artificial substrate, strongly suggesting that presenilins are part of a large γ-secretase complex.

More direct evidence that presenilins are the catalytic components of γ-secretases came from affinity labeling studies using transition-state analog inhibitors. Shearman *et al.* (2000) identified a peptidomimetic γ-secretase inhibitor by rescreening compounds originally designed against HIV protease. The compound blocks γ-secretase activity with an IC$_{50}$ of 0.3 nM in the solubilized protease assay (Li *et al.*, 2000) and contains a hydroxyethyl isostere, a transition-state mimicking moiety found in many aspartyl protease inhibitors. While the transition-state mimicking alcohol directs the compound to aspartyl proteases, flanking substructures determine specificity. Indeed, this compound does not inhibit aspartyl proteases cathepsin D and HIV-1 protease (Shearman *et al.*, 2000).

Photoactivatable versions of this compound bound covalently to presenilin subunits exclusively (Li *et al.*, 2000). Interestingly, installation of the photoreactive group on one end of the inhibitor led to labeling of the N-terminal presenilin sub-unit, while installation on the other end resulted in the tagging of the C-terminal subunit. Moreover, although these agents did not label wild-type PS1 holoprotein, they did tag PS1 ΔE9, which as described earlier is not processed to heterodimers but nevertheless is active. Similarly, Esler *et al.* (2000) identified peptidomimetic inhibitors containing a difluoro alcohol group, another type of transition-state mimicking moiety, and these compounds were developed starting from a substrate-based inhibitor designed from the γ-secretase cleavage site in APP. Conversion of one such analog to a reactive bromoacetamide provided an affinity reagent that likewise bound covalently and specifically to PS1 subunits in cell lysates, isolated microsomes, and whole cells. Either PS1 subunit so labeled could be brought down with antibodies to the other subunit under coimmunoprecipitation condi-tions, demonstrating that the inhibitor bound to heterodimeric PS1. Seiffert and colleagues likewise identified presenilin subunits as the molecular target of novel peptidomimetic γ-secretase inhibitors. The affinity probe, however, does not re-semble known transition-state mimics, so it is not clear whether this compound would be expected to bind to the active site of the protease (Seiffert *et al.,* 2000).

Taken together, these results strongly suggest that heterodimeric presenilin con-tains the catalytic component of γ-secretase: inhibitors in two of the three studies are transition-state analogs targeted to the active site. The wild-type presenilin holoprotein could be an inactive zymogen that requires cleavage into two sub-units for activation. In any event, the active site is likely at the PS heterodimeric interface: both subunits are labeled by γ-secretase affinity reagents, and each con-tributes one critical aspartate. Whether a separate "presenilinase" converts the holoprotein to subunits or presenilins undergo autoproteolysis remains to be de-termined, although the absolute requirement of the two transmembrane aspartates for heterodimer formation suggests the latter.

F. Biological Roles of γ-Secretases

The presenilins are not only involved in the proteolytic processing of APP, they are also critical for processing of the Notch receptor, a signaling molecule crucial for cell-fate determination during embryogenesis (Artavanis-Tsakonas *et al.,* 1999). After translation in the ER, Notch is processed by a furin-like protease, resulting in a heterodimeric receptor that is shuttled to the cell surface (Logeat *et al.,* 1998) (Fig. 5). Upon interaction with a cognate ligand, the ectodomain of Notch is shed by a metalloprotease apparently identical to tumor necrosis factor-α converting enzyme (TACE) (Brou *et al.,* 2000; Mumm *et al.,* 2000). Interestingly, metal-loproteases such as TACE and ADAM-10 are among the identified α-secretases that shed the APP ectodomain (Buxbaum *et al.,* 1998; Lammich *et al.,* 1999) (see Section IV). The membrane associated C terminus is then cut within the pos-tulated transmembrane domain to release the Notch intracellular domain (NICD),

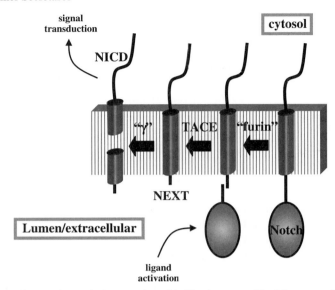

Figure 5 Topology and proteolytic processing of the Notch receptor. Notch is processed in the Golgi by a furin-like convertase and then transported to the cell surface as a heterodimer. Upon ligand binding, the ectodomain is shed by TACE, and the resulting membrane-associated fragment (Notch extracellular truncation; NEXT) is cleaved within its transmembrane domain by a γ-secretase-like protease, releasing the Notch intracellular domain (NICD).

which then translocates to the nucleus where it interacts with and activates the CSL family of transcription factors (Schroeter *et al.*, 1998). NICD formation is absolutely required for signaling from the Notch receptor: knockin of a single point mutation near the transmembrane cleavage site in Notch1 results in an embryonic lethal phenotype in mice virtually identical to that observed upon knockout of the entire Notch1 protein (Huppert *et al.*, 2000).

The parallels between APP and Notch processing are striking. Not only are both cleaved by TACE, but also the transmembrane regions of both proteins are processed by a γ-secretase-like protease that requires presenilins. Deletion of PS1 in mice is embryonic lethal, with a phenotype similar to that observed upon knockout of Notch1 (Wong *et al.*, 1997; Shen *et al.*, 1997), and the PS1/PS2 double knockout phenotype is even more similar (Herreman *et al.*, 1999; Donoviel *et al.*, 1999). Deficiency in PS1 dramatically reduces NICD formation (De Strooper *et al.*, 1999), and the complete absence of presenilins results in total abolition of NICD production (Herreman *et al.*, 2000; Zhang *et al.*, 2000). Treatment of cells with γ-secretase inhibitors designed from the transmembrane cleavage site within APP likewise blocks NICD production (De Strooper *et al.*, 1999) and nuclear translocation (Berezovska *et al.*, 2000) and reduces Notch signaling from a reporter gene (Berezovska *et al.*, 2000). Moreover, the two conserved TM aspartates in presenilins are required for cleavage of the Notch TM domain: as seen with γ-secretase inhibitors, expression of Asp-mutant PS1 or PS2 results in reduction of NICD formation, translocation, and signaling (Ray *et al.,* 1999; Steiner *et al.,* 1999a; Capell *et al.,* 2000; Berezovska *et al.,* 2000). Thus, if presenilins are the catalytic components of the γ-secretases that process APP, they are also likely the catalytic components of the related proteases that clip the transmembrane region of Notch.

Another γ-secretase substrate has been identified: the growth factor-dependent membrane tyrosine kinase receptor ErbB4 (Ni *et al.,* 2001). Ectodomain shedding of ErbB4, but not other ErbB receptors, can be stimulated by activators of protein kinase C or binding with its cognate ligand heregulin. The remaining membrane-bound C-terminal fragment is cleaved to release the cytosolic domain in a PS-dependent process that is blocked by γ-secretase inhibitors. Inhibition of γ-secretase also prevented growth inhibition by heregulin. Thus, γ-secretase proteolysis may be part of a new mechanism for receptor tyrosine kinase signaling.

G. Identification and Role of Nicastrin

In 2000, Yu and colleagues identified another factor apparently critical for γ-secretase activity, a type I integral membrane protein they dubbed nicastrin (Yu *et al.,* 2000b). This protein coimmunoprecipitated with anti-PS1 antibodies from glycerol gradient fractions containing high molecular weight ($\sim$250 kDa) presenilin complexes. RNA interference in *C. elegans* resulted in a Notch-deficient phenotype similar to that seen upon knockout of both worm presenilin genes. Human nicastrin associates with the γ-secretase substrates C83 and C99, and mutations in nicastrin can dramatically reduce Aβ production. More recently, knockout studies in *Drosophila* demonstrated that nicastrin is required for the γ-secretase processing of Notch and subsequent signaling (Chung and Struhl, 2001; Hu *et al.,* 2002; Lopez-Schier and St Johnston, 2002). Moreover, the development of an affinity purification method for γ-secretase using an immobilized transition-state analogue inhibitor led to the isolation of presenilin heterodimers and nicastrin (Esler *et al.,* 2002). Also, γ-secretase activity could be precipitated with anti-nicastrin antibodies. Thus, nicastrin appears to be an obligate member of the γ-secretase complex. The biochemical role nicastrin plays in substrate binding and turnover, however, is completely unknown.

H. An Emerging Family of Polytopic Membrane Proteases

The hypothesis that presenilin is the catalytic component of γ-secretase has generated considerable controversy. The major reason for this is the lack of sequence homology with other known aspartyl proteases. However, another family of putative polytopic aspartyl proteases has been identified: type 4 prepilin peptidases (LaPointe and Taylor, 2000). These proteins also contain eight transmembrane domains and two conserved aspartates required for substrate proteolysis. However, in this case, the two aspartates do not appear to lie within transmembrane domains. Again, the type 4 prepilin peptidases do not bear sequence homology with other known (cytosolic) aspartyl proteases, but no other candidate protease has been proffered. In addition, several other putative polytopic membrane proteases have been identified. These include:

1. The site 2 protease (S2P) that processes the sterol regulatory element binding protein (SREBP), a transcription factor essential for cholesterol biosynthesis (Rawson et al., 1997; Duncan et al., 1998; Ye et al., 2000). An entire family of S2P-like proteins have been identified in bacteria and apparently process a mating factor (Rudner et al., 1999). The S2P family contains a conserved HEXXH motif essential to catalysis, suggesting that they are metalloproteases.

2. The yeast protein Ste24p that processes the CAAX C terminus of certain proteins in conjunction with protein prenylation (Boyartchuk et al., 1997). This protein has been purified to homogeneity and shown to cleave its substrate, demonstrating that Ste24p is a bona fide protease (Tam et al., 2001). The requirement of zinc in this process suggests that Ste24p is a metalloprotease.

3. The rhomboid family of seven-TM proteins. *Drosophila* rhomboid-1 is responsible for the proteolysis of a membrane-anchored TGF-α-like protein called Spitz. Mutagenesis studies suggest that rhomboid-1 is a novel serine protease (Urban et al., 2001).

Other than the identification of conserved and essential residues suggesting a particular proteolytic class, none of these proteins bears sequence homology with other known proteases. This appears to be an example of mechanistic convergence: nature has arrived at the same basic proteolytic mechanisms using polytopic proteins found in soluble proteins, suggesting that there are few other catalytic solutions to the problem of efficient proteolysis.

IV. α-Secretase

A. Substrate Specificity

Alternative processing of APP by α-secretase precludes Aβ production, as this proteolysis occurs within the Aβ sequence (Esch et al., 1990). α-Secretase cleaves APP at Lys-686–Leu-687 (APP770 numbering; see Fig. 2), releasing α-APP$_s$ and concomitantly forming C83, the latter being further processed to the N-terminally truncated Aβ variant called p3. Two missense mutations in APP near the α-secretase cleavage site are pathogenic and increase Aβ production: A692G and E693Q (Haass et al., 1994; De Jonghe et al., 1998; Watson et al., 1999). These mutations occur in families with hereditary cerebral hemorrhage with amyloidosis (Levy et al., 1990; Van Broeckhoven et al., 1990; Hendriks et al., 1992), in which the patients suffer recurrent and ultimately fatal cerebral hemorrhage due to amyloid deposition in the meningeal and cerebral vasculature. Carriers of these mutations may or may not develop early-onset AD. The increased $A\beta$ deposition may be due to elevated $A\beta$ production caused by decreased α-secretase processing (i.e., more APP is shunted into the β-secretase pathway) (Haass *et al.*, 1994); however, a report has suggested that the A692G mutation enhances C99 (and hence $A\beta$) production by preventing alternative processing near this site by BACE2 (Farzan *et al.*, 2000). The principal determinants of APP cleavage by α-secretase appear to be the distance of the hydrolyzed bond from the membrane (12 or 13 residues) and a local helical conformation (Sisodia, 1992). Indeed, whereas a peptide substrate spanning the α-secretase cleavage site representing wild-type APP displays α-helical character, the corresponding substrate from A692G APP is random coiled (Lammich *et al.*, 1999).

B. Pharmacological Profiling and Tentative Identification as TACE and Kuzbanian

Pharmacological studies initially suggested that α-secretase may be a zinc-dependent metalloprotease, as the activity can be blocked by peptide hydroxamates (Parvathy *et al.*, 1998a,b). Interestingly, a series of hydroxamates inhibited APP processing by α-secretase and angiotensin-converting enzyme (ACE) processing by ACE secretase with similar potencies and rank order of potencies (Parvathy *et al.*, 1998a). In contrast, these compounds had a differential effect on TACE (Parvathy *et al.*, 1998b). Nevertheless, knockout of TACE in mice resulted in elimination of inducible α-secretase activity, suggesting that TACE plays a key role in α-secretase processing of APP (Buxbaum *et al.*, 1998). Moreover, TACE cleaved peptide substrates based on the α-secretase cleavage site in APP at Lys-686–Leu-687, suggesting that TACE itself may be one of the α-secretases. TACE apparently processes a spectrum of type I membrane proteins, including TNFα, the p75 TNF receptor, L-selectin adhesion molecule, and transforming growth factor-α (TGFα), and knockout of TACE is lethal, emphasizing the importance of ectodomain shedding during embryonic development (Peschon *et al.*, 1998). As mentioned above, TACE appears to be the sheddase responsible for release of the Notch ectodomain after ligand binding (Brou *et al.*, 2000), a role that would likely be essential for embryonic development; however, the defects seen upon TACE knockout were similar to those of mice lacking TGFα (Peschon *et al.*, 1998). In any event, manipulation of TACE as a therapeutic strategy for AD may lead to toxicity due to effects on the shedding of other membrane proteins.

Another metalloprotease, ADAM-10 (or kuzbanian), also appears to process APP in an α-secretase-like manner (Lammich *et al.*, 1999). Overexpression of ADAM-10 in human cells increased both basal and PKC-inducible α-secretase activity severalfold. Moreover, endogenous α-secretase activity was inhibited by a dominant-negative form of ADAM-10 with a point mutation in the zinc binding site. A peptide substrate based on the α-secretase cleavage site in APP was processed by ADAM-10 at the Lys-686–Leu-687 bond, whereas a substrate based on the disease-causing A692G mutant APP is cleaved less efficiently. Interestingly, ADAM-10 is also implicated in the Notch signaling pathway (Wen *et al.*, 1997). Thus, TACE and ADAM-10 appear to have similar roles with respect to APP and Notch processing. Whether stimulation of ADAM-10 would be a better therapeutic strategy than stimulating TACE is presently unclear, as little is known about the role of ADAM-10 in shedding other membrane proteins.

C. Regulation of α-Secretase Processing of APP

Although cells contain a certain level of basal α-secretase activity, this proteolysis can be substantially induced by classical protein kinase C activators such as phorbol esters (Hung *et al.*, 1993; Buxbaum *et al.*, 1993; Felsenstein *et al.*, 1994; Jacobsen *et al.*, 1994). Moreover, activation of receptors that work through protein kinase C can augment α-secretase cleavage of APP with concomitant reduction in β-secretase processing. For instance, agonists of the metabotropic glutamate receptors can lower Aβ by shunting APP toward the α-secretase pathway (Lee *et al.*, 1995). Muscarinic agonists (M1 and M3) can also decrease Aβ production by this means, and this effect has been observed *in vitro* as well as *in vivo* (Nitsch *et al.*, 1992; Haring *et al.*, 1994; Wolf *et al.*, 1995a; Lin *et al.*, 1999a). Because of this effect on Aβ production, M1 and M3 agonists may be useful agents for treating AD. In addition, evidence suggests that α-APP$_s$ may have a neuroprotective effect and enhance learning and cognition (Meziane *et al.*, 1998), so augmenting α-secretase processing of APP may be beneficial in treating AD for this reason as well.

V. Concluding Remarks

Only a few years ago, the proteases responsible for the processing of APP were completely unknown, hampering efforts to understand the molecular basis of AD and develop effective therapeutic agents. β-Secretase has since been definitively identified as a membrane-tethered aspartyl protease and appears to be a safe drug target based on observation of knockout mice. The protease has been cocrystallized with inhibitors, an advance that allows rational drug design. The normal biological roles of BACE and its homolog BACE2 are unclear and need to be resolved to more fully evaluate BACE as a therapeutic target. γ-Secretase remains incompletely understood, but strong evidence suggests that presenilin is the catalytic component of this membrane-embedded protease and that nicastrin is also an obligate member of the complex. One major function of γ-secretase is the proteolysis and signaling of the Notch receptor during differentiation and development.

This function has also raised concerns about γ-secretase as a therapeutic target, although the role of Notch in the aging adult is not clear. The ADAMs 10 and 17 are apparently responsible for the alternative α-secretase processing of APP that precludes Aβ production, and these proteases are involved in the shedding of a number of other membrane proteins besides APP. Enhancing α-secretase activity via PKC stimulation is under consideration as a therapeutic strategy to lower Aβ. Work toward characterizing and identifying these Alzheimer secretases has led to important insights in biology and will likely continue to do so in the coming years.

References

http://www.alzforum.org/members/resources/app_mutations/app_table.html

http://www.alzforum.org/members/resources/pres_mutations/index.html

Acquati, F., Accarino, M., Nucci, C., Fumagalli, P., Jovine, L., Ottolenghi, S., and Taramelli, R. (2000). The gene encoding DRAP (BACE2), a glycosylated transmembrane protein of the aspartic protease family, maps to the down critical region. *FEBS Lett.* **468,** 59–64.

Artavanis-Tsakonas, S., Rand, M. D., and Lake, R. J. (1999). Notch signaling: cell fate control and signal integration in development. *Science* **284,** 770–776.

Bennett, B. D., Denis, P., Haniu, M., Teplow, D. B., Kahn, S., Louis, J. C., Citron, M., and Vassar, R. (2000). A furin-like convertase mediates propeptide cleavage of BACE, the Alzheimer's β-secretase. *J. Biol. Chem.,* **275,** 37712–37717.

Berezovska, O., Jack, C., McLean, P., Aster, J. C., Hicks, C., Xia, W., Wolfe, M. S., Kimberly, W. T., Weinmaster, G., Selkoe, D. J., and Hyman, B. T. (2000). Aspartate mutations in presenilin and γ-secretase inhibitors both impair Notch1 proteolysis and nuclear translocation with relative preservation of Notch1 signaling. *J. Neurochem.* **75,** 583–593.

Blessed, G., Tomlinson, B. E., and Roth, M. (1968). The association between quantitative measure of dementia and of senile change in the cerebral grey matter of elderly subjects. *Brit. J. Psychiat.* **114,** 797–811.

Borchelt, D. R., Thinakaran, G., Eckman, C. B., Lee, M. K., Davenport, F., Ratovitsky, T., Prada, C. M., Kim, G., Seekins, S., Yager, D., Slunt, H. H., Wang, R., Seeger, M., Levey, A. I., Gandy, S. E., Copeland, N. G., Jenkins, N. A., Price, D. L., Younkin, S. G., and Sisodia, S. S. (1996). Familial Alzheimer's disease-linked presenilin 1 variants elevate Abeta1-42/1-40 ratio in vitro and in vivo. *Neuron* **17,** 1005–1013.

Boyartchuk, V. L., Ashby, M. N., and Rine, J. (1997). Modulation of Ras and a-factor function by carboxyl-terminal proteolysis. *Science* **275,** 1796–1800.

Brou, C., Logeat, F., Gupta, N., Bessia, C., LeBail, O., Doedens, J. R., Cumano, A., Roux, P., Black, R. A., and Israël, A. (2000). A novel proteolytic cleavage involved in notch signaling: The role of the disintegrin-metalloprotease TACE. *Mol. Cell* **5,** 207–216.

Buxbaum, J. D., Koo, E. H., and Greengard, P. (1993). Protein phosphorylation inhibits production of Alzheimer amyloid beta/A4 peptide. *Proc. Natl. Acad. Sci. USA* **90,** 9195–9198.

Buxbaum, J. D., Liu, K. N., Luo, Y., Slack, J. L., Stocking, K. L., Peschon, J. J., Johnson, R. S., Castner, B. J., Cerretti, D. P., and Black, R. A. (1998). Evidence that tumor necrosis factor α converting enzyme is involved in regulated a-secretase cleavage of the Alzheimer's amyloid protein precursor. *J. Biol. Chem.* **273,** 27765–27767.

Cai, H., Wang, Y., McCarthy, D., Wen, H., Borchelt, D. R., Price, D. L., and Wong, P. C. (2001). BACE1 is the major beta-secretase for generation of Abeta peptides by neurons. *Nat. Neurosci.* **4,** 233–234.

Cai, X. D., Golde, T. E., and Younkin, S. G. (1993). Release of excess amyloid beta protein from a mutant amyloid beta protein precursor. *Science* **259**, 514–516.

Campion, D., Dumanchin, C., Hannequin, D., Dubois, B., Belliard, S., Puel, M., Thomas-Anterion, C., Michon, A., Martin, C., Charbonnier, F., Raux, G., Camuzat, A., Penet, C., Mesnage, V., Martinez, M., Clerget-Darpoux, F., Brice, A., and Frebourg, T. (1999). Early-onset autosomal dominant Alzheimer disease: prevalence, genetic heterogeneity, and mutation spectrum. *Am. J. Hum. Genet.* **65**, 664–670.

Capell, A., Grunberg, J., Pesold, B., Diehlmann, A., Citron, M., Nixon, R., Beyreuther, K., Selkoe, D. J., and Haass, C. (1998). The proteolytic fragments of the Alzheimer's disease-associated presenilin-1 form heterodimers and occur as a 100-150-kDa molecular mass complex. *J. Biol. Chem.* **273**, 3205–3211.

Capell, A., Steiner, H., Romig, H., Keck, S., Baader, M., Grim, M. G., Baumeister, R., and Haass, C. (2000). Presenilin-1 differentially facilitates endoproteolysis of the beta-amyloid precursor protein and Notch. *Nat. Cell. Biol.* **2**, 205–211.

Chartier-Harlin, M. C., Crawford, F., Houlden, H., Warren, A., Hughes, D., Fidani, L., Goate, A., Rossor, M., Roques, P., Hardy, J., and Mullan, M. (1991). Early-onset Alzheimer's disease caused by mutations at codon 717 of the beta-amyloid precursor protein gene. *Nature* **353**, 844–846.

Chung, H. M., and Struhl, G. (2001). Nicastrin is required for Presenilin-mediated transmembrane cleavage in *Drosophila. Nat. Cell Biol.* **3**, 1129–1132.

Citron, M., Oltersdorf, T., Haass, C., McConlogue, L., Hung, A. Y., Seubert, P., Vigo-Pelfrey, C., Lieberburg, I., and Selkoe, D. J. (1992). Mutation of the beta-amyloid precursor protein in familial Alzheimer's disease increases beta-protein production. *Nature* **360**, 672–674.

Citron, M., Teplow, D. B., and Selkoe, D. J. (1995). Generation of amyloid beta protein from its precursor is sequence specific. *Neuron* **14**, 661–670.

Citron, M., Vigo-Pelfrey, C., Teplow, D. B., Miller, C., Schenk, D., Johnston, J., Winblad, B., Venizelos, N., Lannfelt, L., and Selkoe, D. J. (1994). Excessive production of amyloid beta-protein by peripheral cells of symptomatic and presymptomatic patients carrying the Swedish familial Alzheimer disease mutation. *Proc. Natl. Acad. Sci. USA* **91**, 11993–11997.

Citron, M., Westaway, D., Xia, W., Carlson, G., Diehl, T., Levesque, G., Johnson-Wood, K., Lee, M., Seubert, P., Davis, A., Kholodenko, D., Motter, R., Sherrington, R., Perry, B., Yao, H., Strome, R., Lieberburg, I., Rommens, J., Kim, S., Schenk, D., Fraser, P., St George Hyslop, P., and Selkoe, D. J. (1997). Mutant presenilins of Alzheimer's disease increase production of 42-residue amyloid beta-protein in both transfected cells and transgenic mice. *Nat. Med.* **3**, 67–72.

Cummings, B. J., and Cotman, C. W. (1995). Image analysis of β-amyloid load in Alzheimer's disease and relation to dementia severity. *Lancet* **346**, 1524–1528.

De Jonghe, C., Zehr, C., Yager, D., Prada, C. M., Younkin, S., Hendriks, L., Van Broeckhoven, C., and Eckman, C. B. (1998). Flemish and Dutch mutations in amyloid beta precursor protein have different effects on amyloid beta secretion. *Neurobiol. Dis.* **5**, 281–286.

De Strooper, B., Annaert, W., Cupers, P., Saftig, P., Craessaerts, K., Mumm, J. S., Schroeter, E. H., Schrijvers, V., Wolfe, M. S., Ray, W. J., Goate, A., and Kopan, R. (1999). A presenilin-1-dependent γ-secretase-like protease mediates release of Notch intracellular domain. *Nature* **398**, 518–522.

De Strooper, B., Saftig, P., Craessaerts, K., Vanderstichele, H., Guhde, G., Annaert, W., Von Figura, K., and Van Leuven, F. (1998). Deficiency of presenilin-1 inhibits the normal cleavage of amyloid precursor protein. *Nature* **391**, 387–390.

Donoviel, D. B., Hadjantonakis, A. K., Ikeda, M., Zheng, H., Hyslop, P. S., and Bernstein, A. (1999). Mice lacking both presenilin genes exhibit early embryonic patterning defects. *Genes Dev.* **13**, 2801–2810.

Duff, K., Eckman, C., Zehr, C., Yu, X., Prada, C. M., Perez-tur, J., Hutton, M., Buee, L., Harigaya, Y., Yager, D., Morgan, D., Gordon, M. N., Holcomb, L., Refolo, L., Zenk, B., Hardy, J., and Younkin, S. (1996). Increased amyloid-beta42(43) in brains of mice expressing mutant presenilin 1. *Nature* **383**, 710–713.

Duncan, E. A., Dave, U. P., Sakai, J., Goldstein, J. L., and Brown, M. S. (1998). Second-site cleavage in sterol regulatory element-binding protein occurs at transmembrane junction as determined by cysteine panning. *J. Biol. Chem.* **273**, 17801–17809.

El Khoury, J., Hickman, S. E., Thomas, C. A., Cao, L., Silverstein, S. C., and Loike, J. D. (1996). Scavenger receptor-mediated adhesion of microglia to beta-amyloid fibrils. *Nature* **382**, 716–719.

Esch, F. S., Keim, P. S., Beattie, E. C., Blacher, R. W., Culwell, A. R., Oltersdorf, T., McClure, D., and Ward, P. J. (1990). Cleavage of amyloid beta peptide during constitutive processing of its precursor. *Science* **248**, 1122–1124.

Esler, W. P., Kimberly, W. T., Ostaszewski, B. L., Diehl, T. S., Moore, C. L., Tsai, J.-Y., Rahmati, T., Xia, W., Selkoe, D. J., and Wolfe, M. S. (2000). Transition-state analogue inhibitors of γ-secretase bind directly to presenilin-1. *Nat. Cell Biol.* **2**, 428–434.

Esler, W. P., Kimberly, W. T., Ostaszewski, B. L., Ye, W., Diehl, T. S., Selkoe, D. J., and Wolfe, M. S. (2002). Activity-dependent isolation of the presenilin/γ-secretase complex reveals nicastrin and a γ substrate. *Proc. Natl. Acad. Sci. USA* **99**, 2720–2725.

Farzan, M., Schnitzler, C. E., Vasilieva, N., Leung, D., and Choe, H. (2000). BACE2, a beta-secretase homolog, cleaves at the beta site and within the amyloid-beta region of the amyloid-beta precursor protein. *Proc. Natl. Acad. Sci. USA* **97**, 9712–9717.

Felsenstein, K. M., Ingalls, K. M., Hunihan, L. W., and Roberts, S. B. (1994). Reversal of the Swedish familial Alzheimer's disease mutant phenotype in cultured cells treated with phorbol 12,13-dibutyrate. *Neurosci. Lett.* **174**, 173–176.

Funato, H., Yoshimura, M., Yamazaki, T., Saido, T. C., Ito, Y., Yokofujita, J., Okeda, R., and Ihara, Y. (1998). Astrocytes containing amyloid beta-protein (Abeta)-positive granules are associated with Abeta40-positive diffuse plaques in the aged human brain. *Am. J. Pathol.* **152**, 983–992.

Ghosh, A. K., Shin, D., Downs, D., Koelsch, G., Lin, X., Ermolieff, J., and Tang, J. (2000). Design of potent inhibitors for human brain memapsin 2 (beta-secretase). *J. Am. Chem. Soc.* **122**, 3522–3523.

Goate, A., Chartier-Harlin, M. C., Mullan, M., Brown, J., Crawford, F., Fidani, L., Giuffra, L., Haynes, A., Irving, N., James, L., Mant, R., Newton, P., Rooke, K., Roques, P., Talbot, C., Pericak-Vance, M., Roses, A., Williamson, R., Rossor, M., Owen, M., and Hardy, J. (1991). Segregation of a missense mutation in the amyloid precursor protein gene with familial Alzheimer's disease. *Nature* **349**, 704–706.

Gruninger-Leitch, F., Schlatter, D., Kung, E., Nelbock, P., and Dobeli, H. (2002). Substrate and inhibitor profile of BACE (beta-secretase) and comparison with other mammalian aspartic proteases. *J. Biol. Chem.* **277**, 4687–4693.

Haass, C., Hung, A. Y., Selkoe, D. J., and Teplow, D. B. (1994). Mutations associated with a locus for familial Alzheimer's disease result in alternative processing of amyloid beta-protein precursor. *J. Biol. Chem.* **269**, 17741–17748.

Haniu, M., Denis, P., Young, Y., Mendiaz, E. A., Fuller, J., Hui, J. O., Bennett, B. D., Kahn, S., Ross, S., Burgess, T., Katta, V., Rogers, G., Vassar, R., and Citron, M. (2000). Characterization of Alzheimer's beta-secretase protein BACE. A pepsin family member with unusual properties. *J. Biol. Chem.* **275**, 21099–21106.

Hardy, J. (1997). Amyloid, the presenilins and Alzheimer's disease. *Trends Neurosci.* **20**, 154–159.

Haring, R., Gurwitz, D., Barg, J., Pinkas-Kramarski, R., Heldman, E., Pittel, Z., Wengier, A., Meshulam, H., Marciano, D., Karton, Y., and Fisher, A. (1994). Amyloid precursor protein secretion via muscarinic receptors: reduced desensitization using the M1-selective agonist AF102B. *Biochem. Biophys. Res. Commun.* **203**, 652–658.

Hendriks, L., van Duijn, C. M., Cras, P., Cruts, M., Van Hul, W., van Harskamp, F., Warren, A., McInnis, M. G., Antonarakis, S. E., Martin, J. J., Hofman, A., and Van Broeckhoven, C. (1992). Presenile dementia and cerebral haemorrhage linked to a mutation at codon 692 of the beta-amyloid precursor protein gene. *Nat. Genet.* **1**, 218–221.

Herreman, A., Hartmann, D., Annaert, W., Saftig, P., Craessaerts, K., Serneels, L., Umans, L., Schrijvers, V., Checler, F., Vanderstichele, H., Baekelandt, V., Dressel, R., Cupers, P.,

Huylebroeck, D., Zwijsen, A., Van Leuven, F., and De Strooper, B. (1999). Presenilin 2 deficiency causes a mild pulmonary phenotype and no changes in amyloid precursor protein processing but enhances the embryonic lethal phenotype of presenilin 1 deficiency. *Proc. Natl. Acad. Sci. USA* **96,** 11872–11877.

Herreman, A., Serneels, L., Annaert, W., Collen, D., Schoonjans, L., and De Strooper, B. (2000). Total inactivation of gamma-secretase activity in presenilin-deficient embryonic stem cells. *Nat. Cell Biol.* **2,** 461–462.

Hong, L., Koelsch, G., Lin, X., Wu, S., Terzyan, S., Ghosh, A. K., Zhang, X. C., and Tang, J. (2000). Structure of the protease domain of memapsin 2 (beta-secretase) complexed with inhibitor. *Science* **290,** 150–153.

Hu, Y., Ye, Y., and Fortini, M. E. (2002). Nicastrin is required for gamma-secretase cleavage of the Drosophila Notch receptor. *Dev. Cell* **2,** 69–78.

Huff, J. R. (1991). HIV protease: a novel chemotherapeutic target for AIDS. *J. Med. Chem.* **34,** 2305–2314.

Hung, A. Y., Haass, C., Nitsch, R. M., Qiu, W. Q., Citron, M., Wurtman, R. J., Growdon, J. H., and Selkoe, D. J. (1993). Activation of protein kinase C inhibits cellular production of the amyloid beta-protein. *J. Biol. Chem.* **268,** 22959–22962.

Huppert, S. S., Le, A., Schroeter, E. H., Mumm, J. S., Saxena, M. T., Milner, L. A., and Kopan, R. (2000). Embryonic lethality in mice homozygous for a processing-deficient allele of Notch1. *Nature* **405,** 966–970.

Huse, J. T., Pijak, D. S., Leslie, G. J., Lee, V. M., and Doms, R. W. (2000b). Maturation and endosomal targeting of beta-site amyloid precursor protein-cleaving enzyme. The Alzheimer's disease β-secretase. *J. Biol. Chem.* **275,** 33729–33737.

Hussain, I., Christie, G., Schneider, K., Moore, S., and Dingwall, C. (2001). Prodomain processing of Asp1 (BACE2) is autocatalytic. *J. Biol. Chem.* **276,** 23322–23328.

Hussain, I., Powell, D., Howlett, D. R., Tew, D. G., Meek, T. D., Chapman, C., Gloger, I. S., Murphy, K. E., Southan, C. D., Ryan, D. M., Smith, T. S., Simmons, D. L., Walsh, F. S., Dingwall, C., and Christie, G. (1999). Identification of a novel aspartic protease (Asp 2) as beta-secretase. *Mol. Cell. Neurosci.* **14,** 419–427.

Iwatsubo, T., Odaka, A., Suzuki, N., Mizusawa, H., Nukina, N., and Ihara, Y. (1994). Visualization of A beta 42(43) and A beta 40 in senile plaques with end-specific A beta monoclonals: evidence that an initially deposited species is A beta 42(43). *Neuron* **13,** 45–53.

Jacobsen, J. S., Spruyt, M. A., Brown, A. M., Sahasrabudhe, S. R., Blume, A. J., Vitek, M. P., Muenkel, H. A., and Sonnenberg-Reines, J. (1994). The release of Alzheimer's disease beta amyloid peptide is reduced by phorbol treatment. *J. Biol. Chem.* **269,** 8376–8382.

James, M. N., Sielecki, A. R., Hayakawa, K., and Gelb, M. H. (1992). Crystallographic analysis of transition state mimics bound to penicillopepsin: difluorostatine- and difluorostatone-containing peptides. *Biochemistry* **31,** 3872–3886.

Jarrett, J. T., Berger, E. P., and Lansbury, P. T., Jr. (1993). The carboxy terminus of the beta amyloid protein is critical for the seeding of amyloid formation: implications for the pathogenesis of Alzheimer's disease. *Biochemistry* **32,** 4693–4697.

Johnson-Wood, K., Lee, M., Motter, R., Hu, K., Gordon, G., Barbour, R., Khan, K., Gordon, M., Tan, H., Games, D., Lieberburg, I., Schenk, D., Seubert, P., and McConlogue, L. (1997). Amyloid precursor protein processing and A beta42 deposition in a transgenic mouse model of Alzheimer disease. *Proc. Natl. Acad. Sci. USA* **94,** 1550–1555.

Kimberly, W. T., Xia, W., Rahmati, T., Wolfe, M. S., and Selkoe, D. J. (2000). The transmembrane aspartates in presenilin 1 and 2 are obligatory for γ-secretase activity and amyloid β-protein generation. *J. Biol. Chem.* **275,** 3173–3178.

Kitazume, S., Tachida, Y., Oka, R., Shirotani, K., Saido, T. C., and Hashimoto, Y. (2001). Alzheimer's beta-secretase, beta-site amyloid precursor protein-cleaving enzyme, is responsible for cleavage secretion of a Golgi-resident sialyltransferase. *Proc. Natl. Acad. Sci. USA* **98,** 13554–13559.

256   Michael S. Wolfe

6666666566666666Lammich, S., Kojro, E., Postina, R., Gilbert, S., Pfeiffer, R., Jasionowski, M., Haass, C., and Fahrenholz, F. (1999). Constitutive and regulated alpha-secretase cleavage of Alzheimer's amyloid precursor protein by a disintegrin metalloprotease. *Proc. Natl. Acad. Sci. USA* **96,** 3922–3927.

LaPointe, C. F., and Taylor, R. K. (2000). The type 4 prepilin peptidases comprise a novel family of aspartic acid proteases. *J. Biol. Chem.* **275,** 1502–1510.

Lee, R. K., Wurtman, R. J., Cox, A. J., and Nitsch, R. M. (1995). Amyloid precursor protein processing is stimulated by metabotropic glutamate receptors. *Proc. Natl. Acad. Sci. USA* **92,** 8083–8087.

Leimer, U., Lun, K., Romig, H., Walter, J., Grunberg, J., Brand, M., and Haass, C. (1999). Zebrafish (*Danio rerio*) presenilin promotes aberrant amyloid beta-peptide production and requires a critical aspartate residue for its function in amyloidogenesis. *Biochemistry* **38,** 13602–13609.

Lemere, C. A., Grenfell, T. J., and Selkoe, D. J. (1999). The AMY antigen co-occurs with abeta and follows its deposition in the amyloid plaques of Alzheimer's disease and Down syndrome. *Am. J. Pathol.* **155,** 29–37.

Levitan, D., Doyle, T. G., Brousseau, D., Lee, M. K., Thinakaran, G., Slunt, H. H., Sisodia, S. S., and Greenwald, I. (1996). Assessment of normal and mutant human presenilin function in Caenorhabditis elegans. *Proc. Natl. Acad. Sci. USA* **93,** 14940–14944.

Levy, E., Carman, M. D., Fernandez-Madrid, I. J., Power, M. D., Lieberburg, I., van Duinen, S. G., Bots, G. T., Luyendijk, W., and Frangione, B. (1990). Mutation of the Alzheimer's disease amyloid gene in hereditary cerebral hemorrhage, Dutch type. *Science* **248,** 1124–1126.

Levy-Lahad, E., Wasco, W., Poorkaj, P., Romano, D. M., Oshima, J., Pettingell, W. H., Yu, C. E., Jondro, P. D., Schmidt, S. D., Wang, K., Crowley, A. C., Fu, Y.-H., Guenette, S. Y., Galas, D., Nenens, E., Wijsman, E. M., Bird, T. D., Schellenberg, G. D., and Tanzi, R. E. (1995). Candidate gene for the chromosome 1 familial Alzheimer's disease locus. *Science* **269,** 973–977.

L'Hernault, S. W., and Arduengo, P. M. (1992). Mutation of a putative sperm membrane protein in Caenorhabditis elegans prevents sperm differentiation but not its associated meiotic divisions. *J. Cell. Biol.* **119,** 55–68.

Li, Y. M., Lai, M. T., Xu, M., Huang, Q., DiMuzio-Mower, J., Sardana, M. K., Shi, X. P., Yin, K. C., Shafer, J. A., and Gardell, S. J. (2000). Presenilin 1 is linked with gamma-secretase activity in the detergent solubilized state. *Proc. Natl. Acad. Sci. USA* **97,** 6138–6143.

Li, Y. M., Xu, M., Lai, M. T., Huang, Q., Castro, J. L., DiMuzio-Mower, J., Harrison, T., Lellis, C., Nadin, A., Neduvelil, J. G., Register, R. B., Sardana, M. K., Shearman, M. S., Smith, A. L., Shi, X. P., Yin, K. C., Shafer, J. A., and Gardell, S. J. (2000). Photoactivated gamma-secretase inhibitors directed to the active site covalently label presenilin 1. *Nature* **405,** 689–694.

Lichtenthaler, S. F., Beher, D., Grimm, H. S., Wang, R., Shearman, M. S., Masters, C. L., and Beyreuther, K. (2002). The intramembrane cleavage site of the amyloid precursor protein depends on the length of its transmembrane domain. *Proc. Natl. Acad. Sci. USA* **99,** 1365–1370.

Lichtenthaler, S. F., Wang, R., Grimm, H., Uljon, S. N., Masters, C. L., and Beyreuther, K. (1999). Mechanism of the cleavage specificity of Alzheimer's disease gamma-secretase identified by phenylalanine-scanning mutagenesis of the transmembrane domain of the amyloid precursor protein. *Proc. Natl. Acad. Sci. USA* **96,** 3053–3058.

Lin, L., Georgievska, B., Mattsson, A., and Isacson, O. (1999a). Cognitive changes and modified processing of amyloid precursor protein in the cortical and hippocampal system after cholinergic synapse loss and muscarinic receptor activation. *Proc. Natl. Acad. Sci. USA* **96,** 12108–12113.

Lin, X., Kaul, S., Rounsley, S., Shea, T. P., Benito, M. I., Town, C. D., Fujii, C. Y., Mason, T., Bowman, C. L., Barnstead, M., Feldblyum, T. V., Buell, C. R., Ketchum, K. A., Lee, J., Ronning, C. M., Koo, H. L., Moffat, K. S., Cronin, L. A., Shen, M., Pai, G., Van Aken, S., Umayam, L., Tallon, L. J., Gill, J. E., Adams, M. D., Carrera, A. J., Creasy, T. H., Goodman, H. M., Somerville, C. R., Copenhaver, G. P., Preuss, D., Nierman, W. C., White, O., Eisen, J. A., Salzberg, S. L., Fraser, C. M., and Venter, J. C. (1999b). Sequence and analysis of chromosome 2 of the plant *Arabidopsis thaliana*. *Nature* **402,** 761–768.

Lin, X., Koelsch, G., Wu, S., Downs, D., Dashti, A., and Tang, J. (2000). Human aspartic protease memapsin 2 cleaves the beta-secretase site of beta-amyloid precursor protein. *Proc. Natl. Acad. Sci. USA* **97**, 1456–1460.

Logeat, F., Bessia, C., Brou, C., LeBail, O., Jarriault, S., Seidah, N. G., and Israel, A. (1998). The Notch1 receptor is cleaved constitutively by a furin-like convertase. *Proc. Natl. Acad. Sci. USA* **95**, 8108–8112.

Lopez-Schier, H., and St Johnston, D. (2002). *Drosophila* nicastrin is essential for the intramembranous cleavage of notch. *Dev. Cell* **2**, 79–89.

Luo, Y., Bolon, B., Kahn, S., Bennett, B. D., Babu-Khan, S., Denis, P., Fan, W., Kha, H., Zhang, J., Gong, Y., Martin, L., Louis, J. C., Yan, Q., Richards, W. G., Citron, M., and Vassar, R. (2001). Mice deficient in BACE1, the Alzheimer's beta-secretase, have normal phenotype and abolished beta-amyloid generation. *Nat. Neurosci.* **4**, 231–232.

Marciniszyn, J., Jr., Hartsuck, J. A., and Tang, J. (1976). Mode of inhibition of acid proteases by pepstatin. *J. Biol. Chem.* **251**, 7088–7094.

McConlogue, L. C., Agard, D. A., Nobuyuki, O., and Tatsuno, G. (2000). Functional analysis of beta-secretase using mutagenesis and structural homology modeling. *Neurobiol. Aging* **21**, S278.

Meda, L., Cassatella, M. A., Szendrei, G. I., Otvos, L., Jr., Baron, P., Villalba, M., Ferrari, D., and Rossi, F. (1995). Activation of microglial cells by beta-amyloid protein and interferon-gamma. *Nature* **374**, 647–650.

Meziane, H., Dodart, J. C., Mathis, C., Little, S., Clemens, J., Paul, S. M., and Ungerer, A. (1998). Memory-enhancing effects of secreted forms of the beta-amyloid precursor protein in normal and amnestic mice. *Proc. Natl. Acad. Sci. USA* **95**, 12683–12688.

Moore, C. L., Leatherwood, D. D., Diehl, T. S., Selkoe, D. J., and Wolfe, M. S. (2000). Difluoro ketone peptidomimetics suggest a large S1 pocket for Alzheimer's γ-secretase: Implications for inhibitor design. *J. Med. Chem.* **43**, 3434–3442.

Mumm, J. S., Schroeter, E. H., Saxena, M. T., Griesemer, A., Tian, X., Pan, D. J., Ray, W. J., and Kopan, R. (2000). A ligand-induced extracellular cleavage regulates-gamma-secretase-like proteolytic activation of Notch1. *Mol. Cell* **5**, 197–206.

Murrell, J., Farlow, M., Ghetti, B., and Benson, M. D. (1991). A mutation in the amyloid precursor protein associated with hereditary Alzheimer's disease. *Science* **254**, 97–99.

Ni, C. Y., Murphy, M. P., Golde, T. E., and Carpenter, G. (2001). gamma-Secretase cleavage and nuclear localization of ErbB-4 receptor tyrosine kinase. *Science* **294**, 2179–2181.

Nitsch, R. M., Slack, B. E., Wurtman, R. J., and Growdon, J. H. (1992). Release of Alzheimer amyloid precursor derivatives stimulated by activation of muscarinic acetylcholine receptors. *Science* **258**, 304–307.

Parvathy, S., Hussain, I., Karran, E. H., Turner, A. J., and Hooper, N. M. (1998a). Alzheimer's amyloid precursor protein alpha-secretase is inhibited by hydroxamic acid-based zinc metalloprotease inhibitors: similarities to the angiotensin converting enzyme secretase. *Biochemistry* **37**, 1680–1685.

Parvathy, S., Karran, E. H., Turner, A. J., and Hooper, N. M. (1998b). The secretases that cleave angiotensin converting enzyme and the amyloid precursor protein are distinct from tumour necrosis factor-alpha convertase. *FEBS Lett.* **431**, 63–65.

Perry, E. K., Tomlinson, B. E., Blessed, G., Bergmann, K., Gibson, P. H., and Perry, R. H. (1978). Correlation of cholinergic abnormalities with senile plaques and mental test scores in senile dementia. *Brit. Med. J.* **2**, 1457–1459.

Peschon, J. J., Slack, J. L., Reddy, P., Stocking, K. L., Sunnarborg, S. W., Lee, D. C., Russell, W. E., Castner, B. J., Johnson, R. S., Fitzner, J. N., Boyce, R. W., Nelson, N., Kozlosky, C. J., Wolfson, M. F., Rauch, C. T., Cerretti, D. P., Paxton, R. J., March, C. J., and Black, R. A. (1998). An essential role for ectodomain shedding in mammalian development. *Science* **282**, 1281–1284.

Podlisny, M. B., Citron, M., Amarante, P., Sherrington, R., Xia, W., Zhang, J., Diehl, T., Levesque, G., Fraser, P., Haass, C., Koo, E. H., Seubert, P., St.George-Hyslop, P., Teplow, D. B., and

Selkoe D. J. (1997). Presenilin proteins undergo heterogeneous endoproteolysis between Thr291 and Ala299 and occur as stable N- and C-terminal fragments in normal and Alzheimer brain tissue. *Neurobiol. Dis.* **3**, 325–337.

Ratovitski, T., Slunt, H. H., Thinakaran, G., Price, D. L., Sisodia, S. S., and Borchelt, D. R. (1997). Endoproteolytic processing and stabilization of wild-type and mutant presenilin. *J. Biol. Chem.* **272**, 24536–24541.

Rawson, R. B., Zelenski, N. G., Nijhawan, D., Ye, J., Sakai, J., Hasan, M. T., Chang, T. Y., Brown, M. S., and Goldstein, J. L. (1997). Complementation cloning of S2P, a gene encoding a putative metalloprotease required for intramembrane cleavage of SREBPs. *Mol. Cell* **1**, 47–57.

Ray, W. J., Yao, M., Mumm, J., Schroeter, E. H., Saftig, P., Wolfe, M., Selkoe, D. J., Kopan, R., and Goate, A. M. (1999). Cell surface presenilin-1 participates in the γ-secretase-like proteolysis of Notch. *J. Biol. Chem.* **274**, 36801–36807.

Roberds, S. L., Anderson, J., Basi, G., Bienkowski, M. J., Branstetter, D. G., Chen, K. S., Freedman, S. B., Frigon, N. L., Games, D., Hu, K., Johnson-Wood, K., Kappenman, K. E., Kawabe, T. T., Kola, I., Kuehn, R., Lee, M., Liu, W., Motter, R., Nichols, N. F., Power, M., Robertson, D. W., Schenk, D., Schoor, M., Shopp, G. M., Shuck, M. E., Sinha, S., Svensson, K. A., Tatsuno, G., Tintrup, H., Wijsman, J., Wright, S., and McConlogue, L. (2001). BACE knockout mice are healthy despite lacking the primary beta-secretase activity in brain: implications for Alzheimer's disease therapeutics. *Hum. Mol. Genet.* **10**, 1317–1324.

Rudner, D. Z., Fawcett, P., and Losick, R. (1999). A family of membrane-embedded metalloproteases involved in regulated proteolysis of membrane-associated transcription factors. *Proc. Natl. Acad. Sci. USA* **96**, 14765–14770.

Sasaki, A., Yamaguchi, H., Ogawa, A., Sugihara, S., and Nakazato, Y. (1997). Microglial activation in early stages of amyloid beta protein deposition. *Acta Neuropathol. (Berl.)* **94**, 316–322.

Saunders, A. J., Kim, T.-W., Tanzi, R. E., Fan, W., Bennett, B. D., Babu-Kahn, S., Luo, Y., Louis, J.-C., McCaleb, M., Citron, M., Vassar, R., and Richards, W. G. (1999). BACE maps to chromosome 11 and a BACE homolog, BACE2, resides in the obligate Down syndrome region of chromosome 21. *Science* **286**, 1255a.

Scheuner, D., Eckman, C., Jensen, M., Song, X., Citron, M., Suzuki, N., Bird, T. D., Hardy, J., Hutton, M., Kukull, W., Larson, E., Levy-Lahad, E., Viitanen, M., Peskind, E., Poorkaj, P., Schellenberg, G., Tanzi, R., Wasco, W., Lannfelt, L., Selkoe, D., and Younkin, S. (1996). Secreted amyloid beta-protein similar to that in the senile plaques of Alzheimer's disease is increased in vivo by the presenilin 1 and 2 and APP mutations linked to familial Alzheimer's disease. *Nat. Med.* **2**, 864–870.

Schroeter, E. H., Kisslinger, J. A., and Kopan, R. (1998). Notch-1 signalling requires ligand-induced proteolytic release of intracellular domain. *Nature* **393**, 382–386.

Seiffert, D., Bradley, J. D., Rominger, C. M., Rominger, D. H., Yang, F., Meredith, J. E., Jr., Wang, Q., Roach, A. H., Thompson, L. A., Spitz, S. M., Higaki, J. N., Prakash, S. R., Combs, A. P., Copeland, R. A., Arneric, S. P., Hartig, P. R., Robertson, D. W., Cordell, B., Stern, A. M., Olson, R. E., and Zaczek, R. (2000). Presenilin-1 and -2 are molecular targets for gamma-secretase inhibitors. *J. Biol. Chem.* **275**, 34086–34091.

Selkoe, D. J. (1994). Cell biology of the amyloid beta-protein precursor and the mechanism of Alzheimer's disease. *Annu. Rev. Cell. Biol.* **10**, 373–403.

Selkoe, D. J. (1999). Translating cell biology into therapeutic advances in Alzheimer's disease. *Nature* **399**, A23–A31.

Seubert, P., Oltersdorf, T., Lee, M. G., Barbour, R., Blomquist, C., Davis, D. L., Bryant, K., Fritz, L. C., Galasko, D., Thal, L. J., Lieberburg, I., and Schenk, D. B. (1993). Secretion of beta-amyloid precursor protein cleaved at the amino terminus of the beta-amyloid peptide. *Nature* **361**, 260–263.

Shearman, M. S., Beher, D., Clarke, E. E., Lewis, H. D., Harrison, T., Hunt, P., Nadin, A., Smith, A. L., Stevenson, G., and Castro, J. L. (2000). L-685,458, an aspartyl protease transition state mimic, is a potent inhibitor of amyloid beta-protein precursor gamma-secretase activity. *Biochemistry* **39**, 8698–8704.

Shen, J., Bronson, R. T., Chen, D. F., Xia, W., Selkoe, D. J., and Tonegawa, S. (1997). Skeletal and CNS defects in Presenilin-1-deficient mice. *Cell* **89,** 629–639.

Sherrington, R., Rogaev, E. I., Liang, Y., Rogaeva, E. A., Levesque, G., Ikeda, M., Chi, H., Lin, C., Li, G., Holman, K., Tsuda, T., Mar, L., Foncin, J.-F., Bruni, A. C., Montesi, M. P., Sorbi, S., Rainero, I., Pinessi, L., Nee, L., Chumakov, I., Pollen, D., Brookes, A., Sanseau, P., Polinsky R. J., Wasco, W., Da Silva, H. A. R., Haines, J. L., Pericak-Vance, M. A., Tanzi, R. E., Roses, A. D., Fraser, P. E., Rommes, J. M., and St. George-Hyslop, P. H. (1995). Cloning of a gene bearing missense mutations in early-onset familial Alzheimer's disease. *Nature* **375,** 754–760.

Shi, X. P., Chen, E., Yin, K. C., Na, S., Garsky, V. M., Lai, M. T., Li, Y. M., Platchek, M., Register, R. B., Sardana, M. K., Tang, M. J., Thiebeau, J., Wood, T., Shafer, J. A., and Gardell, S. J. (2001). The pro domain of beta-secretase does not confer strict zymogen-like properties but does assist proper folding of the protease domain. *J. Biol. Chem.* **276,** 10366–10373.

Silva, A. M., Cachau, R. E., Sham, H. L., and Erickson, J. W. (1996). Inhibition and catalytic mechanism of HIV-1 aspartic protease. *J. Mol. Biol.* **255,** 321–346.

Sinha, S., Anderson, J. P., Barbour, R., Basi, G. S., Caccavello, R., Davis, D., Doan, M., Dovey, H. F., Frigon, N., Hong, J., Jacobson-Croak, K., Jewett, N., Keim, P., Knops, J., Lieberburg, I., Power, M., Tan, H., Tatsuno, G., Tung, J., Schenk, D., Seubert, P., Suomensaari, S. M., Wang, S., Walker, D., and John, V. (1999). Purification and cloning of amyloid precursor protein beta-secretase from human brain. *Nature* **402,** 537–540.

Sisodia, S. S. (1992). Beta-amyloid precursor protein cleavage by a membrane-bound protease. *Proc. Natl. Acad. Sci. USA* **89,** 6075–6079.

Stalder, M., Phinney, A., Probst, A., Sommer, B., Staufenbiel, M., and Jucker, M. (1999). Association of microglia with amyloid plaques in brains of APP23 transgenic mice. *Am. J. Pathol.* **154,** 1673–1684.

Steiner, H., Capell, A., Pesold, B., Citron, M., Kloetzel, P. M., Selkoe, D. J., Romig, H., Mendla, K., and Haass, C. (1998). Expression of Alzheimer's disease-associated presenilin-1 is controlled by proteolytic degradation and complex formation. *J. Biol. Chem.* **273,** 32322–32331.

Steiner, H., Duff, K., Capell, A., Romig, H., Grim, M. G., Lincoln, S., Hardy, J., Yu, X., Picciano, M., Fechteler, K., Citron, M., Kopan, R., Pesold, B., Keck, S., Baader, M., Tomita, T., Iwatsubo, T., Baumeister, R., and Haass, C. (1999a). A loss of function mutation of presenilin-2 interferes with amyloid beta-peptide production and notch signaling. *J. Biol. Chem.* **274,** 28669–28673.

Steiner, H., Romig, H., Grim, M. G., Philipp, U., Pesold, B., Citron, M., Baumeister, R., and Haass, C. (1999b). The biological and pathological function of the presenilin-1 Δexon 9 mutation is independent of its defect to undergo proteolytic processing. *J. Biol. Chem.* **274,** 7615–7618.

Steiner, H., Romig, H., Pesold, B., Philipp, U., Baader, M., Citron, M., Loetscher, H., Jacobsen, H., and Haass, C. (1999c). Amyloidogenic function of the Alzheimer's disease-associated presenilin 1 in the absence of endoproteolysis. *Biochemistry* **38,** 14600–14605.

Suzuki, N., Cheung, T. T., Cai, X. D., Odaka, A., Otvos, L., Jr., Eckman, C., Golde, T. E., and Younkin, S. G. (1994). An increased percentage of long amyloid beta protein secreted by familial amyloid beta protein precursor (beta APP717) mutants. *Science* **264,** 1336–1340.

Tam, A., Schmidt, W. K., and Michaelis, S. (2001). The multispanning membrane protein Ste24p catalyzes CAAX proteolysis and NH2-terminal processing of the yeast a-factor precursor. *J. Biol. Chem.* **276,** 46798–46806.

Teller, J. K., Russo, C., DeBusk, L. M., Angelini, G., Zaccheo, D., Dagna-Bricarelli, F., Scartezzini, P., Bertolini, S., Mann, D. M., Tabaton, M., and Gambetti, P. (1996). Presence of soluble amyloid beta-peptide precedes amyloid plaque formation in Down's syndrome [see comments]. *Nat. Med.* **2,** 93–95.

Thaisrivongs, S., Pals, D. T., Kati, W. M., Turner, S. R., Thomasco, L. M., and Watt, W. (1986). Design and synthesis of potent and specific renin inhibitors containing difluorostatine, difluorostatone, and related analogues. *J. Med. Chem.* **29,** 2080–2087.

Thinakaran, G., Borchelt, D. R., Lee, M. K., Slunt, H. H., Spitzer, L., Kim, G., Ratovitsky, T., Davenport, F., Nordstedt, C., Seeger, M., Hardy, J., Levey, A. I., Gandy, S. E., Jenkins, N. A., Copeland, N. G., Price, D. L., and Sisodia, S. S. (1996). Endoproteolysis of presenilin 1 and accumulation of processed derivatives in vivo. *Neuron* **17,** 181–190.

Thinakaran, G., Harris, C. L., Ratovitski, T., Davenport, F., Slunt, H. H., Price, D. L., Borchelt, D. R., and Sisodia, S. S. (1997). Evidence that levels of presenilins (PS1 and PS2) are coordinately regulated by competition for limiting cellular factors. *J. Biol. Chem.* **272,** 28415–28422.

Tomita, T., Maruyama, K., Saido, T. C., Kume, H., Shinozaki, K., Tokuhiro, S., Capell, A., Walter, J., Grunberg, J., Haass, C., Iwatsubo, T., and Obata, K. (1997). The presenilin 2 mutation (N141I) linked to familial Alzheimer disease (Volga German families) increases the secretion of amyloid beta protein ending at the 42nd (or 43rd) residue. *Proc. Natl. Acad. Sci. USA* **94,** 2025–2030.

Urban, S., Lee, J. R., and Freeman, M. (2001). Drosophila rhomboid-1 defines a family of putative intramembrane serine proteases. *Cell* **107,** 173–182.

Van Broeckhoven, C., Haan, J., Bakker, E., Hardy, J. A., Van Hul, W., Wehnert, A., Vegter-Van der Vlis, M., and Roos, R. A. (1990). Amyloid beta protein precursor gene and hereditary cerebral hemorrhage with amyloidosis (Dutch). *Science* **248,** 1120–1122.

Vassar, R., Bennett, B. D., Babu-Khan, S., Kahn, S., Mendiaz, E. A., Denis, P., Teplow, D. B., Ross, S., Amarante, P., Loeloff, R., Luo, Y., Fisher, S., Fuller, J., Edenson, S., Lile, J., Jarosinski, M. A., Biere, A. L., Curran, E., Burgess, T., Louis, J. C., Collins, F., Treanor, J., Rogers, G., and Citron, M. (1999). Beta-secretase cleavage of Alzheimer's amyloid precursor protein by the transmembrane aspartic protease BACE. *Science* **286,** 735–741.

Watson, D. J., Selkoe, D. J., and Teplow, D. B. (1999). Effects of the amyloid precursor protein Glu693 → Gln 'Dutch' mutation on the production and stability of amyloid beta-protein. *Biochem. J.* **340,** 703–709.

Wen, C., Metzstein, M. M., and Greenwald, I. (1997). SUP-17, a Caenorhabditis elegans ADAM protein related to *Drosophila* KUZBANIAN, and its role in LIN-12/NOTCH signalling. *Development* **124,** 4759–4767.

Wolf, B. A., Wertkin, A. M., Jolly, Y. C., Yasuda, R. P., Wolfe, B. B., Konrad, R. J., Manning, D., Ravi, S., Williamson, J. R., and Lee, V. M. (1995). Muscarinic regulation of Alzheimer's disease amyloid precursor protein secretion and amyloid beta-protein production in human neuronal NT2N cells. *J. Biol. Chem.* **270,** 4916–4922.

Wolfe, M. S. (2001). Secretase targets for Alzheimer's disease: identification and therapeutic potential. *J. Med. Chem.* **44,** 2039–2060.

Wolfe, M. S., De Los Angeles, J., Miller, D. D., Xia, W., and Selkoe, D. J. (1999a). Are presenilins intramembrane-cleaving proteases? Implications for the molecular mechanism of Alzheimer's disease. *Biochemistry* **38,** 11223–11230.

Wolfe, M. S., Xia, W., Moore, C. L., Leatherwood, D. D., Ostaszewski, B., Donkor, I. O., and Selkoe, D. J. (1999b). Peptidomimetic probes and molecular modeling suggest Alzheimer's γ-secretases are intramembrane-cleaving aspartyl proteases. *Biochemistry* **38,** 4720–4727.

Wolfe, M. S., Xia, W., Ostaszewski, B. L., Diehl, T. S., Kimberly, W. T., and Selkoe, D. J. (1999c). Two transmembrane aspartates in presenilin-1 required for presenilin endoproteolysis and γ-secretase activity. *Nature* **398,** 513–517.

Wong, P. C., Zheng, H., Chen, H., Becher, M. W., Sirinathsinghji, D. J., Trumbauer, M. E., Chen, H. Y., Price, D. L., Van der Ploeg, L. H., and Sisodia, S. S. (1997). Presenilin 1 is required for Notch1 and Dll1 expression in the paraxial mesoderm. *Nature* **387,** 288–292.

Yan, R., Bienkowski, M. J., Shuck, M. E., Miao, H., Tory, M. C., Pauley, A. M., Brashier, J. R., Stratman, N. C., Mathews, W. R., Buhl, A. E., Carter, D. B., Tomasselli, A. G., Parodi, L. A., Heinrikson, R. L., and Gurney, M. E. (1999). Membrane-anchored aspartyl protease with Alzheimer's disease beta-secretase activity. *Nature* **402,** 533–537.

Ye, J., Dave, U. P., Grishin, N. V., Goldstein, J. L., and Brown, M. S. (2000). Asparagine-proline sequence within membrane-spanning segment of SREBP triggers intramembrane cleavage by site-2 protease. *Proc. Natl. Acad. Sci. USA* **97,** 5123–5128.

Yu, G., Chen, F., Nishimura, M., Steiner, H., Tandon, A., Kawarai, T., Arawaka, S., Supala, A., Song, Y. Q., Rogaeva, E., Holmes, E., Zhang, D. M., Milman, P., Fraser, P. E., Haass, C., and St George-Hyslop, P. (2000a). Mutation of conserved aspartates affect maturation of both aspartate-mutant and endogenous presenilin 1 and presenilin 2 complexes. *J. Biol. Chem.* **275**, 27348–27353.

Yu, G., Nishimura, M., Arawaka, S., Levitan, D., Zhang, L., Tandon, A., Song, Y. Q., Rogaeva, E., Chen, F., Kawarai, T., Supala, A., Levesque, L., Yu, H., Yang, D. S., Holmes, E., Milman, P., Liang, Y., Zhang, D. M., Xu, D. H., Sato, C., Rogaev, E., Smith, M., Janus, C., Zhang, Y., Aebersold, R., Farrer, L. S., Sorbi, S., Bruni, A., Fraser, P., and St George-Hyslop, P. (2000b). Nicastrin modulates presenilin-mediated notch/glp-1 signal transduction and betaAPP processing. *Nature* **407**, 48–54.

Zhang, J., Kang, D. E., Xia, W., Okochi, M., Mori, H., Selkoe, D. J., and Koo, E. H. (1998). Subcellular distribution and turnover of presenilins in transfected cells. *J. Biol. Chem.* **273**, 12436–12442.

Zhang, Z., Nadeau, P., Song, W., Donoviel, D., Yuan, M., Bernstein, A., and Yankner, B. A. (2000). Presenilins are required for gamma-secretase cleavage of beta-APP and transmembrane cleavage of Notch-1. *Nat. Cell Biol.* **2**, 463–465.

Zhao, J., Paganini, L., Mucke, L., Gordon, M., Refolo, L., Carman, M., Sinha, S., Oltersdorf, T., Lieberburg, I., and McConlogue, L. (1996). Beta-secretase processing of the beta-amyloid precursor protein in transgenic mice is efficient in neurons but inefficient in astrocytes. *J. Biol. Chem.* **271**, 31407–31411.

9

Plasminogen Activation at the Cell Surface

Vincent Ellis
School of Biological Sciences
University of East Anglia
Norwich, NR4 7TJ, United Kingdom

I. Introduction

A. The Plasminogen Activators as Cell-Surface Proteases

The plasminogen activation system essentially consists of three serine proteases. These are the two plasminogen activators, uPA and tPA, both of which can specifically activate plasminogen to the broad specificity protease plasmin. In addition

Current Topics in Developmental Biology, Vol. 54

to its well-described role in hemostasis, i.e., the dissolution of fibrin clots or fibrinolysis, the plasminogen activation system also has important functions in the pericellular environment with physiological and pathological roles in tissues. To an extent these two apparently distinct functions are accounted for by the individual characteristics of the two plasminogen activators, with tPA being associated with fibrinolysis and uPA with pericellular proteolysis. However, it is becoming clear that the plasminogen activators have a degree of functional overlap and that both uPA and tPA have pericellular functions. These functions are regulated by cell surface receptors or binding sites for the proteases, and therefore despite being secreted as soluble proteins, in terms of their function these are true cell surface proteases.

It is widely accepted that the generation of proteolytic activity is a key event in mediating cell migration and invasion, dynamically modulating interactions between the cell and its surrounding extracellular matrix (Werb, 1997). Many cell surface proteases contribute to this proteolytic activity but the characteristics of plasminogen make it ideally suited to a central role in determining these events. Of particular importance are the broad substrate specificity of plasmin, allowing it to target a wide range of potential substrates, and its high concentration, making it the most abundant protease zymogen in the body. Together these characteristics make plasminogen a vast source of proteolytic potential. Therefore the functional regulation of the plasminogen activation system may represent a key step in both the initiation and regulation of the overall process of pericellular proteolysis.

By comparison with other proteolytic cascade systems the plasminogen activation system is rather simple in terms of the number of protease components involved. The complexity in this system arises instead from the different mechanisms that regulate the activity of these proteases, and it is now clear that cells can use a diversity of mechanisms to regulate the function of the plasminogen activation system, as will be emphasized in this chapter. The most well characterized of these mechanisms is provided by uPAR, the specific cellular receptor for uPA.

B. Discovery of the Plasminogen Activators and Their Biological Roles

The protease components of the plasminogen activation system have been known for many years, with the presence of a fibrinolytic activity in tissues ("fibrinolysin") first identified in the early 1900s. In the 1940s it became apparent that this activity was due to a proteolytic conversion of a proenzyme which was subsequently identified as profibrinolysin or plasminogen. A variety of activities that could activate plasminogen were identified at around the same time, from urine, blood, vascular, and other tissues. The understanding that these activities were due to just two distinct plasminogen activators first came from their immunological reactivities;

the protease originally identified in urine was termed urokinase, and now know as uPA, and the protease originally identified in tissues was termed tissue plasminogen activator, tPA. The identity of these proteases was subsequently unequivocally confirmed by their molecular cloning (Pennica et al., 1983; Verde et al., 1984).

The earliest evidence that the plasminogen activators, and in particular uPA, played a role in tissue degradation and cancer were the observations of uPA expression upon oncogenic transformation of cultured cells and the presence of uPA at the invasive fronts of experimental tumors. This and other evidence were covered in depth in the landmark review of Danø et al. (1985). This role has since been corroborated by a vast array of evidence obtained in different systems, and particularly emphasizes a role in tumor metastasis. This evidence includes the immunohistochemical localization of components of the system in human tumors, the positive correlation between levels of uPA and metastatic potential in both cell and animal tumor models, suppression of metastasis in model systems by inhibition of uPA or blocking of uPAR, the effect enhancement of the metastatic phenotype of cell lines by transgenic overexpression of uPA, reduced tumor progression in uPA-deficient mice, and the prognostic value of uPA levels in certain human cancers (Duffy, 2002).

As mentioned earlier, the originally identified function of this proteolytic system was in fibrinolysis. However, the two plasminogen activators proved to have very different efficiencies in mediating this fibrinolytic activity. This is because the formation of fibrin acts as a stimulus for its own subsequent dissolution, i.e., fibrinolysis. The conversion of fibrinogen to fibrin, as well as leading to the formation of an insoluble polymer, also exposes cryptic binding sites for both tPA and plasminogen. This binding increases plasminogen activation dramatically. Although not the topic of this chapter, this effect of fibrin on the activity of tPA introduces the concept that mechanisms to stimulate the activity of the plasminogen activators might represent a general mechanism for the regulation of their activity. Such mechanisms have previously been identified in the blood coagulation cascade where many of the proteolytic reactions are stimulated by interactions of the serine proteases with membrane binding-sites.

The review of Danø et al. (1985) discussed a number of studies that suggested that plasminogen activators, and in particular uPA, may have a subcellular localization consistent with an association with the plasma membrane. However, these studies failed to predict the role that would soon emerge for the cell surface in the regulation of the plasminogen activation system. This came from the observation of the specific binding of uPA to the cell surface and the discovery of uPAR (Stoppelli et al., 1985; Vassalli et al., 1985), the observation that plasminogen could also bind to the cell surface (Plow et al., 1986), and the demonstration that the cell surface was the preferential site for plasminogen activation (Ellis et al., 1989; Stephens et al., 1989).

C. Evolutionary Clues to Plasminogen Activator Function

The cellular function of the plasminogen activation system has often been considered to be secondary to its function in hemostatic fibrinolysis. In this scenario the preexisting activity of this powerful proteolytic system would have been subverted by cellular mechanisms for a role in ECM remodeling. However, among the serine proteases, the plasminogen activators and plasminogen are not closely related and derive from two different phylogenetic lineages. This is unusual as the members of proteolytic cascades are usually closely related and derive from a single lineage, e.g., the blood coagulation proteases. Examination of these two different lineages suggests that both the plasminogen activators and plasminogen are associated with lineages independently involved in developmental functions. The plasminogen activator lineage derives from primordial developmental proteases (Krem *et al.*, 2000), and plasminogen also has a developmental connection and is closely related to the plasminogen-related growth factors (Donate *et al.*, 1994). These proteins lack protease activity, but have signaling developmentally related roles, particularly in epithelial–mesenchymal transitions. Therefore it is possible that the proteases of the plasminogen activation system originally evolved with roles in ECM remodeling and that hemostatic fibrinolysis developed as a specialized form of this remodeling and is therefore the auxiliary function of this system.

Deletion of the genes for the various components of this system in mice has demonstrated that none of them are now required for development, although developmentally related functions are impaired, particularly in the CNS. However, the activity of the system is implicated in both physiological and pathological processes involving cell migration and/or tissue remodeling. In addition to a role in cancer, the plasminogen activation system is also implicated in a variety of other diseases including cardiovascular disease, arthritis, and neurodegeneration. The involvement of this proteolytic system in such a diversity of pathologies may stem, at least in part, from the variety of mechanisms that have been identified which regulate the activity of this system. These mechanisms all involve cell-surface molecules and all lead to a large increase in the generation of plasmin.

II. Structural Chemistry/Biochemistry

A. Modular Construction of Serine Proteases

The three protease components of the plasminogen activation system, uPA, tPA and plasminogen, are mosaic proteins having a modular construction with long N-terminal extensions containing multiple additional domains or modules (Fig. 1), in common with many other serine proteases involved in embryogenesis, immunity, and blood coagulation. These additional protein modules are thought to have been

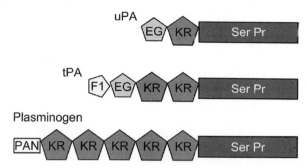

Figure 1   Modular organization of the protease components of the plasminogen activation system. The nomenclature and symbols used in naming protein modules are those proposed by Bork and Bairoch (1995).

acquired as a result of exon shuffling during evolution, a process made possible by the presence of a phase I intron–exon junction at the 5′ end of the protease domain ancestral gene and phase I introns flanking the various modules (Patthy, 1999). Although the two plasminogen activators are only relatively distantly related to plasminogen in evolutionary terms, their modular structure is quite similar, the presence of kringle modules (KR) being the common feature. The only other subgroup of serine proteases that contain this module is the group containing the blood coagulation factors, where it occurs in prothrombin. Kringle modules are not found in any other proteins. The fibronectin type-1 module (F1) found in tPA is only found in the closely related proteases factor XII and hepatocyte growth factor activator and in fibronectin itself. By contrast, the EGF-like module (EG) present in uPA and tPA is found in a wide variety of proteins, often in multiple copies.

The N-terminal regions of the serine proteases mediate interactions with other macromolecules and have critical functional roles, allowing the catalytic domain to interact efficiently with other components of the system. Therefore, although the serine protease domain is responsible for primary recognition of substrate, substrate specificity, and catalysis, it is the N-terminal regions that to a large extent dictate biological function. Not all of the domains in the N-terminal regions have defined functional roles; those that do not appear instead to have structural roles, maintaining the appropriate spacing or orientation of the functional domains.

1. Kringle (KR) Modules

This module is unusual in having just one clearly defined function, which is exemplified by the KR modules of plasminogen. Four of the five KR modules present in plasminogen contain a "lysine binding site," which can bind lysine and certain other aminocarboxylic acids. Each of these sites is somewhat different and therefore they have different affinities and selectivities for their various ligands.

The highest affinity sites are those in the KR-1 and KR-4 modules, which can bind the lysine analog 6-aminohexanoic acid with the K_d values as low as 10 μM (Marti *et al.*, 1997). These sites have two aspartic acid residues that interact with the amino group of the ligand (i.e., the side-chain amino group of lysine), one or two basic residues (lysine or arginine) that interact with the carboxyl group of the ligand (which means that C-terminal lysine residues are preferential ligands), and a hydrophobic groove which interacts with the aliphatic region of the lysine side chain. The KR-5 module of plasminogen has slightly different characteristics. It lacks the basic binding site, has equal affinities for ligands with and without a carboxyl group, and therefore does not discriminate between lysine residues internal to a protein sequence and those at the C terminus.

The KR modules are responsible for two of the important features of plasminogen. The first of these is that the N-terminal PAN module (described below) interacts intramolecularly with a KR module, most likely KR-4 or KR-5, giving the plasminogen molecule a compact, closed conformation which is relatively resistant to activation. In the presence of lysine analogs this interaction is lost and plasminogen adopts an extended, open conformation which is susceptible to activation (Mangel *et al.*, 1990). Plasminogen bound to proteins such as fibrin also adopts this conformation. Removal of the PAN module by proteolysis generates Lys[77]-plasminogen which also has the extended conformation and is activated approximately 10-fold more efficiently than native or Glu[1]-plasminogen.

The plasminogen activators also contain KR modules, but only KR-2 of tPA contains a lysine binding site. This has ligand-binding characteristics most similar to those of plasminogen KR-4. This module appears to play an important role in the binding of tPA to fibrin and some of its cell-surface binding sites. The KR module of uPA, although lacking a lysine binding site, has a characteristic basic sequence motif (Arg[108]-Arg-Arg) that has affinity for heparin (Stephens *et al.*, 1992b). The significance of this binding site has not been demonstrated; it may be involved in the binding of uPA to components of the extracellular matrix such as laminin-nidogen (Stephens *et al.*, 1992a).

2. EGF-Like (EG) Modules

Both of the plasminogen activators contain single EG modules that have conformations close to the consensus structure for these modules. EG modules are found in a wide variety of proteins, including some other serine proteases. The EG module of uPA is at the N-terminus of the protein and is responsible for the binding of uPA to its specific cell-surface receptor uPAR. A flexible, seven-residue Ω-loop connecting the two strands of the module's major β-sheet constitutes the binding epitope (Ploug *et al.*, 1995), and synthetic peptides derived from this sequence are strong binding antagonists (Magdolen *et al.*, 2001). The EG module of tPA is in close juxtaposition to the N-terminal F1 module and they are best considered as a module pair or supermodule as there are extensive interactions between them.

3. Fibronectin Type-1 (F1) Modules

Fibronectin contains 12 repeats of the F1 module, and the only other proteins known to contain the F1 module are tPA, blood coagulation factor XII, and the closely related hepatocyte growth factor activator, each of which contains a single copy. The F1 module of tPA, which shows close similarity to the seventh repeat of fibronectin, has an abundance of hydrophobic residues on the exposed surface of its triple-stranded β-sheet. In the supermodule this surface forms an interface with the major β-sheet of the EG module and forms an extremely stable module pair. This an important functional region of the tPA molecule as it has a key role in binding to fibrin and its cell-surface binding sites.

4. PAN Modules

The N-terminal module of tPA has previously been known as the N-terminal peptide (NTP), which is a misnomer as it is a highly structured region. This structure was thought to be unique to plasminogen and the plasminogen-related growth factors, hepatocyte growth factor/scatter factor (HGF/SF) and macrophage-stimulating proteins (MSP), but has been found to be homologous to the apple domains of the kallikrein/coagulation factor XI family and domains of various nematode proteins. This module superfamily is now termed PAN (plasminogen, apple, nematode) (Tordai et al., 1999). As mentioned previously, this module has an important role in modulating plasminogen function by its intramolecular interaction with KR-modules. Lys-50 and the surrounding region has been proposed to be important for this interaction (An et al., 1998).

B. Catalytic Features

1. Serine Protease Activity

The C-terminal catalytic domains of the serine proteases of the plasminogen activation system share high sequence identity with the archetypal family members trypsin and chymotrypsin. An alignment of the sequences of these serine proteases is shown in Fig. 2. The major differences between these sequences are inserted regions corresponding to surface exposed loops in the three-dimensional structures, often involved in interactions with substrates, inhibitors, or cofactors. The serine proteases are characterized by a catalytic triad of His-57, Asp-102, and Ser-195 (numbered according to the convention of using the sequence of chymotrypsin). They are all synthesized as single-chain precursors or zymogens that require specific proteolytic cleavage to become fully active. This proteolytic cleavage exposes a new N-terminus which is almost invariably an Ile or Val residue (Ile-16 in chymotrypsin). The free amino group of Ile-16 forms a salt bridge with Asp-194, the residue preceding the catalytic Ser residue, to stabilize the structure of four

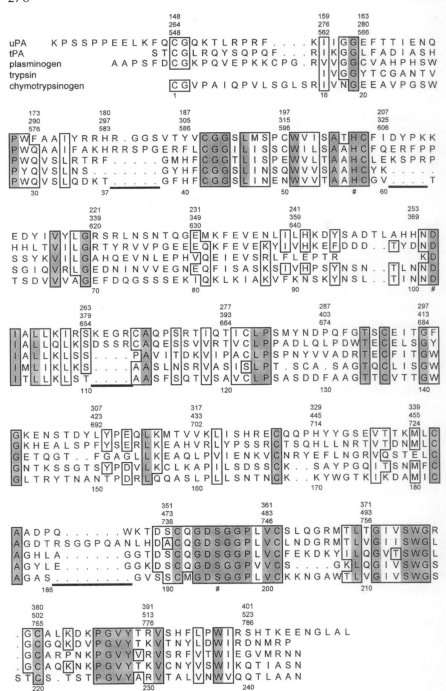

disordered peptide segments collectively known as the "activation domain" (Huber and Bode, 1978). This leads to the formation of a fully mature active site in the protease, by stabilizing the "oxyanion hole" and the substrate binding pocket. It is thought that these two distinct conformations of the protease domain exist in an equilibrium in the absence of the activation cleavage. For most of the serine proteases this equilibrium lies far toward the inactive zymogen form; for example, in the case of trypsinogen the equilibrium constant has been estimated to be 10^7 (Bode, 1979). Plasminogen is an example of a similarly inert zymogen. However, some serine proteases have intrinsic mechanisms that can affect this equilibrium and lead to significant activity in the single-chain zymogens. The plasminogen activators, and in particular tPA, are the principal examples of proteases using this mechanism and it is critical to their function.

2. tPA, an "Active Zymogen"

The single-chain form of tPA can be proteolytically cleaved by plasmin at a canonical activation site of Arg^{275}-Ile. However, the zymogen form has an unusually high level of intrinsic activity, approximately 10–20% that of the two-chain form, and consequently this cleavage has a much smaller effect on the activity of tPA than it does on more typical serine proteases. The structural features of tPA underlying this zymogen activity have been deduced from the crystal structure of the non-activated protease domain (Renatus et al., 1997). The stabilization of the critical Asp-194, which is achieved by formation of a salt bridge with the α-amino group of Ile-16 in the activated two-chain form, is instead achieved by interaction with the side-chain of Lys-156 (Lys-429 in the tPA sequence, Fig. 2).

The high intrinsic activity of single-chain tPA would appear to be contrary to the basic biological principle of zymogens, i.e., that enzymes are synthesized as inactive precursors to prevent premature and possibly harmful activity of the protease. One of the reasons that tPA can do this is that its proteolytic activity is extremely low. At physiological concentrations of plasminogen one molecule of tPA will take approximately 10 min to generate one molecule of plasmin. In addition, despite its poor catalytic activity, single-chain tPA reacts rapidly with its physiological inhibitor PAI-1. tPA overcomes this inherent inefficiency by the

---

Figure 2 Sequence alignment of the catalytic domains of the protease components of the plasminogen activation system. Alignment of the sequences of human uPA, tPA, and plasminogen with the sequences of bovine trypsin and chymotrypsinogen A. The alignment is based on the topological equivalence of residues (Bode and Renatus, 1997). The sequence numbering of uPA, tPA, and plasminogen, respectively, is shown above the alignment and the sequence numbering of chymotrypsinogen below it. Regions of sequence identity and similarity are indicated by gray and white boxes, respectively. Four significant insertion loops in the plasminogen activators are underlined in bold. The sequences include the linker region preceding c-Cys-1 which forms the interchain disulfide bridge with c-Cys-122 in the activated proteases. In the case of uPA the sequence is shown from a proteolytic cleavage site in the linker and therefore lacks four additional N-terminal residues.

use of nonenzymatic cofactors to stimulate its activity. The "classical" cofactor or stimulator of tPA is fibrin, but it is now clear that certain proteins present on the cell surface can also fulfill this function and that tPA may also have an important role in pericellular proteolysis.

Fibrin increases tPA-catalyzed plasminogen activation by up to three orders of magnitude (Hoylaerts *et al.,* 1982). In the presence of fibrin single-chain and two-chain tPAs are equally active demonstrating that fibrin binding shifts the conformational equilibrium of the protease domain to the fully active form. However, the majority of the stimulation comes from the co-assembly of tPA and plasminogen on fibrin, which behaves as a catalytic "template," increasing the local concentration of reactants and also positioning them to react in a catalytically productive manner. Similar template mechanisms appear to underlie increases in the activities of both of the plasminogen activators in the context of the cell surface as will later be described in Section III,A,2.

3. Pro-uPA, a Partially Active Zymogen?

The zymogen pro-uPA also appears to have an intrinsic catalytic activity. There has been much controversy concerning the absolute level of this activity, arising from the sensitivity of pro-uPA to proteolytic activation, particularly by plasmin. A variety of strategies have been used to suppress this activation, leading to estimates that pro-uPA has an activity less than 1% that of the two-chain enzyme (Ellis *et al.,* 1987; Petersen *et al.,* 1988). Experiments using nonactivatable pro-uPA (Lys^{158}>Glu) and active-site mutated plasminogen (Ser^{740}>Ala) confirmed these estimates and demonstrated an intrinsic activity of approximately 0.2%. This is much lower than the activity of the tPA zymogen, and raises the question of whether this activity is of any functional significance. This is also an area of controversy and is discussed further next and in Section III,A,3.

4. Zymogen Activation

The activation of all of these proteases can be catalyzed *in vitro* by a variety of both serine and other proteases, although whether many of these proteases have such a role *in vivo* is uncertain. Plasminogen can be activated by kallikrein, coagulation factors XIa and XIIa, and trypsin (Castellino, 1998); however, there is little doubt that the two plasminogen activators are the only relevant activators *in vivo*. In support of this the phenotype of plasminogen-deficient mice is identical to that of mice with a combined deficiency of tPA and uPA (Bugge *et al.,* 1996a). The activation of tPA by proteases other than plasmin has not been reported and, as it seems that this activation has little functional consequence (for the reasons described above), this is not surprising.

By contrast a number of proteases have been shown to activate pro-uPA and suggested to have a role in its biological activation. In addition to plasmin (Gunzler *et al.,* 1982), these include the serine proteases trypsin (Ichinose *et al.,* 1986), plasma kallikrein (Ichinose *et al.,* 1986), mast cell tryptase (Stack and Johnson, 1994), and matriptase/MT-SP1 (Takeuchi *et al.,* 2000). The last may be of particular interest as it is a member of a recently identified family of type II-transmembrane serine proteases (Hooper *et al.,* 2001); however, it has yet to be shown whether the native transmembrane form of this protease can activate uPAR-bound pro-uPA. Proteases of other classes that can activate pro-uPA include the lysosomal cysteine protease cathepsin B (Kobayashi *et al.,* 1991). This may be of particular significance as cathepsin B is also implicated in disease processes involving pericellular proteolysis as it can associate with the cell surface and may colocalize there with the uPA/uPAR system, as discussed in Chapter 10 by Cavallo-Medved and Sloane. Although plasmin is thought to be the primary physiological activator of pro-uPA, the only protease that has been formally shown to activate pro-uPA *in vivo* is the glandular kallikrein mGK-6, which was found to be responsible for the presence of active uPA in the urine of plasminogen-deficient mice (List *et al.,* 2000).

The main reason to suspect that plasmin is the primary physiological activator is because of the process of reciprocal zymogen activation, i.e., the active protease of each zymogen/protease pair can activate the zymogen of the other pair. Once active protease is present in such a system of two zymogens, there will be a rapid and amplified generation of proteolytic activity. Therefore, plasmin activation of pro-uPA is potentially a powerful mechanism for propagating the activity of this system. This still leaves the question: what is the initial active protease? It is in this context that the proteases mentioned earlier may have a role, i.e., in the initiation of the system. However, an alternative mechanism may involve the intrinsic zymogen activity of pro-uPA, and it has been established, at least under certain conditions *in vitro,* that this can occur with high efficiency and that this could represent the initiation mechanism for this proteolytic system (Ellis and Danø, 1993a). This issue will be discussed in detail in Section III,A,3, when considering the mechanisms involved in uPAR regulation of plasminogen activation on the cell surface.

5. Substrate Specificities

The individual serine proteases demonstrate a remarkably wide range in their substrate specificities, and the proteases of the plasminogen activation system fall at either end of this spectrum. Plasmin has a very broad specificity and is a highly efficient protease, with an activity very similar to that of the digestive enzyme trypsin, cleaving C-terminally of Lys and Arg residues. By contrast the two plasminogen activators are extremely specific, with very few substrates identified.

Apart from plasminogen, only the previously mentioned plasminogen-related growth factor HGF/SF (Naldini *et al.,* 1992) and uPAR (Høyer-Hansen *et al.,* 1992) have been identified as potentially relevant substrates for uPA. The latter, a cleavage at a sequence similar to the activation cleavage site of plasminogen (Ser-Gly-Arg ↓ Ala-Val), leads to loss of uPA binding and therefore represents a potential mechanism for downregulation of the uPA/uPAR interaction. It is not yet known whether HGF/SF activation can take place on the cell surface via uPAR-bound uPA, although this may be possible, as HGF/SF can interact with its receptor c-Met prior to proteolytic activation. tPA has also been reported to activate HGF/SF (Mars *et al.,* 1993) and to cleave and upregulate the activity of the glutamatergic NMDA receptor (Nicole *et al.,* 2001), which may be of relevance to the emerging role for tPA in the brain and particularly in neurodegeneration. The efficiencies of these reactions have not been established and it is not known whether they involve mechanisms to stimulate the otherwise very low catalytic activity of tPA.

The proteins identified as potential substrates for plasmin are far too numerous to list here, but certain groups of proteins stand out as being of particular interest in the context of the cell-surface role of this proteolytic system. These are extra-cellular matrix proteins, the MMPs, and growth factors. Plasmin can cleave many of the nonfibrillar components of the extracellular matrix, including fibronectin, laminin, type-IV collagen (Liotta *et al.,* 1981), and proteoglycans, such as per-lecan (Whitelock *et al.,* 1996). Plasmin can also, of course, degrade fibrin which can be considered as a provisional matrix protein in many tissue repair situations (Loskutoff and Quigley, 2000).

The fibrillar matrix proteins, such as collagen, that cannot be degraded by plas-min are substrates for members of the MMP family. Plasmin is thought to con-tribute to a cascade of proteolytic activity by activation of certain of the MMPs. Plasmin has been shown to fully, or partially, activate interstitial collagenase (MMP-1), stromelysin-1 (MMP-3), matrilysin (MMP-7), gelatinase-B (MMP-9), metalloelastase (MMP-12), and collagenase-3 (MMP-13) *in vitro,* but notably not gelatinase-A (MMP-2) (Murphy *et al.,* 1999). Some of these reactions have also been demonstrated in cell culture systems (He *et al.,* 1989; Mazzieri *et al.,* 1997) and also *in vivo* (Carmeliet *et al.,* 1997a).

The processing of growth factors by plasmin is another important mechanism by which the plasminogen activation system can influence cellular behavior. The best studied examples of this include the activation of latent TGF-β (Odekon *et al.,* 1994) and the mobilization of matrix-sequestered FGF-2 (Saksela and Rifkin, 1990). Interestingly in a cell culture system the activation of TGF-β was shown to be dependent on the presence of uPAR (Odekon *et al.,* 1994). Plasmin has been shown to proteolytically activate pro-forms of members of the neurotrophic nerve growth factor family members, brain-derived neurotrophic factor (BDNF) and β-NGF, converting them to neuronal cell-survival factors (Lee *et al.,* 2001). None of these reactions has yet been shown to occur *in vivo,* although there is indirect evidence for the activation of TGF-β (Grainger *et al.,* 1993, 1995).

C. Inhibitors

1. Physiological Inhibitors

The physiological inhibitors of the proteases of the plasminogen activation system are all members of the large group of proteins known as the serpins (serine protease inhibitor superfamily). These proteins have a conserved structure and adopt a metastable conformation that is required for their inhibitory activity. Serpins inhibit their target proteases by an irreversible suicide substrate mechanism; the initial formation of a noncovalent Michaelis-type complex leads to cleavage of the scissile bond (Arg-Met) and the formation of a covalent ester linkage between the active-site Ser residue of the protease and the backbone carbonyl of the serpin P1 residue in the reactive center loop. Subsequent structural rearrangements in the serpin leads to the transport of the protease to the opposite end of the serpin molecule, and distortion of the protease structure leads to its inactivation and the kinetic trapping of the covalent acyl intermediate (Silverman *et al.*, 2001).

The main inhibitor of the two plasminogen activators is the serpin plasminogen activator inhibitor-1 (PAI-1). This is both a very specific and a very efficient inhibitor of these enzymes with association rate constants of greater than $2 \times 10^7 \ M^{-1}s^{-1}$ for both of the activated proteases, and with single-chain tPA inhibited approximately 5-fold slower (Thorsen *et al.*, 1988). PAI-1 is expressed by many cell types in a highly regulated manner, although low concentrations (less than 1 nM) can also be found in plasma. PAI-2, which falls within a different subclass of the serpins, has a much more restricted expression pattern, being found primarily in monocytes and in placenta. It lacks a signal sequence and is poorly secreted, giving rise to speculation that it may have an intracellular role rather than being an inhibitor of the plasminogen activators. PAI-2 is a less efficient inhibitor than PAI-1 and displays a differential inhibition of the plasminogen activators. It is a very poor inhibitor of tPA ($5 \times 10^3 \ M^{-1}s^{-1}$ for single-chain tPA, greater than 1000-fold lower than with PAI-1) and only a moderate inhibitor of uPA ($2 \times 10^6 \ M^{-1}s^{-1}$) (Thorsen *et al.*, 1988). Other serpins are less efficient than the PAIs, but nevertheless may have important functional roles. Among these is the neuronally expressed and axonally secreted inhibitor neuroserpin that is found in most regions of the brain. Neuroserpin is a relatively efficient and specific inhibitor of both uPA and tPA, with association rate constants of $5 \times 10^4 \ M^{-1}s^{-1}$ and $1.5 \times 10^5 \ M^{-1}s^{-1}$, respectively (Osterwalder *et al.*, 1998). Because of their metastable nature, mutations in serpins can cause polymerization of the protein, which in the case of neuroserpin leads to the formation of neuronal inclusion bodies and is associated with familial dementia (Davis *et al.*, 1999).

The activity of plasmin is regulated by α_2-antiplasmin. This serpin is expressed in the liver and present at a high concentration in plasma (1 μM). It is possibly the most efficient of the serpins, with an association constant approaching $10^8 \ M^{-1}s^{-1}$ (Wiman and Collen, 1979), approaching the diffusionally controlled rate, i.e., the theoretical maximum. Part of the reason for this high efficiency is the unique C-terminal extension present in α_2-antiplasmin. This binds to the lysine-binding sites in the KR modules of plasmin (and plasminogen) and greatly facilitates the subsequent inhibitory reaction. Therefore, in the presence of lysine analogs, inhibition by α_2-antiplasmin is greatly reduced. Despite plasmin's broad substrate specificity, it reacts very specifically with α_2-antiplasmin, an effect which has been shown to involve interactions of the inhibitor with at least two surface-exposed loops on the protease domain of plasmin (Turner *et al.*, 2002).

2. Synthetic and Nonphysiological Inhibitors

A variety of natural and synthetic inhibitors of the plasminogen activation system have been identified. The activity of plasmin is most readily inhibited by aprotinin (bovine basic pancreatic trypsin inhibitor, BPTI), a member of the Kunitz family of protease inhibitors by the formation of a very stable noncovalent complex (K_i approximately 1 nM). Aprotinin does not inhibit the plasminogen activators to any appreciable extent, but it does inhibit a number of other serine proteases. Because of its broad substrate specificity, small synthetic inhibitors tend to have very little selectivity toward plasmin. Peptide-based inhibitors such as Val-Leu-Lys-chloromethane have more selectivity, but still not sufficient to be of general use. It is worth noting that the same lysine analogs that bind to the KR modules of plasminogen are also weak inhibitors of plasmin activity, binding in the S1 pocket of the active site. However, compounds with a methylated carboxyl group are much more effective than the corresponding carboxylic acids, e.g., K_i of 1 mM for lysine methyl ester compared to 100 mM for lysine.

The much more restricted substrate specificity of the plasminogen activators should allow the identification of relatively specific inhibitors, although it has proven to be difficult to obtain inhibitors that are both specific and efficient. The diuretic drug amiloride (3,5-diamino-N-(aminoiminomethyl)-6-chloropyrazinecarboxamide) has been shown to selectively inhibit the activity of uPA, but not tPA, with a K_i of 7 μM (Vassalli and Belin, 1987). Some plant-derived Kunitz inhibitors, such as *Erythrina* trypsin inhibitor, have been found to be relatively efficient inhibitors of tPA with a K_i of 0.1 μM (Joubert and Dowdle, 1987). Because of its potential as an therapeutic target in cancer, there are significant efforts to synthesize effective low-molecular-weight inhibitors of uPA. Promising groups of compounds include 4-substituted benzo[*b*]thiophene-2-carboxamidines ($K_i < 0.2 \mu M$) (Towle *et al.*, 1993) and (4-aminomethyl)phenylguanidine derivatives ($K_i < 3 \mu M$), which have an unusual binding mode involving interactions with both the S1 and S1′ subsites of the protease, which may facilitate the design of inhibitors with further improvements in potency (Sperl *et al.*, 2000).

D. Expression of Components of the Plasminogen Activation System

1. Plasminogen

Plasminogen is constitutively expressed in the liver, primarily by hepatocytes. Despite the high concentration that this leads to in plasma and other extracellular fluids, it has often been suggested that there are also extrahepatic sites of plasminogen expression. A study using sensitive RT-PCR methods has demonstrated a quiet broad extrahepatic distribution of plasminogen mRNA in the mouse (Zhang *et al.*, 2002), but whether this expression is of functional significance is not known. One site where extrahepatic expression is thought to be important is the brain, which is normally deprived of circulating plasminogen by the blood–brain barrier. Plasminogen has been shown to be expressed in hippocampal neurons (Tsirka et al., 1997b), and surprisingly it is highly regulated in these cells, as its expression is greatly upregulated by stimulation of these cells with the glutamate analog kainic acid (Matsuoka et al., 1998). The role of plasminogen in the brain will be discussed in Section V,D in the context of both neuronal function and neurodegeneration.

2. tPA

Both of the plasminogen activators are the products of highly regulated genes. tPA is expressed in a very limited number of cell types, primarily endothelial cells, vascular smooth muscle cells, neurons, and cells derived from the neural crest, including glial cells and melanocytes. Its expression in the both the central and peripheral nervous systems is remarkably widespread and has been mapped both in rodents (Sappino et al., 1993a; Davies et al., 1998) and in humans (Teesalu et al., 2002). Endothelial cells are thought to be the main source of circulating tPA. In cultured endothelial cells, tPA synthesis is transcriptionally upregulated in response to fluid shear stress, thrombin, histamine, retinoic acid, VEGF, and sodium butyrate (Lansink and Kooistra, 1996; Kawai et al., 1996). The promoter region of the tPA gene varies quite significantly between species; for example, the human gene, but not those of mouse and rat, contains a TATA box, although a TATA-box-independent transcription of the human gene from a downstream initiation site has been observed after PMA stimulation (Henderson and Sleigh, 1992). These and other observations point to a variation in the regulation of tPA expression between species (Feng et al., 1990; Holmberg et al., 1995). tPA also appear to be translationally regulated, at least in the mouse, by an active adenylation–deadenylation process. This has been observed in oocytes (Huarte et al., 1987) and may explain the presence of tPA mRNA but not activity in some areas of the brain (Sappino et al., 1993).

Regulated secretion of tPA is another aspect of the control of tPA, as it is stored in some cell types, notably endothelial cells and neurons. Endothelial cells have been shown to acutely release tPA from preformed stores in a regulated manner both in vitro and in vivo. A rapid increase in circulating tPA is observed in response to adrenergic agents, exercise, shear stress, and vasopressin analogs (Stein et al., 1998). This increase is mainly attributed to release from endothelial cells in which tPA has been shown to be stored in Weibel–Palade bodies and released by exocytosis together with von Willebrand factor (Huber et al., 2002). tPA is also targeted to the regulated secretory pathway in neuroendocrine cells, in which it accumulates in catecholamine storage vesicles (Parmer et al., 1997). Its release from neuroendocrine chromaffin cells upon sympathoadrenal activation may contribute to acute increases in plasma tPA levels. tPA in neurons of the CNS is also stored in granules and can be released in a calcium-dependent manner by membrane depolarization (Gualandris et al., 1996).

3. uPA

uPA is expressed in many different cell types, including leukocytes, fibroblasts, keratinocytes, and epithelial and endothelial cells. The expression of uPA in these cells is often absent under normal physiological conditions, but is upregulated in pathological conditions, e.g., activated leukocytes, migrating keratinocytes, and most solid tumors, and its expression is under the control of diverse extracellular signals (reviewed by Irigoyen *et al.*, 1999). The promoter region of the uPA gene contains elements characteristic of both regulated and housekeeping genes. A TATA box, found in many regulated genes, is found 25 bp upstream of the transcription initiation site, but also present are GC-rich stretches in the first 200 bp upstream of the TATA-box, a characteristic of constitutively expressed genes. Consensus *cis*-acting elements present in the promoter region include those for transcription factors PEA3/AP-1, NF-κB, and a cAMP response element, and have been shown to be active in driving uPA transcription in response to tumor promoters, growth factors, cytokines, and hormones (Nerlov *et al.*, 1992; D'Orazio *et al.*, 1997). uPA expression is also influenced by mRNA turnover due to instability determinants in the 3′ untranslated region (Nanbu *et al.*, 1994). The induction of uPA expression can be downregulated by antiinflammatory glucocorticoids (Medcalf *et al.*, 1986), an effect which may be indirect as there appear to be no sites for the glucocorticoid receptor in the uPA promoter region. Interestingly, uPA expression is also induced by disruption of the cytoskeleton, via a *Ras*/ERK signaling pathway (Irigoyen *et al.*, 1997).

In many biological situations leukocytes are an important source of uPA activity. A number of studies have shown that uPA can be stored in these cells and secreted on cellular activation. In neutrophils uPA has been shown to be stored primarily in specific granules and also in easily mobilizable secretory vesicles, and uPA can be released by exocytosis induced by both strong and weak activators, e.g., TNFα and fMLP (Plesner *et al.*, 1994). Interestingly, uPAR is also found in both of these subcellular compartments, suggesting that the receptor may be occupied prior to exocytosis of these vesicles.

4. uPAR

uPAR is expressed by a range of cell types similar to those expressing uPA, in many cases leading to autocrine binding of uPA. The promoter region of the human uPAR gene lacks conventional TATA and CAAT boxes but contains a CpG-rich island and sequences related to consensus *cis*-acting elements for the AP-1, NF-κB, and Sp-1 transcription factors. Most of the promoter activity is contained within the first 180 bp upstream of the transcriptional start site (Soravia *et al.*, 1995). uPAR gene expression is influenced by tumor promoters, growth factors, cytokines, and hormones, indicating the potentially complex nature of uPAR gene regulation

(Lund *et al.*, 1991, 1995; Mandriota *et al.*, 1995), consistent with its putative role in diverse biological settings.

III. Cellular Binding Sites and the Regulation of the Plasminogen Activation System

The protease components of the plasminogen activation system are all soluble proteins and for many years were studied in that context, even in situations where cellular roles were implicated. However, the observation in the mid-1980s that uPA could bind specifically to the surface of a number of cell types (Stoppelli *et al.*, 1985; Vassalli *et al.*, 1985) and that plasminogen could do the same (Plow *et al.*, 1986) led to the demonstration that the cell surface was the preferential site for plasminogen activation (Ellis *et al.*, 1989; Stephens *et al.*, 1989). It has since been shown that the binding of uPA to its specific cellular receptor uPAR regulates the proteolytic activity of this system at many different levels, as well as linking this proteolytic activity to two other fundamental systems involved in the regulation of cellular behavior, namely signal transduction and cell adhesion. Although the activity of tPA is clearly regulated by fibrin, it, too, has been found to interact with cellular binding sites. A variety of such sites have been identified, all of which appear to be less specific than uPAR as they bind tPA with a lower affinity and also bind other ligands. Nevertheless, these sites may have an important function in regulating tPA activity in certain situations.

The parallels between the cellular regulation of plasminogen activation and the regulation of other serine proteases by similar mechanisms, e.g., the proteases involved in blood coagulation and complement activation, suggested that this may be a general mechanism for the regulation of protease activity. This has subsequently proved to be the case as other classes of proteases, many of which are also described in this book and include other membrane-associated soluble proteases as well as integral membrane proteases, are now known to be regulated by events at the cell surface.

The binding of plasminogen to the cell surface is essential for its high-efficiency activation by cell-associated plasminogen activators, as will described in detail later, yet the molecules responsible for this binding are not well defined. One reason for this is that they are almost certainly heterogeneous and, in contrast to uPAR and many other cell-surface proteins that are found at an abundance of 10^3–10^4 per cell, greater than 10^7 molecules of plasminogen can be bound to some cells types (Plow *et al.*, 1986). Consistent with this, a variety of molecules have been shown to bind plasminogen at the cell surface, including both proteins and nonproteins, such as gangliosides (Miles *et al.*, 1989). The proteins include α-enolase (Miles *et al.*, 1991), annexin II (Hajjar *et al.*, 1994) and its heterotetrameric form (Kassam *et al.*, 1998a), TIP49a (Hawley *et al.*, 2001), cytokeratin-8 (Hembrough *et al.*, 1995), and other proteins exposing a C-terminal Lys residue (Hawley *et al.*, 2000). Plasminogen appears to bind to all of these molecules via a similar mechanism involving the lysine binding sites in the KR modules of plasminogen, with K_d values of approximately 1–2 μM, and all of these interactions are competed similarly by various lysine analogs. A question that will be addressed in Section III,A,2 is, how can such a diverse group of proteins be involved in facilitating cell-surface plasminogen activation, or indeed are they involved?

A. Cellular Mechanisms Regulating uPA Function

1. uPAR

The specific cellular binding site for uPA, uPAR, is a GPI-anchored membrane protein (Ploug *et al.*, 1991). uPAR is composed of three extracellular domains which, although having less than 20% sequence identity, are homologous and belong to the module family known as Ly-6/uPAR (LU) (Ploug and Ellis, 1994). Although the three-dimensional structure of uPAR or any of its individual domains has yet to be solved, the structures of many single-domain LU proteins are known. The overall topology of the module is characterized by a three-finger fold with a central β-sheet projecting three loops from a globular core that contains a conserved scaffold of disulfide bonds. Mapping of various epitopes on uPAR by site-directed mutagenesis is consistent with its individual domains having this structure (Gårdsvoll *et al.*, 1999).

uPAR binds uPA with high-affinity (K_d 0.3 nM). This binding requires the intact multidomain structure of uPAR and cleavage at a protease-sensitive region between the first two domains results in a greater than 10^3-fold reduction in affinity (Ploug *et al.*, 1994). The molecular basis for this is not fully elucidated but may be due to a multidomain binding site or extensive interdomain interactions in uPAR. Studies using a decapeptide antagonist of uPA binding have shown that it can span the N- and C-terminal domains, compatible with both of these models (Ploug *et al.*, 1998). Interestingly, this antagonist of human uPAR is not an antagonist of the murine protein, thus mimicking the species specificity observed for the binding of uPA. Although uPA is the only high-affinity ligand for uPAR, it has been shown that the adhesive glycoprotein vitronectin can also bind with a K_d of approximately 0.1 μM (Okumura *et al.*, 2002). This binding occurs only with the "denatured" form of vitronectin that is found in the extracellular matrix and is mediated by the N-terminal somatamedin B domain of vitronectin, which also binds PAI-1. This may be of significance to the potential role of uPAR in cell adhesion, as will be discussed in detail in Section IV,B,1.

uPA binds to uPAR via its N-terminal EG-module. An important feature of this interaction is that the isolated EG module and the intact protein bind to uPAR equivalently, as determined by both equilibrium binding constants and kinetic rate constants (Ploug *et al.,* 1998). This strongly suggests that the other domains of uPA have no interaction with uPAR, consistent with the known structural and dynamic independence of the domains of uPA. An important corollary of this is that uPAR would not be expected to directly influence the catalytic properties of uPA. However, as described in detail later, uPAR strongly influences the activity of uPA, but this is due to the assembly of specific complexes with plasminogen on the cell surface and not to direct effects on uPA catalytic activity.

2. uPAR-Mediated Plasminogen Activation

Binding of pro-uPA to uPAR on the cell surface leads to a greatly increased rate of plasmin generation (Ellis *et al.,* 1989) and does so by maximizing the inherent efficiency of the pro-uPA/plasminogen reciprocal zymogen activation system. Both of the primary reactions of this system, pro-uPA activation by plasmin and plasminogen activation by uPA, are greatly accelerated. This is most easily demonstrated for plasminogen activation by uPAR-bound uPA, the efficiency of which is increased by up to 200-fold by a large reduction in the apparent K_m from greater than 20 μM to 0.1 μM (Ellis *et al.,* 1991). This decrease in K_m is of particular importance as it falls well below the physiological concentration of plasminogen (2 μM in plasma) and therefore further favors the cell-associated reaction. The activation of uPAR-bound pro-uPA is also accelerated by at least 50-fold.

Both of these kinetic effects are achieved without directly influencing the catalytic activity of either of the proteases. This is highlighted by two observations. Firstly, the overall increase in plasmin generation is absolutely dependent on the cellular binding of plasminogen. No increase in plasmin generation is observed in the presence of lysine analogs which compete the cellular binding of plasminogen (Ellis *et al.,* 1989). Secondly, the activity of the plasminogen activation system is not enhanced by non-membrane-associated forms of uPAR, either purified native uPAR (Ellis *et al.,* 1991) or recombinant C-terminally truncated uPAR (Ellis, 1996). The large acceleration of plasmin generation therefore requires concomitant binding of pro-uPA and plasminogen and appears to be due to the cell-surface binding sites providing a catalytic template for the assembly of the reactants.

Despite the importance of plasminogen binding and the variety of proteins that contribute to the high plasminogen-binding capacity of the cell surface, the precise role of these various molecules in plasminogen activation is not yet clear. For example, they could contribute equally to plasminogen activation, or a discrete subset could be functionally involved. The simplest interpretation is that the high surface density of plasminogen favors catalysis, and mechanistic models based on this assumption have been made (Longstaff *et al.,* 1999). However, experimental evidence is not compatible with such models but instead supports a model involving specific stoichiometric interactions between plasminogen and the uPA–uPAR complex. This evidence includes the following: (i) the effect of uPAR on plasminogen activation can be quantitatively mimicked in 1:1 complexes of uPA and plasminogen assembled on a monoclonal antibody (Ellis and Danø, 1993a). (ii) Cell-surface plasminogen activation can be specifically blocked by certain low-molecular-weight anionic compounds in the absence of any effect on the cellular binding of uPA or plasminogen or on the catalytic activity of uPA or plasmin (Ellis and Danø, 1993b). (iii) Plasminogen can bind to uPA via an "exosite" interaction, i.e., independent of the active site, with an affinity (K_d 50 nM as determined by surface plasmon resonance) that is close to the K_m for plasminogen activation (100 nM) and much higher than the overall affinity for the cellular binding of plasminogen (K_d 1–2 μM) (Ellis et al., 1999). Together, these results suggest that the exosite interaction between uPA and plasminogen is necessary for the assembly of plasminogen activation complexes on the cell surface, with the interaction possibly being responsible for orientating the membrane-immobilized protease/zymogen components. A provocative speculation that can be made from these observations is that uPAR-bound uPA itself constitutes the functional cellular binding site for plasminogen. Interestingly, circumstantial evidence supports this hypothesis as it has been observed that a proportion of plasminogen co-localizes with uPA on a breast cancer cell line (Andronicos and Ranson, 2001), although there are several alternative interpretations of this observation. Surprisingly, the role of uPAR in plasminogen activation appears to be relatively passive and limited to localizing uPA at the cell surface, as a directly GPI-anchored mutant of uPA displays an activity on the cell surface indistinguishable from that of uPAR-bound uPA (Lee et al., 1994; Vines et al., 2000).

3. Initiation of uPAR-Mediated Plasminogen Activation

The catalytically favorable environment provided by assembly of the plasminogen activation complexes described earlier provides a mechanism for the amplification and propagation of an initial proteolytic stimulus by reciprocal zymogen activation, in which plasmin is responsible for the activation of pro-uPA. But the question remains of what is that initial stimulus, i.e., what initiates plasminogen activation? A variety of proteases capable of activating pro-uPA have been mentioned (Section II,B,4.), any of which could fulfill this function. However, the most plausible mechanism is initiation by the intrinsic activity of pro-uPA. In situations where the reactions of this reciprocal zymogen activation system are kinetically favored, e.g., in the presence of uPAR on the cell surface, the low intrinsic activity of pro-uPA will be favored to an equal extent. This is difficult to demonstrate in a cellular system, where exogenous proteases may be present, but it has been demonstrated in an in vitro model of receptor-mediated plasminogen activation, a model which also showed that any discrete level of intrinsic activity will lead to efficient activation of the system (Ellis and Danø, 1993a). Therefore pro-uPA itself appears to be the initiator of the system, generating the initial plasmin activity that will then proteolytically activate pro-uPA (i.e., the initiation phase), with further plasmin and uPA being generated by reciprocal zymogen activation (i.e., the propagation phase). This means that plasmin is the principal activator of pro-uPA in the context of both initiating and then propagating the activity of this cell-surface plasminogen activation system, although it cannot be excluded that exogenous proteases may have an additional influence in certain pathological situations.

4. Regulation of uPAR-Mediated Plasminogen Activation by Serpins

The serpin inhibitors of the plasminogen activation system play critical, and distinct, roles in regulating its activity on the cell surface. The plasminogen activator inhibitors PAI-1 and PAI-2 inhibit uPAR-bound uPA with kinetics almost equivalent to those with the free protease (Ellis *et al.*, 1990) and uPA/PAI-1 complexes bind efficiently to uPAR (Cubellis *et al.*, 1990). Therefore, these proteins are efficient inhibitors of this proteolytic system and act to suppress its activity. By contrast, plasmin generated on the cell surface is almost completely resistant to inhibition by α_2-antiplasmin (Ellis *et al.*, 1991; Hall *et al.*, 1991). However, plasmin is slowly inhibited, but this is due to its dissociation from the cell surface, which occurs quite readily because of its relatively low binding affinity. Therefore, this appears to be a mechanism to restrict the activity of generated plasmin to the cell surface or its immediate environment, but which still allows a dynamic turnover of plasmin so that its activity will not persist on the cell surface. These and other interactions and reactions involved in the cell-surface plasminogen activation system are shown in Fig. 3 (see color insert).

5. LRP and uPAR recycling

The final fate of uPAR-bound uPA once activated and subsequently inhibited by PAI-1 is removal from the cell surface by internalization. The molecule responsible for this internalization is the multiligand clearance receptor LRP (low-density lipoprotein-related receptor). LRP has been shown to internalize uPA/PAI-1 complexes leading to their subsequent degradation (Nykjær *et al.*, 1992), and this internalization is promoted when these complexes are bound to uPAR, which is also then internalized (Conese *et al.*, 1995). However, it has been demonstrated that uPAR is not degraded, but instead recycled to the plasma membrane (Nykjaer *et al.*, 1997). This recycling of uPAR may be a mechanism, in some respects analogous to that described earlier for the inhibition of plasmin by α_2-antiplasmin, to allow dynamic turnover of the components of this proteolytic system on the cell surface.

B. Cellular Mechanisms Regulating tPA Function

Subsequent to the discovery of uPAR and its role in regulating plasminogen activation, there has been significant interest in identifying cellular binding sites for tPA.

These studies originally concentrated on endothelial cells, on the basis of a presumed hemostatic role, but have since been extended to other cell types. A number of binding sites for tPA have been identified, although most do not have the high level of specificity and affinity displayed by uPAR. This may suggest that these interactions are relatively unimportant, but the existence of non-fibrin-dependent roles for tPA is supported by a number of considerations. It has been observed that the fibrin stimulation of tPA is not optimal and it has proven relatively easy to design mutants with a superior fibrin specificity, and mutations at multiple sites in the protease domain have this effect (Paoni et al., 1993). This may suggest that tPA has evolved with suboptimal fibrin specificity to enable other functions. This certainly appears to be the case in the brain, an organ normally devoid of fibrinogen or fibrin, but where tPA is expressed quite extensively and has been shown to activate locally synthesized plasminogen (Tsirka et al., 1997a).

1. Endothelial Binding Sites for tPA

tPA has been shown to bind specifically to cultured endothelial cells. A very high-affinity component of this binding proved to be due to matrix-associated PAI-1 (Russell et al., 1990), but a lower affinity binding was shown to be due to the presence of the Ca^{2+}/phospholipid binding protein annexin II, peripherally associated with the plasma membrane (Hajjar et al., 1994, 1996). This binding required generation of an internal C-terminal Lys residue in annexin II by plasmin cleavage, and led to an approximate 20-fold increase in plasminogen activation (Cesarman et al., 1994). This Lys-dependent binding, presumably via KR-2 of tPA, does not discriminate between tPA and plasminogen, and annexin II binds both. Annexin II has also been shown to stimulate plasminogen activation by another mechanism, involving formation of a heterotetrameric complex consisting of two molecules of annexin II and two molecules of a p11 subunit, which has a native C-terminal Lys residue (Kassam et al., 1998a). Plasminogen binding to p11 in the tetramer leads to a greater stimulation of plasminogen activation than annexin II alone (Kassam et al., 1998b).

Annexin II is not endothelial cell-specific and has also been shown to mediate plasminogen activation on macrophages (Falcone et al., 2001). However, as these cells do not express tPA, annexin II is likely to be acting solely as a plasminogen binding site in these cells. Differentiation and neurite growth in the neuroendocrine cell line PC-12, a plasmin-dependent process, is also influenced by annexin II (Jacovina et al., 2001). Interestingly, the activation of microglial cells by tPA, a process which is independent of tPA proteolytic activity, has also been suggested to involve annexin II (Siao and Tsirka, 2002). In addition to a variety of other non-proteolytically related functions, annexin II has also been shown to bind procathepsin B on certain tumor cell lines (Mai et al., 2000). The potential significance of this is discussed in Chapter 10 by Cavallo-Medved & Sloane.

2. Vascular Smooth Muscle Cells

Vascular smooth muscle cells are another cell type that expresses tPA, particularly in response to vascular injury (Clowes *et al.,* 1990), suggesting that tPA is involved in the migration and phenotypic changes of these cells that lead to intimal hyperplasia and thickening of the arterial wall. These cells have been shown to specifically bind tPA with a greater than 100-fold increase in plasminogen activation (Ellis and Whawell, 1997). As with the uPAR system this effect was absolutely dependent on the concomitant cellular binding of plasminogen. Plasminogen activation on these cells is further favored as bound tPA is protected from inhibition by PAI-1, with a 30-fold reduction in inhibition kinetics (Werner *et al.,* 1999).

Both of these effects were found to correlate with a conformational change in the protease domain of tPA, and complexes of tPA with both PAI-1 and certain low-molecular-weight inhibitors are unable to interact with the cellular binding site, indicating that this domain is intimately involved in the binding interaction. Consistent with these observations, and in contrast to the situation with endothelial cells, this binding is not mediated by annexin II and requires neither C-terminal Lys residues or KR-2 of tPA (Ellis and Whawell, 1997). The protein responsible for this binding has been identified as a 63-kDa type-II transmembrane protein, although this has yet to be fully characterized (T. Razzaq and V. Ellis, unpublished observations).

The protection of tPA bound to these cells from inhibition by PAI-1 possibly has a greater functional significance than might otherwise be the case when it is considered that uPA bound to uPAR, on these or other cells, is not similarly protected (Ellis *et al.,* 1990; Werner *et al.,* 1999). Therefore, at least on vascular smooth muscle cells, tPA and uPA may have either differential or complementary roles in plasminogen activation dependent, respectively, on the presence or absence of PAI-1. These and other comparisons between the functional regulation of the two proteases are summarized in Table I.

Table I   Comparison of Functional Regulation of Plasminogen Activators by Pericellular Binding Sites

	uPAR	tPAR[a]	apo-PrP
Stimulation of plasminogen activation	√	√	√
Dependence on cellular binding of plasminogen	√	√	n.d.[b]
Dependent on specific interaction with plasminogen	√	—	√
Direct effect on catalytic activity	—	√	√
Protection from inhibition	—	√	n.d.

[a] tPAR, specific tPA receptor on vascular smooth muscle cells.
[b] n.d., not determined.

3. Prion Protein

The brain has been mentioned in Section II,D,2 as an organ with an unexpectedly widespread expression of tPA, and in some regions it is the most abundant protease (Davies *et al.,* 1998). Various experimental models have shown that it has an important functional role there, both in normal brain function (Seeds *et al.,* 1995, 1999; Huang *et al.,* 1996) and in neurodegenerative events (Tsirka *et al.,* 1995), and ECM targets for generated plasmin have been identified (Chen and Strickland, 1997). How tPA is regulated in these situations is not known, but it is likely that its activity is stimulated in these tissues and that this may involve cell-surface mechanisms. It has been known for many years that tPA binds specifically to the surface of neuronal cells (Verrall and Seeds, 1989), although the binding site has yet to be identified. A potential candidate for a functional tPA binding site on neuronal cells is the prion protein (PrP). This is a glycolipid-anchored membrane protein, which subsequent to a conformational transition becomes the infectious agent in the group of diseases known as the transmissible spongiform encephalopathies or prion diseases (Prusiner, 1998). It has been shown that a recombinant soluble form of PrP with the disease-associated conformation can specifically bind tPA with a K_d below 2.5 nM (Ellis *et al.,* 2002). This is a much higher affinity than observed for binding to fibrin or the cellular binding sites described earlier, suggesting an important role for this interaction, and in the presence of plasminogen a greater than 300-fold increase in activation is observed. This effect requires the concomitant binding of plasminogen to what appears to be an independent site on the same molecule of PrP. The latter observation is consistent with the report that plasminogen can bind to the disease-associated form of PrP isolated from the brains of infected animals (Fischer *et al.,* 2000). Although it has yet to be demonstrated whether the membrane-associated form of PrP can also mediate these effects on plasminogen activation, it seems likely that PrP-stimulated plasminogen activation may play a role in the pathogenesis of the prion diseases, a hypothesis that is further discussed in Section V,D.

IV. uPAR as a Regulator of Cell Migration

A. Proteolysis and Cell Migration

The fundamental basis for the study of proteases in invasive cell migration is their role in proteolytic remodeling of the extracellular matrix. Although these proteases are now known to have many other functions that affect cellular behavior, important questions remain regarding this basic proteolytic function. Central among these is the issue of *how much* proteolysis is necessary for migration and invasion. Although in simplistic terms the ECM, particularly in the form of basement membranes, is seen as a barrier to cell movement, it is also clear that excessive proteolysis will remove the substratum necessary for cell motility. Therefore it would appear that very precisely controlled levels of proteolytic activity are required. The multiplicity of mechanisms described here for the functional regulation of the plasminogen activation system on the cell surface in all probability contribute to this precise control. This contention would appear to be borne out by *in vivo* observations, and highlighted in particular by the seemingly paradoxical role of the inhibitor PAI-1 in angiogenesis *in vivo*. Angiogenesis is a complex process with a requirement for proteolytic activity; however, in a model of tumor angiogenesis it has been observed that PAI-1-deficient mice have decreased rather than increased local invasion and vascularization of the tumor (Bajou *et al.*, 1998). The requirement for PAI-1 has been shown to be due to its protease inhibitory activity, as the phenotype of PAI-deficient mice can only be rescued by adenoviral expression of inhibitory-competent PAI-1 (Bajou *et al.*, 2001). Although seemingly obvious, this is in fact an important confirmation, as *in vitro* PAI-1 is also implicated in cell adhesion independently of its protease inhibitory activity, as will be discussed later. The role of the plasminogen activation system in this model appears to have close parallels in human cancer, as high levels of both uPA and PAI-1 are correlated with poor clinical outcome in a variety of tumors (see Section V,B).

B. uPAR and Cell Adhesion

Cell migration is a process that requires constant modulation of the adhesive properties of the cell, and this is achieved in a directed manner through complex connections between the ECM and the actin cytoskeleton mediated by transmembrane adhesion molecules such as integrins. The question arises: can migrating cells utilize the activity of the plasminogen activation system in a directed manner in the absence of a transmembrane linkage between the glycolipid-anchored protein uPAR and the cytoskeleton? The polarization of uPAR to the leading edge of migrating cells was first observed in monocytes (Estreicher *et al.*, 1990) and has since been demonstrated in a variety of other cell types, such as vascular smooth muscle cells (Okada *et al.*, 1995). In stationary cells uPAR is found both in integrin-rich focal adhesions (Pollanen *et al.*, 1988) and in lipid raft membrane microdomains including caveolae (Stahl and Mueller, 1995). The presence of uPAR in focal adhesions and its polarization in migrating cells are thought to be dependent on interactions with other proteins and a number of potential mechanisms have emerged.

1. uPAR and Vitronectin

Central to these mechanisms is the adhesive protein vitronectin. The first observation of uPA-dependent cell adhesion was during PMA-stimulated differentiation of myeloid cell lines (Nusrat and Chapman, 1991). This adhesion was later shown to involve interactions between uPAR and vitronectin, which were promoted by uPA binding (Wei *et al.*, 1994). Interactions between these proteins have been confirmed in purified systems and shown to require the intact three-domain form of uPAR (Høyer-Hansen *et al.*, 1997) and the N-terminal somatomedin-B domain of vitronectin (Okumura *et al.*, 2002). The direct interaction between uPAR and vitronectin has also been shown to mediate a reorganization of the actin cytoskeleton in murine fibroblasts overexpressing human uPAR (Kjøller and Hall, 2001). This effect, which led to an increase in cell motility, was not dependent on uPA binding (due to the species specificity of the uPA/uPAR interaction) or the interaction of vitronectin with integrins. The cytoskeletal reorganization was shown to be dependent on activation Rac, a member of the Rho family of small GTPases. Rac is known to regulate the assembly of lamellipodia and cell ruffles and have a critical role in regulating protrusive activity at the leading edge of cells (Nobes and Hall, 1995). Although uPAR-mediated signaling to the cytoskeleton via Rac may be a direct consequence of the adhesive properties of uPAR, it is also possible that these effects are independent. The mechanisms by which uPAR induces this signaling are as yet unresolved, but will be discussed in Section IV,C in the context of other proposed signaling activities of uPAR.

Another vitronectin-binding protein, PAI-1, has also been implicated in the regulation of cell adhesion. PAI-1 binds with high affinity to the somatomedin-B domain of vitronectin (Seiffert *et al.*, 1994), i.e., the same domain that binds uPAR, and the majority of PAI-1 in plasma occurs in complex with vitronectin, stabilizing its inhibitory activity. This has led to the suggestion that PAI-1 may regulate cell adhesion by competing with uPAR for binding to matrix-associated vitronectin (Deng *et al.*, 1996). An alternative mechanism by which PAI-1 influences cell adhesion has also been proposed in which the inhibitor masks the binding site on vitronectin for the integrin receptor $\alpha_v\beta_3$ and, as PAI-1 is released from vitronectin when it reacts with proteases, the presence of uPA activity would unmask these otherwise cryptic integrin binding sites (Stefansson and Lawrence, 1996).

2. uPAR and Integrins

Although the effect of uPAR on cell adhesion discussed above have all been shown to be integrin-independent, interactions between uPAR and integrins have also been demonstrated and proposed to have important roles in cell motility. The first evidence for these interactions was the observation of clustering of uPAR and the β_2 integrin Mac-1 ($\alpha_M\beta_2$ or CD11b/CD18) on neutrophils (Xue *et al.*, 1994), which was later shown to be related to ability of these cells to respond to chemotactic stimuli (Gyetko *et al.*, 1995). There are conflicting data regarding the potential effect of such interactions on integrin function. Downregulation of uPAR expression in monocytes has been shown to lead to a reduction in Mac-1-mediated adhesion, suggesting that uPAR facilitates integrin function (Sitrin *et al.*, 1996). However, uPAR has also been shown to inhibit the function of both β_1 and β_2 integrins (Wei *et al.*, 1996). It is possible that variables such as the activation state of the integrins and the presence or absence of bound uPA may influence these effects. Many studies have also demonstrated correlations between the presence of uPA/uPAR complex and integrin-mediated cell migration. An example of this is the observation that upregulation of uPA and uPAR expression leads to an increase in $\alpha_V\beta_5$-dependent, but not $\alpha_V\beta_3$-dependent, migration on vitronectin (Yebra *et al.,* 1996). This effect on migration, but not adhesion, was dependent on both the binding and activity of uPA. Other studies have also demonstrated a uPA-dependent interaction between uPAR and $\alpha_V\beta_5$ leading to cytoskeletal rearrangements mediated by the protein kinase C pathway, whereas interaction of vitronectin with $\alpha_V\beta_5$ led to protein kinase C-independent rearrangements, again suggesting an alteration in integrin function on association with uPAR (Carriero *et al.,* 1999).

Compared to the data just discussed there are very few examples of direct demonstrations of physical interactions between uPAR and integrins, and those that do exist are either in cells overexpressing uPAR or interactions between integrins and recombinant soluble uPAR. An interesting example of the latter is the observation of the binding of soluble uPAR to $\alpha_4\beta_1$, $\alpha_6\beta_1$, $\alpha_9\beta_1$, and $\alpha_V\beta_3$ integrins transfected into CHO cells (Tarui *et al.,* 2000). The characteristics of some of these interactions suggested that uPAR might be a ligand for integrins as well as, or in addition to, being an integrin-associated protein, i.e., uPAR may be engaged in both *cis* and *trans* interactions with integrins. Consistent with this, interactions between GPI-anchored uPAR and integrins on apposing cells were also demonstrated. The difficulty in demonstrating such interactions directly without overexpression of the components may be that the interactions are transitory, at least in motile cells. It has been shown by FRET techniques that uPAR interactions with β_2 integrins oscillate in migrating neutrophils (Kindzelskii *et al.,* 1996, 1997), and such transitory interactions might be necessary for the dynamic process of cell migration.

The role of uPA in many of the scenarios presented here for the involvement of uPAR in cell migration is not fully resolved, but it is clear that in some circumstances both cell adhesion and migration are affected by the binding of uPA. However, it is important to return to the original question, which was, how do migrating cells utilize the activity of the plasminogen activation system in a directed manner? These various mechanisms clearly have the potential to direct uPAR to relevant areas of the cell surface during migration, but do they also dynamically affect the activity of the proteolytic system? There is evidence that they can, as vitronectin-dependent relocalization of uPAR to focal adhesions has been reported to reduce cell-surface uPA activity (Wilcox *et al.,* 1996), and it has been observed that certain members of the tetraspan family of integrin-interacting proteins can influence the binding of uPA to uPAR (R. Bass, F. Berditchevski and V. Ellis, unpublished observations).

C. uPAR Signaling

The binding of uPA to uPAR has been shown to activate a variety of different signaling pathways in various cell types. In almost all cases these effects have been shown to be independent of uPA catalytic activity, by using either inactivated uPA or isolated receptor-binding, N-terminal domains of uPA. Examples of this are protein kinase C-mediated phosphorylation in epithelial cells (Busso *et al.*, 1994), activation of the MAP kinases ERK-1 and ERK-2 in the breast cancer cell-line MCF-7 (Nguyen *et al.*, 1998), activation of the Jak/STAT signaling pathway in vascular smooth muscle cells (Dumler *et al.*, 1998), and activation of nonreceptor tyrosine kinases such as *Hck* (Resnati *et al.*, 1996). It is thought that these various signals are transduced by uPAR, in the absence of a transmembrane or cytoplasmic region, via interactions with transmembrane "adapter" proteins. Identification of such proteins has proven to be difficult, but two potential mechanisms have emerged.

The interactions between uPAR and integrins, discussed earlier, represent a potential mechanism for transducing uPA-dependent signals. Although they are not signaling receptors in their own right, i.e., they do not have tyrosine kinase activity and are not coupled to G-proteins, integrins are known to transduce signals from the ECM. They do this by both direct and indirect interactions with a number of major signaling pathways, in particular the FAK and *Fyn/Shc* pathways (reviewed in Schwartz, 2001). Integrins can therefore activate some, if not all, of the signaling pathways implicated in uPAR-dependent signaling. Strong evidence for uPAR signaling via integrins has been obtained from the observation that tumor dormancy is induced in Hep 3 carcinoma cells *in vivo* by downregulation of uPAR. This has been shown to be due to a reduction in uPAR interactions with $\alpha_5\beta_1$ integrin and a consequent reduction ERK activation status signaled through FAK and *Src* (Aguirre Ghiso *et al.*, 1999; Aguirre Ghiso, 2002). In these experiments reexpression of uPAR led to a restoration of FAK phosphorylation and cell growth. uPA strongly stimulated ERK activation in this model, presumably by increasing the avidity of uPAR for $\alpha_5\beta_1$.

A second potential mechanism has also been identified. The identification of this mechanism stems from the observation that not only uPA, but also recombinant soluble uPAR, is chemotactic for monocytes (Resnati *et al.*, 1996). However, this effect of soluble uPAR is dependent on cleavage of the previously discussed protease-sensitive region between its first and second domains. This was subsequently shown to be due to exposure of a peptide sequence in this linker region, and corresponding synthetic peptides reproduce the chemotactic effect (Fazioli *et al.*, 1997). These signals have been shown to be transduced in leukocytes by the G protein-coupled chemotactic receptor FPRL1/LXA4R, as desensitization of this receptor desensitizes uPAR signaling and a physical association between the two proteins could be demonstrated (Resnati *et al.*, 2002). Therefore it appears that both proteolytic and nonproteolytic mechanisms can lead to uPAR-mediated activation of this G-protein-coupled receptor, which is also expressed in other cell types and may therefore be more widely involved in uPA/uPAR signaling.

The question of how uPA binding leads to activating associations with adapter proteins remains. A potential common pathway for activation of both integrins and G-protein-coupled receptors, such as FPRL1/LXA4R, may involve a conformational change in uPAR upon binding of uPA. The observation that both uPA binding and proteolytic cleavage in the uPAR linker region lead to activation of FPRL1/LXA4R suggests that uPA binding may expose the peptide sequence necessary for receptor activation. Therefore ligand-induced conformational changes in uPAR may also influence uPAR/integrin interactions and again lead to signaling. Biochemical studies suggest that uPAR undergoes conformational changes upon ligand binding and are therefore compatible with such a model (Ploug and Ellis, 1994; Higazi et al., 1996).

Redistribution of uPAR on the cell surface could also be involved in promoting its associations with transmembrane adapter proteins. uPAR has been reported to be present in caveolae (Stahl and Mueller, 1995), a specialized cholesterol and glycosphingolipid-rich membrane invagination characterized by the presence of the structural protein caveolin-1. The potential role of caveolae in regulating protease function, due to the presence of multiple proteases in these structures, including cathepsin B, MMP-2, and MT1-MMP, is discussed extensively in Chapter 10 by Cavallo-Medved and Sloane. However, these structures are widely considered to be platforms for the assembly of signal-transducing complexes (Simons and Toomre, 2000). Therefore clustering of uPAR in caveolae may lead to activation of resident signaling molecules which include *Src* family kinases and heterotrimeric G-proteins. Caveolin-1 itself has been shown to be required for integrin-dependent signaling to the ERK pathway by direct interactions with β_1 integrins and the *Src*-family kinase *Fyn* in WI-38 lung fibroblasts (Wary et al., 1998). Consistent with this, antisense depletion of caveolin-1 in HEK293 cells disrupted complexes between *Src* and β_1 integrins, but these complexes could be stabilized by uPAR (Wei et al., 1999). These data suggest that uPAR regulates integrin signaling in caveolae, but the role of uPA in this process has yet to be elucidated.

V. Biological and Pathological Roles

The proteases of the plasminogen activation system are implicated in wide range of physiological and pathological processes, some of which have already been mentioned in the context of the mechanisms regulating the activity of this proteolytic system. There are too many of these processes to discuss in detail here, so the focus will be on four areas where there is strong evidence from more than one experimental system for the functional involvement of the plasminogen activation system. These are wound healing, cancer, cardiovascular disease, and neurodegeneration.

A. Wound Healing

Skin wound healing has been adopted as a model for the study of tissue remodeling and repair that recapitulates many of the processes involved in pathological situations, including tumor invasion. Histological studies of wound healing in mice have shown that both uPA and uPAR are expressed by keratinocytes at the leading edge of regenerating epidermis but are absent from adjacent normal skin (Rømer et al., 1994), suggesting that the generation of plasmin activity is involved in the migration of these keratinocytes into the wound field. Similar experiments in plasminogen-deficient mice confirm this involvement, as wound healing is severely impaired in these animals (Rømer et al., 1996). This plasminogen dependency of keratinocyte migration could involve the action of plasmin on a variety of its potential substrates, but the observation that combined deficiency of plasminogen and fibrinogen resulted in "phenotypic rescue" and the restoration of macroscopically normal wound healing (Bugge et al., 1996b) points to fibrin being the major substrate for plasmin in this setting. Therefore, in the absence of plasmin activity to solubilize fibrin, cellular migration into, and reorganization of, the fibrin-rich wound matrix is severely impeded. However, despite the impaired wound repair in plasminogen-deficient mice the wounds do eventually heal with an outcome that is generally comparable to that of control mice, suggesting that fibrin is eventually cleared in the absence of plasmin activity. Experiments in which these mice were treated systemically with a broad-spectrum MMP inhibitor have shown that MMPs are responsible for this residual activity as wound healing was completely abolished in these animals (Lund et al., 1999). The MMP(s) involved in this fibrin degradation are not known, but membrane-type MMPs, and in particular MT1-MMP, have been shown to have fibrinolytic activity in other situations (Hiraoka et al., 1998). What should be emphasized here is that these observations suggest that, at least in wound healing, these two proteolytic systems appear to be acting independently and plasmin is not necessary for MMP activation, although if MT1-MMP is indeed involved this is not surprising as this protease is thought to be activated intracellularly by a furin-like enzyme.

These observations in fibrin-rich healing wounds do not exclude the possibility that nonfibrin substrates, both matrix and nonmatrix, are the major substrates in other situations. Nevertheless, it should be noted that fibrin may also be an important substrate in other matrices, and for example is a common component of tumor stroma (Dvorak, 1986) and atherosclerotic lesions (Smith and Thompson, 1994).

B. Cancer

Early studies of transplanted animal tumors, such as the Lewis lung carcinoma model, showed that uPA is expressed by the cancer cells in the tumor, with the highest expression associated with cells in the invasive areas (Skriver et al., 1984). uPAR has a similar expression pattern, but PAI-1, which is also is also expressed in the cancer cells, is lowest in the invasive areas (Kristensen et al., 1990). These and other observations are consistent with a relatively simple model in which high uPA activity and low PAI-1 contribute to a proteolytic environment at the invasive regions of tumors. However, later studies in various human cancers have shown a much more complex situation in many cases. These studies have demonstrated expression of the components of this system in cells which form the stroma of the tumor, including fibroblasts, macrophages, and endothelial cells. An example of this is in colon cancer where virtually all uPA is expressed by fibroblast-like cells in the tumor stroma but uPAR is expressed by adjacent cancer cells (Pyke et al., 1991b), suggesting paracrine binding of uPA to the cancer cells to promote proteolysis. In these tumors the expression of PAI-1 is restricted to endothelial cells in the tumor stroma and was not detected in surrounding normal vasculature (Pyke et al., 1991a), where, as previously discussed, it appears to play a permissive role in tumor angiogenesis. This collaboration between different cell types is not confined to colon cancer and has also been observed for other proteolytic systems, in particular the MMPs (reviewed in MacDougall and Matrisian, 1995; Danø et al., 1999).

The components of the plasminogen activation system have proven to be robust prognostic markers for various types of cancer (Duffy, 2002), strongly suggesting that the detection of their expression in the histological studies reflects a functional role in tumor progression. An excellent example of the prognostic value of these components is the recent meta-analysis of uPA and PAI-1 levels in primary breast cancer showing that high levels of both uPA and PAI-1 are strong and independent prognostic indicators of both poor overall and risk-free survival (Look et al., 2002). The apparent paradox of high levels of both enzyme and inhibitor being risk factors has been mentioned in Section IV,A in the context of tumor vascularization, as PAI-1 is required for this process in mouse models (Bajou et al., 1998), and likely reflects the need for carefully regulated proteolysis in this angiogenic process.

The evidence that uPAR plays an important role in mediating plasminogen activation in cancer is mainly derived from experimental tumor models in which the binding of uPA to uPAR has been reduced. Metastasis of PC3 prostate carcinoma cells transfected with a catalytically inactive, but uPAR-binding, mutant of uPA was dramatically reduced in nude mice, with growth of the primary tumor unaffected (Crowley et al., 1993), and invasive primary tumor growth was substantially suppressed in a B16 melanoma model in syngeneic mice by treatment with the noncatalytic N-terminal fragment of murine uPA (Min et al., 1996). Using the chick chorioallantoic membrane (CAM) model of cellular invasion, antisense inhibition of uPAR expression in squamous carcinoma cells showed that invasiveness was directly proportional to the level of uPAR and that cells with the lowest uPAR expression were not tumorigenic on CAMs of chick embryos (Kook et al., 1994). Surprisingly it has been more difficult to obtain strong evidence for the involvement of the plasminogen activation system in tumors in gene-ablated mice; for example, both primary tumor growth and metastasis of Lewis lung carcinoma were found to be only moderately affected in plasminogen-deficient mice (Bugge et al., 1998). However, genetically induced breast tumors in these mice displayed a significant suppression of both the number and total burden of lung metastases, while the growth of the primary tumor was unaffected (Bugge et al.,

1997). Although the pharmaceutical industry has developed effective inhibitors of the plasminogen activation system for use in patients with cancer, their efficacy has yet to be determined in clinical trials.

C. Cardiovascular Disease and Atherosclerosis

Atherosclerosis is another complex disease in which proteases are implicated, potentially having roles both in the development of the atherosclerotic plaque and also in its clinical complications, including the rupture of unstable plaques and aneurysm formation due to thinning of the vessel wall (Lijnen, 2002). As with the situation described for cancer, evidence for the involvement of the plasminogen activation system was initially gained from the histological analysis of the components and subsequently further evidence was gained experimentally in various animal models.

Both of the plasminogen activators have been found to be upregulated in human advanced atherosclerotic plaques, with tPA increased in vascular smooth muscle cells present in the intima and in macrophage-derived foam cells, and uPA also increased in vascular smooth muscle cells and particularly in macrophages located at the edges of the necrotic core (Lupu et al., 1995). The expression of both activators has also been observed in experimental arterial injury in rats, where they were found to be differentially regulated in vascular smooth muscle cells as uPA was expressed during the early proliferative phase of the response and tPA at the later phase when these cells migrate into the intima (Clowes et al., 1990).

Experiments in gene-ablated mice have contributed greatly to the understanding of the role of these proteases in atherosclerosis. Initial observations using models of vascular injury showed that arterial repair and formation of a neointima, largely involving smooth muscle cell migration, are severely compromised in uPA-deficient, but not tPA-deficient, mice (Carmeliet et al., 1997a), and PAI-1 deficiency was found to have the opposite effect (Carmeliet et al., 1997c). A model more closely reflecting human disease, using apolipoprotein E-deficient mice genetically susceptible to atherosclerosis, confirmed these findings. Although early development of the lesions was unaffected in uPA-deficient mice, protection against destruction of the underlying media was observed, as well as a reduction in subsequent erosion, smooth muscle cell necrosis, and resultant aneurysm formation (Carmeliet et al., 1997b; Heymans et al., 1999). Studies in this more complex model also revealed a counterintuitive effect of PAI-1 deficiency, somewhat analogous to that seen in tumor angiogenesis, as plaque growth was found to be promoted in advanced lesions due to an accumulation of matrix, possibly due to an increased level of TGF-β activation (Luttun et al., 2002).

A particularly interesting aspect of these studies is that they provide strong evidence for the *in vivo* activation of MMPs by plasmin, thus supporting the evidence gained from biochemical and cell biological studies. The observed vascular erosion observed in wild-type, but not uPA-deficient, mice involves degradation of elastin and collagen. These are substrates for MMPs but not plasmin, and macrophages from these mice were found to be unable to activate MMP-3, MMP-9, MMP-12, and MMP-13 (Carmeliet *et al.,* 1997b). Subsequent observations in MMP-3 (stromelysin-1)-deficient mice are consistent with this interpretation (Heymans *et al.,* 1999).

The complexity of the role of the plasminogen activation system in disease is further highlighted by these studies when cardiac rupture, a fatal complication of acute myocardial infarction, is considered. uPA-deficient mice are completely protected from the occurrence of this event (although in this case MMP-3-deficient mice were not), but opposing this the infarcted areas failed to heal, predisposing them to subsequent cardiac failure (Heymans *et al.,* 1999). Observations such as these, in complex disease processes, again underscore the multiplicity of cell types that can utilize this proteolytic system as well as the multiplicity of its molecular targets.

D. Neurodegeneration

Recent observations in gene-ablated mice have confirmed earlier speculation that tPA has an important role in the central nervous system, in both normal function and neurodegeneration. These studies showed that tPA is expressed in neuroectoderm-derived tissues during rodent development (Friedman and Seeds, 1994) and widely detected in the adult brain, predominantly in the hippocampus, hypothalamus, cerebellum, and amygdala (Sappino *et al.,* 1993). It was also observed to be induced as an immediate-early gene during a screen of genes upregulated during seizure, kindling, and long-term potentiation, suggesting that increased expression of tPA may play a role in the structural changes that accompany activity-dependent plasticity (Qian *et al.,* 1993). The effect of tPA on long-term potentiation, the long-lasting increase in synaptic efficiency thought to be involved in the development of long-term memory, has been shown to be compromised in tPA-deficient mice (Huang *et al.,* 1996) and increased in mice expressing a tPA transgene in postnatal neurons (Madani *et al.,* 1999). Neuronal migration, critical for brain development, is also retarded in tPA-deficient mice (Seeds *et al.,* 1999). The upregulation of tPA during seizure, excessive and synchronous synaptic activity, has attracted much interest as seizure is associated with neuronal cell death in the hippocampus, suggesting a link between tPA and neurodegeneration.

Neuronal cell death can be induced by excitotoxic injury in the hippocampus by injection of glutamate analogs, such as kainic acid, and tPA-deficient mice have been shown to be protected from this neuronal degeneration (Tsirka *et al.,* 1995). It has been shown that the mechanism by which tPA contributes to neuronal death is via activation of plasminogen (Tsirka *et al.,* 1997b), and furthermore that laminin, an extracellular matrix protein highly expressed in this hippocampus, is the target substrate for the generated plasmin (Chen and Strickland, 1997). Degradation of laminin presumably affects the viability of the neurons in some way, possibly by anoikis.

Excitotoxic injury is thought to contribute to many neurodegenerative diseases, which may therefore be influenced by tPA. What is not yet known is whether the activity of tPA in the brain is regulated by binding to cofactor molecules or cellular binding sites, although it has been reported that tPA binds to neurons (Verrall and Seeds, 1989). The observation discussed in Section III,B,3 that the disease-associated form of the prion protein specifically binds tPA with high affinity and greatly stimulates plasminogen activation (Ellis *et al.,* 2002) therefore represents a possible mechanism by which tPA activity in the brain could be increased in pathological situations. From the neurodegenerative effects of tPA discussed here, it would be predicted that generation of abnormal prion protein would contribute to the neurodegeneration in the prion diseases. However, in the brain, as in other biological situations, plasmin may have multiple effects, both deleterious and beneficial. For example, it has been mentioned previously that plasmin can activate pro-forms of members of the neurotrophic nerve growth factor family members, brain-derived neurotrophic factor (BDNF) and β-NGF, converting them to neuronal cell-survival factors (Lee *et al.,* 2001), an effect that could be protective in the presence of neurodegenerative stimuli. Interestingly, preliminary experiments have shown that tPA-deficient mice have an accelerated disease progression subsequent to scrapie infection (S. Tsirka, personal communication), an observation that is compatible with a neuroprotective role for the potentially increased generation of plasmin in this prion disease.

Whether tPA also has a potential functional role in other neurodegenerative diseases has yet to be determined. However, plasminogen activation has been reported to be stimulated by analogs of the β-amyloid peptide (Kingston *et al.,* 1995), the amyloid-plaque-forming peptide in Alzheimer's disease. The mechanism of this stimulation has not been elucidated, but is increased on aggregation of this 40-residue peptide into fibrillar, amyloid-like structures which is facilitated by plasmin cleavage of the β-amyloid peptide (Van Nostrand and Porter, 1999). Whether this stimulation occurs *in vivo* and is involved in the disease is not known, but again the effects of plasmin may be complex as it has been shown that plasmin can also act as an α-secretase, i.e., generate nonamyloidogenic peptides from amyloid precursor protein, and plasmin levels are reported to be reduced in Alzheimer's disease brains (Ledesma *et al.,* 2000).

E. uPAR-Mediated Proteolysis *in Vivo*

From much of the evidence discussed here it would be supposed that uPA would have a severely compromised function in the absence of uPAR, and many observations *in vivo* support this contention. However, it cannot be ignored that uPAR-deficient mice do not have the same phenotype as uPA-deficient mice, and in fact do not display an overt physiological phenotype (Bugge *et al.,* 1995). Nevertheless, proteolysis can be demonstrated to be affected by uPAR deficiency. An example of this is the observation that mice with a combined deficiency of uPAR and tPA display major fibrin deposits, unlike mice deficient in tPA alone (Bugge *et al.*, 1996a), demonstrating that uPA function is compromised by the absence of uPAR. By contrast, an example where uPA has been shown to function effectively in the absence of uPAR is during vascular wound healing (Carmeliet *et al.*, 1998). Much of this is probably related to the amount of plasmin that needs to be generated in each of these situations, and it seems that in many cases the presumably reduced level of plasminogen activation in the absence of uPAR is still sufficient to mediate biological effects *in vivo*. The redundancy that is evident in matrix-degrading proteolytic systems no doubt also has an important contribution.

Pathological proteolysis involving upregulation of the activity of the uPAR system may also not be effectively modeled by experiments in uPAR-deficient mice. An example of this is the recent observation that transgenic mice overexpressing either uPA or uPAR in the skin have no cutaneous pathology, but mice crossed to overexpress both uPA and uPAR display extensive epidermal thickening, alopecia, and subepidermal blistering (Zhou *et al.*, 2000). Another recent observation supporting the proteolytic function of uPAR is the that the toxicity of an engineered uPA-activatable anthrax toxin is dependent on the presence of uPAR both in cell culture (Liu *et al.*, 2001) and *in vivo* (Liu *et al.*, 2002). This may prove to be an effective strategy to target the uPAR system therapeutically, as the toxin would not be active simply in the presence of uPAR and uPA, but would require the assembly of a fully functional cell-surface plasminogen activation system.

VI. Concluding Remarks

A. Summary

It has been emphasized here that plasminogen represents an enormous reservoir of proteolytic potential and that its activation is consequently a highly regulated process. This begins at the level of transcriptional regulation of the two plasminogen activators and in some cell types also their regulated secretion. However, the functional regulation of the plasminogen activators is the level at which the complexity of this proteolytic system is most apparent. It is now clear that a diversity of mechanisms exist to regulate the activation and activity of the plasminogen activators, and that these include the utilization of these soluble proteins as cell-surface proteases.

uPA is clearly the primary plasminogen activator in this context, binding uPAR and forming catalytic complexes that generate plasmin activity with very high efficiency, and furthermore cooperating with protease inhibitors to confine this activity to the cell surface or pericellular environment. The activity of this proteolytic system, which is thought to be involved in mediating invasive cellular migration, appears to be functionally linked to other systems fundamental to this cellular behavior. A variety of mechanisms have been proposed, but a common feature is direct or indirect interactions with integrins potentially leading to signaling to the cellular cytoskeletal machinery, suggesting a cooperativity between these systems in mediating invasive migration. tPA clearly also has the potential to act as a cell-surface protease, in addition to its fibrin-dependent role in vascular hemostasis. A variety of cellular binding sites for tPA have been identified, some of which appear to regulate plasminogen activation as efficiently as uPAR. In contrast to uPAR, these binding sites may be cell-type specific, although their roles *in vivo* have yet to be established.

B. Concluding Remarks

It was mentioned in the Introduction (Section I,C) that the phylogenetic lineage of the plasminogen activators and plasminogen suggests that they may have evolved with developmentally related functions in extracellular matrix degradation and remodeling. This putative developmental role has now been lost, or at least is redundant with other protease systems, but the role in extracellular matrix degradation remains. There is no doubt that fibrin is the major target substrate of both uPA- and tPA-catalyzed plasminogen activation, but the fact that the two activators are an evolutionarily stable feature among mammalian species argues that these proteins have important, distinct biological roles.

These roles could be cell-mediated and cell-independent fibrinolysis, for uPA and tPA, respectively, but the situation is unquestionably more complex. In addition to the evidence for cell-mediated, tPA-catalyzed plasminogen activation, tPA has fibrin-independent roles in the CNS. Furthermore, experiments in fibrinogen-deficient mice demonstrate that fibrin is not the only target substrate in pathological conditions, e.g., liver repair (Bezerra *et al.,* 1999) and pulmonary fibrosis (Hattori *et al.,* 2000), and other *in vivo* studies have now clearly demonstrated roles for plasmin in the degradation of extracellular matrix proteins, activation of MMPs, and the activation of growth factors. These observations substantiate predictions from biochemical and cell biological studies that plasmin would have such roles *in vivo*. Therefore, although the two plasminogen activators are functionally regulated by a variety of mechanisms, the multiplicity of their target substrates means that their roles, although clearly different, cannot be considered to be distinct. It is this combination of multiple target substrates, together with the diversity of mechanisms regulating plasmin generation, that results in this proteolytic system being involved in such a wide range of pathological situations.

C. Future Directions

The advances made in the understanding of the function of the plasminogen activation system in recent years, particularly those arising from studies in gene-ablated mice, have reinforced certain established concepts regarding its biological roles but have also opened up new and previously unexpected areas. Many of these observations underscore the necessity for further, and more sophisticated, studies of the mechanisms underlying the various functions of the cell-surface plasminogen activation system.

Key issues that need to be understood in much more detail are manifold. Are other components that regulate the function of the plasminogen activation system yet to be identified? The regulation of tPA activity and function both in normal brain function and in neurodegeneration is an area that will be of particular interest, as will the potential contribution of plasminogen activation to neurodegenerative diseases. What are the true biological substrates of this proteolytic system, and are they targeted specifically? Many potential plasmin substrates have been identified *in vitro,* but validating these *in vivo* is a huge challenge. As mentioned earlier, some important nonfibrin substrates have already been confirmed, largely by detailed analysis in gene-ablated mice, but others need to be either validated or dismissed. These substrates may include proteins yet to be identified, in particular growth factors, that may arise from the human genome sequencing project.

Plasmin activation of various latent MMPs is a long-standing observation that has been confirmed by the type of detailed *in vivo* analysis just mentioned. This is a clear example of cooperativity between two distinct protease systems to form a flexible and adaptable proteolytic cascade. The plasminogen activation system, the MMPs, and the other protease systems described in this book display an astonishing array of cellular tactics to effect the spatial localization of their proteolytic activities. However, an area that has only begun to be addressed is how the activities of these cell-surface proteolytic systems are integrated by cellular mechanisms to achieve concerted effects on cellular function. Some of these have been touched on in this chapter, e.g., colocalization of different proteases in plasma membrane microdomains, and others have been reviewed (Ellis and Murphy, 2001). What may prove to be a key step in such a putative integration mechanism, at least in the context of invasive cell migration, is the relationship between uPAR and integrins.

The relationship between uPAR-mediated proteolysis and cell migration arising from interactions between uPAR and integrins is a particularly intriguing theme to arise from recent cell biological studies. Identifying which are the important components mediating these interactions from among the many that are potentially involved, how they interact, how specific those interactions are, and what the consequences of these interactions are will represent a major advance in the understanding of the extremely dynamic process of cell migration. Much of the emphasis in these studies has been on how uPAR could modulate the activity of these other systems. An area not yet considered in any detail is whether the flow of information between the components of these systems can also go in the opposite direction, i.e., can dynamic changes in the cytoskeleton during cell migration lead to alterations in the activity of the plasminogen activation system at the cell surface? This is a particularly interesting possibility, as such a mechanism would complement the "inside-out" signaling already known to be mediated by integrins.

Understanding the functional regulation of the plasminogen activation system in the context of the diversity of the pathological situations in which it is involved is central to efforts to therapeutically target the activity of this proteolytic system. Targeting the various regulatory mechanisms of the proteolytic system rather than the common end point of plasmin activity may allow the activity of this system to be modulated in different pathological situations, including the major diseases of cancer, atherosclerosis, and neurodegeneration, while leaving other functions of the system unhindered.

Acknowledgments

Research in the author's laboratory is supported by the British Heart Foundation and the Medical Research Council.

References

Aguirre Ghiso, J. A. (2002). Inhibition of FAK signaling activated by urokinase receptor induces dormancy in human carcinoma cells in vivo. *Oncogene* **21,** 2513–2524.

Aguirre Ghiso, J. A., Kovalski, K., and Ossowski, L. (1999). Tumor dormancy induced by downregulation of urokinase receptor in human carcinoma involves integrin and MAPK signaling. *J. Cell Biol.* **147,** 89–104.

An, S. S., Carreno, C., Marti, D. N., Schaller, J., Albericio, F., and Llinas, M. (1998). Lysine-50 is a likely site for anchoring the plasminogen N-terminal peptide to lysine-binding kringles. *Protein Sci.* **7,** 1960–1969.

Andronicos, N. M., and Ranson, M. (2001). The topology of plasminogen binding and activation on the surface of human breast cancer cells. *Br. J. Cancer* **85,** 909–916.

Bajou, K., Masson, V., Gerard, R. D., Schmitt, P. M., Albert, V., Praus, M., Lund, L. R., Frandsen, T. L., Brunner, N., Dano, K., Fusenig, N. E., Weidle, U., Carmeliet, G., Loskutoff, D., Collen, D., Carmeliet, P., Foidart, J. M., and Noel, A. (2001). The plasminogen activator inhibitor PAI-1 controls in vivo tumor vascularization by interaction with proteases, not vitronectin. Implications for antiangiogenic strategies. *J. Cell Biol.* **152,** 777–784.

Bajou, K., Noel, A., Gerard, R. D., Masson, V., Brunner, N., Holst-Hansen, C., Skobe, M., Fusenig, N. E., Carmeliet, P., Collen, D., and Foidart, J. M. (1998). Absence of host plasminogen activator inhibitor 1 prevents cancer invasion and vascularization. *Nat. Med.* **4,** 923–928.

Bezerra, J. A., Bugge, T. H., Melin-Aldana, H., Sabla, G., Kombrinck, K. W., Witte, D. P., and Degen, J. L. (1999). Plasminogen deficiency leads to impaired remodeling after a toxic injury to the liver. *Proc. Natl. Acad. Sci. USA* **96,** 15143–15148.

Bode, W. (1979). The transition of bovine trypsinogen to a trypsin-like state upon strong ligand binding. II. The binding of the pancreatic trypsin inhibitor and of isoleucine-valine and of sequentially related peptides to trypsinogen and to *p*-guanidinobenzoate-trypsinogen. *J. Mol. Biol.* **127,** 357–374.

Bode, W., and Renatus, M. (1997). Tissue-type plasminogen activator: variants and crystal/solution structures demarcate structural determinants of function. *Curr. Opin. Struct. Biol.* **7,** 865–872.

Bork, P., and Bairoch, A. (1995). Extracellular protein modules: A proposed nomenclature. *Trends Biochem. Sci.* **20,** poster C02.

Bugge, T. H., Flick, M. J., Danton, M. J., Daugherty, C. C., Rømer, J., Danø, K., Carmeliet, P., Collen, D., and Degen, J. L. (1996a). Urokinase–type plasminogen activator is effective in fibrin clearance in the absence of its receptor or tissue-type plasminogen activator. *Proc. Natl. Acad. Sci. USA* **93,** 5899–5904.

Bugge, T. H., Kombrinck, K. W., Flick, M. J., Daugherty, C. C., Danton, M. J. S., and Degen, J. L. (1996b). Loss of fibrinogen rescues mice from the pleiotropic effects of plasminogen deficiency. *Cell* **87,** 709–719.

Bugge, T. H., Kombrinck, K. W., Xiao, Q., Holmback, K., Daugherty, C. C., Witte, D. P., and Degen, J. L. (1997). Growth and dissemination of Lewis lung carcinoma in plasminogen-deficient mice. *Blood* **90,** 4522–4531.

Bugge, T. H., Lund, L. R., Kombrinck, K. K., Nielsen, B. S., Holmback, K., Drew, A. F., Flick, M. J., Witte, D. P., Dano, K., and Degen, J. L. (1998). Reduced metastasis of Polyoma virus middle T antigen-induced mammary cancer in plasminogen-deficient mice. *Oncogene* **16,** 3097–3104.

Bugge, T. H., Suh, T. T., Flick, M. J., Daugherty, C. C., Rømer, J., Solberg, H., Ellis, V., Danø, K., and Degen, J. L. (1995). The receptor for urokinase-type plasminogen activator is not essential for mouse development or fertility. *J. Biol. Chem.* **270,** 16886–16894.

Busso, N., Masur, S. K., Lazega, D., Waxman, S., and Ossowski, L. (1994). Induction of cell migration by pro-urokinase binding to its receptor: possible mechanism for signal transduction in human epithelial cells. *J. Cell Biol.* **126,** 259–270.

Carmeliet, P., Moons, L., Dewerchin, M., Rosenberg, S., Herbert, J.-M., Lupu, F., and Collen, D. (1998). Receptor-independent role of urokinase-type plasminogen activator in pericellular plasmin and matrix metalloproteinase proteolysis during vascular wound healing in mice. *J. Cell Biol.* **140,** 233–245.

Carmeliet, P., Moons, L., Herbert, J.-M., Crawley, J., Lupu, F., Lijnen, H. R., and Collen, D. (1997a). Urokinase but not tissue plasminogen activator mediates arterial neointima formation in mice. *Circ. Res.* **81,** 829–839.

Carmeliet, P., Moons, L., Lijnen, R., Baes, M., Lemaître, V., Tipping, P., Drew, A., Eeckhout, Y., Shapiro, S., Lupu, F., and Collen, D. (1997b). Urokinase-generated plasmin activates matrix metalloproteinases during aneurysm formation. *Nat. Genet.* **17,** 439–444.

Carmeliet, P., Moons, L., Lijnen, R., Janssens, S., Lupu, F., Collen, D., and Gerard, R. D. (1997c). Inhibitory role of plasminogen activator inhibitor-1 in arterial wound healing and neointima formation—A gene targeting and gene transfer study in mice. *Circulation* **96,** 3180–3191.

Carriero, M. V., Del Vecchio, S., Capozzoli, M., Franco, P., Fontana, L., Zannetti, A., Botti, G., D'Aiuto, G., Salvatore, M., and Stoppelli, M. P. (1999). Urokinase receptor interacts with alpha(v)beta5 vitronectin receptor, promoting urokinase-dependent cell migration in breast cancer. *Cancer Res.* **59,** 5307–5314.

Castellino, F. J. (1998). Plasmin. *In* "Handbook of Proteolytic Enzymes" (A. J. Barrett, N. D. Rawlings, J. F. Woessner, Eds.), pp. 190–199. Academic Press, London.

Cesarman, G. M., Guevara, C. A., and Hajjar, K. A. (1994). An endothelial cell receptor for plasminogen/tissue plasminogen activator (t-PA). II. Annexin II-mediated enhancement of t-PA-dependent plasminogen activation. *J. Biol. Chem.* **269,** 21198–21203.

Chen, Z. L., and Strickland, S. (1997). Neuronal death in the hippocampus is promoted by plasmin-catalyzed degradation of laminin. *Cell* **91,** 917–925.

Clowes, A. W., Clowes, M. M., Au, Y. P. T., Reidy, M. A., and Belin, D. (1990). Smooth muscle cells express urokinase during mitogenesis and tissue-type plasminogen activator during migration in injured rat carotid artery. *Circ. Res.* **67,** 61–67.

Conese, M., Nykjær, A., Petersen, C. M., Cremona, O., Pardi, R., Andreasen, P. A., Gliemann, J., Christensen, E. I., and Blasi, F. (1995). α_2-Macroglobulin receptor/LDL receptor-related protein(LRP)-dependent internalization of the urokinase receptor. *J. Cell Biol.* **131,** 1609–1622.

Crowley, C. W., Cohen, R. L., Lucas, B. K., Liu, G., Shuman, M. A., and Levinson, A. D. (1993). Prevention of metastasis by inhibition of the urokinase receptor. *Proc. Natl. Acad. Sci. USA* **90,** 5021–5025.

Cubellis, M. V., Wun, T. C., and Blasi, F. (1990). Receptor-mediated internalization and degradation of urokinase is caused by its specific inhibitor PAI-1. *EMBO J.* **9,** 1079–1085.

D'Orazio, D., Besser, D., Marksitzer, R., Kunz, C., Hume, D. A., Kiefer, B., and Nagamine, Y. (1997). Cooperation of two PEA3/AP1 sites in uPA gene induction by TPA and FGF-2. *Gene* **201,** 179–187.

Danø, K., Andreasen, P. A., Grøndahl-Hansen, J., Kristensen, P., Nielsen, L. S., and Skriver, L. (1985). Plasminogen activators, tissue degradation and cancer. *Adv. Cancer Res.* **44,** 139–266.

Danø, K., Rømer, J., Nielsen, B. S., Bjorn, S., Pyke, C., Rygaard, J., and Lund, L. R. (1999). Cancer invasion and tissue remodeling—cooperation of protease systems and cell types. *APMIS* **107,** 120–127.

Davies, B. J., Pickard, B. S., Steel, M., Morris, R. G., and Lathe, R. (1998). Serine proteases in rodent hippocampus. *J. Biol. Chem.* **273,** 23004–23011.

Davis, R. L., Shrimpton, A. E., Holohan, P. D., Bradshaw, C., Feiglin, D., Collins, G. H., Sonderegger, P., Kinter, J., Becker, L. M., Lacbawan, F., Krasnewich, D., Muenke, M., Lawrence, D. A., Yerby, M. S., Shaw, C. M., Gooptu, B., Elliott, P. R., Finch, J. T., Carrell, R. W., and Lomas, D. A. (1999). Familial dementia caused by polymerization of mutant neuroserpin. *Nature* **401,** 376–379.

Deng, G., Curriden, S. A., Wang, S. J., Rosenberg, S., and Loskutoff, D. J. (1996). Is plasminogen activator inhibitor-1 the molecular switch that governs urokinase receptor-mediated cell adhesion and release? *J. Cell Biol.* **134,** 1563–1571.

Donate, L. E., Gherardi, E., Srinivasan, N., Sowdhamini, R., Aparicio, S., and Blundell, T. L. (1994). Molecular evolution and domain structure of plasminogen-related growth factors (HGF/SF and HGF1/MSP). *Protein Sci.* **3,** 2378–2394.

Duffy, M. J. (2002). Urokinase-type plasminogen activator: a potent marker of metastatic potential in human cancers. *Biochem. Soc. Trans.* **30,** 207–210.

Dumler, I., Weis, A., Mayboroda, O. A., Maasch, C., Jerke, U., Haller, H., and Gulba, D. C. (1998). The Jak/Stat pathway and urokinase receptor signaling in human aortic vascular smooth muscle cells. *J. Biol. Chem.* **273,** 315–321.

Dvorak, H. F. (1986). Tumors: wounds that do not heal. Similarities between tumor stroma generation and wound healing. *N. Engl. J. Med.* **315,** 1650–1659.

Ellis, V. (1996). Functional analysis of the cellular receptor for urokinase in plasminogen activation. *J. Biol. Chem.* **271,** 14779–14784.

Ellis, V., and Danø, K. (1993a). Potentiation of plasminogen activation by an anti-urokinase monoclonal antibody due to ternary complex formation. A mechanistic model for receptor-mediated plasminogen activation. *J. Biol. Chem.* **268,** 4806–4813.

Ellis, V., and Danø, K. (1993b). Specific inhibition of the activity of the urokinase receptor-mediated cell-surface plasminogen activation system by suramin. *Biochem. J.* **296,** 505–510.

Ellis, V., and Murphy, G. (2001). Cellular strategies for proteolytic targeting during migration and invasion. *FEBS Lett.* **506,** 1–5.

Ellis, V., and Whawell, S. A. (1997). Vascular smooth muscle cells potentiate plasmin generation by both urokinase and tissue plasminogen activator dependent mechanisms: Evidence for a specific tPA receptor on these cells. *Blood* **90,** 2312–2322.

Ellis, V., Behrendt, N., and Danø, K. (1991). Plasminogen activation by receptor-bound urokinase. A kinetic study with both cell-associated and isolated receptor. *J. Biol. Chem.* **266,** 12752–12758.

Ellis, V., Daniels, M., Misra, R., and Brown, D. R. (2002). Plasminogen activation is stimulated by prion protein and regulated in a copper-dependent manner. *Biochemistry* **41,** 6891–6896.

Ellis, V., Scully, M. F., and Kakkar, V. V. (1987). Plasminogen activation by single-chain urokinase in functional isolation. A kinetic study. *J. Biol. Chem.* **262,** 14998–15003.

Ellis, V., Scully, M. F., and Kakkar, V. V. (1989). Plasminogen activation initiated by single-chain urokinase-type plasminogen activator. Potentiation by U937 monocytes. *J. Biol. Chem.* **264,** 2185–2188.

Ellis, V., Whawell, S. A., Werner, F., and Deadman, J. J. (1999). Assembly of urokinase receptor-mediated plasminogen activation complexes involves direct, non-active-site interactions between urokinase and plasminogen. *Biochemistry* **38,** 651–659.

Ellis, V., Wun, T.-C., Behrendt, N., Rønne, E., and Danø, K. (1990). Inhibition of receptor-bound urokinase by plasminogen-activator inhibitors. *J. Biol. Chem.* **265,** 9904–9908.

Estreicher, A., Muhlhauser, J., Carpentier, J. L., Orci, L., and Vassalli, J. D. (1990). The receptor for urokinase type plasminogen activator polarizes expression of the protease to the leading edge of migrating monocytes and promotes degradation of enzyme inhibitor complexes. *J. Cell Biol.* **111,** 783–792.

Falcone, D. J., Borth, W., Khan, K. M., and Hajjar, K. A. (2001). Plasminogen-mediated matrix invasion and degradation by macrophages is dependent on surface expression of annexin II. *Blood* **97,** 777–784.

Fazioli, F., Resnati, M., Sidenius, N., Higashimoto, Y., Appella, E., and Blasi, F. (1997). A urokinase-sensitive region of the human urokinase receptor is responsible for its chemotactic activity. *EMBO J.* **16,** 7279–7286.

Feng, P., Ohlsson, M., and Ny, T. (1990). The structure of the TATA-less rat tissue-type plasminogen activator gene. Species-specific sequence divergences in the promoter predict differences in regulation of gene expression. *J. Biol. Chem.* **265,** 2022–2027.

Fischer, M. B., Roeckl, C., Parizek, P., Schwarz, H. P., and Aguzzi, A. (2000). Binding of disease-associated prion protein to plasminogen. *Nature* **408,** 479–483.

Friedman, G. C., and Seeds, N. W. (1994). Tissue plasminogen activator expression in the embryonic nervous system. *Dev. Brain Res.* **81,** 41–49.

Gårdsvoll, H., Dano, K., and Ploug, M. (1999). Mapping part of the functional epitope for ligand binding on the receptor for urokinase-type plasminogen activator by site-directed mutagenesis. *J. Biol. Chem.* **274,** 37995–38003.

Grainger, D. J., Kirschenlohr, H. L., Metcalfe, J. C., Weissberg, P. L., Wade, D. P., and Lawn, R. M. (1993). Proliferation of human smooth muscle cells promoted by lipoprotein(a). *Science* **260,** 1655–1658.

Grainger, D. J., Wakefield, L., Bethell, H. W., Farndale, R. W., and Metcalfe, J. C. (1995). Release and activation of platelet latent TGF-β in blood clots during dissolution with plasmin. *Nature Med.* **1,** 932–937.

Gualandris, A., Jones, T. E., Strickland, S., and Tsirka, S. E. (1996). Membrane depolarization induces calcium-dependent secretion of tissue plasminogen activator. *J. Neurosci.* **16,** 2220–2225.

Gunzler, W. A., Steffens, G. J., Otting, F., Buse, G., and Flohe, L. (1982). Structural relationship between human high and low molecular mass urokinase. *Hoppe Seylers Z. Physiol. Chem.* **363,** 133–141.

Gyetko, M. R., Sitrin, R. G., Fuller, J. A., Todd, R. F. 3rd, Petty, H., and Standiford, T. J. (1995). Function of the urokinase receptor (CD87) in neutrophil chemotaxis. *J. Leukocyte Biol.* **58,** 533–538.

Hajjar, K. A., Guevara, C. A., Lev, E., Dowling, K., and Chacko, J. (1996). Interaction of the fibrinolytic receptor, annexin II, with the endothelial cell surface. Essential role of endonexin repeat 2. *J. Biol. Chem.* **271,** 21652–21659.

Hajjar, K. A., Jacovina, A. T., and Chacko, J. (1994). An endothelial cell receptor for plasminogen/tissue plasminogen activator. I. Identity with annexin II. *J. Biol. Chem.* **269,** 21191–21197.

Hall, S. W., Humphries, J. E., and Gonias, S. L. (1991). Inhibition of cell surface receptor-bound plasmin by alpha 2-antiplasmin and alpha 2-macroglobulin. *J. Biol. Chem.* **266,** 12329–12336.

Hattori, N., Degen, J. L., Sisson, T. H., Liu, H., Moore, B. B., Pandrangi, R. G., Simon, R. H., and Drew, A. F. (2000). Bleomycin-induced pulmonary fibrosis in fibrinogen-null mice. *J. Clin. Invest.* **106,** 1341–1350.

Hawley, S. B., Green, M. A., and Miles, L. A. (2000). Discriminating between cell surface and intracellular plasminogen-binding proteins: heterogeneity in profibrinolytic plasminogen-binding proteins on monocytoid cells. *Thromb. Haemost.* **84,** 882–890.

Hawley, S. B., Tamura, T., and Miles, L. A. (2001). Purification, cloning, and characterization of a profibrinolytic plasminogen-binding protein, TIP49a. *J. Biol. Chem.* **276,** 179–186.

He, C., Wilhelm, S. M., Pentland, A. P., Marner, B. L., Grant, G. A., Eisen, A. Z., and Goldberg, G. I. (1989). Tissue cooperation in a proteolytic cascade activating human interstitial collagenase. *Proc. Natl. Acad. Sci. USA* **86,** 2632–2636.

Hembrough, T. A., Vasudevan, J., Allietta, M. M., Glass, W. F. 2nd, and Gonias, S. L. (1995). A cytokeratin 8-like protein with plasminogen-binding activity is present on the external surfaces of hepatocytes, HepG2 cells and breast carcinoma cell lines. *J. Cell Sci.* **108,** 1071–1082.

Henderson, B. R., and Sleigh, M. J. (1992). TATA box-independent transcription of the human tissue plasminogen activator gene initiates within a sequence conserved in related genes. *FEBS Lett.* **309,** 130–134.

Heymans, S., Luttun, A., Nuyens, D., Theilmeier, G., Creemers, E., Moons, L., Dyspersin, G. D., Cleutjens, J. P., Shipley, M., Angellilo, A., Levi, M., Nube, O., Baker, A., Keshet, E., Lupu, F., Herbert, J. M., Smits, J. F., Shapiro, S. D., Baes, M., Borgers, M., Collen, D., Daemen, M. J., and Carmeliet, P. (1999). Inhibition of plasminogen activators or matrix metalloproteinases prevents cardiac rupture but impairs therapeutic angiogenesis and causes cardiac failure. *Nat. Med.* **5,** 1135–1142.

Higazi, A. A., Upson, R. H., Cohen, R. L., Manuppello, J., Bognacki, J., Henkin, J., McCrae, K. R., Kounnas, M. Z., Strickland, D. K., Preissner, K. T., Lawler, J., and Cines, D. B. (1996). Interaction of single-chain urokinase with its receptor induces the appearance and disappearance of binding epitopes within the resultant complex for other cell surface proteins. *Blood* **88,** 542–551.

Hiraoka, N., Allen, E., Apel, I. J., Gyetko, M. R., and Weiss, S. J. (1998). Matrix metalloproteinases regulate neovascularization by acting as pericellular fibrinolysins. *Cell* **95,** 365–377.

Holmberg, M., Leonardsson, G., and Ny, T. (1995). The species-specific differences in the cAMP regulation of the tissue-type plasminogen activator gene between rat, mouse and human is caused by a one-nucleotide substitution in the cAMP-responsive element of the promoters. *Eur. J. Biochem.* **231,** 466–474.

Hooper, J. D., Clements, J. A., Quigley, J. P., and Antalis, T. M. (2001). Type II transmembrane serine proteases. Insights into an emerging class of cell surface proteolytic enzymes. *J. Biol. Chem.* **276,** 857–860.

Høyer-Hansen, G., Behrendt, N., Ploug, M., Danø, K., and Preissner, K. T. (1997). The intact urokinase receptor is required for efficient vitronectin binding: receptor cleavage prevents ligand interaction. *FEBS Lett.* **420,** 79–85.

Høyer-Hansen, G., Rønne, E., Solberg, H., Behrendt, N., Ploug, M., Lund, L. R., Ellis, V., and Danø, K. (1992). Urokinase plasminogen activator cleaves its cell surface receptor releasing the ligand-binding domain. *J. Biol. Chem.* **267,** 18224–18229.

Hoylaerts, M., Rijken, D. C., Lijnen, H. R., and Collen, D. (1982). Kinetics of the activation of plasminogen by human tissue plasminogen activator. Role of fibrin. *J. Biol. Chem.* **257,** 2912–2919.

Huang, Y. Y., Bach, M. E., Lipp, H. P., Zhuo, M., Wolfer, D. P., Hawkins, R. D., Schoonjans, L., Kandel, E. R., Godfraind, J. M., Mulligan, R., Collen, D., and Carmeliet, P. (1996). Mice lacking the gene encoding tissue-type plasminogen activator show a selective interference with late-phase long-term potentiation in both Schaffer collateral and mossy fiber pathways. *Proc. Natl. Acad. Sci. USA* **93,** 8699–8704.

Huarte, J., Belin, D., Vassalli, A., Strickland, S., and Vassalli, J. D. (1987). Meiotic maturation of mouse oocytes triggers the translation and polyadenylation of dormant tissue-type plasminogen activator mRNA. *Genes Dev.* **1,** 1201–1211.

Huber, D., Cramer, E. M., Kaufmann, J. E., Meda, P., Masse, J. M., Kruithof, E. K., and Vischer, U. M. (2002). Tissue-type plasminogen activator (t-PA) is stored in Weibel–Palade bodies in human endothelial cells both in vitro and in vivo. *Blood* **99,** 3637–3645.

Huber, R., and Bode, W. (1978). Structural basis of the activation and action of trypsin. *Acc. Chem. Research* **11,** 114–122.

Ichinose, A., Fujikawa, K., and Suyama, T. (1986). The activation of pro-urokinase by plasma kallikrein and its inactivation by thrombin. *J. Biol. Chem.* **261,** 3486–3489.

Irigoyen, J. P., Besser, D., and Nagamine, Y. (1997). Cytoskeleton reorganization induces the urokinase-type plasminogen activator gene via the Ras/extracellular signal-regulated kinase (ERK) signaling pathway. *J. Biol. Chem.* **272,** 1904–1909.

Irigoyen, J. P., Munoz-Canoves, P., Montero, L., Koziczak, M., and Nagamine, Y. (1999). The plasminogen activator system: biology and regulation. *Cell Mol. Life Sci.* **56,** 104–132.

Jacovina, A. T., Zhong, F., Khazanova, E., Lev, E., Deora, A. B., and Hajjar, K. A. (2001). Neuritogenesis and the nerve growth factor-induced differentiation of PC-12 cells requires annexin II-mediated plasmin generation. *J. Biol. Chem.* **276,** 49350–49358.

Joubert, F. J., and Dowdle, E. B. (1987). The primary structure of the inhibitor of tissue plasminogen activator found in the seeds of Erythrina caffra. *Thromb. Haemost.* **57,** 356–360.

Kassam, G., Choi, K. S., Ghuman, J., Kang, H. M., Fitzpatrick, S. L., Zackson, T., Zackson, S., Toba, M., Shinomiya, A., and Waisman, D. M. (1998a). The role of annexin II tetramer in the activation of plasminogen. *J. Biol. Chem.* **273,** 4790–4799.

Kassam, G., Le, B. H., Choi, K. S., Kang, H. M., Fitzpatrick, S. L., Louie, P., and Waisman, D. M. (1998b). The p11 subunit of the annexin II tetramer plays a key role in the stimulation of t-PA-dependent plasminogen activation. *Biochemistry* **37,** 16958–16966.

Kawai, Y., Matsumoto, Y., Watanabe, K., Yamamoto, H., Satoh, K., Murata, M., Handa, M., and Ikeda, Y. (1996). Hemodynamic forces modulate the effects of cytokines on fibrinolytic activity of endothelial cells. *Blood* **87,** 2314–2321.

Kindzelskii, A. L., Eszes, M. M., Todd, R. F., and Petty, H. R. (1997). Proximity oscillations of complement type 4 (alphaX beta2) and urokinase receptors on migrating neutrophils. *Biophys. J.* **73,** 1777–1784.

Kindzelskii, A. L., Laska, Z. O., Todd, R. F., and Petty, H. R. (1996). Urokinase-type plasminogen activator receptor reversibly dissociates from complement receptor type 3 (alpha M beta 2' CD11b/CD18) during neutrophil polarization. *J. Immunol.* **156,** 297–309.

Kingston, I. B., Castro, M. J., and Anderson, S. (1995). In vitro stimulation of tissue-type plasminogen activator by Alzheimer amyloid beta-peptide analogues. *Nat. Med.* **1,** 138–142.

Kjøller, L., and Hall, A. (2001). Rac mediates cytoskeletal rearrangements and increased cell motility induced by urokinase-type plasminogen activator receptor binding to vitronectin. *J. Cell Biol.* **152,** 1145–1158.

Kobayashi, H., Schmitt, M., Goretzki, L., Chucholowski, N., Calvete, J., Kramer, M., Gunzler, W. A., Janicke, F., and Graeff, H. (1991). Cathepsin B efficiently activates the soluble and the tumor cell receptor-bound form of the proenzyme urokinase-type plasminogen activator (Pro-uPA). *J. Biol. Chem.* **266,** 5147–5152.

Kook, Y. H., Adamski, J., Zelent, A., and Ossowski, L. (1994). The effect of antisense inhibition of urokinase receptor in human squamous cell carcinoma on malignancy. *EMBO J.* **13,** 3983–3991.

Krem, M. M., Rose, T., and Di Cera, E. (2000). Sequence determinants of function and evolution in serine proteases. *Trends Cardiovasc. Med.* **10,** 171–176.

Kristensen, P., Pyke, C., Lund, L. R., Andreasen, P. A., and Dano, K. (1990). Plasminogen activator inhibitor-type 1 in Lewis lung carcinoma. *Histochemistry* **93**, 559–566.

Lansink, M., and Kooistra, T. (1996). Stimulation of tissue-type plasminogen activator expression by retinoic acid in human endothelial cells requires retinoic acid receptor beta 2 induction. *Blood* **88**, 531–541.

Ledesma, M. D., Da Silva, J. S., Crassaerts, K., Delacourte, A., De Strooper, B., and Dotti, C. G. (2000). Brain plasmin enhances APP alpha-cleavage and Abeta degradation and is reduced in Alzheimer's disease brains. *EMBO Rep.* **1**, 530–535.

Lee, R., Kermani, P., Teng, K. K., and Hempstead, B. L. (2001). Regulation of cell survival by secreted proneurotrophins. *Science* **294**, 1945–1948.

Lee, S. W., Ellis, V., and Dichek, D. A. (1994). Characterization of plasminogen activation by glycosylphosphatidylinositol-anchored urokinase. *J. Biol. Chem.* **269**, 2411–2418.

Lijnen, H. R. (2002). Extracellular proteolysis in the development and progression of atherosclerosis. *Biochem. Soc. Trans.* **30**, 163–167.

Liotta, L. A., Goldfarb, R. H., Brundage, R., Siegal, G. P., Terranova, V., and Garbisa, S. (1981). Effect of plasminogen activator (urokinase), plasmin, and thrombin on glycoprotein and collagenous components of basement membrane. *Cancer Res.* **41**, 4629–4636.

List, K., Jensen, O. N., Bugge, T. H., Lund, L. R., Ploug, M., Dano, K., and Behrendt, N. (2000). Plasminogen-independent initiation of the pro-urokinase activation cascade in vivo. Activation of pro-urokinase by glandular kallikrein (mGK-6) in plasminogen-deficient mice. *Biochemistry* **39**, 508–515.

Liu, S., Aaronson, H., Mitola, D. J., Leppla, S. H., and Bugge, T. H. (2002). Potent anti-tumor activity of a urokinase-activated engineered anthrax toxin. *Proc. Natl. Acad. Sci. USA.* **In Press.**

Liu, S., Bugge, T. H., and Leppla, S. H. (2001). Targeting of tumor cells by cell surface urokinase plasminogen activator-dependent anthrax toxin. *J. Biol. Chem.* **276**, 17976–17984.

Longstaff, C., Merton, R. E., Fabregas, P., and Felez, J. (1999). Characterization of cell-associated plasminogen activation catalyzed by urokinase-type plasminogen activator, but independent of urokinase receptor (uPAR, CD87). *Blood* **93**, 3839–3846.

Look, M. P., van Putten, W. L., Duffy, M. J., Harbeck, N., Christensen, I. J., Thomssen, C., Kates, R., Spyratos, F., Ferno, M., Eppenberger-Castori, S., Sweep, C. G., Ulm, K., Peyrat, J. P., Martin, P. M., Magdelenat, H., Brunner, N., Duggan, C., Lisboa, B. W., Bendahl, P. O., Quillien, V., Daver, A., Ricolleau, G., Meijer-van Gelder, M. E., Manders, P., Fiets, W. E., Blankenstein, M. A., Broet, P., Romain, S., Daxenbichler, G., Windbichler, G., Cufer, T., Borstnar, S., Kueng, W., Beex, L. V., Klijn, J. G., O'Higgins, N., Eppenberger, U., Janicke, F., Schmitt, M., and Foekens, J. A. (2002). Pooled analysis of prognostic impact of urokinase-type plasminogen activator and its inhibitor PAI-1 in 8377 breast cancer patients. *J. Natl. Cancer Inst.* **94**, 116–128.

Loskutoff, D. J., and Quigley, J. P. (2000). PAI-1, fibrosis, and the elusive provisional fibrin matrix. *J. Clin. Invest.* **106**, 1441–1443.

Lund, L. R., Ellis, V., Rønne, E., Pyke, C., and Danø, K. (1995). Transcriptional and post-transcriptional regulation of the receptor for urokinase-type plasminogen activator by cytokines and tumour promoters in the human lung carcinoma cell line A549. *Biochem. J.* **310**, 345–352.

Lund, L. R., Rømer, J., Bugge, T. H., Nielsen, B. S., Frandsen, T. L., Degen, J. L., Stephens, R. W., and Danø, K. (1999). Functional overlap between two classes of matrix-degrading proteases in wound healing. *EMBO J.* **18**, 4645–4656.

Lund, L. R., Rømer, J., Rønne, E., Ellis, V., Blasi, F., and Danø, K. (1991). Urokinase-receptor biosynthesis, mRNA level and gene transcription are increased by transforming growth factor beta 1 in human A549 lung carcinoma cells. *EMBO J.* **10**, 3399–3407.

Lupu, F., Heim, D. A., Bachmann, F., Hurni, M., Kakkar, V. V., and Kruithof, E. K. O. (1995). Plasminogen activator expression in human atherosclerotic lesions. *Arterioscler. Thromb. Vasc. Biol.* **15**, 1444–1455.

Luttun, A., Lupu, F., Storkebaum, E., Hoylaerts, M. F., Moons, L., Crawley, J., Bono, F., Poole, A. R., Tipping, P., Herbert, J. M., Collen, D., and Carmeliet, P. (2002). Lack of plasminogen activator inhibitor-1 promotes growth and abnormal matrix remodeling of advanced atherosclerotic plaques in apolipoprotein E-deficient mice. *Arterioscler. Thromb. Vasc. Biol.* **22,** 499–505.

MacDougall, J. R., and Matrisian, L. M. (1995). Contributions of tumor and stromal matrix metalloproteinases to tumor progression, invasion and metastasis. *Cancer Metastasis Rev.* **14,** 351–362.

Madani, R., Hulo, S., Toni, N., Madani, H., Steimer, T., Muller, D., and Vassalli, J. D. (1999). Enhanced hippocampal long-term potentiation and learning by increased neuronal expression of tissue-type plasminogen activator in transgenic mice. *EMBO J.* **18,** 3007–3012.

Magdolen, V., Burgle, M., de Prada, N. A., Schmiedeberg, N., Riemer, C., Schroeck, F., Kellermann, J., Degitz, K., Wilhelm, O. G., Schmitt, M., and Kessler, H. (2001). Cyclo19,31[D-Cys19]-uPA19-31 is a potent competitive antagonist of the interaction of urokinase-type plasminogen activator with its receptor (CD87). *Biol. Chem.* **382,** 1197–1205.

Mai, J., Finley, R. L., Jr., Waisman, D. M., and Sloane, B. F. (2000). Human procathepsin B interacts with the annexin II tetramer on the surface of tumor cells. *J. Biol. Chem.* **275,** 12806–12812.

Mandriota, S. J., Seghezzi, G., Vassalli, J. D., Ferrara, N., Wasi, S., Mazzieri, R., Mignatti, P., and Pepper, M. S. (1995). Vascular endothelial growth factor increases urokinase receptor expression in vascular endothelial cells. *J. Biol. Chem.* **270,** 9709–9716.

Mangel, W. F., Lin, B. H., and Ramakrishnan, V. (1990). Characterization of an extremely large, ligand-induced conformational change in plasminogen. *Science* **248,** 69–73.

Mars, W. M., Zarnegar, R., and Michalopoulos, G. K. (1993). Activation of hepatocyte growth factor by the plasminogen activators uPA and tPA. *Am. J. Pathol.* **143,** 949–958.

Marti, D. N., Hu, C.-K., An, S. S. A., Von Haller, P., Schaller, J., and Llinás, M. (1997). Ligand preferences of kringle 2 and homologous domains of human plasminogen: Canvassing weak, intermediate, and high-affinity binding sites by ^{1}H-NMR. *Biochemistry* **36,** 11591–11604.

Matsuoka, Y., Kitamura, Y., and Taniguchi, T. (1998). Induction of plasminogen in rat hippocampal pyramidal neurons by kainic acid. *Neurosci. Lett.* **252,** 119–122.

Mazzieri, R., Masiero, L., Zanetta, L., Monea, S., Onisto, M., Garbisa, S., and Mignatti, P. (1997). Control of type IV collagenase activity by components of the urokinase–plasmin system: a regulatory mechanism with cell-bound reactants. *EMBO J.* **16,** 2319–2332.

Medcalf, R. L., Richards, R. I., Crawford, R. J., and Hamilton, J. A. (1986). Suppression of urokinase-type plasminogen activator mRNA levels in human fibrosarcoma cells and synovial fibroblasts by anti-inflammatory glucocorticoids. *EMBO J.* **5,** 2217–2222.

Miles, L. A., Dahlberg, C. M., Levin, E. G., and Plow, E. F. (1989). Gangliosides interact directly with plasminogen and urokinase and may mediate binding of these fibrinolytic components to cells. *Biochemistry* **28,** 9337–9343.

Miles, L. A., Dahlberg, C. M., Plescia, J., Felez, J., Kato, K., and Plow, E. F. (1991). Role of cell-surface lysines in plasminogen binding to cells: identification of alpha-enolase as a candidate plasminogen receptor. *Biochemistry* **30,** 1682–1691.

Min, H. Y., Doyle, L. V., Vitt, C. R., Zandonella, C. L., Stratton-Thomas, J. R., Shuman, M. A., and Rosenberg, S. (1996). Urokinase receptor antagonists inhibit angiogenesis and primary tumor growth in syngeneic mice. *Cancer Res.* **56,** 2428–2433.

Murphy, G., Stanton, H., Cowell, S., Butler, G., Knauper, V., Atkinson, S., and Gavrilovic, J. (1999). Mechanisms for pro matrix metalloproteinase activation. *APMIS* **107,** 38–44.

Naldini, L., Tamagnone, L., Vigna, E., Sachs, M., Hartmann, G., Birchmeier, W., Daikuhara, Y., Tsubouchi, H., Blasi, F., and Comoglio, P. M. (1992). Extracellular proteolytic cleavage by urokinase is required for activation of hepatocyte growth factor/scatter factor. *EMBO J.* **11,** 4825–4833.

Nanbu, R., Menoud, P. A., and Nagamine, Y. (1994). Multiple instability-regulating sites in the 3′ untranslated region of the urokinase-type plasminogen activator mRNA. *Mol. Cell Biol.* **14**, 4920–4928.

Nerlov, C., De Cesare, D., Pergola, F., Caracciolo, A., Blasi, F., Johnsen, M., and Verde, P. (1992). A regulatory element that mediates co-operation between a PEA3-AP-1 element and an AP-1 site is required for phorbol ester induction of urokinase enhancer activity in HepG2 hepatoma cells. *EMBO J.* **11**, 4573–4582.

Nguyen, D. H., Hussaini, I. M., and Gonias, S. L. (1998). Binding of urokinase-type plasminogen activator to its receptor in MCF-7 cells activates extracellular signal-regulated kinase 1 and 2 which is required for increased cellular motility. *J. Biol. Chem.* **273**, 8502–8507.

Nicole, O., Docagne, F., Ali, C., Margaill, I., Carmeliet, P., MacKenzie, E. T., Vivien, D., and Buisson, A. (2001). The proteolytic activity of tissue-plasminogen activator enhances NMDA receptor-mediated signaling. *Nat. Med.* **7**, 59–64.

Nobes, C. D., and Hall, A. (1995). Rho, rac, and cdc42 GTPases regulate the assembly of multimolecular focal complexes associated with actin stress fibers, lamellipodia, and filopodia. *Cell* **81**, 53–62.

Nusrat, A. R., and Chapman, H. A. (1991). An autocrine role for urokinase in phorbol ester-mediated differentiation of myeloid cell lines. *J. Clin. Invest.* **87**, 1091–1097.

Nykjær, A., Conese, M., Christensen, E. I., Olson, D., Cremona, O., Gliemann, J., and Blasi, F. (1997). Recycling of the urokinase receptor upon internalization of the uPA:serpin complexes. *EMBO J.* **16**, 2610–2620.

Nykjær, A., Petersen, C. M., Moller, B., Jensen, P. H., Moestrup, S. K., Holtet, T. L., Etzerodt, M., Thogersen, H. C., Munch, M., Andreasen, P. A., and Gliemann, J. (1992). Purified alpha 2-macroglobulin receptor/LDL receptor-related protein binds urokinase.plasminogen activator inhibitor type-1 complex. Evidence that the alpha 2-macroglobulin receptor mediates cellular degradation of urokinase receptor-bound complexes. *J. Biol. Chem.* **267**, 14543–14546.

Odekon, L. E., Blasi, F., and Rifkin, D. B. (1994). Requirement for receptor-bound urokinase in plasmin-dependent cellular conversion of latent TGF-beta to TGF-beta. *J. Cell. Physiol.* **158**, 398–407.

Okada, S. S., Tomaszewski, J. E., and Barnathan, E. S. (1995). Migrating vascular smooth muscle cells polarize cell surface urokinase receptors after injury in vitro. *Exp. Cell Res.* **217**, 180–187.

Okumura, Y., Kamikubo, Y., Curriden, S. A., Wang, J., Kiwada, T., Futaki, S., Kitagawa, K., and Loskutoff, D. J. (2002). Kinetic analysis of the interaction between vitronectin and the urokinase receptor. *J. Biol. Chem.* **277**, 9395–9404.

Osterwalder, T., Cinelli, P., Baici, A., Pennella, A., Krueger, S. R., Schrimpf, S. P., Meins, M., and Sonderegger, P. (1998). The axonally secreted serine proteinase inhibitor, neuroserpin, inhibits plasminogen activators and plasmin but not thrombin. *J. Biol. Chem.* **273**, 2312–2321.

Paoni, N. F., Chow, A. M., Pena, L. C., Keyt, B. A., Zoller, M. J., and Bennett, W. F. (1993). Making tissue-type plasminogen activator more fibrin specific. *Protein Eng.* **6**, 529–534.

Parmer, R. J., Mahata, M., Mahata, S., Sebald, M. T., O'Connor, D. T., and Miles, L. A. (1997). Tissue plasminogen activator (t-PA) is targeted to the regulated secretory pathway. Catecholamine storage vesicles as a reservoir for the rapid release of t-PA. *J. Biol. Chem.* **272**, 1976–1982.

Patthy, L. (1999). Genome evolution and the evolution of exon-shuffling—a review. *Gene* **238**, 103–114.

Pennica, D., Holmes, W. E., Kohr, W. J., Harkins, R. N., Vehar, G. A., Ward, C. A., Bennett, W. F., Yelverton, E., Seeburg, P. H., Heyneker, H. L., Goeddel, D. V., and Collen, D. (1983). Cloning and expression of human tissue-type plasminogen activator cDNA in E. coli. *Nature* **301**, 214–221.

Petersen, L. C., Lund, L. R., Nielsen, L. S., Danø, K., and Skriver, L. (1988). One-chain urokinase-type plasminogen activator from human sarcoma cells is a proenzyme with little or no intrinsic activity. *J. Biol. Chem.* **263**, 11189–11195.

Plesner, T., Ploug, M., Ellis, V., Rønne, E., Høyer-Hansen, G., Wittrup, M., Pedersen, T. L., Tscherning, T., Danø, K., and Hansen, N. E. (1994). The receptor for urokinase-type plasminogen activator and urokinase is translocated from two distinct intracellular compartments to the plasma membrane on stimulation of human neutrophils. *Blood* **83,** 808–815.

Ploug, M., and Ellis, V. (1994). Structure–function relationships in the receptor for urokinase-type plasminogen activator. Comparison to other members of the Ly-6 family and snake venom α-neurotoxins. [Review]. *FEBS Lett.* **349,** 163–168.

Ploug, M., Ellis, V., and Danø, K. (1994). Ligand interaction between urokinase-type plasminogen activator and its receptor probed with 8-anilino-1-naphthalenesulfonate. Evidence for a hydrophobic binding site exposed only on the intact receptor. *Biochemistry* **33,** 8991–8997.

Ploug, M., Ostergaard, S., Hansen, L. B., Holm, A., and Dano, K. (1998). Photoaffinity labeling of the human receptor for urokinase-type plasminogen activator using a decapeptide antagonist. Evidence for a composite ligand-binding site and a short interdomain separation. *Biochemistry* **37,** 3612–3622.

Ploug, M., Rahbek-Nielsen, H., Ellis, V., Roepstorff, P., and Danø, K. (1995). Chemical modification of the urokinase-type plasminogen activator and its receptor using tetranitromethane. Evidence for the involvement of specific tyrosine residues in both molecules during receptor–ligand interaction. *Biochemistry* **34,** 12524–12534.

Ploug, M., Rønne, E., Behrendt, N., Jensen, A. L., Blasi, F., and Danø, K. (1991). Cellular receptor for urokinase plasminogen activator. Carboxyl-terminal processing and membrane anchoring by glycosyl-phosphatidylinositol. *J. Biol. Chem.* **266,** 1926–1933.

Plow, E. F., Freaney, D. E., Plescia, J., and Miles, L. A. (1986). The plasminogen system and cell surfaces: evidence for plasminogen and urokinase receptors on the same cell type. *J. Cell Biol.* **103,** 2411–2420.

Pollanen, J., Hedman, K., Nielsen, L. S., Dano, K., and Vaheri, A. (1988). Ultrastructural localization of plasma membrane-associated urokinase-type plasminogen activator at focal contacts. *J. Cell Biol.* **106,** 87–95.

Prusiner, S. B. (1998). Prions. *Proc. Natl. Acad. Sci. USA* **95,** 13363–13383.

Pyke, C., Kristensen, P., Ralfkiaer, E., Eriksen, J., and Dano, K. (1991a). The plasminogen activation system in human colon cancer: messenger RNA for the inhibitor PAI-1 is located in endothelial cells in the tumor stroma. *Cancer Res.* **51,** 4067–4071.

Pyke, C., Kristensen, P., Ralfkiaer, E., Grøndahl-Hansen, J., Eriksen, J., Blasi, F., and Danø, K. (1991b). Urokinase-type plasminogen activator is expressed in stromal cells and its receptor in cancer cells at invasive foci in human colon adenocarcinomas. *Am. J. Pathol.* **138,** 1059–1067.

Qian, Z., Gilbert, M. E., Colicos, M. A., Kandel, E. R., and Kuhl, D. (1993). Tissue-plasminogen activator is induced as an immediate-early gene during seizure, kindling and long-term potentiation. *Nature* **361,** 453–457.

Renatus, M., Engh, R. A., Stubbs, M. T., Huber, R., Fischer, S., Kohnert, U., and Bode, W. (1997). Lysine 156 promotes the anomalous proenzyme activity of tPA: X-ray crystal structure of single-chain human tPA. *EMBO J.* **16,** 4797–4805.

Resnati, M., Guttinger, M., Valcamonica, S., Sidenius, N., Blasi, F., and Fazioli, F. (1996). Proteolytic cleavage of the urokinase receptor substitutes for the agonist-induced chemotactic effect. *EMBO J.* **15,** 1572–1582.

Resnati, M., Pallavicini, I., Wang, J. M., Oppenheim, J., Serhan, C. N., Romano, M., and Blasi, F. (2002). The fibrinolytic receptor for urokinase activates the G protein-coupled chemotactic receptor FPRL1/LXA4R. *Proc. Natl. Acad. Sci. USA* **99,** 1359–1364.

Rømer, J., Bugge, T. H., Pyke, C., Lund, L. R., Flick, M. J., Degen, J. L., and Danø, K. (1996). Impaired wound healing in mice with a disrupted plasminogen gene. *Nat. Med.* **2,** 287–292.

Rømer, J., Lund, L. R., Eriksen, J., Pyke, C., Kristensen, P., and Danø, K. (1994). The receptor for urokinase-type plasminogen activator is expressed by keratinocytes at the leading edge during re-epithelialization of mouse skin wounds. *J. Invest. Dermatol.* **102,** 519–522.

Russell, M. E., Quertermous, T., Declerck, P. J., Collen, D., Haber, E., and Homcy, C. J. (1990). Binding of tissue-type plasminogen activator with human endothelial cell monolayers. Characterization of the high affinity interaction with plasminogen activator inhibitor-1. *J. Biol. Chem.* **265,** 2569–2575.

Saksela, O., and Rifkin, D. B. (1990). Release of basic fibroblast growth factor–heparan sulphate complexes from endothelial cells by plasminogen activator–mediated proteolytic activity. *J. Cell Biol.* **110,** 767–775.

Sappino, A. P., Madani, R., Huarte, J., Belin, D., Kiss, J. Z., Wohlwend, A., and Vassalli, J. D. (1993). Extracellular proteolysis in the adult murine brain. *J. Clin. Invest.* **92,** 679–685.

Schwartz, M. A. (2001). Integrin signaling revisited. *Trends Cell Biol.* **11,** 466–470.

Seeds, N. W., Basham, M. E., and Haffke, S. P. (1999). Neuronal migration is retarded in mice lacking the tissue plasminogen activator gene. *Proc. Natl. Acad. Sci. USA* **96,** 14118–14123.

Seeds, N. W., Williams, B. L., and Bickford, P. C. (1995). Tissue plasminogen activator induction in Purkinje neurons after cerebellar motor learning. *Science* **270,** 1992–1994.

Seiffert, D., Ciambrone, G., Wagner, N. V., Binder, B. R., and Loskutoff, D. J. (1994). The somatomedin B domain of vitronectin. Structural requirements for the binding and stabilization of active type 1 plasminogen activator inhibitor. *J. Biol. Chem.* **269,** 2659–2666.

Siao, C. J., and Tsirka, S. E. (2002). Tissue plasminogen activator mediates microglial activation via its finger domain through annexin II. *J. Neurosci.* **22,** 3352–3358.

Silverman, G. A., Bird, P. I., Carrell, R. W., Church, F. C., Coughlin, P. B., Gettins, P. G., Irving, J. A., Lomas, D. A., Luke, C. J., Moyer, R. W., Pemberton, P. A., Remold-O'Donnell, E., Salvesen, G. S., Travis, J., and Whisstock, J. C. (2001). The serpins are an expanding superfamily of structurally similar but functionally diverse proteins. Evolution, mechanism of inhibition, novel functions, and a revised nomenclature. *J. Biol. Chem.* **276,** 33293–33296.

Simons, K., and Toomre, D. (2000). Lipid rafts and signal transduction. *Nat. Rev. Mol. Cell Biol.* **1,** 31–39.

Sitrin, R. G., Todd, R. F. 3rd, Petty, H. R., Brock, T. G., Shollenberger, S. B., Albrecht, E., and Gyetko, M. R. (1996). The urokinase receptor (CD87) facilitates CD11b/CD18-mediated adhesion of human monocytes. *J. Clin. Invest.* **97,** 1942–1951.

Skriver, L., Larsson, L. I., Kielberg, V., Nielsen, L. S., Andresen, P. B., Kristensen, P., and Danø, K. (1984). Immunocytochemical localization of urokinase-type plasminogen activator in Lewis lung carcinoma. *J. Cell Biol.* **99,** 752–757.

Smith, E. B., and Thompson, W. D. (1994). Fibrin as a factor in atherogenesis. *Thromb. Res.* **73,** 1–19.

Soravia, E., Grebe, A., De Luca, P., Helin, K., Suh, T. T., Degen, J. L., and Blasi, F. (1995). A conserved TATA-less proximal promoter drives basal transcription from the urokinase-type plasminogen activator receptor gene. *Blood* **86,** 624–635.

Sperl, S., Jacob, U., Arroyo, de Prada, Sturzebecher, J., Wilhelm, O. G., Bode, W., Magdolen, V., Huber, R., and Moroder, L. (2000). (4-Aminomethyl)phenylguanidine derivatives as nonpeptidic highly selective inhibitors of human urokinase. *Proc. Natl. Acad. Sci. USA* **97,** 5113–5118.

Stack, M. S., and Johnson, D. A. (1994). Human mast cell tryptase activates single-chain urinary-type plasminogen activator (pro-urokinase). *J. Biol. Chem.* **269,** 9416–9419.

Stahl, A., and Mueller, B. M. (1995). The urokinase-type plasminogen activator receptor, a GPI-linked protein, is localized in caveolae. *J. Cell Biol.* **129,** 335–344.

Stefansson, S., and Lawrence, D. A. (1996). The serpin PAI-1 inhibits cell migration by blocking integrin $\alpha_v\beta_3$ binding to vitronectin. *Nature* **383,** 441–443.

Stein, C. M., Brown, N., Vaughan, D. E., Lang, C. C., and Wood, A. J. (1998). Regulation of local tissue-type plasminogen activator release by endothelium-dependent and endothelium-independent agonists in human vasculature. *J. Am. Coll. Cardiol.* **32,** 117–122.

Stephens, R. W., Aumailley, M., Timpl, R., Reisberg, T., Tapiovaara, H., Myohanen, H., Murphy-Ullrich, J., and Vaheri, A. (1992a). Urokinase binding to laminin-nidogen. Structural requirements and interactions with heparin. *Eur. J. Biochem.* **207,** 937–942.

Stephens, R. W., Bokman, A. M., Myohanen, H. T., Reisberg, T., Tapiovaara, H., Pedersen, N., Grøndahl-Hansen, J., Llinas, M., and Vaheri, A. (1992b). Heparin binding to the urokinase kringle domain. *Biochemistry* **31,** 7572–7579.

Stephens, R. W., Pollanen, J., Tapiovaara, H., Leung, K. C., Sim, P. S., Salonen, E. M., Rønne, E., Behrendt, N., Danø, K., and Vaheri, A. (1989). Activation of pro-urokinase and plasminogen on human sarcoma cells: a proteolytic system with surface-bound reactants. *J. Cell Biol.* **108,** 1987–1995.

Stoppelli, M. P., Corti, A., Soffientini, A., Cassani, G., Blasi, F., and Assoian, R. K. (1985). Differentiation-enhanced binding of the amino-terminal fragment of human urokinase plasminogen activator to a specific receptor on U937 monocytes. *Proc. Natl. Acad. Sci. USA* **82,** 4939–4943.

Takeuchi, T., Harris, J. L., Huang, W., Yan, K. W., Coughlin, S. R., and Craik, C. S. (2000). Cellular localization of membrane-type serine protease 1 and identification of protease-activated receptor-2 and single-chain urokinase-type plasminogen activator as substrates. *J. Biol. Chem.* **275,** 26333–26342.

Tarui, T., Mazar, A. P., Cines, D. B., and Takada, Y. (2001). Urokinase receptor (uPAR/CD87) is a ligand for integrins and mediates cell-cell interaction. *J. Biol. Chem.* **276,** 3983–3990.

Teesalu, T., Kulla, A., Asser, T., Koskiniemi, M., and Vaheri, A. (2002). Tissue plasminogen activator as a key effector in neurobiology and neuropathology. *Biochem. Soc. Trans.* **30,** 183–189.

Thorsen, S., Philips, M., Selmer, J., Lecander, I., and Astedt, B. (1988). Kinetics of inhibition of tissue-type and urokinase-type plasminogen activator by plasminogen-activator inhibitor type 1 and type 2. *Eur. J. Biochem.* **175,** 33–39.

Tordai, H., Banyai, L., and Patthy, L. (1999). The PAN module: the N-terminal domains of plasminogen and hepatocyte growth factor are homologous with the apple domains of the prekallikrein family and with a novel domain found in numerous nematode proteins. *FEBS Lett.* **461,** 63–67.

Towle, M. J., Lee, A., Maduakor, E. C., Schwartz, C. E., Bridges, A. J., and Littlefield, B. A. (1993). Inhibition of urokinase by 4-substituted benzo[b]thiophene-2-carboxamidines: an important new class of selective synthetic urokinase inhibitor. *Cancer Res.* **53,** 2553–2559.

Tsirka, S. E., Bugge, T. H., Degen, J. L., and Strickland, S. (1997a). Neuronal death in the central nervous system demonstrates a non-fibrin substrate for plasmin. *Proc. Natl. Acad. Sci. USA* **94,** 9779–9781.

Tsirka, S. E., Gualandris, A., Amaral, D. G., and Strickland, S. (1995). Excitotoxin-induced neuronal degeneration and seizure are mediated by tissue plasminogen activator. *Nature* **377,** 340–344.

Tsirka, S. E., Rogove, A. D., Bugge, T. H., Degen, J. L., and Strickland, S. (1997b). An extracellular proteolytic cascade promotes neuronal degeneration in the mouse hippocampus. *J. Neurosci.* **17,** 543–552.

Turner, R. B., Liu, L., Sazonova, I. Y., and Reed, G. L. (2002). Structural elements that govern the substrate specificity of the clot-dissolving enzyme plasmin. *J. Biol. Chem.* **277,** 33068–33074.

Van Nostrand, W. E., and Porter, M. (1999). Plasmin cleavage of the amyloid beta-protein: alteration of secondary structure and stimulation of tissue plasminogen activator activity. *Biochemistry* **38,** 11570–11576.

Vassalli, J. D., and Belin, D. (1987). Amiloride selectively inhibits the urokinase-type plasminogen activator. *FEBS Lett.* **214,** 187–191.

Vassalli, J. D., Baccino, D., and Belin, D. (1985). A cellular binding site for the M_r 55,000 form of the human plasminogen activator, urokinase. *J. Cell Biol.* **100,** 86–92.

Verde, P., Stoppelli, M. P., Galeffi, P., Di Nocera, P., and Blasi, F. (1984). Identification and primary sequence of an unspliced human urokinase poly(A)+ RNA. *Proc. Natl. Acad. Sci. USA* **81,** 4727–4731.

Verrall, S., and Seeds, N. W. (1989). Characterization of ^{125}I-tissue plasminogen activator binding to cerebellar granule neurons. *J. Cell Biol.* **109,** 265–271.

312   Vincent Ellis

Vines, D. J., Lee, S. W., Dichek, D. A., and Ellis, V. (2000). Receptor-mediated regulation of plasminogen activator function: plasminogen activation by two directly membrane-anchored forms of urokinase. *J. Pept. Sci.* **6,** 432–439.

Wary, K. K., Mariotti, A., Zurzolo, C., and Giancotti, F. G. (1998). A requirement for caveolin-1 and associated kinase Fyn in integrin signaling and anchorage-dependent cell growth. *Cell* **94,** 625–634.

Wei, Y., Lukashev, M., Simon, D. I., Bodary, S. C., Rosenberg, S., Doyle, M. V., and Chapman, H. A. (1996). Regulation of integrin function by the urokinase receptor. *Science* **273,** 1551–1555.

Wei, Y., Waltz, D. A., Rao, N., Drummond, R. J., Rosenberg, S., and Chapman, H. A. (1994). Identification of the urokinase receptor as an adhesion receptor for vitronectin. *J. Biol. Chem.* **269,** 32380–32388.

Wei, Y., Yang, X., Liu, Q., Wilkins, J. A., and Chapman, H. A. (1999). A role for caveolin and the urokinase receptor in integrin-mediated adhesion and signaling. *J. Cell Biol.* **144,** 1285–1294.

Werb, Z. (1997). ECM and cell surface proteolysis: Regulating cellular ecology. *Cell* **91,** 439–442.

Werner, F., Razzaq, T. M., and Ellis, V. (1999). Tissue plasminogen activator binds to human vascular smooth muscle cells by a novel mechanism—Evidence for a reciprocal linkage between inhibition of catalytic activity and cellular binding. *J. Biol. Chem.* **274,** 21555–21561.

Whitelock, J. M., Murdoch, A. D., Iozzo, R. V., and Underwood, P. A. (1996). The degradation of human endothelial cell-derived perlecan and release of bound basic fibroblast growth factor by stromelysin, collagenase, plasmin, and heparanases. *J. Biol. Chem.* **271,** 10079–10086.

Wilcox, S. A., Reho, T., Higgins, P. J., Tominna-Sebald, E., and McKeown-Longo, P. J. (1996). Localization of urokinase to focal adhesions by human fibrosarcoma cells synthesizing recombinant vitronectin. *Biochem. Cell Biol.* **74,** 899–910.

Wiman, B., and Collen, D. (1979). On the mechanism of the reaction between human alpha 2-antiplasmin and plasmin. *J. Biol. Chem.* **254,** 9291–9297.

Xue, W., Kindzelskii, A. L., Todd, R. F. 3rd, and Petty, H. R. (1994). Physical association of complement receptor type 3 and urokinase-type plasminogen activator receptor in neutrophil membranes. *J. Immunol.* **152,** 4630–4640.

Yebra, M., Parry, G. C. N., Strömblad, S., Mackman, N., Rosenberg, S., Mueller, B. M., and Cheresh, D. A. (1996). Requirement for receptor-bound urokinase-type plasminogen activator for integrin $\alpha_v\beta_5$-directed cell migration. *J. Biol. Chem.* **271,** 29393–29399.

Zhang, L., Seiffert, D., Fowler, B. J., Jenkins, G. R., Thinnes, T. C., Loskutoff, D. J., Parmer, R. J., and Miles, L. A. (2002). Plasminogen has a broad extrahepatic distribution. *Thromb. Haemost.* **87,** 493–501.

Zhou, H. M., Nichols, A., Meda, P., and Vassalli, J. D. (2000). Urokinase-type plasminogen activator and its receptor synergize to promote pathogenic proteolysis. *EMBO J.* **19,** 4817–4826.

10

Cell-Surface Cathepsin B: Understanding Its Functional Significance

Dora Cavallo-Medved[1] *and Bonnie F. Sloane*[1,2]
[1]Department of Pharmacology,
Wayne State University,
School of Medicine,
Detroit, Michigan 48201

[2]Barbara Ann Karmanos Cancer Institute,
Wayne State University,
School of Medicine,
Detroit, Michigan 48201

I. Introduction

Cathepsin B is a lysosomal cysteine protease of the papain family of enzymes, which is traditionally believed to degrade proteins that have entered the lysosomal system (Kirsche *et al.,* 1995; Mort and Buttle, 1997). This bilobal protein is initially synthesized on the surface of the rough endoplasmic reticulum (RER) as a preproenzyme containing a signal sequence. In the RER, the signal sequence is cleaved and cathepsin B becomes an inactive 46-kDa proenzyme (procathepsin B). Procathepsin B is further processed into two active forms: (1) a 31-kDa

Current Topics in Developmental Biology, Vol. 54
Copyright 2003, Elsevier Science (USA). All rights reserved.
0070-2153/03 $35.00

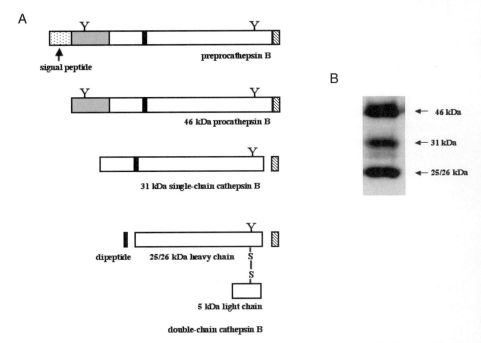

Figure 1 A schematic diagram illustrating the various isoforms of cathepsin B. (A) Cathepsin B is synthesized as a preproenzyme containing a signal sequence (dotted area). In the RER, the signal sequence is cleaved and cathepsin B becomes a latent 46-kDa proenzyme. In the late endosomes, the proregion (shaded gray) of procathepsin B is cleaved converting the enzyme into an active 31-kDa single chain form. The striped region represents the C-terminal removal of 6 amino acids that does not affect activity. Further processing of cathepsin B occurs in the lysosomes by cleavage of a dipeptide (shaded black) resulting in an active double-chain form of cathepsin B. The double-chain form consists of a 25-/26-kDa heavy chain linked by a disulfide bond to a 5-kDa light chain. "Y" depicts potential glycosylation sites. (B) Immunoblot illustrating the 46-kDa procathepsin B and the 31-kDa and 25-/26-kDa active cathepsin B secreted into the media of DU145 human prostate cancer cells. The 5-kDa light chain cannot be observed under these conditions.

single-chain form found in late endosomes and (2) a double-chain form consisting of a 25-/26-kDa heavy chain and 5-kDa light chain found in lysosomes (Fig. 1).

The role of cathepsin B does not appear to be limited to the degradation of proteins within lysosomes but instead this enzyme is hypothesized to be involved in a number of disease processes and is a candidate for participation in tumor progression. In synovial tissue from the joints of patients with rheumatoid arthritis, transcription of cathepsin B is enhanced in synovial cells when compared with normal fibroblasts. Localization of cathepsin B is restricted mainly to the synovial cells that are attached to cartilage and bone at sites of rheumatoid joint erosion

(Trabandt *et al.*, 1991). In osteoarthritis, increased extracellular cathepsin B activity is observed in areas around clefts and in zones of hypercellularity and is involved in cartilage catabolism (Baici *et al.*, 1995). Upregulation of cathepsin B expression, secretion, and activity in osteoarthritis suggest that this enzyme plays a leading role in the progression of the disease (Berardi *et al.*, 2001). Increased cathepsin B levels are also present in the gingival fluid of patients suffering from periodontal disease. Distribution of cathepsin B in periodontal tissue revealed that this enzyme is present mainly in the area in between the epithelium and the adjacent subepithelial connective tissue (Trabandt *et al.*, 1995). Macrophages and fibroblasts in both the inflamed and to a lesser extent the noninflamed areas of the gingival tissue expressed cathepsin B on the surface of these cells. Thus, extracellular cathepsin B may be involved in the destruction of the collagenous matrix of the surrounding connective tissue in gingivitis and periodontitis (Kennett *et al.*, 1997). Extracellular cathepsin B is also observed in normal cell processes. For example, normal thyroid epithelial cells secrete both pro- and active cathepsin B (Brix *et al.*, 1996). Cathepsin B is also transported to the apical plasma membrane of these cells and is proteolytically active at the cell surface. The biological function of cell-surface cathepsin B in thyroid epithelial cells has been reported to be proteolysis of thyroglobulin, the precursor of thyroid hormones (Brix *et al.*, 2001).

The majority of research involving cell-surface cathepsin B has stemmed from studies in cancer. Observations that implicate cathepsin B in malignant progression are its overexpression in tumors as compared to normal tissues, its redistribution from perinuclear lysosomes to peripheral vesicles, and its association with the plasma membrane. Secretion and relocalization of cathepsin B to the cell surface in tumor cells are believed to correlate with tumor progression and clinical outcome for cancer patients. Because of the extensive research conducted on cell-surface cathepsin B and cancer, in this chapter we will focus on the relationship between this enzyme and malignancy with respect to cathepsin B expression, trafficking, and localization. We will further characterize the association of cathepsin B with the plasma membrane by discussing potential cathepsin B binding proteins and introducing caveolae as specific regions for the localization of cell surface cathepsin B. Finally, we will attempt to provide insight into the functional significance of cell surface cathepsin B as an active member of a proteolytic cascade that is postulated to be involved in tumor invasion.

II. Cathepsin B Expression and Cancer

A. Upregulation of Cathepsin B Expression in Tumors and Cancer Cells

Several early studies reported an association of cathepsin B with human and murine malignancies (Poole *et al.*, 1978; Mort *et al.*, 1981; Sloane *et al.*, 1981, 1982; Recklies *et al.*, 1982). Reports by Poole and co-workers (1978) revealed a significantly higher level of cathepsin B secretion from malignant breast tissues as compared to nonmalignant breast tissues. Since these initial studies, several other groups have shown an upregulation of cathepsin B expression and/or activity in breast (Castiglioni *et al.*, 1994), colorectal (Campo *et al.*, 1994), bladder (Eijan *et al.*, 2000; Staack *et al.*, 2002), gastric (Watanabe *et al.*, 1989), lung (Sukoh *et al.*, 1994) and prostate (Sinha *et al.*, 1995) carcinomas, melanomas (Sloane *et al.*, 1981), gliomas (Rempel *et al.*, 1994), osteoclastomas (Page *et al.*, 1992), and chondrosarcomas (Hackel *et al.*, 2000). Work by Frohlich *et al.* (2001) revealed an increase in cathepsin B mRNA, protein expression, and activity levels in human melanoma specimens as compared to normal skin tissues. They further suggested a positive correlation between the upregulation of cathepsin B gene transcription in the tumor cells and the malignancy of the melanoma (Frohlich *et al.*, 2001). Sloane *et al.* (1981) found a positive correlation between cathepsin B activity and the metastatic potential of murine melanoma cell lines. The relationship between upregulation of cathepsin B expression and metastasis may suggest an involvement of this enzyme in tumor invasion. Studies in glioblastomas revealed that an increase in cathepsin B expression and altered localization of the enzyme is correlated to an increase in both histomorphological and clinical evidence of invasion as shown by magnetic resonance imaging (Rempel *et al.*, 1994). *In situ* hybridization in prostate cancer revealed an increase in cathepsin B mRNA at the invasive edges of these tumors (Sinha *et al.*, 1993). In colorectal carcinoma tissues, cathepsin B mRNA levels are enhanced, particularly in tumors that are in the process of invasion (Murnane *et al.*, 1991). This correlation between cathepsin B expression and progression of colon cancer also appears to be associated with the shorter survival of the patients (Campo *et al.*, 1994).

Several regulatory mechanisms have been suggested to contribute to the modulation of cathepsin B expression and localization in tumor and cancer cells. Gene amplification is one of the modes of overexpression of cathepsin B message. In esophageal adenocarcinoma, an amplicon at 8p22, the locus of the cathepsin B gene, has been found to be associated with the amplification and overexpression of the cathepsin B gene (Hughes *et al.*, 1998). Amplification of cathepsin B has also been detected in transformed ovarian cells (Abdollahi *et al.*, 1999). Other regulatory mechanisms including stability of the cathepsin B message (Berquin *et al.*, 1999), alternative mRNA splicing, and use of multiple promoters (Berquin *et al.*, 1995) are also believed to play significant roles in the overexpression of cathepsin B. Konduri *et al.* (2001) reported that the increase in cathepsin B mRNA, protein, and activity levels in glioblastoma cell lines is associated with an increase in cathepsin B promoter activity. Increased cathepsin B activity in thyroid papillary carcinomas as compared to normal thyroid tissue is also correlated with an increase in cathepsin B mRNA content (Shuja and Murnane, 1996). These papillary carcinomas also revealed a prominent cathepsin B staining close to the underlying basement membranes of these cells (Shuja and Murnane, 1996). Cathepsin B staining at the inner surface of the basal plasma membrane immediately adjacent to the basement membrane has also been observed during the progression of colon cancer (Campo *et al.*, 1994). In this case, the redistribution of cathepsin B appears to precede the increase in expression of cathepsin B protein thus suggesting that the altered trafficking of the enzyme is independent of its expression. Immunohistochemical staining for cathepsin B in normal colon tissue and colon carcinoma is illustrated in Fig. 2 (see color insert).

B. Expression of Cathepsin B at the Invasive Edges of Tumors

A role for cathepsin B during tumor invasion has been demonstrated in many human carcinomas. Expression of cathepsin B protein is intense at the invasive edges of tumors including prostate (Sinha *et al.*, 1995), colon (Emmert-Buck *et al.*, 1994; Visscher *et al.*, 1994a) and bladder (Visscher *et al.*, 1994b) carcinomas and infiltrating glioblastomas (Mikkelsen *et al.*, 1995; Rempel *et al.*, 1994). *In situ* hybridization has revealed increases in cathepsin B mRNA levels at the invasive edges of prostate tumors (Sinha *et al.*, 1993). In addition, cathepsin B activity has been shown to be elevated at the invasive or infiltrating edges of microdissected specimens from human colon and breast carcinomas (Frosch *et al.*, 1999; Emmert-Buck *et al.*, 1994). In colorectal carcinomas, elevated cathepsin B activity is detected at the invasive fronts of the cancer cells (Hazen *et al.*, 2000). Staining for cathepsin B in stromal cells associated with colon tumors also has been shown to correlate with increased malignancy and reduced patient survival (Campo *et al.*, 1994). Expression of active cathepsin B at the invading regions of tumors and their associated stroma suggests a functional role for cathepsin B in the degradation of the extracellular matrix during tumor invasion and progression.

C. Expression of Cathepsin B at the Cell Surface

The characteristic localization of cathepsin B in tumor cells appears to be at the cell surface. Subcellular fractionation of metastatic B16 murine melanoma cells identified an association of cathepsin B with both the lysosomal and plasma membrane/endosomal fractions of these cells (Sloane *et al.*, 1986). In both these fractions, only the 31-kDa single-chain cathepsin B was detected. The catalytic activities of cathepsin B from these fractions did not differ nor did the ability of E64, a cysteine protease inhibitor, to reduce these activities (Moin *et al.*, 1998). In the lysosomal fractions, three isozymes of the single chain cathepsin B with pIs of 5.33, 5.2, and 5.1 could be detected. These isozymes were also present in the plasma membrane/endosomal fractions in addition to a fourth cathepsin B isozyme with a pI of 5.64 (Moin *et al.*, 1998). An increase in specific activity of cathepsin B in the plasma membrane fraction positively correlated with the metastatic potential of the three melanoma cell lines (Sloane *et al.*, 1986). Subcellular fractionation of prostate cancer cells also showed an increase in cathepsin B activity in plasma membrane/endosome fractions of malignant prostate as compared to those of benign prostatic hyperplasia (Sinha *et al.,* 2001). Immunogold microscopy further confirmed the presence of cathepsin B in these plasma membrane/endosome fractions (Sinha *et al.,* 2001). Cathepsin B has also been identified in the plasma membrane of lung (Erdel *et al.,* 1990), colorectal (Campo *et al.,* 1994; Hazen *et al.,* 2000) and ovarian (Warwas *et al.,* 1997) carcinomas. Immunofluorescent staining of MCF-10AneoT, MCF-7, and BT20 breast carcinoma cell lines has confirmed the localization of cathepsin B on the outer surface of these cells (Sameni *et al.,* 1995). In MCF-10AneoT cells, cathepsin B staining can also be detected in the cell processes and on the extracellular basal surface of these cells (Sameni *et al.,* 1995). More specifically, in colorectal carcinomas active cathepsin B and the annexin II heterotetramer (AIIt) colocalize to the caveolar membrane regions of these cells (Cavallo-Medved, unpublished data). AIIt has been shown to also interact with components of the extracellular matrix such as collagen I (Wirl and Scwartz-Albiez, 1990) and tenascin-C (Chung and Erikson, 1994), thus suggesting a role for cathepsin B in the degradation of these extracellular matrix proteins. Cell-surface cathepsin B staining also has been shown to be localized to cell processes in regions resembling focal adhesions (Sloane *et al.,* 1994; Rempel *et al.,* 1994) where contact with the extracellular matrix may occur. Earlier studies by Buck *et al.* (1992) revealed that cathepsin B is able to degrade extracellular matrix proteins including laminin, type IV collagen, and fibronectin. In gastric cancer, a significant correlation is observed between increased tumor-associated cathepsin B, decreased tumor-associated laminin, and higher tumor stage (Khan *et al.,* 1998). The presence of active cathepsin B on the cell surface has been demonstrated by continuous cathepsin B activity assays in living cells (Linebaugh *et al.,* 1999). In addition, a novel confocal imaging assay has revealed the participation of cathepsin B in the pericellular degradation of quenched fluorescent extracellular substrates by living human breast cancer cells (Sameni *et al.,* 2000). In both these assay systems the highly selective cathepsin B inhibitor, CA074 (Murata *et al.,* 1991), reduces the degradation of substrates. Hence, localization of cathepsin B at the cell surface and in particular in areas of extracellular matrix degradation indicates a functional role for this protease in tumor invasion.

III. Compartmentalization of Cathepsin B

A. Classical Pathway: Trafficking of Cathepsin B to Lysosomes

In normal systems, trafficking of cathepsin B appears to follow the conventional mechanism found for most soluble lysosomal enzymes (Fig. 3). Cathepsin B is synthesized as a preproenzyme on membrane ribosomes and is recognized by the signal recognition particle, which directs its translation into the RER. In the RER,

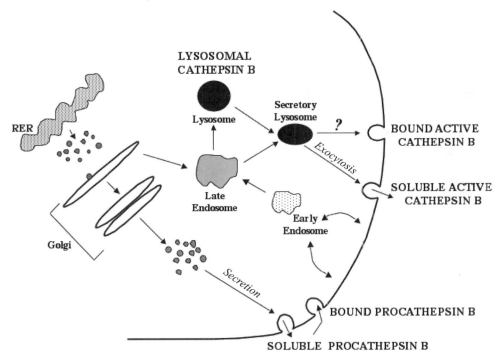

Figure 3 Classical and alternative trafficking pathways of cathepsin B. This diagram illustrates the potential pathways of cathepsin B trafficking in either normal or tumor cells. Procathepsin B is synthesized in the RER and transported through the Golgi. In normal cells, procathepsin B is transported from the Golgi to late endosomes where it is converted to an active 31-kDa single-chain form. Cathepsin B is then transported into lysosomes where it is further processed into an active double-chain form of 25/26 and 5 kDa. This classical cathepsin B trafficking mechanism is also found in tumor cells with the addition of the following alternative pathways: (1) procathepsin B is secreted from the cells via the secretory pathway where it may exist extracellularly as soluble procathepsin B or it may become associated with the plasma membrane as bound procathepsin B; and (2) cathepsin B may be trafficked to secretory lysosomes that either undergo exocytosis thereby secreting soluble active cathepsin B or fuse with the plasma membrane giving rise to bound active cathepsin B.

procathepsin B is cotranslationally glycosylated at two potential N-glycosylation sites: one in the propeptide region and the other in the mature region of the enzyme. Following translation, the presequence is cleaved and cathepsin B proenzyme is then transported to the Golgi complex for modification of oligosaccharides to high-mannose carbohydrates and phosphorylation of the mannose residues. These modifications allow for the binding of procathepsin B (46 kDa) to membrane-bound mannose 6-phosphate receptors (MPRs), which then transport the enzyme to prelysosomal compartments (i.e., late endosomes). Two MPRs involved in this event are the 300-kDa cation-independent CI-MPR and the 46-kDa cation-dependent CD-MPR. Fibroblasts established from mouse embryos defective in either or both CI-MPR and CD-MPR show that both receptors are required for efficient trafficking of lysosomal enzymes (Ludwig *et al.*, 1994). Acidification of the endosomal compartments to pH 5.0 allows for the dissociation of enzymes from their receptors and the MPRs are then recycled back to the Golgi apparatus or to the cell surface. Within these prelysosomal compartments the proregion of cathepsin B is cleaved, thus converting the enzyme to a single-chain 31-kDa active form. Cathepsin B is then transported to the lysosomes where in some tissues the enzyme is further processed to its double-chain form that consists of a 25-/26-kDa heavy chain and a 5-kDa light chain (Nishimura *et al.*, 1988; Rowan *et al.*, 1992). Under normal conditions, cathepsin B is stored in perinuclear lysosomes as a housekeeping enzyme and is involved in protein degradation, although secretion of approximately 5–10% procathepsin B has been observed in human skin fibroblasts (Hanewinkel *et al.*, 1987) and human hepatoma cells (Mach *et al.*, 1992).

B. Alternative Pathway: Trafficking of Cathepsin B to the Cell Surface and Secretion

In cancer cells, cathepsin B is localized in perinuclear lysosomes, in vesicles located in the peripheral cytoplasm and in cell processes, and on the plasma membrane. As well, extracellular secretion of pro- and mature cathepsin B has been observed. Currently, there are few data illustrating the mode of cathepsin B trafficking to the cell surface and beyond (Fig. 3). In breast-cancer cell lines, staining for cathepsin B and two other lysosomal cysteine proteases, cathepsin D and L, has revealed that the three lysosomal enzymes are localized in separate vesicles, thus suggesting that these proteases are shuttled intracellularly via discrete pathways (Sameni *et al.*, 1995). In I-cell fibroblasts, trafficking of cathepsin D is believed to occur via both MPR-dependent and MPR-independent pathways (von Figura and Hasilik, 1986). An MPR-independent pathway for the trafficking of both cathepsin D and L has been identified and a 43-kDa lysosomal proenzyme receptor was proposed to be involved (McIntyre and Erikson, 1991, 1993). This lysosomal proenzyme receptor is an integral membrane protein that at pH 5.0 binds specifically to a 9-residue sequence in the N-terminal region of procathepsin L (McIntyre *et al.*, 1994). Procathepsin D has also been proposed to traffic via a lysosomal proenzyme receptor (McIntyre and Erikson, 1991); however, to date these receptors have not been shown to bind procathepsin B.

The vesicles containing active cathepsin B at the cell periphery do not stain for either Rab 11 (a marker for post-Golgi membranes including the trans-Golgi network and secretory vesicles) or CI-MPR (a marker for late endosomes). As well, these vesicles do not stain for early endosomal markers thus indicating that active cathepsin B is unlikely to have been engulfed from an external source. Instead these vesicles appear to be lysosomes since they do stain for lysosomal markers such as LAMP-1 and Lysotracker (Sameni and Sloane, unpublished data).

This suggests that secretion of active cathepsin B is directly from exocytosis of lysosomes. Linebaugh *et al.* (1999) reported that secretion of active cathepsin B in tumor cells is independent of procathepsin B secretion. Retrograde trafficking of lysosomes and calcium-regulated exocytosis of lysosomes at the cell surface has been suggested in several cell types (Rodriguez *et al.,* 1997). Traditionally, lysosomes have been defined as dead-end organelles; however, more recent studies depict lysosomes as dynamic compartments possessing molecular machinery that allows them, under certain conditions, to fuse with the plasma membrane (Andrews, 2000). Secretory lysosome is the term initially used to define lysosomes that were thought to be transformed into regulated secretory organelles by mechanisms unique to hematopoietic cells. Interestingly, other cell types including fibroblasts have also been shown to have the ability to induce transformation of lysosomes into secretory organelles, thus suggesting that this event is more common than originally proposed (Andrews, 2000). In osteoclasts, a reorganization of lysosomes results in the translocation of lysosomal enzymes and proton pumps to the ruffled border membrane and the release of these enzymes at the site of bone resorption (Mostov and Werb, 1997). Calcium-dependent exocytosis of secretory lysosomes in dendritic cells is induced by specific binding of these cells to CD8(+) T cells resulting in the polarized secretion of IL-1β and active lysosomal cathepsin D toward the interacting T lymphocytes (Gardella *et al.,* 2001). These results expose the existence of cross-talk between the two cell types. Although the mechanism(s) involved in cathepsin B transport to the tumor cell surface remains to be elucidated, it is tempting to suggest that exocytosis of secretory lysosomes may be responsible for the trafficking and secretion of active cathepsin B to the cell surface and into the extracellular milieu (Fig. 3). In turn, it is intriguing to suggest that interactions between tumor cells and the underlying stroma may induce transformation of lysosomes into secretory lysosomes. Although there has been no direct evidence that the altered trafficking and secretion of cathepsin B in tumor cells is via calcium-dependent exocytosis of secretory lysosomes, recent studies suggest that 12-*O*-tetradecanoylphorbol-13-acetate (PMA) activates signaling processes that trigger exocytosis of a subpopulation of active cathepsin-B-containing lysosomes/endosomes. In turn, secreted cathepsin B appears to initiate a proteolytic cascade involving urokinase-type plasminogen activator (uPA), plasminogen, and plasmin resulting in the activation of serum-derived latent transforming growth factor-β (TGF-β) (Guo *et al.,* 2002).

Altered localization of cathepsin B to the plasma membrane appears to be influenced by several factors. Sloane *et al.* (1986) initially found that cathepsin B was associated with the plasma membrane of B16 metastatic murine melanoma cells and that cathepsin B activity in the plasma membrane fractions correlated with the metastatic potential of these cells. Further studies in these melanoma cells revealed that membrane-associated cathepsin B appears to be stable at physiological extracellular pH and has decreased inhibitor binding potential as compared to non-membrane-associated cathepsin B. The release of cathepsin B from the surface of these cells was shown to occur only with treatment using detergents (Rozhin *et al.*, 1987). Although these studies did not identify a specific cell-surface receptor for cathepsin B they did exclude the possibility of cathepsin B binding to mannose 6-phosphate receptors on the membrane (Sloane *et al.*, 1986).

The MCF-10 model system, which has been used to analyze early stages of human breast-cancer progression, has also been used in the study of cathepsin B trafficking and localization to the cell surface. Analysis of cathepsin B localization in the parental nontumorigenic cell line MCF-10A and its oncogenic c-H-*ras* transfected subline MCF-10AneoT revealed differential regulation and alterations in the trafficking of cathepsin B between the two cell lines (Sloane *et al.*, 1994). MCF-10AneoT cells displayed an increase in membrane-associated cathepsin B protein and activity without an observed increase in the level of cathepsin B mRNA. Immunocytochemical analysis of neoplastic MCF-10AneoT cells revealed perinuclear staining for cathepsin B as well as staining at the cell periphery, including in cell processes. This pattern of cathepsin B staining was similarly observed in human breast carcinoma cell lines MCF-7 and BT20 (Sameni *et al.*, 1995). Extracellular staining for cathepsin B in MCF-10AneoT cells was seen as discrete patches on the basal surface of the cells. In contrast, there was no surface staining for cathepsin B in MCF-10A cells, and instead cathepsin B staining was shown to be exclusively perinuclear (Sloane *et al.*, 1994; Sameni *et al.*, 1995). Although *ras*-transfected MCF-10A cells also show trafficking of cathepsin D to the cell periphery, studies using this model system suggest that these enzymes are trafficked in separate vesicles (Sameni *et al.*, 1995). The role of ras and its related proteins is still being investigated with respect to cathepsin B trafficking and activity in cancer cells. We have found that colorectal carcinoma cells expressing active Ki-*ras* exhibited elevated levels of cell surface cathepsin B activity compared to a Ki-*ras* disrupted subline (Cavallo-Medved, unpublished data). Since expression of cathepsin B protein was not upregulated in these active Ki-*ras* expressing cells, we suggest that the enhanced cell surface cathepsin B activity is a result of increased cathepsin B trafficking to the cell periphery. That alterations in the trafficking lead to elevated expression and activity of cathepsin B on the membrane suggests a functional role for cathepsin B at the cell surface. Considering the positive correlation between cell surface cathepsin B and cancer progression the enzyme is likely involved in the tumor cells' ability to degrade the extracellular matrix.

Secretion of both active and procathepsin B from various cancer cells has been observed. Although secretion of procathepsin B may indicate a saturation of the mannose 6-phosphate pathway, secretion of active cathepsin B would indicate a more complex event. Two mechanisms may exist: (1) cathepsin B is processed intracellularly, transported through lysosomes, and secreted in its active forms and/or (2) cathepsin is processed and activated on the cell surface and released extracellularly (Fig. 3).

Although the mechanism of cathepsin B secretion remains unclear, our laboratory has shown various factors that appear to induce this process. Increased secretion of both latent and active cathepsin B is characteristic of many tumor types such as breast (Poole *et al.*, 1978) and colon (Maciewicz *et al.*, 1989) carcinomas, gliomas (McCormick, 1993), and murine melanomas (Qian *et al.*, 1989). Studies on B16 melanoma cells revealed that under slightly acidic conditions (pH 6.5) cathepsin-B-containing vesicles migrate toward the cell periphery and secretion of active cathepsin B is enhanced (Rozhin *et al.*, 1994). A reduction in cytoplasmic pH has been shown to induce movement of lysosomes to the cell surface along microtubules (Heuser, 1989). Distribution of cathepsin B along microtubules has also been observed in MCF-10AneoT cells (Sloane *et al.*, 1994). The effects of pH may be physiologically significant considering the acidic microenvironments that exist beneath osteoclasts and inflammatory macrophages (Silver *et al.*, 1988) and surrounding tumors (Young and Spevacek, 1992). The *in vitro* modulation of pH may mimic the *in vivo* environment of these systems.

Another method in which mature cathepsin B secretion is enhanced is via the treatment of cells with 12-(S)-hydroxyeicosatetraenoic acid (12-S-HETE), a product of arachidonic acid metabolism (Honn *et al.*, 1994). In B16a cells, 12-S-HETE treatment triggers the secretion of active cathepsin B from the cells (Honn *et al.*, 1994). In addition, vesicular staining for cathepsin B is observed at the cell periphery. This pattern of peripheral staining for cathepsin B is also detected in lung tumors treated with 12-S-HETE (Ulbricht *et al.*, 1997). It was suggested that the effects of this compound on cathepsin B secretion might be via modulation of the cytoskeleton. Along with its effects on active cathepsin B secretion, 12-S-HETE has been shown to reduce the secretion of endogenous cysteine protease inhibitors from B16a cells (Sloane *et al.*, 1991) and upregulate the expression of a cell surface integrin, $\alpha_{IIb}\beta_3$ (Chopra *et al.*, 1991) and autocrine motility factor receptor (Silletti *et al.*, 1994). Changes in the expression of these proteins may modulate the composition of the extracellular matrix and favor conditions for cathepsin B participation in tumor invasion.

IV. Association of Cathepsin B with the Cell Surface

The mechanism(s) by which cathepsin B binds to the extracellular surface of tumor cells still remains to be elucidated. Cathepsin B has been shown to bind alpha 2 macroglobulin and its receptor, the low-density lipoprotein receptor-related protein/alpha 2-macroglobulin on tumor cell surfaces (Arkona and Wiederanders, 1996). This receptor has also been shown to bind both pro-uPA and urokinase/plasminogen activator inhibitor-1 complexes (Zhang *et al.*, 1998). Using a yeast two-hybrid system several other proteins have been identified as potential cathepsin B cell surface binding proteins (Mai *et al.*, 2000). The most studied of these proteins is p11, the light chain of the AIIt, which has also been shown to bind plasminogen (Kassam *et al.*, 1998a). The heavy chain of the AIIt, p36, binds tissue plasminogen activator (tPA) (Hajjar *et al.*, 1994). Although a physiological effect of AIIt at the cell surface has yet to be established, it has been suggested that this complex may facilitate the interactions between proteases and the extracellular matrix proteins. Other potential cathepsin B binding proteins are: (1) MAGE-3, a melanoma surface antigen expressed in colon cancer and recognized by specific cytotoxic T lymphocytes (Kocher *et al.*, 1995); (2) preprogalanin, a precursor of galanin that inhibits the development of chemically induced colon cancers in rats (Iishi *et al.*, 1995); (3) DP1, a gene of the familial adenomatous polyposis locus encoding a protein of unknown function; and (4) a calcium and integrin binding protein that interacts with the α_{IIb} domain of the platelet fibrinogen receptor $\alpha_{IIb}\beta_3$ (Naik *et al.*, 1997). Similar to cell surface cathepsin B, the expression of $\alpha_{IIb}\beta_3$ can also be upregulated by 12-S-HETE (Timar *et al.*, 1995).

A. Annexin II Heterotetramer and Cathepsin B

AIIt is a member of a family of calcium-dependent, phospholipid binding proteins. It is associated with the plasma membrane (Thiel *et al.*, 1992) and is composed of two copies of the 36-kDa heavy chain, p36, and two copies of the 11-kDa light chain, p11 (Gerke and Weber, 1984; Erikson *et al.*, 1984). The physiological function of AIIt at the cell surface may include a role in calcium-dependent exocytosis (Burgoyne, 1998), endocytosis (Emans *et al.*, 1993), cell–cell adhesion (Robitzki *et al.*, 1990), and most recently as a membrane binding site for extracellular pro-cathepsin B (Mai *et al.*, 2000).

AIIt expression on the cell surface appears to be positively correlated with the tumorigenicity and malignancy of metastatic sublines of human colon cancer and rodent breast cancer and lymphomas (Yeatman *et al.*, 1993). p11 is a member of the S100 family of proteins, which are associated with neoplasia via deregulation of their gene expression. Overexpression of p36 has been associated with gliomas (Reeves *et al.*, 1992; Roseman *et al.*, 1994), hepatomas (Frohlich *et al.*, 1990), lymphomas (Chiang *et al.*, 1996), and pancreas (Kumble *et al.*, 1992; Vishwanatha *et al.*, 1993) and lung (Cole *et al.*, 1992) cancers. As a membrane protein, the role of AIIt in cancer metastasis may be linked to its ability to act as a receptor for many different ligands such as extracellular matrix proteins collagen I (Wirl and Schwartz-Albiez, 1990) and tenascin-C (Chung and Erikson, 1994) and proteases including plasminogen (Kassam *et al.*, 1998a), tPA (Hajjar *et al.*, 1994) and procathepsin B (Mai *et al.*, 2000). AIIt may function to localize both extracellular matrix proteins and enzymes in order to organize and mediate interactions between these proteins at the cell surface. In tumor cells, the AIIt-dependent associations between matrix proteins and enzymes that degrade them may facilitate regulated modification and turnover of the extracellular matrix, essential for tumor invasion.

It was reported that p36 binds plasminogen and stimulates the activation of plasminogen via a tPA-dependent event (Hajjar and Hamel, 1990). Inhibition of plasminogen binding to p36 in macrophages results in the loss of cell surface matrix degradation and cell invasion through the extracellular matrix (Falcone *et al.,* 2001). In contrast, Kassam *et al.* (1998b) established the p11 subunit of AIIt as the plasminogen-binding site and revealed that p11 is responsible for most of the AIIt stimulation of tPA-mediated plasminogen activation. The association of AIIt with the uPA/plasmin system is interesting since p11 was found to be the binding site for membrane associated procathepsin B (Mai *et al.,* 2000). It was speculated that activation of procathepsin B at the cell surface may involve its binding to AIIt. This hypothesis is attractive since active cathepsin B initiates the conversion of both soluble and membrane-bound pro-uPA to uPA that in turn can activate plasminogen (Kobayashi *et al.,* 1991; Ikeda *et al.,* 2000). In ovarian cancer cells, inhibition of cell surface cathepsin B has been shown to prevent both the activation of pro-uPA to uPA and the invasion of these cells through Matrigel (Kobayashi *et al.,* 1993). Inhibition of cathepsin B in MCF-10A cells also suppresses activation of pro-uPA and the cytostatic effects of PMA (Guo *et al.,* 2002). Thus, the association of procathepsin B, AIIt, and the uPA/plasmin system may stimulate early proteolytic cascade events leading to tumor cell invasion.

B. Caveolae as Sites for Localization of Cell-Surface Cathepsin B

AIIt has been identified in the lipid-rich regions of the plasma membrane termed caveolae (Sargiacomo *et al.,* 1993; Harder and Gerke, 1994). Caveolae are 50- to 100-nm vesicular invaginations that represent highly lipid-rich subdomains of the plasma membrane. They are found in most cells and are abundant in adipocytes, endothelial cells, fibroblasts, epithelial cells, and muscle cells (Sargiacomo *et al.,* 1993; Lisanti *et al.,* 1994). Caveolae have unique lipid compositions that are rich in cholesterol and glycosphingolipids and as such they are insoluble to Triton X-100, resistant to high-salt conditions, and lighter in buoyant density than other cellular membranes (Sargiacomo *et al.,* 1993). They are believed to play roles in endocytosis, cholesterol transport, and cell signaling events (Smart *et al.,* 1999). Caveolae are also believed to be involved in cell-surface proteolysis (Fig. 4). The clustering of urokinase-plasminogen activator receptor (uPAR) and its ligand uPA in caveolae enhances the activation of plasminogen at the cell surface (Stahl and Mueller, 1995) and in turn may lead to cell invasion. In addition, the colocalization of MMP-2, its activator MT1-MMP, its proposed receptor $\alpha_v\beta_3$, and its inhibitor TIMP-2 within caveolae implicate this membrane region as a site for restricted cell invasion (Annabi *et al.,* 2001; Puyraimond *et al.,* 2001). More recently, both the 31-kDa and the 25-/26-kDa active cathepsin B isoforms have been identified in the caveolae of colorectal carcinoma cells. The localization of cathepsin B to caveolae is not surprising considering that cathepsin B activates soluble and membrane-bound pro-uPA (Kobayashi *et al.,* 1991), an early event associated with the proteolytic cascade involved in tumor invasion.

The major structural protein of caveolae, caveolin-1, has been shown to be a negative regulator of many signaling proteins such as heterotrimeric G proteins, protein kinase C, adenylyl cyclase, nitric oxide synthase, and *src*-family tyrosine

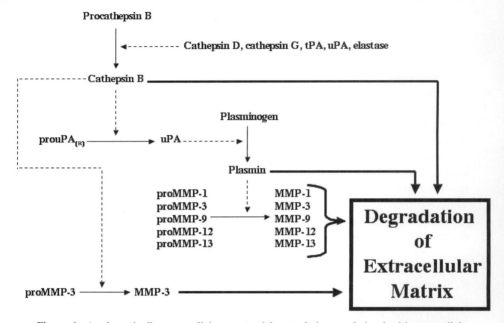

Figure 4   A schematic diagram outlining a potential proteolytic cascade involved in extracellular matrix degradation and tumor invasion. This extracellular proteolytic cascade begins with the activation of procathepsin B by other proteases. Active cathepsin B then activates proMMP-3 to MMP-3 and both soluble and receptor-bound pro-uPA$_{(R)}$ to uPA, which leads to the activation of plasminogen to plasmin. Plasmin in turn can activate a series of metalloproteinases that can be directly involved in degradation of the extracellular matrix. In addition, both active cathepsin B and plasmin can directly participate in degradation of the extracellular matrix. The cascade of proteolytic events involved in degradation of the extracellular matrix leads to tumor invasion and metastasis. (Dotted lines) Participation of an enzyme(s) in the activation of other enzymes; (thin solid lines) conversion of an enzyme from its inactive to active form; (thick solid lines) direct proteolysis of the extracellular matrix.

kinases that may participate in cell adhesion, migration, and invasion (Smart *et al.,* 1999). Caveolin-1 is also involved in the trafficking of cholesterol between the cell surface and the ER (Smart *et al.,* 1996) and participates in the organization and activation of signaling molecules concentrated on the cytoplasmic side of caveolae (Smart *et al.,* 1999). Interestingly, it has been revealed that caveolin-1 plays a role in cell growth and cancer development, and more specifically, it has been suggested that this protein possess tumor suppressor activities. Evidence shows that transformation of NIH 3T3 cells via v-*abl* or H-*ras* activation reduces the levels of caveolin-1 expression which in turn leads to the loss of anchorage-dependent growth and the ability of these cells to form foci on soft agar (Koleske *et al.,* 1995). Although in certain cell types a reduction in caveolin-1 expression is sufficient to mediate cellular transformation, the hypothesis that caveolin-1 may be a tumor suppressor still remains controversial. The caveolin-1 gene is located on human chromosome 7q31.1 (Engelman *et al.*, 1997), a region frequently deleted in human breast cancer (Zenklusen *et al.*, 1994), and it was identified as one of 26 gene products that are downregulated in breast cancer cells (Sager *et al.*, 1994). Caveolin-1 protein levels are also reduced in mammary (Lee *et al.*, 1998) and lung (Racine *et al.*, 1999) carcinomas. On the other hand, several observations argue against caveolin-1 as a tumor suppressor. Caveolin-1 overexpression, observed in primary and metastatic prostate carcinomas (Yang *et al.*, 1998, 1999), colon adenocarcinoma (Fine *et al.*, 2001), invasive ductal breast carcinoma (Yang *et al.*, 1998), and high-grade bladder cancer (Rajjayabun *et al.*, 2001), has been suggested to be positively correlated with tumorigenesis. A correlation has been observed between caveolin-1 expression and multidrug resistance of cancer cell lines (Lavie *et al.*, 1998). Furthermore, the overexpression of caveolin-1 in esophageal squamous cell carcinoma appears to correlate with lymph node metastasis and a decrease in the overall patient survival rate (Kato *et al.*, 2002).

Recent findings suggest that the expression of caveolin-1 is upregulated in HCT 116 colorectal carcinoma cells expressing active Ki-*ras* in comparison to HKh-2 colorectal carcinoma cells in which the mutant Ki-*ras* allele has been disrupted (Cavallo-Medved, unpublished data). Altered trafficking of caveolin-1 to various other subcellular compartments is also observed in HCT 116 cells. This is interesting considering that the trafficking to and activity of cathepsin B at the cell surface is also upregulated in these Ki-*ras* mutated colon cells. Other studies also revealed a correlation between mutations in Ki-*ras* and upregulation in cathepsin B expression and activity in colorectal carcinoma cell lines (Yan *et al.*, 1997) as well as primary human colorectal carcinoma cells (Kim *et al.*, 1998). As previously mentioned, both the expression and activity of mature cathepsin B at the plasma membrane increases without any alteration in cathepsin B mRNA levels in H-*ras* transformed MCF-10 breast epithelial cells. Thus, modulation of Ki-*ras* and H-*ras* expression appears to increase the trafficking of active cathepsin B to the cell surface, which in turn contributes to the invasive potential of these cells. Kim *et al.* (1998) reported that a mutation in Ki-*ras* or expression of altered forms of N-*ras* protein increases the tumorigenicity of colorectal carcinomas by inducing both cathepsin B and cathepsin L expression. Increases in cathepsin B trafficking to and activity at the surface of colon cancer cells may enhance tumor cell invasion. Studies using Matrigel invasion assays revealed that HCT 116 cells are more invasive than HKh-2 cells (Cavallo-Medved, unpublished data). Perhaps the enhanced invasive potential of HCT 116 cells is influenced by the increase in caveolin-1 expression. A contribution to the invasive potential of colon cancer cells by an increased caveolin expression has previously been suggested (Fine *et al.*, 2001). Furthermore, the fact that active cathepsin B was detected in the caveolae of both HCT 116 and HKh-2 cells suggests a functional role for caveolae-associated cathepsin B. This role may include tumor invasion in coordination with the uPA/uPAR system and matrix metalloproteinases.

V. Functional Significance of Cell-Surface Cathepsin B

The main question in regard to the localization of cathepsin B to the cell surface of tumor cells is its functional significance. From the evidence accumulated thus far many scientists hypothesize that cathepsin B plays a role in tumor invasion and metastasis. These processes involve the participation of cathepsin B: (1) directly in the degradation of extracellular matrix proteins; and (2) indirectly by activating other proteases, thereby initiating proteolytic cascades. Our investigations into the functional significance of cell surface cathepsin B in tumor invasion have mostly consisted of *in vitro* activity assays. These assays have been established to determine the ability of living cells to degrade substrates in the presence and absence of protease inhibitors. They include: (1) a "real-time" assay which measures hydrolysis of substrates by cathepsin B both pericellularly (i.e., cell-surface-associated and secreted cathepsin B) and intracellularly; and (2) a confocal imaging assay that colocalizes the hydrolysis of quenched fluorescent proteins and cathepsin B activity in living cells grown on a three-dimensional matrix.

A. Cathepsin B Activity at the Cell Surface

Clues toward understanding the role of cathepsin B at the cell surface have been found via the measurement of both secreted and membrane-associated cathepsin B. The "real-time" cathepsin B activity assay has been used to measure the hydrolysis of synthetic substrates by cathepsin B in living cells over a continuous time period (Linebaugh *et al.,* 1999). This assay allows for the quantitation of pericellular levels of cathepsin B activity in comparison to both intracellular and total cellular cathepsin B activities. Cathepsin B activity has been measured in various tumor cell lines including breast carcinoma, colon adenocarcinoma, and glioblastoma cells, and in some of these cell lines the pericellular cathepsin B activity was shown to be 40–50% membrane-associated (Linebaugh *et al.,* 1999). In *ras*-transfected MCF-10AneoT cells, the pericellular cathepsin B activity was shown to be twice as much as the pericellular cathepsin B activity exhibited by their parental MCF-10A counterparts (Linebaugh *et al.,* 1999). Although it remains to be determined whether secreted cathepsin B is activated at the cell surface or becomes associated with the cell surface following activation, in some cell lines membrane-associated cathepsin B activity is independent of constitutive secretion of procathepsin B. This would suggest that cathepsin B is transported to the cell surface directly by lysosomes and/or late endosomes and thus supports the observations of calcium-mediated exocytosis of lysosomes by fibroblasts and epithelial cells. Although the "real-time" activity assay is beneficial in that it can monitor changes in the kinetic rate of pericellular cathepsin B activity under various conditions, it is incapable of determining the ability of cell-surface cathepsin B to degrade specific matrix proteins.

Optical imaging, using near-infrared probes, has been used to visualize proteolytic activity of tumors *in vivo* (Mahmood *et al.*, 1999; Tung *et al.*, 2000). These studies have shown tumor-associated activity of cathepsin B in mouse models and suggest that cathepsin B is a potential biomarker for the detection of such lesions (Marten *et al.*, 2002). A recently developed *in vitro* optical imaging assay uses confocal microscopy to visualize degradation of specific quenched fluorescent substrates by living cells (Sameni *et al.*, 2000). In this case, living cells are grown on matrix that is mixed with a protein substrate cross-linked to a quenched fluorescein iso thio cyanate (FITC) derivative (e.g., DQ-collagen IV). When this protein substrate is cleaved the product formed is fluorescent and is qualitatively observed under confocal microscopy. This technology allows for visualization of proteolysis of specific extracellular matrix proteins both extracellularly and intracellularly by optical sectioning through both the cells and the surrounding extracellular matrix.

Collagen and laminin are basement membrane proteins that can be cleaved by cathepsin B. Localization of cathepsin B is found at the basal surface of breast carcinoma and glioma cells in areas of laminin degradation (Sloane, 1996). Moreover, inhibition of cathepsin B activity partially reduces laminin degradation. Cathepsin B staining is also localized in the pseudopodia of lung tumor cells invading into a collagen gel (Strohmaier *et al.*, 1997). In the confocal imaging assays, DQ-collagen IV mixed in a three-dimensional matrix is used as a proteolytic substrate. This assay was employed to compare the degradation pattern between two human breast cancer cell lines, BT20 and BT549 (Sameni *et al.*, 2000). Interestingly, as these cell lines invaded into the matrix they revealed distinct differences in their ability to degrade DQ-collagen IV. BT20 cells appeared to degrade the DQ-collagen IV extracellularly and more specifically in the finger-like pits or tunnels directly underneath the cells. In BT549 cells, however, fluorescent degradation products are localized intracellularly in perinuclear vesicles that were identified as lysosomes. Considering there was no pericellular accumulation of degradation products in BT549 cells, the DQ-collagen IV likely enters the cells by an endocytic mechanism and is degraded in the lysosomes. Degradation of cresyl-violet substrates specifically by cathepsin B revealed that this enzyme is present in both the pericellular DQ-collagen IV degradation sites in BT20 cells and the intracellular DQ-collagen IV degradation sites in BT549 cells, thus suggesting a role for cathepsin B in the degradation of this matrix protein (Sameni *et al.*, 2000). Furthermore, a significant reduction in the pericellular and intracellular proteolysis of DQ-collagen IV in the presence of specific cathepsin B inhibitors, CA-074 and CA-074Me (membrane-permeable derivative of CA-074), was also revealed. Pericellular degradation is also partially inhibited by both serine and metalloproteinase inhibitors. These findings suggest that cathepsin B is the primary enzyme responsible for the degradation of DQ-collagen IV intracellularly in BT549, and in BT20 cells cathepsin B is a participant in pericellular DQ-collagen IV degradation in addition to other proteases. Similar results were also found in U87 glioma cells grown on a matrix mixed with DQ-collagen IV (Sameni *et al.*, 2001). In this case, degradation of the protein substrate occurs solely intracellularly and colocalizes with cathepsin B in perinuclear vesicles and in vesicles within long pseudopodia. Cathepsin B has previously been shown to participate in the invasion by glioma cells through Matrigel and in the invasion of glioma spheroids into an aggregate of normal rat brain cells (Demchik *et al.,* 1999). Inhibition of cathepsin B by CA074 and transfection of antisense cathepsin B were also shown to inhibit Matrigel invasion of glioma cells (Mohanam *et al.,* 2001).

Intracellular degradation of extracellular matrix proteins by cathepsin B is interesting and has been shown to occur not only in breast cancer cells and gliomas, but also in human colon and prostate cancer cells (Sameni and Sloane, unpublished results) and *rac*-transformed rat fibroblasts (Ahram *et al.,* 2000). Intracellular degradation of extracellular matrix proteins, particularly type IV collagen, by lysosomal cathepsin B suggests an important role for lysosomal proteases in this process. Although it is traditionally believed that pericellular enzymes such as matrix metalloproteinases are responsible for degradation of extracellular matrix proteins, several pieces of evidence support the participation of intracellular lysosomal proteases. For example, electron microscopy studies by Levine *et al.* (1978) revealed tumors that contained fragments of collagen within intracellular vesicles. Based on observations of large acidic vesicles that internalized extracellular matrix and contained cathepsin D, Montcourrier *et al.* (1990, 1994) hypothesized that the lysosomal aspartic protease cathepsin D is responsible for the degradation of extracellular matrix proteins in breast cancer cells. Thus, these data along with results from confocal imaging assays reinforce the possibility that intracellular proteolysis in conjunction with extracellular proteolysis contributes to the invasion of cancer cells.

B. Cell-Surface Cathepsin B as a Member of a Proteolytic Cascade

The concept of various types and classes of proteases working together to allow cancer cells to invade is very appealing. Although classically it was believed that matrix metalloproteinases are the primary class of proteases involved in tumor invasion, several other classes including serine and cysteine proteases have also been shown to play significant roles in this process. Inhibitor studies using the confocal imaging assays have revealed that inhibitors to serine and cysteine protease and metalloproteinase activities in BT20 cells all contribute to the reduction of pericellular DQ-collagen IV degradation (Sameni *et al.,* 2000). This suggests that cathepsin B is a key component in a proteolytic cascade involving the extracellular degradation of matrix proteins. However, as yet no clear mechanism(s) and/or regulation process(es) have been established for this proteolytic cascade. Proteases likely participate by activating proenzymes in a sequential order resulting in the degradation of extracellular matrix and the invasion and migration of tumor cells through basement membranes and interstitial stroma. The role of cathepsin B in this enzymatic cascade is currently being investigated.

A potential proteolytic cascade illustrating the role of cathepsin B is depicted in Fig. 4. Procathepsin B can be activated by cathepsins D (van der Stappen *et al.*, 1996) and G, tPA, uPA, and elastases (Dalet-Fumeron *et al.*, 1993, 1996). Active cathepsin B can activate proMMP-3 to MMP-3 (Murphy *et al.*, 1992) and prouPA to uPA (Kobayashi *et al.*, 1991). uPA can convert plasminogen to plasmin (Dano *et al.*, 1985), a broad-spectrum protease that in turn can activate several latent metalloproteinases including MMP-1, MMP-3, MMP-9, MMP-12, and MMP-13 (He *et al.*, 1989; Matrisian and Bowden, 1990; Baramova *et al.*, 1997; Carmeliet *et al.*, 1997; Festuccia *et al.*, 1998). Plasmin along with these metalloproteinases can degrade fibronectin, collagen IV, and laminin, components of the extracellular matrix (Dano *et al.*, 1985). In ovarian cancer, activation of pro-uPA and invasion of these carcinoma cells through Matrigel are prevented by the inhibition of cathepsin B (Kobayashi *et al.*, 1993). The relationship between cathepsin B and the uPA/plasmin system is quite intriguing since these proteases have been linked to cell surface AIIt. Thus, AIIt may provide a structural linkage between proteases and the extracellular matrix to facilitate degradation of the matrix proteins by these proteases.

Where and how this pathological process occurs on the cell surface remains to be determined. Recent evidence points to the specialized caveolae regions of the plasma membrane that have been shown to contain several classes of proteases (Fig. 5). Activation of procathepsin B on the cell surface may initiate a cascade that activates MMP-2 in the caveolae. Increased levels of pro-MMP-2 and active MMP-2 and enhanced cathepsin B activity have all been localized to regions of colon tumor invasion (Emmert-Buck *et al.*, 1994). The caveolae may also be the site for cathepsin B activation of the prouPA/plasminogen system. uPAR, which interacts with matrix proteins such as vitronectin (Carriero *et al.*, 1997), also forms a complex with caveolin and β_1 integrin on the cell surface and enhances the migration of cells on fibronectin (Wei *et al.*, 1999). The association of uPAR with the β_1 integrin and caveolin complex has also been observed in human chondrocytes (Schwab *et al.*, 2001). Integrins have been shown to be present in invadopodia that play an important role in malignant progression (Nakahara *et al.*, 1996). In human 12T breast fibroblast cells, treatment with blocking antibodies against β_1 integrins appears to reduce secreted cathepsin B activity (Koblinski and Sloane, unpublished data). Thus, it is possible that caveolae are organizing centers for the localization of proteases, their receptors, and substrates and for the facilitation of extracellular remodeling processes. In addition, caveolin protein may modulate signal transduction events that regulate interactions between proteases and matrix proteins and in turn mediate cytoskeletal reorganization and cell adhesion and migration events. The fact that active cathepsin B has been localized to caveolae suggests a functional role for this enzyme in these specialized compartments.

Expression of cathepsin B also appears to be regulated by the extracellular matrix. In mesangial cells, cathepsin B expression is upregulated when these cells are grown on Matrigel (Singhal *et al.*, 1998). Activity and expression of cathepsin B

CAVEOLAE

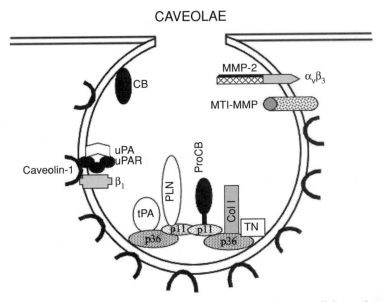

Figure 5 A schematic illustration of caveolae as an organizing center for extracellular matrix proteins and various proteases. Cathepsin B (CB) is shown both as a proenzyme (proCB) bound to p11 of the AIIt and as an active caveolae-associated enzyme. In caveolae, active cathepsin B may convert pro-uPA to uPA. uPA is shown bound to its receptor uPAR, which is associated with the caveolin-1-β_1 integrin complex. Caveolin-1 protein is shown as a structural scaffolding protein for caveolae. The binding of procathepsin B to AIIt shows an association of this enzyme with other proteases such as tPA and plasminogen (PLN) and extracellular matrix proteins, collagen I (Col I) and tenascin C (TN). This diagram also depicts the caveolae-associated metalloproteinases MT1-MMP and MMP-2 and the MMP-2 receptor, $\alpha_v\beta_3$ integrin, which may coordinate with caveolae-associated cathepsin B in the activation of proteolytic cascades leading to extracellular matrix degradation and tumor invasion.

was also enhanced in human colorectal carcinoma cells grown on Matrigel (Sameni and Sloane, unpublished data). Human breast fibroblasts show an increase in secreted cathepsin B activity when grown on collagen I as compared to plastic surfaces (Koblinski and Sloane, unpublished data). Both *in vivo* and *in vitro* data also suggest that cathepsin B expression is regulated by tumor–stromal cell interactions. For example, cathepsin B expression is observed in fibroblasts and invading macrophages at the invasive edge of breast (Castiglioni *et al.*, 1994) and colon carcinomas (Campo *et al.*, 1994). Rabbit V2 carcinomas produce an unidentified cytokine that increases the expression of cathepsin B and its secretion in rabbit skin fibroblasts (Baici *et al.*, 1988). These observed increases in cathepsin B expression and activity at the tumor–stromal interface suggest that communications between the tumor and the surrounding stroma exist that may coordinate events associated with tumor cell invasion. It is evident from these studies that cell-surface cathepsin B is a key component of these events.

VI. Conclusions

In summary, it is obvious that the role(s) for cell-surface cathepsin B are far more extensive than originally proposed. The association of this protease with various malignancies is a testament to its importance, particularly in the events leading up to and involving the transition to malignancy, tumor invasion, and metastasis. Cell-surface cathepsin B is of special interest in that it is involved in a number of interactions between proteins and proteolytic systems. Although more research is needed to elucidate the mechanism(s) by which cathepsin B is trafficked to and associated with the cell surface, future studies will likely focus on: (1) the physical binding properties between cathepsin B and the AIIt; (2) the characterization of caveolae-associated cathepsin B; (3) the interactions between cathepsin B and matrix proteins; (4) the interactions between tumor and stromal cells with respect to cathepsin B expression and activity; and (5) the role of cathepsin B in initiating proteolytic cascades. These studies will likely provide more insight into the functional significance of cathepsin B at the cell surface and can lead to the identification of new targets for therapeutic intervention.

References

Abdollahi, A., Getts, L. A., Sonoda, G., Miller, P. D., Taguchi, T., Godwin, A. K., Testa, J. R., and Hamilton, T. C. (1999). Genome scanning detects amplification of the cathepsin B gene (CtsB) in transformed rat ovarian surface epithelial cells. *J. Soc. Gynecol. Investig.* **6,** 32–40.

Ahram, M., Sameni, M., Qiu, R. G., Linebaugh, B., Kirn, D., and Sloane, B. F. (2000). Racl-induced endocytosis is associated with intracellular proteolysis during migration through a three-dimensional matrix. *Exp. Cell Res.* **260,** 292–303.

Andrews, N. W. (2000). Regulated secretion of conventional lysosomes. *Trends Cell Biol.* **10,** 316–321.

Annabi, B., Lachambre, M., Bousquet-Gagnon, N., Page, M., Gingras, D., and Beliveau, R. (2001). Localization of membrane-type 1 matrix metalloproteinase in caveolae membrane domains. *Biochem J.* **353,** 547–553.

Arkona, C., and Wiederanders, B. (1996). Expression, subcellular distribution and plasma membrane binding of cathepsin B and gelatinases in bone metastatic tissue. *Biol. Chem.* **377,** 695–702.

Baici, A., Horler, D., Lang, A., Merlin, C., and Kissling, R. (1995). Cathepsin B in osteoarthritis: zonal variation of enzyme activity in human femoral head cartilage. *Ann. Rheum. Dis.* **54,** 281–288.

Baici, A., Knopfel, M., and Keist, R. (1988). Tumor–host interactions in the rabbit V2 carcinoma: stimulation of cathepsin B in host fibroblasts by a tumor-derived cytokine. *Invasion Metastasis* **8,** 143–158.

Baramova, E. N., Bajou, K., Remacle, A., L'Hoir, C., Krell, H. W., Weidle, U. H., Noel, A., and Foidart, J. M. (1997). Involvement of PA/plasmin system in the processing of pro-MMP-9 and in the second step of pro-MMP-2 activation. *FEBS Lett.* **405,** 157–162.

Berardi, S., Lang, A., Kostoulas, G., Horler, D., Vilei, E. M., and Baici, A. (2001). Alternative messenger RNA splicing and enzyme forms of cathepsin B in human osteoarthritic cartilage and cultured chondrocytes. *Arthritis Rheum.* **44,** 1819–1831.

Berquin, I. M., Cao, L., Fong, D., and Sloane, B. F. (1995). Identification of two new exons and multiple transcription start points in the 5'-untranslated region of the human cathepsin-B-encoding gene. *Gene* **159,** 143–149.

Berquin, I. M., Yan, S., Katiyar, K., Huang, L., Sloane, B. F., and Troen, B. R. (1999). Differentiating agents regulate cathepsin B gene expression in HL-60 cells. *J. Leukoc. Biol.* **66,** 609–616.

Brix, K., Lemansky, P., and Herzog, V. (1996). Evidence for extracellularly acting cathepsins mediating thyroid hormone liberation in thyroid epithelial cells. *Endocrinology* **137,** 1963–1974.

Brix, K., Linke, M., Tepel, C., and Herzog, V. (2001). Cysteine proteinases mediate extracellular prohormone processing in the thyroid. *Biol. Chem.* **382,** 717–725.

Buck, M. R., Karustis, D. G., Day, N. A., Honn, K. V., and Sloane, B. F. (1992). Degradation of extracellular-matrix proteins by human cathepsin B from normal and tumour tissues. *Biochem. J.* **282,** 273–278.

Burgoyne, R. D. (1988). Calpactin in exocytosis. *Nature* **331,** 20.

Campo, E., Munoz, J., Miquel, R., Palacin, A., Cardesa, A., Sloane, B. F., and Emmert-Buck, M. R. (1994). Cathepsin B expression in colorectal carcinomas correlates with tumor progression and shortened patient survival. *Am. J. Pathol.* **145,** 301–309.

Carmeliet, P., Moons, L., Lijnen, R., Baes, M., Lemaitre, V., Tipping, P., Drew, A., Eeckhout, Y., Shapiro, S., Lupu, F., and Collen, D. (1997). Urokinase-generated plasmin activates matrix metalloproteinases during aneurysm formation. *Nat. Genet.* **17,** 439–444.

Carriero, M. V., Del Vecchio, S., Franco, P., Potena, M. I., Chiaradonna, F., Botti, G., Stoppelli, M. P., and Salvatore, M. (1997). Vitronectin binding to urokinase receptor in human breast cancer. *Clin. Cancer Res.* **3,** 1299–1308.

Castiglioni, T., Merino, M. J., Elsner, B., Lah, T. T., Sloane, B. F., and Emmert-Buck, M. R. (1994). Immunohistochemical analysis of cathepsins D, B, and L in human breast cancer. *Hum. Pathol.* **25,** 857–862.

Chiang, Y., Davis, R. G., and Vishwanatha, J. K. (1996). Altered expression of annexin II in human B-cell lymphoma cell lines. *Biochim. Biophys. Acta* **1313,** 295–301.

Chopra, H., Timar, J., Chen, Y. Q., Rong, X. H., Grossi, I. M., Fitzgerald, L. A., Taylor, J. D., and Honn, K. V. (1991). The lipoxygenase metabolite 12(S)-HETE induces a cytoskeleton-dependent increase in surface expression of integrin alpha IIb beta 3 on melanoma cells. *Int. J. Cancer* **49,** 774–786.

Chung, C. Y., and Erickson, H. P. (1994). Cell surface annexin II is a high affinity receptor for the alternatively spliced segment of tenascin-C. *J. Cell Biol.* **126,** 539–548.

Cole, S. P., Pinkoski, M. J., Bhardwaj, G., and Deeley, R. G. (1992). Elevated expression of annexin II (lipocortin II, p36) in a multidrug resistant small cell lung cancer cell line. *Br. J. Cancer* **65,** 498–502.

Dalet-Fumeron, V., Boudjennah, L., and Pagano, M. (1996). Competition between plasminogen and procathepsin B as a probe to demonstrate the *in vitro* activation of procathepsin B by the tissue plasminogen activator. *Arch. Biochem. Biophys.* **335,** 351–357.

Dalet-Fumeron, V., Guinec, N., and Pagano, M. (1993). *In vitro* activation of pro-cathepsin B by three serine proteinases: leucocyte elastase, cathepsin G, and the urokinase-type plasminogen activator. *FEBS Lett.* **332,** 251–254.

Dano, K., Andreasen, P. A., Grondahl-Hansen, J., Kristensen, P., Nielsen, L. S., and Skriver, L. (1985). Plasminogen activators, tissue degradation, and cancer. *Adv. Cancer Res.* **44,** 139–266.

Demchik, L. L., Sameni, M., Nelson, K., Mikkelsen, T., and Sloane, B. F. (1999). Cathepsin B and glioma invasion. *Int. J. Dev. Neurosci.* **17,** 483–494.

Eijan, A. M., Sandes, E., Puricelli, L., Bal De Kier Joffe, E., and Casabe, A. R. (2000). Cathepsin B levels in urine from bladder cancer patients. *Oncol. Rep.* **7,** 1395–1399.

Emans, N., Gorvel, J. P., Walter, C., Gerke, V., Kellner, R., Griffiths, G., and Gruenberg, J. (1993). Annexin II is a major component of fusogenic endosomal vesicles. *J. Cell Biol.* **120,** 1357–1369.

Emmert-Buck, M. R., Roth, M. J., Zhuang, Z., Campo, E., Rozhin, J., Sloane, B. F., Liotta, L. A., and Stetler-Stevenson, W. G. (1994). Increased gelatinase A (MMP-2) and cathepsin B activity in invasive tumor regions of human colon cancer samples. *Am. J. Pathol.* **145**, 1285–1290.

Engelman, J. A., Wykoff, C. C., Yasuhara, S., Song, K. S., Okamoto, T., and Lisanti, M. P. (1997). Recombinant expression of caveolin-1 in oncogenically transformed cells abrogates anchorage-independent growth. *J. Biol. Chem.* **272**, 16374–16381.

Erdel, M., Trefz, G., Spiess, E., Habermaas, S., Spring, H., Lah, T., and Ebert. W. (1990). Localization of cathepsin B in two human lung cancer cell lines. *J. Histochem. Cytochem.* **38**, 1313–1321.

Erikson, E., Tomasiewicz, H. G., and Erikson, R. L. (1984). Biochemical characterization of a 34-kilodalton normal cellular substrate of pp60v-src and an associated 6-kilodalton protein. *Mol. Cell Biol.* **4**, 77–85.

Falcone, D. J., Borth, W., Khan, K. M., and Hajjar, K. A. (2001). Plasminogen-mediated matrix invasion and degradation by macrophages is dependent on surface expression of annexin II. *Blood* **97**, 777–784.

Festuccia, C., Dolo, V., Guerra, F., Violini, S., Muzi, P., Pavan, A., and Bologna, M. (1998). Plasminogen activator system modulates invasive capacity and proliferation in prostatic tumor cells. *Clin. Exp. Metastasis* **16**, 513–528.

Fine, S. W., Lisanti, M. P., Galbiati, F., and Li, M. (2001). Elevated expression of caveolin-1 in adenocarcinoma of the colon. *Am. J. Clin. Pathol.* **115**, 719–724.

Frohlich, M., Motte, P., Galvin, K., Takahashi, H., Wands, J., and Ozturk, M. (1990). Enhanced expression of the protein kinase substrate p36 in human hepatocellular carcinoma. *Mol. Cell Biol.* **10**, 3216–3223.

Frohlich, E., Schlagenhauff, B., Mohrle, M., Weber, E., Klessen, C., and Rassner, G. (2001). Activity, expression, and transcription rate of the cathepsins B, D, H, and L in cutaneous malignant melanoma. *Cancer* **91**, 972–982.

Frosch, B. A., Berquin, I., Emmert-Buck, M. R., Moin, K., and Sloane, B. F. (1999). Molecular regulation, membrane association and secretion of tumor cathepsin B. *APMIS* **107**, 28–37.

Gardella, S., Andrei, C., Lotti, L. V., Poggi, A., Torrisi, M. R., Zocchi, M. R., and Rubartelli, A. (2001). CD8(+) T lymphocytes induce polarized exocytosis of secretory lysosomes by dendritic cells with release of interleukin-1beta and cathepsin D. *Blood* **98**, 2152–2159.

Gerke, V., and Weber, K. (1984). Identity of p36K phosphorylated upon Rous sarcoma virus transformation with a protein purified from brush borders; calcium-dependent binding to non-erythroid spectrin and F-actin. *EMBO J.* **3**, 227–233.

Guo, M., Mathieu, P. A., Linebaugh, B., Sloane, B. F., and Reiners, J. J. Jr. (2002). Phorbol ester activation of a proteolytic cascade capable of activating latent TGF-beta : A process initiated by the exocytosis of cathepsin B. *J. Biol. Chem.* **277**, 4829–4837.

Hackel, C. G., Krueger, S., Grote, H. J., Oshiro, Y., Hodges, S., Johnston, D. A., Johnson, M. E., Roessner, A., Ayala, A. G., and Czerniak, B. (2000). Overexpression of cathepsin B and urokinase plasminogen activator is associated with increased risk of recurrence and metastasis in patients with chondrosarcoma. *Cancer* **89**, 995–1003.

Hajjar, K. A., and Hamel, N. M. (1990). Identification and characterization of human endothelial cell membrane binding sites for tissue plasminogen activator and urokinase. *J. Biol. Chem.* **265**, 2908–2916.

Hajjar, K. A., Jacovina, A. T., and Chacko, J. (1994). An endothelial cell receptor for plasminogen/tissue plasminogen activator. I. Identity with annexin II. *J. Biol. Chem.* **269**, 21191–21197.

Hanewinkel, H., Glossl, J., and Kresse, H. (1987). Biosynthesis of cathepsin B in cultured normal and I-cell fibroblasts. *J. Biol. Chem.* **262**, 12351–12355.

Harder, T., and Gerke, V. (1994). The annexin II2p11(2) complex is the major protein component of the Triton X-100-insoluble low-density fraction prepared from MDCK cells in the presence of Ca^{2+}. *Biochim. Biophys. Acta* **1223**, 375–382.

Hazen, L. G., Bleeker, F. E., Lauritzen, B., Bahns, S., Song, J., Jonker, A., Van Driel, B. E., Lyon, H., Hansen, U., Kohler, A., and Van Noorden, C. J. (2000). Comparative localization of cathepsin B protein and activity in colorectal cancer. *J. Histochem. Cytochem.* **48,** 1421–1430.

He, C. S., Wilhelm, S. M., Pentland, A. P., Marmer, B. L., Grant, G. A., Eisen, A. Z., and Goldberg, G. I. (1989). Tissue cooperation in a proteolytic cascade activating human interstitial collagenase. *Proc. Natl. Acad. Sci. USA* **86,** 2632–2636.

Heuser, J. (1989). Changes in lysosome shape and distribution correlated with changes in cytoplasmic pH. *J. Cell Biol.* **108,** 855–864.

Honn, K. V., Timar, J., Rozhin, J., Bazaz, R., Sameni, M., Ziegler, G., and Sloane, B. F. (1994). A lipoxygenase metabolite, 12-(S)-HETE, stimulates protein kinase C-mediated release of cathepsin B from malignant cells. *Exp. Cell Res.* **214,** 120–130.

Hughes, S. J., Glover, T. W., Zhu, X. X., Kuick, R., Thoraval, D., Orringer, M. B., Beer, D. G., and Hanash, S. (1998). A novel amplicon at 8p22-23 results in overexpression of cathepsin B in esophageal adenocarcinoma. *Proc. Natl. Acad. Sci. USA* **95,** 12410–12415.

Iishi, H., Tatsuta, M., Baba, M., Uehara, H., Yano, H., and Nakaizumi, A. (1995). Chemoprevention by galanin against colon carcinogenesis induced by azoxymethane in Wistar rats. *Int. J. Cancer* **61,** 861–863.

Ikeda, Y., Ikata, T., Mishiro, T., Nakano, S., Ikebe, M., and Yasuoka, S. (2000). Cathepsins B and L in synovial fluids from patients with rheumatoid arthritis and the effect of cathepsin B on the activation of pro-urokinase. *J. Med. Invest.* **47,** 61–75.

Kassam, G., Choi, K. S., Ghuman, J., Kang, H. M., Fitzpatrick, S. L., Zackson, T., Zackson, S., Toba, M., Shinomiya, A., and Waisman, D. M. (1998a). The role of annexin II tetramer in the activation of plasminogen. *J. Biol. Chem.* **273,** 4790–4799.

Kassam, G., Le, B. H., Choi, K. S., Kang, H. M., Fitzpatrick, S. L., Louie, P., and Waisman, D. M. (1998b). The p11 subunit of the annexin II tetramer plays a key role in the stimulation of t-PA-dependent plasminogen activation. *Biochemistry* **37,** 16958–16966.

Kato, K., Hida, Y, Miyamoto, M., Hashida, H., Shinohara, T., Itoh, T., Okushiba, S., Kondo, S., and Katoh, H. (2002). Overexpression of caveolin-1 in esophageal squamous cell carcinoma correlates with lymph-node metastasis and pathologic stage. *Cancer* **94,** 929–933.

Kennett, C. N., Cox, S. W., and Eley, B. M. (1997). Ultrastructural localization of cathepsin B in gingival tissue from chronic periodontitis patients. *Histochem. J.* **29,** 727–734.

Khan, A., Krishna, M., Baker, S. P., and Banner, B. F. (1998). Cathepsin B and tumor-associated laminin expression in the progression of colorectal adenoma to carcinoma. *Mod. Pathol.* **11,** 704–708.

Kim, K., Cai, J., Shuja, S., Kuo, T., and Murnane, M. J. (1998). Presence of activated ras correlates with increased cysteine proteinase activities in human colorectal carcinomas. *Int. J. Cancer* **79,** 324–333.

Kirsche, H., Barrett, A. J., and Rawlings, N. D. (1995). Proteinases 1: lysosomal cysteine proteinases. *Protein Profile* **2,** 1587–1643.

Kobayashi, H., Moniwa, N., Sugimura, M., Shinohara, H., Ohi, H., and Terao, T. (1993). Increased cell-surface urokinase in advanced ovarian cancer. *Jpn. J. Cancer Res.* **84,** 633–640.

Kobayashi, H., Schmitt, M., Goretzki, L., Chucholowski, N., Calvete, J., Kramer, M., Gunzler, W. A., Janicke, F., and Graeff, H. (1991). Cathepsin B efficiently activates the soluble and the tumor cell receptor-bound form of the proenzyme urokinase-type plasminogen activator (Pro-uPA). *J. Biol. Chem.* **266,** 5147–5152.

Kocher, T., Schultz-Thater, E., Gudat, F., Schaefer, C., Casorati, G., Juretic, A., Willimann, T., Harder, F., Heberer, M., and Spagnoli, G. C. (1995). Identification and intracellular location of MAGE-3 gene product. *Cancer Res.* **55,** 2236–2239.

Koleske, A. J., Baltimore, D., and Lisanti, M. P. (1995). Reduction of caveolin and caveolae in oncogenically transformed cells. *Proc. Natl. Acad. Sci. USA* **92,** 1381–1385.

Konduri, S., Lakka, S. S., Tasiou, A., Yanamandra, N., Gondi, C. S., Dinh, D. H., Olivero, W. C.,

Gujrati, M., and Rao, J. S. (2001). Elevated levels of cathepsin B in human glioblastoma cell lines. *Int. J. Oncol.* **19,** 519–524.

Kumble, K. D., Hirota, M., Pour, P. M., and Vishwanatha, J. K. (1992). Enhanced levels of annexins in pancreatic carcinoma cells of Syrian hamsters and their intrapancreatic allografts. *Cancer Res.* **52,** 163–167.

Lavie, Y., Fiucci, G., and Liscovitch, M. (1998). Up-regulation of caveolae and caveolar constituents in multidrug-resistant cancer cells. *J. Biol. Chem.* **273,** 32380–32383.

Lee, S. W., Reimer, C. L., Oh, P., Campbell, D. B., and Schnitzer, J. E. (1998). Tumor cell growth inhibition by caveolin re-expression in human breast cancer cells. *Oncogene* **16,** 1391–1397.

Levine, A. M., Reddick, R., and Triche, T. (1978). Intracellular collagen fibrils in human sarcomas. *Lab Invest.* **39,** 531–540.

Linebaugh, B. E., Sameni, M., Day, N. A., Sloane, B. F., and Keppler, D. (1999). Exocytosis of active cathepsin B enzyme activity at pH 7.0, inhibition and molecular mass. *Eur. J. Biochem.* **264,** 100–109.

Lisanti, M. P., Scherer, P. E., Vidugiriene, J., Tang, Z., Hermanowski-Vosatka, A., Tu, Y. H., Cook, R. F., and Sargiacomo, M. (1994). Characterization of caveolin-rich membrane domains isolated from an endothelial-rich source: implications for human disease. *J. Cell Biol.* **126,** 111–126.

Ludwig, T., Munier-Lehmann, H., Bauer, U., Hollinshead, M., Ovitt, C., Lobel, P., and Hoflack, B. (1994). Differential sorting of lysosomal enzymes in mannose 6-phosphate receptor-deficient fibroblasts. *EMBO J.* **13,** 3430–3437.

Mach, L., Stuwe, K., Hagen, A., Ballaun, C., and Glossl, J. (1992). Proteolytic processing and glycosylation of cathepsin B. The role of the primary structure of the latent precursor and of the carbohydrate moiety for cell-type-specific molecular forms of the enzyme. *Biochem. J.* **282,** 577–582.

Maciewicz, R. A., Wardale, R. J., Etherington, D. L., and Paraskev, C. (1989). Immunodetection of cathepsin B and L present in and secreted from human premalignant and malignant colorectal tumor cell lines. *Int. J. Cancer* **43,** 478–486.

Mahmood, U., Tung, C. H., Bogdanov, A. Jr., and Weissleder, R. (1999). Near-infrared optical imaging of protease activity for tumor detection. *Radiology* **213,** 866–870.

Mai, J., Finley, R. L. Jr., Waisman, D. M., and Sloane, B. F. (2000). Human procathepsin B interacts with the annexin II tetramer on the surface of tumor cells. *J. Biol. Chem.* **275,** 12806–12812.

Marten, K., Bremer, C., Khazaie, K., Sameni, M., Sloane, B., Tung, C. H., and Weissleder, R. (2002). Detection of dysplastic intestinal adenomas using enzyme-sensing molecular beacons in mice. *Gastroenterology* **122,** 406–414.

Matrisian, L. M., and Bowden, G. T. (1990). Stromelysin/transin and tumor progression. *Sem. Cancer Biol.* **1,** 107–115.

McCormick, D. (1993). Secretion of cathepsin B by human gliomas *in vitro. Neuropathol. Appl. Neurobiol.* **19,** 146–151.

McIntyre, G. F., and Erickson, A. H. (1991). Procathepsins L and D are membrane-bound in acidic microsomal vesicles. *J. Biol. Chem.* **266,** 15438–15445.

McIntyre, G. F., and Erickson, A. H. (1993). The lysosomal proenzyme receptor that binds procathepsin L to microsomal membranes at pH 5 is a 43-kDa integral membrane protein. *Proc. Natl. Acad. Sci. USA* **90,** 10588–10592.

McIntyre, G. F., Godbold, G. D., and Erickson, A. H. (1994). The pH-dependent membrane association of procathepsin L is mediated by a 9-residue sequence within the propeptide. *J. Biol. Chem.* **269,** 567–572.

Mikkelsen, T., Yan, P. S., Ho, K. L., Sameni, M., Sloane, B. F., and Rosenblum, M. L. (1995). Immunolocalization of cathepsin B in human glioma: implications for tumor invasion and angiogenesis. *J. Neurosurg.* **83,** 285–290.

Mohanam, S., Jasti, S. L., Kondraganti, S. R., Chandrasekar, N., Lakka, S. S., Kin, Y., Fuller, G. N., Yung, A. W., Kyritsis, A. P., Dinh, D. H., Olivero, W. C., Gujrati, M., Ali-Osman, F., and Rao,

J. S. (2001). Down-regulation of cathepsin B expression impairs the invasive and tumorigenic potential of human glioblastoma cells. *Oncogene* **20,** 3665–3673.

Moin, K., Cao, L., Day, N. A., Koblinski, J. E., and Sloane, B. F. (1998). Tumor cell membrane cathepsin B. *Biol. Chem.* **379,** 1093–1099.

Montcourrier, P., Mangeat, P. H., Salazar, G., Morisset, M., Sahuquet, A., and Rochefort, H. (1990). Cathepsin D in breast cancer cells can digest extracellular matrix in large acidic vesicles. *Cancer Res.* **50,** 6045–6054.

Montcourrier, P., Mangeat, P. H., Valembois, C., Salazar, G., Sahuquet, A., Duperray, C., and Rochefort, H. (1994). Characterization of very acidic phagosomes in breast cancer cells and their association with invasion. *J. Cell Sci.* **107,** 2381–2391.

Mort, J. S., and Buttle, D. J. (1997). Cathepsin B. *Int. J. Biochem. Cell Biol.* **29,** 715–720.

Mort, J. S., Leduc, M., and Recklies, A. D. (1981). A latent thiol proteinase from ascitic fluid of patients with neoplasia. *Biochim. Biophys. Acta* **662,** 173–180.

Mostov, K., and Werb, Z. (1997). Journey across the osteoclast. *Science* **276,** 219–220.

Murata, M., Miyashita, S., Yokoo, C., Tamai, M., Hanada, K., Hatayama, K., Towatari, T., Nikawa, T., and Katunuma, N. (1991). Novel epoxysuccinyl peptides. Selective inhibitors of cathepsin B, *in vitro. FEBS Lett.* **280,** 307–310.

Murnane, M. J., Sheahan, K., Ozdemirli, M., and Shuja, S. (1991). Stage-specific increases in cathepsin B messenger RNA content in human colorectal carcinoma. *Cancer Res.* **51,** 1137–1142.

Murphy, G., Ward, R., Gavrilovic, J., and Atkinson, S. (1992). Physiological mechanisms for metalloproteinase activation. *Matrix Suppl.* **1,** 224–230.

Naik, U. P., Patel, P. M., and Parise, L. V. (1997). Identification of a novel calcium-binding protein that interacts with the integrin alphaIIb cytoplasmic domain. *J. Biol. Chem.* **272,** 4651–4654.

Nakahara, H., Nomizu, M., Akiyama, S. K., Yamada, Y., Yeh, Y., and Chen, W. T. (1996). A mechanism for regulation of melanoma invasion. Ligation of alpha6beta1 integrin by laminin G peptides. *J. Biol. Chem.* **271,** 27221–27224.

Nishimura, Y., Kawabata, T., and Kato, K. (1988). Identification of latent procathepsins B and L in microsomal lumen: characterization of enzymatic activation and proteolytic processing in vitro. *Arch. Biochem. Biophys.* **261,** 64–71.

Page, A. E., Warburton, M. J., Chambers, T. J., Pringle, J. A., and Hayman, A. R. (1992). Human osteoclastomas contain multiple forms of cathepsin B. *Biochim. Biophys. Acta* **1116,** 57–66.

Poole, A. R., Tiltman, K. J., Recklies, A. D., and Stoker, T. A. (1978). Differences in secretion of the proteinase cathepsin B at the edges of human breast carcinomas and fibroadenomas. *Nature* **273,** 545–547.

Puyraimond, A., Fridman, R., Lemesle, M., Arbeille, B., and Menashi, S. (2001). MMP-2 colocalizes with caveolae on the surface of endothelial cells. *Exp. Cell Res.* **262,** 28–36.

Qian, F., Bajkowski, A. S., Steiner, D. F., Chan, S. J., and Frankfater, A. (1989). Expression of five cathepsins in murine melanomas of varying metastatic potential and normal tissues. *Cancer Res.* **49,** 4870–4875.

Racine, C., Belanger, M., Hirabayashi, H., Boucher, M., Chakir, J., and Couet, J. (1999). Reduction of caveolin 1 gene expression in lung carcinoma cell lines. *Biochem. Biophys. Res. Commun.* **255,** 580–586.

Rajjayabun, P. H., Garg, S., Durkan, G. C., Charlton, R., Robinson, M. C., and Mellon, J. K. (2001). Caveolin-1 expression is associated with high-grade bladder cancer. *Urology* **58,** 811–814.

Recklies, A. D., Mort, J. S., and Poole, A. R. (1982). Secretion of a thiol proteinase from mouse mammary carcinomas and its characterization. *Cancer Res.* **42,** 1026–1032.

Reeves, S. A., Chavez-Kappel, C., Davis, R., Rosenblum, M., and Israel, M. A. (1992). Developmental regulation of annexin II (Lipocortin 2) in human brain and expression in high grade glioma. *Cancer Res.* **52,** 6871–6876.

Rempel, S. A., Rosenblum, M. L., Mikkelsen, T., Yan, P. S., Ellis, K. D., Golembieski, W. A., Sameni, M., Rozhin, J., Ziegler, G., and Sloane, B. F. (1994). Cathepsin B expression and localization in glioma progression and invasion. *Cancer Res.* **54,** 6027–6031.

Robitzki, A., Schroder, H. C., Ugarkovic, D., Gramzow, M., Fritsche, U., Batel, R., and Muller, W. E. (1990). cDNA structure and expression of calpactin, a peptide involved in Ca^{2+}-dependent cell aggregation in sponges. *Biochem. J.* **271**, 415–420.

Rodriguez, A., Webster, P., Ortego, J., and Andrews, N. W. (1997). Lysosomes behave as Ca^{2+}-regulated exocytic vesicles in fibroblasts and epithelial cells. *J. Cell Biol.* **137**, 93–104.

Roseman, B. J., Bollen, A., Hsu, J., Lamborn, K., and Israel, M. A. (1994). Annexin II marks astrocytic brain tumors of high histologic grade. *Oncol. Res.* **6**, 561–567.

Rowan, A. D., Mason, P., Mach, L., and Mort, J. S. (1992). Rat procathepsin B. Proteolytic processing to the mature form *in vitro. J. Biol. Chem.* **267**, 15993–15999.

Rozhin, J., Robinson, D., Stevens, M. A., Lah, T. T., Honn, K. V., Ryan, R. E., and Sloane, B. F. (1987). Properties of a plasma membrane-associated cathepsin B-like cysteine proteinase in metastatic B16 melanoma variants. *Cancer Res.* **47**, 6620–6628.

Rozhin, J., Sameni, M., Zielger, G., and Sloane, B. F. (1994). Pericellular pH affects distribution and secretion of cathepsin B in malignant cells. *Cancer Res.* **54**, 6517–6525.

Sager, R., Sheng, S., Anisowicz, A., Sotiropoulou, G., Zou, Z., Stenman, G., Swisshelm, K., Chen, Z., Hendrix, M. J., and Pemberton, P. (1994). RNA genetics of breast cancer: maspin as paradigm. *Cold Spring Harb. Symp. Quant. Biol.* **59**, 537–546.

Sameni, M., Dosescu, J., and Sloane, B. F. (2001). Imaging proteolysis by living human glioma cells. *Biol. Chem.* **382**, 785–788.

Sameni, M., Elliott, E., Ziegler, G., Fortgens, P. H., Dennison, C., and Sloane, B. F. (1995). Cathepsin B and D are localized at the surface of human breast cancer cells. *Pathol. Oncol. Res.* **1**, 43–53.

Sameni, M., Moin, K., and Sloane, B. F. (2000). Imaging proteolysis by living human breast cancer cells. *Neoplasia* **2**, 496–504.

Sargiacomo, M., Sudol, M., Tang, Z., and Lisanti, M. P. (1993). Signal transducing molecules and glycosyl-phosphatidylinositol-linked proteins form a caveolin-rich insoluble complex in MDCK cells. *J. Cell Biol.* **122**, 789–807.

Schwab, W., Gavlik, J. M., Beichler, T., Funk, R. H., Albrecht, S., Magdolen, V., Luther, T., Kasper, M., and Shakibaei, M. (2001). Expression of the urokinase-type plasminogen activator receptor in human articular chondrocytes: association with caveolin and beta 1-integrin. *Histochem. Cell Biol.* **115**, 317–323.

Shuja, S., and Murnane, M. J. (1996). Marked increases in cathepsin B and L activities distinguish papillary carcinoma of the thyroid from normal thyroid or thyroid with non-neoplastic disease. *Int. J. Cancer* **66**, 420–426.

Silletti, S., Timar, J., Honn, K. V., and Raz, A. (1994). Autocrine motility factor induces differential 12-lipoxygenase expression and activity in high- and low-metastatic K1735 melanoma cell variants. *Cancer Res.* **54**, 5752–5756.

Silver, I. A., Murrills, R. J., and Etherington, D. J. (1988). Microelectrode studies on the acid microenvironment beneath adherent macrophages and osteoclasts. *Exp. Cell Res.* **175**, 266–276.

Singhal, P. C., Gibbons, N., Franki, N., Reddy, K., Sharma, P., Mattana, J., Wagner, J. D., and Bansal, V. (1998). Simulated glomerular hypertension promotes mesangial cell apoptosis and expression of cathepsin-B and SGP-2. *J. Invest. Med.* **46**, 42–50.

Sinha, A. A., Gleason, D. F., Deleon, O. F., Wilson, M. J., and Sloane, B. F. (1993). Localization of a biotinylated cathepsin B oligonucleotide probe in human prostate including invasive cells and invasive edges by *in situ* hybridization. *Anat. Rec.* **235**, 233–240.

Sinha, A. A., Jamuar, M. P., Wilson, M. J., Rozhin, J., and Sloane, B. F. (2001). Plasma membrane association of cathepsin B in human prostate cancer: biochemical and immunogold electron microscopic analysis. *Prostate* **49**, 172–184.

Sinha, A. A., Wilson, M. J., Gleason, D. F., Reddy, P. K., Sameni, M., and Sloane, B. F. (1995). Immunohistochemical localization of cathepsin B in neoplastic human prostate. *Prostate* **26**, 171–178.

Sloane, B. F. (1996). Suicidal tumor proteases. *Nat. Biotechnol.* **14**, 826–827.

Sloane, B. F., Dunn, J. R., and Honn, K. V. (1981). Lysosomal cathepsin B: correlation with metastatic potential. *Science* **212,** 1151–1153.

Sloane, B. F., Honn, K. V., Sadler, J. G., Turner, W. A., Kimpson, J. J., and Taylor, J. D. (1982). Cathepsin B activity in B16 melanoma cells: a possible marker for metastatic potential. *Cancer Res.* **42,** 980–986.

Sloane, B. F., Moin, K., Sameni, M., Tait, L. R., Rozhin, J., and Ziegler, G. (1994). Membrane association of cathepsin B can be induced by transfection of human breast epithelial cells with c-Ha-*ras* oncogene. *J. Cell Sci.* **107,** 373–384.

Sloane, B. F., Rozhin, J., Johnson, K., Taylor, H., Crissman, J. D., and Honn, K. V. (1986). Cathepsin B: association with plasma membrane in metastatic tumors. *Proc. Natl. Acad. Sci. USA* **83,** 2483–2487.

Sloane, B. F., Rozhin, J., Krepela, E., Ziegler, G., and Sameni, M. (1991). The malignant phenotype and cysteine proteinases. *Biomed. Biochim. Acta.* **50,** 549–554.

Smart, E. J., Ying, Y. S., Donzell, W. C., and Anderson, R. G. (1996). A role for caveolin in transport of cholesterol from endoplasmic reticulum to plasma membrane. *J. Biol. Chem.* **271,** 29427–29435.

Smart, E. J., Graf, G. A., McNiven, M. A., Sessa, W. C., Engelman, J. A., Scherer, P. E., Okamoto, T., and Lisanti, M. P. (1999). Caveolins, liquid-ordered domains, and signal transduction. *Mol. Cell Biol.* **19,** 7289–7304.

Staack, A., Koenig, F., Daniltchenko, D., Hauptmann, S., Loening, S. A., Schnorr, D., and Jung, K. (2002). Cathepsins B, H, and L activities in urine of patients with transitional cell carcinoma of the bladder. *Urology* **59,** 308–312.

Stahl, A., and Mueller, B. M. (1995). The urokinase-type plasminogen activator receptor, a GPI-linked protein, is localized in caveolae. *J. Cell Biol.* **129,** 335–344.

Strohmaier, A. R., Porwol, T., Acker, H., and Spiess, E. (1997). Tomography of cells by confocal laser scanning microscopy and computer-assisted three-dimensional image reconstruction: localization of cathepsin B in tumor cells penetrating collagen gels *in vitro. J. Histochem. Cytochem.* **45,** 975–983.

Sukoh, N., Abe, S., Ogura, S., Isobe, H., Takekawa, H., Inoue, K., and Kawakami, Y. (1994). Immunohistochemical study of cathepsin B. Prognostic significance in human lung cancer. *Cancer* **74,** 46–51.

Thiel, C., Osborn, M., and Gerke, V. (1992). The tight association of the tyrosine kinase substrate annexin II with the submembranous cytoskeleton depends on intact p11- and Ca^{2+}-binding sites. *J. Cell Sci.* **103,** 733–742.

Timar, J., Bazaz, R., Kimler, V., Haddad, M., Tang, D. G., Robertson, D., Tovari, J., Taylor, J. D., and Honn, K. V. (1995). Immunomorphological characterization and effects of 12-(S)-HETE on a dynamic intracellular pool of the alpha IIb beta 3-integrin in melanoma cells. *J. Cell Sci.* **108,** 2175–2186.

Trabandt, A., Gay, R. E., Fassbander, H. G., and Gay, S. (1991). Cathepsin B in synovial cells at the site of joint destruction in rheumatoid arthritis. *Arthritis Rheum.* **34,** 1444–1451.

Trabandt, A., Muller-Ladner, U., Kriegsmann, J., Gay, R. E., and Gay, S. (1995). Expression of proteolytic cathepsins B, D, and L in periodontal gingival fibroblasts and tissues. *Lab Invest.* **73,** 205–212.

Tung, C. H., Gerszten, R. E., Jaffer, F. A., and Weissleder, R. (2000). A novel near-infrared fluorescence sensor for detection of thrombin activation in blood. *Chembiochem* **3,** 207–211.

Ulbricht, B., Henny, H., Horstmann, H., Spring, H., Faigle, W., and Spiess, E. (1997). Influence of 12(S)-hydroxyeicosatetraenoic acid (12(S)-HETE) on the localization of cathepsin B and cathepsin L in human lung tumor cells. *Eur. J. Cell Biol.* **74,** 294–301.

van der Stappen, J. W., Williams, A. C., Maciewicz, R. A., and Paraskeva, C. (1996). Activation of cathepsin B, secreted by a colorectal cancer cell line requires low pH and is mediated by cathepsin D. *Int. J. Cancer.* **67,** 547–554.

Vishwanatha, J. K., Chiang, Y., Kumble, K. D., Hollingsworth, M. A., and Pour, P. M. (1993). Enhanced expression of annexin II in human pancreatic carcinoma cells and primary pancreatic cancers. *Carcinogenesis* **14**, 2575–2579.

Visscher, D. W., Sloane, B., Sakr, W., Sameni, M., Weaver, D., Bouwman, D., and Crissman, J. D. (1994a). Clinocopathological significance of cathepsin B and urokinase-type plasminogen activator immunostaining in colorectal adenocarcinoma. *Int. J. Surg. Pathol.* **1**, 227–234.

Visscher, D. W., Sloane, B. F., Sameni, M., Babiarz, J. W., Jacobson, J., and Crissman, J. D. (1994b). Clinicopathologic significance of cathepsin B immunostaining in transitional neoplasia. *Mod. Pathol.* **7**, 76–81.

von Figura, and Hasilik, A. (1986). Lysosomal enzymes and their receptors. *Annu. Rev. Biochem.* **55**, 167–193.

Warwas, M., Haczynska, H., Gerber, J., and Nowak, M. (1997). Cathepsin B-like activity as a serum tumour marker in ovarian carcinoma. *Eur. J. Clin. Chem. Clin. Biochem.* **35**, 301–304.

Watanabe, M., Higashi, T., Watanabe, A., Osawa, T., Sato, Y., Kimura, Y., Tominaga, S., Hashimoto, N., Yoshida, Y., Morimoto, S., Shito, S., Hasinoto, M., Kobayashi, M., Tomoda, J., and Tsuji, T. (1989). Cathepsin B and L activities in gastric cancer tissue: correlation with histological findings. *Biochem. Med. Metab. Biol.* **42**, 21–29.

Wei, Y., Yang, X., Liu, Q., Wilkins, J. A., and Chapman, H. A. (1999). A role for caveolin and the urokinase receptor in integrin-mediated adhesion and signaling. *J. Cell Biol.* **144**, 1285–1294.

Wirl, G., and Schwartz-Albiez, R. (1990). Collagen-binding proteins of mammary epithelial cells are related to Ca^{2+}- and phospholipid-binding annexins. *J. Cell Physiol.* **144**, 511–522.

Yan, Z., Deng, X., Chen, M., Xu, Y., Ahram, M., Sloane, B. F., and Friedman, E. (1997). Oncogenic c-Ki-*ras* but not oncogenic c-Ha-*ras* up-regulates CEA expression and disrupts basolateral polarity in colon epithelial cells. *J. Biol. Chem.* **272**, 27902–27907.

Yang, G., Truong, L. D., Timme, T. L., Ren, C., Wheeler, T. M., Park, S. H., Nasu, Y., Bangma, C. H., Kattan, M. W., Scardino, P. T., and Thompson, T. C. (1998). Elevated expression of caveolin is associated with prostate and breast cancer. *Clin. Cancer Res.* **4**, 1873–1880.

Yang, G., Truong, L. D., Wheeler, T. M., and Thompson, T. C. (1999). Caveolin-1 expression in clinically confined human prostate cancer: a novel prognostic marker. *Cancer Res.* **59**, 5719–5123.

Yeatman, T. J., Updyke, T. V., Kaetzel, M. A., Dedman, J. R., and Nicolson, G. L. (1993). Expression of annexins on the surfaces of non-metastatic and metastatic human and rodent tumor cells. *Clin. Exp. Metastasis* **11**, 37–44.

Young, P. R., and Spevacek, S. M. (1992). Substratum acidification by murine B16F10 melanoma cultures. *Biochim. Biophys. Acta* **1139**, 163–166.

Zenklusen, J. C., Bieche, I., Lidereau, R., and Conti, C. J. (1994). (C-A)n microsatellite repeat D7S522 is the most commonly deleted region in human primary breast cancer. *Proc. Natl. Acad. Sci. USA* **91**, 12155–12158.

Zhang, J. C., Sakthivel, R., Kniss, D., Graham, C. H., Strickland, D. K., and McCrae, K. R. (1998). The low density lipoprotein receptor-related protein/alpha2-macroglobulin receptor regulates cell surface plasminogen activator activity on human trophoblast cells. *J. Biol. Chem.* **273**, 32273–32280.

11

Protease-Activated Receptors

Wadie F. Bahou
Division of Hematology
State University of New York
Stony Brook, New York 11794-8151

I. Introduction and General Overview

A. Overview

Distinct proteases—best exemplified by serine proteases such as thrombin—are potent cellular activators. Many of these cellular effects are mediated by a group of seven transmembrane G-protein-coupled receptors, uniquely activated by proteolytic cleavage (protease-activated receptors; PARs). PARs are functionally expressed on a large number of normal and malignant cells and pathogenetically involved in tumor metastases, thrombosis, atherosclerosis, and inflammation. Furthermore, murine transgenic knockout studies implicate PAR1 in a developmentally important role during embryogenesis, suggesting a role for thrombin-generating

Current Topics in Developmental Biology, Vol. 54

pathways or thrombin-responsive PARs in development. Targeted strategies focusing on design of PAR-specific inhibitors based on the unique mechanisms of PAR activation have been initiated with preliminary success.

B. Serine Proteases

Serine proteases contain a typical active-site catalytic triad in close proximity that is responsible for the charge relay system evident in all serine proteases. In the case of the potent multifunctional serine protease α-thrombin, for example, the catalytic triad is composed of His-365, Asp-419, and Ser-527 (Neurath, 1984)]. The precursor protein prothrombin is synthesized exclusively in the liver as an inactive zymogen with a tightly regulated plasmatic concentration of 100 μg/ml and a circulating half-life of about 72 h. The conversion of prothrombin to its 38-kDa active protease (α-thrombin) requires the assembly of a prothrombinase complex comprised of prothrombin, coagulation factor Xa, and the active cofactor Va on the surface of a cellular phospholipid membrane requiring calcium ions. An anion binding exosite confers binding specificity to various thrombin substrates (Maraganore et al., 1990). Whereas thrombin substrate specificity is similar to that of trypsin for small peptides, thrombin's action on larger proteins is more highly selective. Virtually all thrombin substrates contain an arginine or lysine adjacent to the sessile bond, a representative cleavage site containing the consensus X-Pro-Arg-↓-X (Blomback et al., 1967); for example, critical cleavages of fibrinogen at Arg^{16}-Gly^{17} within the Aα chain and Arg^{14}-Gly^{15} within the Bβ chain regulate fibrin polymerization and stabilization of the fibrin clot. Activated thrombin circulates briefly in the circulation, rapidly neutralized by antithrombin III, with less physiologically relevant inhibition by α_2-macroglobulin and α_1-antiprotease (Downing et al., 1978).

C. Protease-Activated Receptors

Serine proteases such as thrombin communicate with cells through a unique class of cell-surface PARs that are members of a larger family of G-protein-coupled receptors that include α- and β-adrenergic receptors and muscarinic and serotonin receptors, among others (Dohlman et al., 1991; Vu et al., 1991a). Unlike the latter group of receptors, which signal through standard receptor/ligand interactions, PARs are activated by a unique proteolytic cleavage within the first extracellular loop, or by synthetic peptidomimetics corresponding to the new N termini generated after receptor cleavage (Chen et al., 1994; Vu et al., 1991b). At the molecular level, the PARs share a common structural organization, suggesting evolution from a common ancestral gene. Evolutionary cross-species differences between murine and human receptor systems suggest functional divergences yet to be fully elucidated (see later discussion).

II. Structural Biochemistry and Classification

A. PAR Classification

To date, the cDNAs for four PARs have been isolated and characterized, enumerated chronologically by their date of initial identification (see Table I).

Table I Classification and Key Properties of Protease-Activated Receptors

	PAR1	PAR2	PAR3	PAR4
Amino acids	425	397	374	385
Cleavage site	Arg^{41}/Ser^{42}	Arg^{42}/Asp^{43}	Lys^{38}/Thr^{39}	Arg^{47}/Gly^{48}
Agonist	Thrombin (EC3.4.21.5)	Trypsin	Thrombin	Thrombin
	Granzyme A (EC3.4.21.78)	Tryptase (EC3.4.21.59)		Trypsin
	Trypsin (EC3.4.21.4)	TF/FVIIa (EC)		Cathepsin G
	Factor Xa (EC3.4.21.6)	Factor Xa		
(Ant)agonist	• Plasmin (EC3.4.21.7)			
	• Cathepsin G (EC3.4.21.20)			
	• Elastase (EC 3.4.21.37)			
	• Proteinase 3 (EC3.4.21.76)			
Peptide agonist[a]	SFLLR[b]	SLIGK	—	GYPGQV
Hirudin-like sequence	DKEYPF	—	FEEFP	—
Gene expression	Wide expression	Liver, kidneys, intestine, colon, pancreas, lung, keratinocytes	Spleen, skin, megakaryocytes	Megakaryocytes
Endothelium	Yes	Yes	Yes	No
Platelets				
Human	Yes	No	Minimal	Yes
Murine	No	No	Yes	Yes
Disease(s)	Thrombosis	Inflammation	Thrombosis	Thrombosis
	Tumor metastases/angiogenesis			
	Inflammation			

[a] Refers to endogenous agonist.
[b] Can directly activate PAR2.

1. Protease Activated Receptor-1 (PAR1)

Initially isolated and characterized in 1991, human PAR1 is the predominant receptor for thrombin-mediated platelet activation, and the prototypic and best-characterized PAR to date (Vu *et al.*, 1991a). PAR1 mRNA is differentially and widely expressed in human tissues and is readily detected in cells intimately involved in hemostatic regulation, i.e., platelets, vascular endothelial cells, and vascular smooth muscle cells (Hung *et al.*, 1992; Vu *et al.*, 1991a). The PAR1 predicted translation product encodes a 425-amino-acid backbone that is heavily glycosylated resulting in a molecular weight approximating 70 kDa by sodium dodecyl sulfate (SDS)–polyacrylamide gel electrophoresis. PAR1 has a relatively long amino-terminal exodomain consisting of 99 amino acid residues uniquely evolved to facilitate interaction with its specific protease and enhance receptor recognition and cleavage (Brass *et al.*, 1992; Vu *et al.*, 1991b). PAR1 is efficiently cleaved and activated by α-thrombin with an EC_{50} for thrombin-stimulated phosphoinositide hydrolysis approximating 0.2 nM. At concentrations approximating 10 nM, α-thrombin cleaves almost all of the cell-surface PAR1 within 1 min. Other receptor-activating proteases include granzyme A (Suidan *et al.*, 1994), trypsin (Brass *et al.*, 1991), and factor Xa (Molino *et al.*, 1997b; Riewald and Ruff, 2001); inactivating proteases (i.e., those that cleave elsewhere within the receptor) include cathepsin G (Molino *et al.*, 1995), plasmin (Kimura *et al.*, 1996), elastase (Renesto *et al.*, 1997), and proteinase 3 (Renesto *et al.*, 1997).

2. Protease Activated Receptor-2 (PAR2)

Human PAR2 was identified and characterized during a search for substance-K receptor homologs (Nystedt *et al.*, 1994). Unlike the other PARs which are cleaved by thrombin, PAR2 is unique in that it is cleaved by trypsin or by mast-cell-derived tryptases but not by thrombin (Mirza *et al.*, 1997; Nystedt *et al.*, 1994). Northern blot analysis demonstrated that the PAR2 gene is widely expressed in human tissues with especially prominent expression patterns in the liver, kidney, pancreas, small intestine, and colon. It is also detected in human endothelial cells, keratinocytes, and smooth muscles of the aorta and coronary artery, but not in the brain or skeletal muscles (Molino *et al.*, 1998; Nystedt *et al.*, 1994). PAR2 is not expressed in human platelets (Mirza *et al.*, 1996). Human PAR2 is cleaved at a specific ^{33}SKGR↓SLIGK^{41} cleavage site; in addition to its physiological cleavage and activation by mast cell tryptases, PAR2 is also activated by coagulation factors Xa and VIIa, although activation by the latter appears to occur optimally in the presence of the tissue factor/factor VIIa complex and factor X (Camerer *et al.*, 2000). Like PAR1, PAR2 appears to mediate proliferative responses as evaluated in primary cultures of vascular endothelial cells (Mirza *et al.*, 1997). Additional functional roles for PAR2 in embryonic development (Jenkins *et al.*, 2000), inflammation (Cocks *et al.*, 1999; Steinhoff *et al.*, 2000), vascular hemodynamic responses

(Sobey *et al.*, 1999), and thrombosis have been suggested (Camerer *et al.*, 2000) (see later discussion).

3. Protease Activated Receptor-3 (PAR3)

In humans, PAR3 mRNA is expressed in the bone marrow and vascular endothelial cells, but minimally in platelets with no more than 150–200 PAR3 receptors per platelet (Cupit *et al.*, 1999; Ishihara *et al.*, 1997; Schmidt *et al.*, 1998). *In situ* hybridization revealed that PAR3 is highly expressed in murine splenic and bone marrow megakaryocytes, cells which do not express murine PAR1. Whereas PAR1 is the primary thrombin receptor in human platelets, PAR3 does not appear to have a physiological function in human platelet activation, but rather represents the predominant thrombin receptor in murine platelets. Like PAR1, PAR3 is also a thrombin substrate, with 20 nM α-thrombin cleaving up to 80% of PAR3-transfected COS cells within 5 min (Ishihara *et al.*, 1997). The EC_{50} for thrombin-induced phosphoinositide hydrolysis is comparable to that of PAR1 (0.2 nM). Thrombin inactivation by its proteolytic inhibitor D-phenylalanyl-L-prolyl-L-arginylchloromethyl ketone (PPACK) abrogates signaling in PAR3-transfected cells, even at concentrations as high as 1 μM, confirming that proteolytic cleavage is necessary for receptor activation. Human PAR1 and PAR3 have about 27% amino acid sequence similarity. The specific thrombin cleavage recognition sequence found in the amino-terminal exodomain of human PAR3 is located at the [35]LPIK↓TFRGAP[44] junction. Similar to the structure of PAR1 is the presence within PAR3 of a hirudin-like sequence ([48]FEEFP[52]) which is also identified carboxyl to the cleavage site and facilitates the protease–receptor interaction (Mathews *et al.*, 1994). Like PAR1, PAR3 is postulated to utilize the hirudin-like domain for thrombin interaction, supported by mutagenesis studies within this sequence which shift the dose-response curve for thrombin activation rightwards by about 10-fold.

The human PAR3 cleavage site [35]LPIK↓TFRGAP[44] is specific and responsive to thrombin alone, and PAR3 cleavage is prevented by substitution of proline for threonine at amino acid residue 39. Other arginine/lysine-specific serine proteases (factor Xa, trypsin, factor VIIa, tissue plasminogen activator or plasmin) demonstrate little or no ability to cleave PAR3. Unlike PAR1, synthetic peptidomimetics based on the N-terminal sequence of the newly generated tethered ligand (TFR-GAP and TFRGAPPNS) resulted in little or no activation of PAR3 even at concentrations as high as 100 μM. Despite this observation, evidence suggests that PAR3's molecular mechanism of activation is comparable to that of PAR1, i.e., signaling mediated via a tethered ligand mechanism (see later discussion). Thus, substitution of Ala-40 for Phe-40 generated a receptor that failed to signal upon thrombin cleavage (Ishihara *et al.*, 1997). Phe-40 would be postulated to be critical in PAR3's intramolecular interaction and signaling, analogous to Phe-43 in PAR1 (Scarborough *et al.*, 1992).

4. Protease Activated Receptor-4 (PAR4)

PAR4 is the most recently identified member of the protease-activated receptor family and a third thrombin receptor in humans (Xu *et al.,* 1998). Alignment of the human PAR4's 397-amino acid sequence with other known PARs indicates that it has about 33% amino acid homology with PAR1, PAR2, and PAR3 (Cupit *et al.,* 1999; Xu *et al.,* 1998). However, PAR4's amino-terminal exodomain and intracellular cytoplasmic domain have very little or no amino acid sequence similarity to the corresponding regions in the three other PARs. Northern blot analysis demonstrated that the PAR4 gene is widely expressed in human tissues with especially high expression patterns in lung, pancreas, thyroid, testis, and small intestine. No PAR4 expression is detected in the brain, kidney, spinal cord, human umbilical vein endothelial cells, or peripheral blood leukocytes. Although less abundant than PAR1, PAR4 mRNA is readily detected in human platelets by reverse transcription–polymerase chain reaction (RT-PCR).

Like PAR1 and PAR3, thrombin is a primary activator of PAR4 although it is 1–2 logs less responsive to thrombin when compared to PAR1 and PAR3 (Kahn *et al.,* 1998; Xu *et al.,* 1998). The EC_{50} for thrombin- and trypsin-induced phosphoinositide hydrolysis mediated by PAR4 activation is about 5 n*M*. The likely explanation for this higher thrombin requirement for PAR4 activation is the absence of the thrombin-interactive hirudin-like exodomain that is present within PAR1 and PAR3. Other arginine-lysine specific serine proteases including factors VIIa, IXa, XIa, plasmin, and urokinase demonstrate little or no ability to activate PAR4. Minimal activation responses were evident using high (100 n*M*) concentrations of factor Xa. Embedded within its first exodomain is the thrombin-cleavage recognition site, [44]PAPR↓GYPGQV[53] specifically cleaved at the Arg-47 and Gly-48 bond. Mutagenesis of Arg-47 to Ala-47 generates a receptor unable to respond to either thrombin or trypsin. Evidence exists that PAR4 is activatable by neutrophil-derived cathepsin G, with cleavage at the identical $Arg^{47} \downarrow Gly^{48}$ sessile bond (Sambrano *et al.,* 2000).

Comparable to human PAR1 and PAR2 (but not to PAR3), human PAR4 may be activated by peptidomimetics that are homologous to the N-terminal amino acid sequences of the tethered ligand. The Tyr-49 in the second position of the tethered ligand is important in the function of this domain (Faruqi *et al.,* 2000). Its peptidomimetic GYPGQV is able to activate PAR4 although at a higher EC_{50} (100 μM) compared to that of SFLLRN for PAR1 (EC_{50} ~5 μM) (Vu *et al.,* 1991a). The mutagenized peptidomimetic AYPGKF is 10-fold more potent than GYPGQV (Faruqi *et al.,* 2000).

B. Molecular Mechanisms of Receptor Activation

The mechanisms regulating PAR cleavage are best characterized using PAR1 as the prototype (refer to Fig. 1). The PAR1 N terminus interacts with thrombin's anion

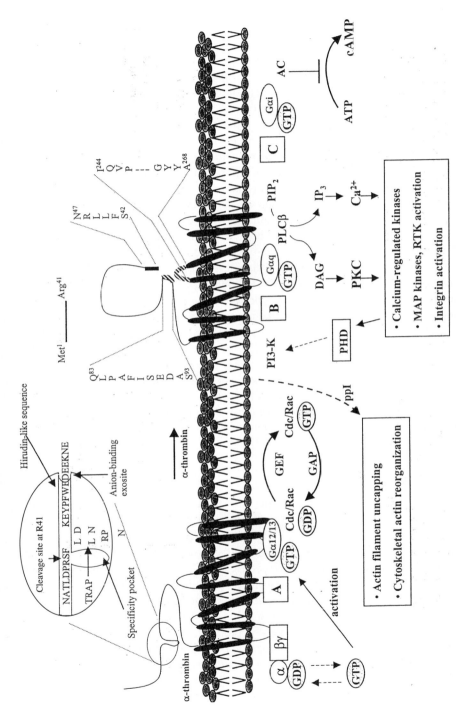

Figure 1 Molecular mechanism of PAR1 activation and intracellular signaling cascades. See text for details.

exosite through an acidic hirudin-like binding domain sequence ^{50}DKYEPF^{55} (found carboxyl to the cleavage site) (Esmon, 1987; Liu *et al.*, 1991), thereby facilitating cleavage at the LDPR^{41}/S^{42}FLLR sessile bond. A comparable ^{38}LDPR$\downarrow$S^{42} sequence is also utilized as the thrombin cleavage site in protein C when thrombin is specifically bound to thrombomodulin (Esmon, 1987). Free thrombin unattached to thrombomodulin is unable to cleave protein C because of the inhibitory effect of the P3 aspartate residue in ^{39}DPR^{41} (LeBonniec and Esmon, 1991). Formation of the thrombin–thrombomodulin complex results in a conformational change in thrombin's active center which can accommodate the ^{39}DPR^{41} sequence with subsequent cleavage of protein C. By extrapolation from the thrombin–thrombomodulin model, thrombin binding to the hirudin-like domain would predictably cause a similar conformational change, with subsequent receptor cleavage (Liu *et al.*, 1991; Mathews *et al.*, 1994). This irreversible cleavage generates a new amino terminus leading to "self-activation" via a tethered ligand mechanism. Critical intramolecular interactions involving the second extracellular loop (hatched) (Gerstzen *et al.*, 1994), the tethered ligand (solid rectangle), and an 11-mer peptide within the long extracellular domain (hatched) (Bahou *et al.*, 1993a; Nanevicz *et al.*, 1996) may serve as the "binding pocket" for downstream receptor-coupling events, although intermolecular activating mechanisms may exist with much less efficiency (Chen *et al.*, 1994). Synthetic peptides containing at least the first five amino acids (SFLLR) of the tethered ligand (referred to as thrombin receptor activating peptides; TRAP) are able to effect receptor activation independently of receptor proteolysis with evidence that the protonated Ser-42 amino group and the Phe-43 side chain appear to be especially critical for peptidomimetic function (Scarborough *et al.*, 1992). After thrombin cleavage the Met^1-Arg^{41} peptide fragment is released from the cellular surface (Furman *et al.*, 1998).

1. Receptor Processing

The intracellular processing and fate of PARs appears to be cell-dependent. On a resting platelet approximately two-thirds of the 1500–2000 PAR1 receptors are located on the plasma membrane (Brass *et al.*, 1992). Prior to activation the remainder are present in the membranes of the intracellular surface connecting system, a structure that is contiguous with the platelet plasma membrane and that is exposed upon platelet activation. Platelet activation by nonthrombin agonists exposes those PAR1 receptors initially present in the canalicular system, thereby increasing the number of cell-surface receptors available for cleavage. Thrombin activation efficiently cleaves the majority of the receptors, with subsequent internalization or receptor shedding into platelet microparticles during platelet aggregation (Molino *et al.*, 1997a). Unlike the situation seen in vascular endothelial cells (see later discussion), platelets do not appear to have a storage pool of receptors and have minimal capacity for protein synthesis. Platelets are thus essentially capable of responding once to thrombin, not an unexpected evolutionary adaptation given their short circulatory half-life and functionally terminal roles in hemostasis. Although it is evident that other proteases can disable PAR1 by cleavage at alternative sites within the receptor (refer to Table I), their physiological role as regulators of thrombin-induced platelet activation remain unclear.

The fate of PARs on generally quiescent and longer-surviving vascular endothelial cells (ECs) is different than that of platelets. PAR1 is readily detectable on ECs (Bahou *et al.*, 1993a), with estimates as high as 10^6 receptors/cell (Tiruppathi *et al.*, 1992). Stimulation of ECs with thrombin results in receptor internalization and cellular desensitization. Activated PAR1 is rapidly uncoupled from signaling and subsequent receptor activation, although nearly half of the receptors remain present on the cell surface. The uncoupling is most likely related to intracytoplasmic phosphorylation events, possibly involving one or more members of the GRK family of receptor kinases. Unlike other G-protein coupled receptors (GPCRs) (refer to Table I), the receptor does not recycle, but rather is delivered to the lysosomal compartment for degradation; this is not unexpected given the proteolytic cleavage which disables further receptor function. Like platelets, endothelial cell PAR1 is used once, but unlike platelets, the EC cell-surface compartment is readily restored within 2 h from a preformed intracellular pool (Molino *et al.*, 1997b). Subsequent protein synthesis replenishes both the cell-surface and intracellular storage pools of receptors. The availability of a rapidly mobilized intracellular storage pool of receptors ensures that stationary and long-lived endothelial cells are able to respond to subsequent thrombin stimuli within a relatively short period, a situation of less relevance for an activated platelet with a terminal function. Comparable mechanisms regulating PAR2 activation and desensitization appear evident in ECs, with suggestions that heterologous cross-desensitization of PAR1 and PAR2 may represent an additional method for regulating PAR function (Mirza *et al.*, 1996; Molino *et al.*, 1997b) (see later discussion).

2. Receptor cocommunication

Evidence has been presented that PARs function in concert at the cell surface and intracellularly to regulate cell activation events. A dual-receptor platelet signaling system exists in both humans and mice, although evidence for receptor cofunction is restricted to the murine system. In mice, PAR3 and PAR4 are the primary thrombin receptors, whereas in humans PAR1 and PAR4 are the primary receptor system (Kahn *et al.*, 1998). PAR4 is responsible for the delayed residual signaling seen at high thrombin concentrations ($> 10 \, nM$) in both murine and human platelets. In mice, PAR3 may actually function as a cofactor for PAR4 cleavage and activation at low thrombin concentrations since PAR3 activation in the absence of PAR4 does not result in signaling (Nakanishi-Matsui *et al.*, 2000). It remains unclear whether this phenomenon occurs via receptor heterodimerization, or whether a comparable mechanism is present in thrombin-mediated human activation systems.

Molecular mechanisms regulating PAR1 activation and desensitization have been extensively studied and include a combination of receptor internalization and phosphorylation, the former mechanism playing a predominant role in endothelial cells (ECs). Evidence for heterologous desensitization between two PARs has also been demonstrated in ECs, known to coexpress PAR1 and PAR2. Thus, whereas internalization of EC PAR1 occurs upon thrombin activation, PAR1 internalization (and functional desensitization) similarly occurs with specific activation of PAR2 (Mirza *et al.*, 1996). Reciprocity is also evident since PAR1 activation results in functional PAR2 desensitization, although the distinction between intracellular "shutoff" and PAR2 internalization is less well established. These heterologous responses are not related to cell-surface PAR1/PAR2 interactions as has been demonstrated for the murine PAR3/PAR4 system, although evidence that cleaved PAR1 receptors can signal through PAR2 trans-activation has been presented (O'Brien *et al.*, 2000).

C. PAR Intracellular Signaling

Thrombin demonstrates profound effects on diverse cells; it is mitogenic for vascular smooth muscle cells and fibroblasts, and a potent secretagogue and stimulus for vascular endothelial cells. Direct effects on monocyte chemotaxis (Bar-Shavit *et al.*, 1983), neutrophil activation, and adhesion link thrombin activity with inflammatory responses (Zimmerman *et al.*, 1985). Known thrombin cellular sequelae involving other diverse tissues include effects on neurons (Turgeon and Hoenou, 2001), tumor metastases (Nierodzik *et al.*, 1991; Nierodzik and Korpatkin, 1992), and myocardial cells (Steinberg *et al.*, 1991).

Upon PAR activation, thrombin-induced intermediate signaling pathways involve phosphoinositide hydrolysis, protein phosphorylation, an increase in intracytosolic free calcium, and suppression of cAMP synthesis, signals that are generally identifiable at thrombin concentrations approximating 100 pM (Fig. 1). These distinct but converging pathways ultimately lead to cytoskeletal actin reorganization and integrin activation, the critical common pathways mediating cellular responsiveness. In quiescent cells, PARs are associated with heterotrimeric $\alpha\beta\gamma$ G proteins, with the α-subunit maintained in the inactive (GDP-bound) state. G-protein activation is regulated by the binding and hydrolysis of GTP. Agonist stimulation by thrombin promotes the release of GDP with replacement by cytosolic GTP. A conformational change of the α-subunit results in its dissociation from $G_{\beta\gamma}$, leaving both in their active states. The subsequent hydrolysis of the GTP-bound form to GDP is mediated by GTPase activating proteins (GAPs), which result in reassociation with $G_{\beta\gamma}$, pending subsequent cycles of receptor-mediated signaling. Upon thrombin stimulation, PARs couple to members of the $G_{\alpha12/13}$ (A), $G_{\alpha q}$ (B), and $G_{\alpha i}$ (C) G-protein families, with downstream activation to a host of intracellular effectors, associated with G_α-charging of GTP. $G_{\alpha i}$ inhibits adenyl cyclase (AC) resulting in diminished levels of cAMP and enhanced cellular responsiveness. The $G_{\alpha q}$-subunit activates phospholipase Cβ (PLCβ), resulting in generation of 1,4,5-inositol trisphosphates (IP$_3$) from phosphatidylinositol 4,5-bisphosphate (PIP$_2$) and release of intracytosolic calcium. Both of these signaling pathways are partially (if not fully) mediated through PAR1 activation (Hung *et al.,* 1992). Concomitant activation of protein kinase C (PKC) from diacylglycerol (DAG) provides for a pathway linked to activation of calcium-regulated kinases, MAP kinases, receptor tyrosine kinases (RTK), and integrins, among others. The $G_{\alpha 12/13}$-subunits are coupled to guanine nucleotide exchange factor(s) (GEF) such as p115RhoGEF (Kozasa *et al.,* 1998), although the upstream signals leading to thrombin-induced GTP-charging of Rac1 and Cdc42 remain poorly characterized. GTP-bound rac1 can directly bind and stimulate phosphatidylinositol-3-kinase (PI3-K), which is also coactivated in thrombin-stimulated platelets. Activated phosphoinositide kinases (such as PI3-K) (Stoyanov *et al.,* 1995) facilitate recruitment and attachment of various signaling proteins (including those containing pleckstrin homology domains [PHD]) to the platelet inner membrane. The generation of D3-, D4-, or D5-phosphorylated phosphoinositides (ppI) appears to be terminal events for PAR1-mediated actin filament uncapping and assembly (Hartwig *et al.,* 1995).

III. PAR Molecular and Developmental Genetics

A. PAR Molecular Genetics

The genes for three PARs (PAR1, PAR2, PAR3) cocluster in the human genome at 5q13, whereas the PAR4 gene is found as a single copy at 19p12 (Bahou *et al.,* 1993b; Schmidt *et al.,* 1996). The gene structures share a common characteristic in that the coding regions of all these genes are contained within two exons. The first of the two exons is smaller, encompassing approximately 30 amino acids. The majority of the coding sequence is located in the larger second exon, which contains the protease cleavage site for all gene products (see Fig. 2). These similarities among the PAR genes suggest a relatively recent gene duplication event and evolution from a common ancestral gene. Homology searches of the near-complete human genome databases provide no strong evidence for the presence of other PARs at either chromosomal location or elsewhere in the genome.

B. PAR Developmental Genetics

Murine model systems have provided powerful insights into the role of serine proteases and PARs in cellular and embryonic development. There is emerging evidence supporting a key role for thrombin-generating pathways (and/or their

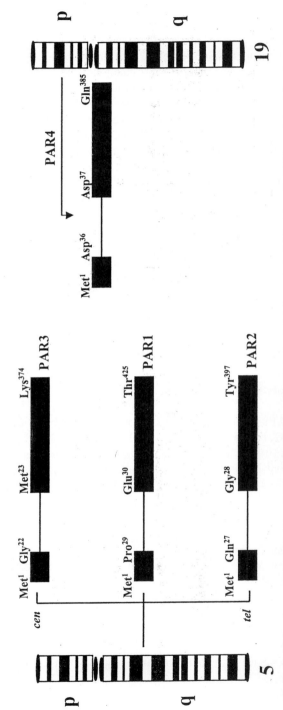

Figure 2 Molecular genetics of PARs. The genes for PAR1, PAR2, and PAR3 are located on chromosome 19p12, whereas the PAR4 gene is located at chromosome 5q13. The interorder chromosomal assignments for PARs 1–3 are updated from previously published reports to conform with recently generated consensus data from the human genome databases (Schmidt et al., 1997; Venter et al., 1998). The amino acid designations are from published cDNA sequences for PAR1 (Vu et al., 1991), PAR2 (Nystedt et al., 1995), PAR3 (Ishihara et al., 1997), and PAR4 (Xu et al., 1998). Smaller boxes represent exons 1 and larger boxes, exons 2 of each PAR gene. AA, Amino acids; cen, centromere; tel, telomere; p, short arm; q, long arm.

cellular receptors) in functions beyond normal hemostasis, with relevance to normal embryogenesis—and in the case of PAR1 imputing a role in vascular wall integrity. Thus, a fundamental role for PAR1 in embryonic development had originally been suggested by generating thrombin receptor-deficient (PAR1$^{-/-}$) knockout mice. Disruption of the murine gene resulted in approximately 50% lethality of homozygous (PAR1$^{-/-}$) mice at embryonic day 9–10, although the other half developed normally, apparently with no hemorrhagic diatheses (Connolly et al., 1996; Darrow et al., 1996). Whereas PAR1$^{-/-}$ fibroblasts lost thrombin responsiveness, PAR1$^{-/-}$ platelets continued to respond normally to thrombin, now known to be related to PAR3 signaling as the predominant murine thrombin receptor (Kinlough-Rathbone et al., 1993; Ishihara et al., 1997). These embryonic deaths corresponded to development of the circulatory system, and the growth deficits of PAR1$^{-/-}$ mice displayed delayed vascularization of the yolk sac, possibly intimating a role for PAR1 in vasculogenesis. Furthermore, these initial observations provided strong presumptive evidence that thrombin-induced platelet activation was not causally implicated in the defective hemostasis evident in these mice.

Other transgenic knockout studies relevant to thrombin generation and PAR signaling are outlined in Fig. 3. The conversion of prothrombin to its active protease α-thrombin is dependent upon factor Xa generation. At sites of vascular injury, plasma coagulation factor VIIa contacts extravascular tissue factor which function to convert the factor X zymogen to activated factor X (FXa). FXa in the presence of the Factor Va cofactor then converts prothrombin to thrombin. It is intriguing that the timing and phenotype of embryonic lethality in PAR1$^{-/-}$ embryos corresponded to that of mice deficient in factor V (Cui et al., 1996) and prothrombin (Sun et al., 1998; Xue et al., 1998), consistent with a role for thrombin generation or cellular signaling in vascular remodeling or its integrity. More pronounced degrees of embryonic lethality were evident in tissue factor-deficient knockout mice because of fatal hemorrhage, presumably associated with defects in vascular integrity (Bugge et al., 1996; Toomey et al., 1996). In contrast, targeted disruption of the factor VII gene (the protease cofactor for tissue factor) had no effect on embryonic development, although FVII$^{-/-}$ embryos did display postpartum hemorrhage (Rosen et al., 1997). One explanation for the normal development of FVII$^{-/-}$ mice is possible trace amounts of transplacental (maternal) factor VII transmission from heterozygote mothers. More recently, similar results have been seen with targeted disruption of the factor X gene, with FX$^{-/-}$ mice displaying partial embryonic lethality (Dewerchin et al., 2000). Nearly one-third of FX$^{-/-}$ pups died by embryonic (E) day 11.5–12.5, although no histologic defects in the vasculature of affected embryos or their yolk sacs were observed. Targeted disruption of the thrombomodulin gene results in an embryonic lethal mutation by E9.5 not clearly related to defective vasculogenesis, but apparently due to dysfunctional maternal–embryonic interactions in the parietal yolk sac (Rosenberg, 1996). These collective observations clearly demonstrate that the serine proteases involved in thrombin-generating pathways play key roles in embryonic development, presumably related to defects

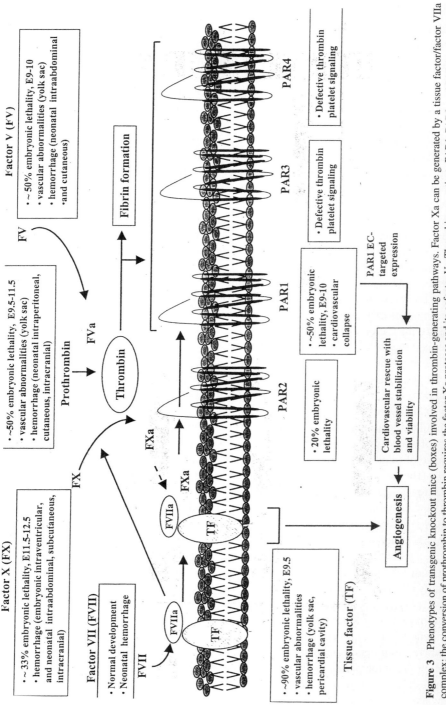

Figure 3 Phenotypes of transgenic knockout mice (boxes) involved in thrombin-generating pathways. Factor Xa can be generated by a tissue factor/factor VIIa complex; the conversion of prothrombin to thrombin requires the factor Xa protease and its cofactor Va. Thrombin can activate PAR1, PAR3, and PAR4 and generate fibrin from fibrinogen cleavage. The TF/FVIIa/FXa complex can directly activate PAR2 and PAR1 unrelated to thrombin generation, whereas the TF/FVIIa complex is directly implicated in angiogenesis. EC-targeted PAR1 expression rescues the PAR1-null phenotype implicating thrombin and/or its endothelial cell PAR1 receptor in vessel wall viability. Refer to text for details.

Text within figure:

Factor X (FX)
- ~33% embryonic lethality, E11.5–12.5
- hemorrhage (embryonic intraventricular, and neonatal intraabdominal, subcutaneous, intracranial)

Factor VII (FVII)
- Normal development
- Neonatal hemorrhage

Tissue factor (TF)
- ~90% embryonic lethality, E9.5
- vascular abnormalities
- hemorrhage (yolk sac, pericardial cavity)

FVII

FVIIa   TF

FVIIa   TF

FX

FXa

FXa

FXa

Angiogenesis

Prothrombin
- ~50% embryonic lethality, E9.5–11.5
- vascular abnormalities (yolk sac)
- hemorrhage (neonatal intraperitoneal, cutaneous, intracranial)

FV

FVa

Thrombin

Fibrin formation

Factor V (FV)
- ~50% embryonic lethality, E9–10
- vascular abnormalities (yolk sac)
- hemorrhage (neonatal intraabdominal and cutaneous)

PAR2
- 20% embryonic lethality

PAR1
- ~50% embryonic lethality, E9–10
- cardiovascular collapse

PAR3
- Defective thrombin platelet signaling

PAR4
- Defective thrombin platelet signaling

Cardiovascular rescue with blood vessel stabilization and viability

PAR1 EC-targeted expression

in vascular integrity or development. Furthermore, the minimal coordinate expression of prothrombin and PAR1 during embryonic development suggested that other endogenous PAR1 activators may exist during development (Soifer *et al.*, 1994).

A unifying explanation for these collective defects strongly implicated coagulation proteases in the maintenance of (endothelial cell) vascular wall integrity, a hypothesis that has now been confirmed. Targeted expression of PAR1 to the vascular endothelial compartment of PAR1 null mice rescued the lethal phenotype, with evidence for normal vascular development, and resolution of the abnormal bleeding causally implicated in intrauterine death. Thus, loss of PAR1 signaling in endothelial cells probably explains the death at midgestation seen with knockout of tissue factor, factor V, or prothrombin (and possibly factor X). In the latter cases, intrauterine death is due to defective thrombin generation, whereas in PAR1-null mice, the intrauterine collapse is related to loss of the endothelial cellular receptor mediating the thrombin response (Griffin *et al.*, 2001).

IV. PARs In Disease and Pathological Processes

A. Angiogenesis

The murine knockout studies outlined above strongly suggest that PARs have a fundamental role in vascular development and/or the maintenance of vessel integrity. At this point, the evidence implicates PAR1 with less clear roles for PARs 2/3/4. Newly forming blood vessels consist primarily of fragile endothelial cells (ECs). Vessel stabilization of such "naked" ECs requires the recruitment of mesenchymal cells which differentiate into vascular smooth muscle cells, with production of an extracellular matrix that strengthens and matures newly forming vessels. Vascular endothelial growth factor (VEGF) initiates endothelial growth, while platelet-derived growth factor (PDGF) and angiopoietin are necessary for vessel wall stabilization. The abnormal vascular fragility evident in PAR1$^{-/-}$ embryos is not dissimilar to that seen in embryos lacking angiopoietin-1 or its Tie2 receptor (Schlaeger *et al.*, 1997), suggesting an interaction between coagulation proteases/PAR1 axis and this angiogenic axis involved in vascular stabilization and integrity (Carmeliet *et al.* 1996). Viewed in concert, coagulation and angiogenesis could cooperate to stabilize new blood vessels, repair injured vessels, and stimulate sprouting of new vessels.

Although the vascular fragility seen in PAR1$^{-/-}$ embryos is likely related to defective thrombin signaling, ancillary role(s) of the tissue factor/FVIIa/FXa complex remain possible. Thrombin's effects are protean: it (1) generates a provisional fibrin scaffold that attracts migrating ECs, (2) increases expression of VEGF receptors, (3) promotes production and extravasation of matrix proteins, (4) loosens ECs from the extracellular matrix by upregulating and activating metalloproteinases

(Zucker *et al.,* 1995), (5) induces EC proliferation and migration, and (6) activates the hypoxia-inducible transcription factor HIF-1α, thereby upregulating expression and production of numerous angiogenic molecules (Rao and Pendurthi, 2001). Similarly, tissue factor is known to stabilize newly forming fragile vessels by recruitment of supporting pericytes (Carmeliet *et al.,* 1996), and the TF/FVIIa complex could stimulate angiogenesis by downregulating the angiogenic inhibitor thrombospondin-1, while upregulating VEGF, fibroblast growth factor-5, collagenases, and their receptors (u-PAR) (Rao and Pendurthi, 2001). The proteolytic activity of FVIIa (independent of thrombin formation) is also involved in angiogenic signaling. Evidence suggests that PAR2 is likely to be activated by FVIIa in the vessel wall. This activation is most evident when sufficient tissue factor and FXa are present, the anticipated scenario in the angiogenic endothelium. Interestingly, the TF/FVIIa/Xa complex has been demonstrated to activate PAR1 (Riewald and Ruf, 2001); thus, although thrombin is likely to be the primary PAR1 activator for vessel stabilization, upstream coagulation proteases may also be involved (Riewald *et al.,* 2001). At this point, it remains unknown whether angiogenesis in pathological conditions is impaired in PAR1$^{-/-}$ mice.

B. Cancer and Tumor Metastases

The tumor microenvironment encompasses an array of signaling molecules designed to facilitate tumor invasion by degrading and remodeling the extracellular matrix. Both thrombin and trypsin are among these activated components, and their expression correlates with the stage and type of carcinoma, thereby associated with cellular invasion and extracellular matrix degradation. Thus, enzymatically active thrombin has been identified on surgically removed tumor tissue, including malignant melanoma. Similarly, tumors express cell-surface tissue factor, providing another means for thrombin generation. Tissue factor expression correlates with hematogenous metastases in melanoma cells and is associated with the leading edge in invasive breast carcinoma; as mentioned earlier, TF can stimulate tumor growth independently of thrombin generation by induction of VEGF and angiogenesis (Mueller *et al.,* 1992; Rao and Pendurthi, 2001).

PAR1 appears to be primarily involved in mediating the enhanced thrombin responsiveness evident in cancer and tumor metastases. PAR1 is expressed in pancreatic tumor cells and melanoma cell lines, and its expression has been correlated with invasiveness in breast carcinoma cells lines and human breast metastatic tissue (Even-Ram *et al.,* 1998). Furthermore, B16F10 melanoma cells transfected with the PAR1 cDNA demonstrated enhanced thrombin-inducible adhesion to fibronectin, and pulmonary metastases nearly 40-fold greater than those of wild-type controls using murine *in vivo* model systems (Nierodzik *et al.,* 1998). More recent evidence implicates PAR1 expression not only on the malignant carcinoma cells, but also on the cell types forming the microenvironment, including endothelial and vascular smooth muscle cells, stromal fibroblasts, and macrophages; interestingly, macrophages expressed both PAR1 and PAR2. Similarly, in stromal fibroblasts, both PAR1 and PAR2 expression were seen immediately surrounding the malignant cells, although their expression was *absent* from the surrounding stromal fibroblasts of normal and benign breast epithelial cells (D'Andrea *et al.*, 2001). Since PAR1 and PAR2 are known to facilitate tumor-cell adhesion, cell proliferation, and release of cytokines and growth factors, these collective observations suggest an autocrine and/or paracrine mechanism linking protease generation to the balance of tumor containment and metastases.

C. Thrombosis

Although thrombin is clearly one of the most potent activators of human platelets *in vitro,* distinguishing thrombin's procoagulant effects (mediated through the coagulation cascade) from its diverse cellular effects (i.e., platelet, endothelial, vascular smooth muscle cell activation, etc.) in the regulation of thrombosis remains a concerted challenge. This issue is relevant for strategies designed to inhibit PAR cellular activation (see later discussion). Local concentrations of thrombin generation *in vivo* are diluted by normal blood flow and by endogenous plasma inhibitors. Nonetheless, thrombin concentrations within the vicinity of a thrombus may be as high as 140 nM, with persistence for up to 10 days, and blood coagulation monitoring suggests that the effective thrombin concentration involved in fibrinogen cleavage approximates 2.5 nM (Brummel *et al.,* 1999); these concentrations are greater than the EC_{50} for human platelet activation.

In vivo, a critical role for thrombin-associated platelet activation in the generation and propagation of arterial thromboses has been suggested, based on the efficacy of heparin to reduce platelet and fibrinogen deposition in models of arterial injury (Chesebro *et al.,* 1991). That this represents an antithrombin effect—as opposed to an anti-factor Xa effect—has been suggested by the use of low-molecular-weight heparins (Heras *et al.,* 1990). More selective antithrombin therapy using the synthetic competitive thrombin inhibitor argatroban or hirudin—which does not directly block other mediators of platelet aggregation such as thromboxane A$_2$, serotonin, ADP, or collagen—abolishes the development of mural thrombosis and markedly limits platelet deposition to a single layer *in vivo* after arterial injury. As importantly, evidence exists that thrombin inhibition is associated with diminished platelet activation markers *in vivo* (Practico *et al.,* 1997) and that direct PAR1 inhibition can abrogate arterial thrombosis in a platelet-dependent primate animal model (Cook *et al.,* 1995).

D. Atherosclerotic Coronary Artery Disease

PARs may be involved in atherosclerotic pathogenesis, specifically coronary artery disease and to some extent restenosis postangioplasty. PAR1 expression is upregulated in advanced human atherosclerotic vessels (Nelken *et al.*, 1992), after experimental injury in animal models (Baykal *et al.*, 1995), and is induced by balloon angioplasty in the rat and baboon (Wilcox *et al.*, 1994). *In situ* hybridization of atherosclerotic vessels demonstrated high levels of PAR1 expression in areas rich in macrophages and in areas of proliferating smooth muscle cells and mesenchymal-like intimal cells (Baykal *et al.*, 1995). Earlier lesions consisting of fatty streaks demonstrated PAR1 expression in the intima alone without significant expression in the underlying media (Nelken *et al.*, 1992). In addition to its rapid upregulation during mechanical vascular injury, PAR1 expression is transiently upregulated by low—but not high—shear stress (Papadaki *et al.*, 1998). Whereas PAR1 antisense oligodeoxynucleotides can inhibit vascular smooth muscle cell thrombin responsiveness *in vitro,* they appear ineffective in blocking neointimal hyperplasia following intimal injury *in vivo* (Herbert *et al.*, 1997). In contrast, neointimal smooth muscle cell accumulation can be attenuated using anti-PAR1 antibodies in a rat angioplasty model, confirming that PAR1 activation is involved in the proliferation and accumulation of neointimal smooth muscle cells induced by balloon injury, and that strategies directed at PAR1 blockade may have clinical relevance for postangioplasty restenosis (Takada *et al.*, 1998). Similar beneficial results using neointimal hyperplasia as the end point have been suggested using a carotid injury model in PAR1$^{-/-}$ mice, although the results did not reach statistical significance (Cheung *et al.*, 1999).

E. Inflammation

PARs provide an important link between hemostatic and inflammatory pathways (Esmon, 2000), best exemplified by thrombin's well-recognized proinflammatory effects, and reinforced by the regulatory functions of kininogens and bradykinin in PAR1 activation (see later discussion). Thrombin activation of endothelial cells facilitates leukocyte adhesion by stimulation of cell-surface adhesion molecules and enhances platelet activating factor formation, a potent neutrophil stimulant (Carveth *et al.*, 1992). Thrombin also stimulates IL-6 and IL-8 release from endothelial cells (Johnson *et al.*, 1998) and increases endothelial cell permeability, in part by gap-junction regulation (Cirino *et al.*, 1996). Thrombin is also a mitogen for lymphocytes and is chemotactic for monocytes. Linking coagulant protease generation with proinflammatory stimuli is not limited to thrombin effects, however. It is known that factor XII (Schmeidler-Sapiro *et al.*, 1991), factor VIIa, and factor Xa also activate various cellular types, with evidence that some of factor X's effects are mediated by binding to effector protease receptor-1 (Nicholson *et al.*, 1996). As outlined earlier, FXa and the tissue factor/VIIa complex have been shown to activate PAR2 (Camerer *et al.*, 2000), known to be expressed on vascular endothelial cells (Mirza *et al.*, 1996), and upregulated during endothelial cell stimulation using various inflammatory stimuli (Nystedt *et al.*, 1996).

Disseminated intravascular coagulation with small-vessel thrombosis can develop in the setting of strong systemic inflammatory stimuli such as bacteremia, and/or with congenital deficiencies that modulate thrombin production, such as antithrombin III deficiency or factor V Leiden. Furthermore, proinflammatory roles for coagulant proteins *in vivo* have been suggested by studies demonstrating amelioration of bacterial-induced endotoxic shock after treatment with activated protein C (Taylor *et al.,* 1987) or anti–tissue factor antibodies (Taylor *et al.,* 1991). Similarly, in a murine model of injury-induced crescentic glomerulonephritis, hirudin was shown to attenuate renal injury and glomerular crescent formation by reducing T-cell and macrophage infiltration and fibrin deposition. Comparable results were evident in PAR1$^{-/-}$ mice, suggesting that this receptor was involved in inflammatory cell-mediated renal injury (Cunningham *et al.,* 2000). Similar supportive observations implicate PARs and/or the coagulant pathways in pathogenesis of bronchial inflammation (Cocks *et al.,* 1999; Hauck *et al.,* 1999), mast-cell-induced neurogenic inflammation (Steinhoff *et al.,* 2000), periodontal disease (Hou *et al.,* 1998; Lourbakos *et al.,* 1998), and myocardial infarction (Ehrlich *et al.,* 2000).

V. Development and Application of PAR Inhibitors

Agents that block thrombin-induced cellular responsiveness could potentially be useful therapeutic agents for a wide variety of diseases, although the evidence for this remains largely inferential and based on efficacy of thrombin inhibition. Successful generation of monoclonal or polyclonal antibodies directed against the PAR1 and/or PAR4 cleavage sites inhibits *in vitro* platelet thrombin responsiveness at variable concentrations of thrombin (Bahou *et al.,* 1993a; Norton *et al.,* 1993). Likewise, antisense oligonucleotides directed against PAR1 can effectively modulate cellular thrombin responses, although such strategies are not applicable for anucleate human platelets. Finally, the presence of dual thrombin receptor systems on various cells creates a unique problem for thrombin blockade. Since inhibition of PAR1 attenuates platelet thrombin responses at low—but not high—thrombin concentrations (Kahn *et al.,* 1999), isolated use of PAR1 inhibitors would expectedly be incomplete. Whether such inhibition would be effective as an antithrombotic in situations of "low thrombin generation" remains to be established.

Selective approaches to development of PAR-specific inhibition have been described, using strategies targeting both endogenous regulators, or those targeting the unique intramolecular binding mechanisms of PAR tethered ligands. Previous studies had indicated that bradykinin (Shima *et al.,* 1992) and both high- (Puri *et al.,* 1991) and low-molecular-weight kininogens (Meloni and Schmaier, 1991) inhibit thrombin-induced platelet activation. A core pentapeptide derived from these sequences (RPPGF) was subsequently shown to inhibit platelet activation and thrombin cleavage of PAR1 by binding to the PAR1 tethered ligand

sequence NATLDPRSFLLR (refer to Fig. 1). This appeared to function as a selective platelet thrombin inhibitor since it failed to interfere with ADP, collagen,
or U46619-induced platelet activation (Hasan *et al.*, 1996). *In vivo* confirmation
of its potential efficacy has been established in an electrically induced canine
coronary thrombosis model, an effect apparently brought about by inhibition of
PAR-mediated platelet activation (Mosher *et al.*, 1979). Inhibitory peptide antagonists have also been designed by applying iterations of the PAR1 peptide sequence
(Bernatowicz *et al.*, 1996). For example, based on analysis of key functional and
spatial groups of the PAR1 agonist peptide SFLLRN, a prototype PAR1 antagonist containing a rigid indole template has been developed (Andrade-Gordon
et al., 1999). It was found to selectively bind and inhibit PAR1 activation without
displaying PAR1-agonist or thrombin inhibitory activity. The peptide inhibits intracellular calcium mobilization and platelet aggregation and displayed no activity
against PAR2, PAR3, or PAR4. Flow cytometry confirmed that this compound directly inhibited PAR1 activation and internalization, without affecting exodomain
cleavage. At high concentration of α-thrombin and SFLLRN, it failed to completely
block platelet activation, most likely related to PAR4 saturation. Subsequent *in vivo*
studies using two guinea-pig platelet-dependent thrombosis models revealed
modest effects presumably related to incomplete PAR4 inhibition. These results
contrast with more beneficial effects seen in a rat restenosis model after balloon angioplasty (Andrade-Gordon *et al.*, 2001). Although these reagents remain in early
phases of development, they represent logical and targeted developmental strategy
based on the known molecular mechanisms of PAR1 activation. Depending on
results of further *in vivo* testing, the application of comparable strategies designed
to inhibit PAR4 function may be necessary for optimal inhibitory strategies.

VI. Concluding Remarks

A family of highly adapted seven-transmembrane receptors have evolved for the
transmission of extracellular signals from serine proteases generated by inflammatory, fibrinolytic, or hemostatic pathways. To date, four PARs have been identified
and characterized with no emerging evidence for the presence of other PARs
in the human genome. Two of the PARs (PAR3 and PAR4) appear to function
solely as platelet receptors for thrombin-induced aggregation, whereas PAR1 and
PAR2 have more extensive functions. In the cascade of coagulation proteases, the
evolution of a family of protease-activated receptors uniquely adapted for protease signaling would intuitively represent a logical adaptation for physiological
hemostasis. Similarly, during angiogenesis, the involvement of PAR1 as a sta-
bilizer of blood vessels could be envisioned as a unique adaptation linking the
coagulation system (locally activated in sites of new vessel formation) to vessel
generation and stabilization. Likewise, malignant cells are usually embedded in
cell-surface proteases which may explain the adaptive response for enhanced PAR1

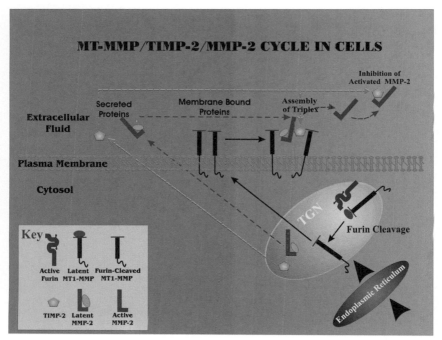

Figure 1.2 Tri-molecular formation between MT1-MMP, TIMP-2, and MMP-2 on the cell surface leading to activation of pro-MMP-2. Two molecules of activated MT1-MMP are required to activate one molecule of proMMP-2 (72 kDa). Following binding of the N-terminal domain of TIMP-2 to the catalytic site of MT1-MMP, the C-terminal domain of TIMP-2 binds to the hemopexin domain of proMMP-2, thereby immobilizing it for activation by a second active MT1-MMP molecule. It has been proposed (not shown in figure) that an intermediate form of activated MMP-2 (64 kDa) binds to integrin αvβ3 on the cell surface before conversion to activated MMP-2 (62 kDa). Following release from the cell surface, activated MMP-2 becomes inhibited by excess TIMP-2. Proprotein convertases (paired basic amino acid cleaving enzymes, i.e., furin) are required for cleavage of the MT1-MMP propeptide in the trans Golgi network (TGN).

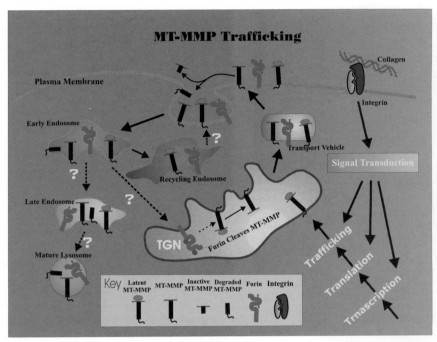

Figure 1.3. Trafficking of the MT1-MMP molecule to and from the cell surface (hypothesis). Following translation of latent MT1-MMP (63 kDa) in the endoplasmic reticulum and transport to the Golgi apparatus, furin cleaves the prodomain of MT1-MMP in the trans Golgi network (TGN) leading to an active enzyme (57/60 kDa). MT1-MMP and furin then enter a transport vesicle destined for trafficking to the cell surface. Latent MT1-MMP appears to be less efficiently trafficked to the cell surface. On the cell surface, MT1-MMP undergoes further cleavage to a 43/45 kDa fragment consisting of the hemopexin-like, transmembrane, and cytoplasmic domains; the catalytic domain and other fragments are released from the cell surface. Active MT1-MMP and the inactive 42 to 45-kDa MT1-MMP fragment enter into clathrin-coated vesicles, presumably destined for intracellular degradation. The possibility of recycling of latent or active MT1-MMP to the cell surface after endocytosis remains to be evaluated. The signal transduction mechanism affecting MT1-MMP transcription, translation, and trafficking is depicted involving collagen:integrin interactions on the cell surface; other outside—inside signaling mechanisms remain to be illuminated.

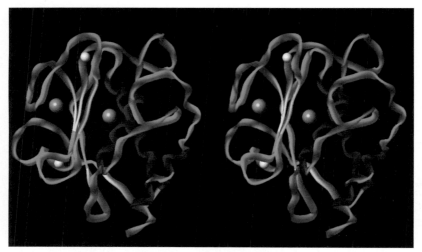

Figure 1.4. Stereo view of the ribbon representation of the three-dimensional X-ray structure of the catalytic domain of the MT1-MMP (see reference). Variable loops and α-helices are colored in cyan and magenta, respectively. Metals are depicted in space-filled rendering with zinc ions colored in orange and structural calcium ions are in white. The catalytic zinc is located in the center of the image and structural zinc is located at 9 o'clock. Reference from Fernandez-Catalan, C., Bode, W., Huber, R., Turk, D., Calvete, J. J., Lichte, A., Tschesche, H., and Maskos, K. (1998). Crystal structure of the complex formed by the membrane type 1 matrix metalloproteinase with the tissue inhibitor of metalloproteinases-2, the soluble progelatinase a receptor. *Embo J.* **17,** 5238.

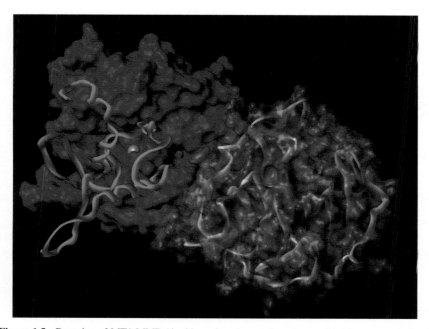

Figure 1.5. Domains of MT1-MMP (the hinge domain not shown) assembled in a computational model for the enzyme (unpublished results from the Mobashery group). A solvent-accessible surface area is constructed around the catalytic domain and colored in magenta. Covering the active site is a ribbon representation of the propeptide domain of MT1-MMP in green. The hemopexin domain is represented by a yellow ribbon representation and an orange translucent solvent-accessible surface area colored in orange. The catalytic zinc ion is shown in an orange space-filled representation.

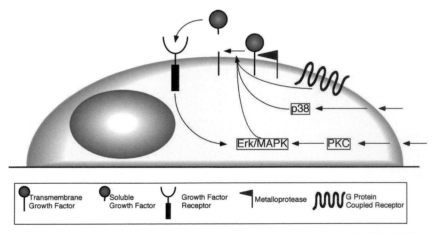

Figure 4.3. Different stimuli and intracellular factors regulate protein ectodomain shedding. Different extracellular stimuli such as growth factors or stress stimulate ectodomain shedding through the action of several intracellular mediators: the Erk MAP kinases, the p38 MAP kinase, or PKC. The stimulation of the shedding of growth factors creates a positive feedback loop and, in addition, it explains the cross-talk between G-protein-coupled receptors and tyrosine kinase receptors, because the former receptors are also able to activate the shedding of growth factors.

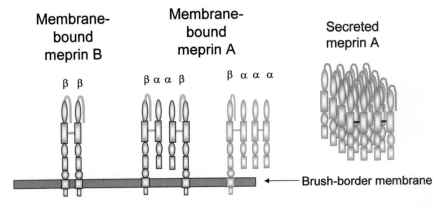

Figure 5.4. Oligomeric structure of meprin A and B. The meprin subunits are shown schematically with domains as rectangles or oval units. Both subunits contain protease, MAM, MATH, and AM domains; the mature meprin β subunit also contains EGF-like, transmembrane, and cytoplasmic domains. Intersubunit disulfide bonds are indicated by horizontal bars. Meprin B is a meprin β homodimer. Meprin A at the membrane is a heterotetramer consisting of at least one anchored β subunit; dimers associate by noncovalent interactions. The secreted form of meprin A, homooligomeric meprin α dimers, tends to self-associate and form large multimers extracellularly. The secreted isoform of meprin A in mouse kidney retains the prosequence and is therefore primarily in the latent enzyme; the meprin α subunit in the kidney membrane is activated, whereas the meprin β subunit retains the prosequence and is latent at the membrane (Kounnas *et al.*, 1991).

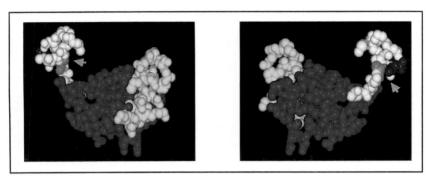

Figure 5.9. Three-dimensional model of monocyte chemoattractant protein-1 (MCP-1). The orange arrow indicates the meprin A cleavage site between Asn-6 (red) and Ala-7 (green). White indicates the N terminus, blue indicates the globular domain, and yellow indicates the α-helical C terminus.

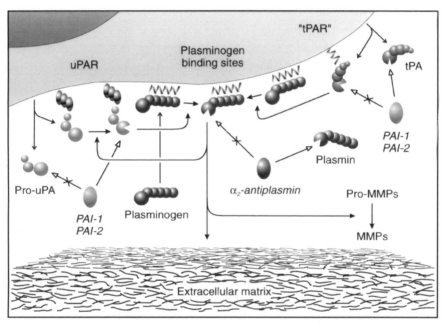

Figure 9.3. Receptor-mediated plasminogen activation. Two independent pathways for pericellular plasmin generation are shown with the principal reactions between membrane-bound components. Closed arrows denote proteolytic activation (or other cleavage) reactions; open arrows denote inhibition; and crossed arrows denote lack of an inhibitory effect. The binding of secreted pro-uPA to its cellular receptor, uPA, leads to efficient reciprocal zymogen activation in which uPA preferentially activates cell-bound plasminogen, and cell-bound plasmin can efficiently activate uPAR-bound pro-uPA. Similarly, the binding of secreted tPA to its receptor(s) leads directly to the efficient activation of cell-associated plasminogen, as tPA does not require proteolytic activation because of its unique "active zymogen" characteristics. Generated plasmin can degrade many nonfibrillar proteins of the extracellular matrix, activate certain matrix metalloproteases, and activate or release matrix-bound growth factors. These pathways are modulated by physiological inhibitors. α₂-Antiplasmin acts to further focus plasmin activity at the cell surface, as plasmin that is generated is fully protected from inhibition while it remains bound but is rapidly inhibited on dissociation from the cell surface. The plasminogen activator inhibitors, in particular PAI-1, appear to have contrasting functions in the regulation of uPA and tPA activities. Pro-uPA is not inhibited by PAI-1, but once activated to uPA the receptor-bound enzyme is fully available for inhibition. By contrast, because of its inherent activity, nascent tPA is susceptible to inhibition by PAI-1, but is protected to a degree when bound to cellular binding sites.

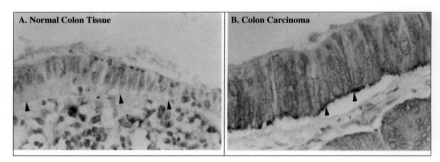

Figure 10.2. Immunohistochemical staining of cathepsin B in normal colon tissue verses colorectal carcinoma. (A) In epithelial cells of normal colon tissue, cathepsin B staining is in vesicles in the apical region of the cytoplasm. There is no cathepsin B staining at the basal pole of the cells (*arrowheads*). (B) In early colorectal carcinoma, cathepsin B staining is polarized to the basal pole of the epithelial cells (*arrowheads*) directly adjacent to the underlying basement membrane (*arrowheads*). [Campo and Sloane, unpublished data].

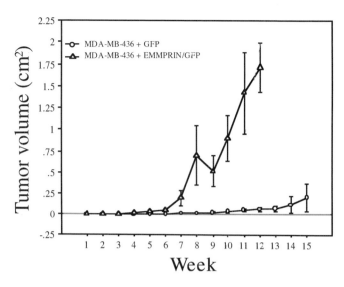

Figure 12.3. Stimulation of tumor growth by increased expression of emmprin. MDA-MB-436 cells were transfected with constructs expressing emmprin or with vector alone. The transfectants were then injected into the fat pad of *nu/nu* mice (1 x 10⁶ cells per animal), and growth was assessed by weekly measurement of tumor size (± S.E. for 10 animals per group). The numbers of animals remaining alive at various time points are indicated on the graph. [Adapted from Zucker *et al.,* (2001).] Examples of tumors obtained in this experiment are shown in the lower panel (left, emmprin; right, control).

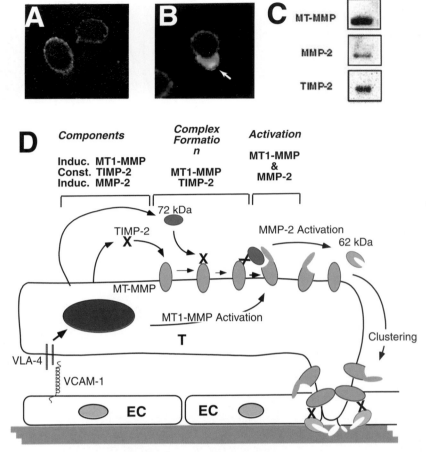

Figure 13.3. Clustering of MT1-MMP, TIMP-2, and MMP-2 on the "invadopodia" of transmigrating T lymphocytes. (A, B) Representative confocal microscopic micrographs of α4β1 high expressing T cells 60 min (A) and 8 h (B) after plating on a monolayer of VCAM-1 positive endothelial cells. (A) The T cells (identified by fluorescein-labeled CD4) exhibit a diffuse peripheral plasma membrane localization of TIMP-2 (rhodamine-labeled) as well as low levels of MT1-MMp (not shown) before and 60 min after plating onto an endothelial cell monolayer. (B) After 8 h, the T cells (identified by fluorescein-labeled CD4) have clustered TIMP-2 to specific membrane surface regions (rhodamine-labeled). This staining colocalizes with robust MT1-MMP and MMP-2 staining (data not shown), consistent with a tertiary complex formation and clustering (arrow). (C) Immunoprecipitation of MT1-MMP reveals that MT1-MMP is complexed with TIMP-2 *and* MMP-2 (as determined by Western blotting) following engagement of T cell α4β1 intefrin by endothelial cell VCAM-1. (D) Schematic representation of the proposed α4β1-mediated induction, assembly, and activation of MT-MMP, TIMP-2, and MMP-2 during T-cell migration through the endothelium and its underlying basement membrane. Engagement of T-cell α4β1 elicits the induction of MT1-MMP and pro-MMP-2, followed by clustering and activation of MT-MMP and MMP-2. Following complex assembly, activation, and clustering, the activated MT1-MMP and MMP-2 on the surface of the T cell proteolyze ECM components, facilitating migration of the T cell into the perivascular tissues. Based on data from Graesser *et al.,*(1998, 2000, 2002); Romanic *et al.,* (1997); Romanic and Madri (1994).

expression. Given these observations, a logical question is the role of PAR1 modulators in regulating the balance in cancer metastatic potential; theoretically, PAR1 blockade could be envisioned to simultaneously inhibit tumor cell PAR1 and the angiogenic response required for initiation of metastatic tumor growth. Comparable approaches have previously been pursued by inhibiting the coagulation cascade using anticoagulants, generally with mixed results. Whether specific PAR1 (or other PAR) antagonists can be adapted for this indication remains unclear. For inhibition of thrombosis, it is clear that PAR blockade will be more complicated given the known coexpression of a dual-PAR system on human platelets. Nonetheless, given the preliminary successes in development of PAR1 antagonists, it is likely that progress in this area will add a new dimension to experimental therapeutics applicable to a large number of disease phenotypes.

References

Andrade-Gordon, P., Derian, C. K., Maryanoff, B. E., Zhang, H.-C., Addo, M. F., Cheung, W. M., Damiano, B. P., D'Andrea, M. R., Darrow, A. L., de Gavavilla, L., Eckardt, A. J., Giardino, E., Haertlein, B., and McComsey, D. F. (2001). Administration of a potent antagonist of protease-activated receptor-1 (PAR-1) attenuates vascular restenosis following balloon angioplasty in rats. *J. Pharmacol. Exp. Ther.* **298**, 34–42.

Andrade-Gordon, P., Maryanoff, B. E., Derian, C. K., Zhang, H.-C., Addo, M. F., Darrow, A. L., Eckardt, A. J., Hoekstra, W. J., McComsey, D. F., Oksenberg, D., Reynolds, E. E., Santulli, R. J., Scarborough, R., Smit, E. M., and White, K. B. (1999). Design, synthesis, and biological characterization of a peptide-mimetic antagonist for a tethered-ligand receptor. *Proc. Natl. Acad. Sci. USA* **96**, 12257–12262.

Bahou, W., Coller, B., Potter, C., Norton, K., Kutok, J., and Goligorsky, M. (1993a). The thrombin receptor extracellular domain contains sites crucial for peptide ligand-induced activation. *J. Clin. Invest.* **91**, 1405–1413.

Bahou, W. F., Nierman, W. C., Durkin, A. S., Potter, C. L., and Demetrick, D. J. (1993b). Chromosomal assignment of the human thrombin receptor gene: localization to region q13 of chromosome 5. *Blood* **82**, 1532–1537.

Bar-Shavit, R., Kahn, A., Wilner, G. D., and Fenton, J. W. (1983). Monocyte chemotaxis: stimulation by specific exosite region in thrombin. *Science* **220**, 728–731.

Baykal, D., Schmedtje, Jr., J., and Runge, M. (1995). Role of the thrombin receptor in restenosis and atherosclerosis. *Am. J. Cardiol.* **75**, 82B–87B.

Bernatowicz, M., Klimas, C., Hartl, K., Peluso, M., Alegretto, N., and Seiler, S. M. (1996). Development of potent thrombin receptor antagonist peptides. *J. Med. Chem.* **39**, 4879–4887.

Blomback, B., Blomback, M., Hessel, B., and Iwanaga, S. (1967). Structure of N-terminal fragments of fibrinogen and specificity of thrombin. *Nature* **215**, 1445–1448.

Brass, L. F., Manning, D. R., Williams, A. G., Woolkalis, M. J., and Poncz, M. (1991). Receptor and G protein-mediated responses to thrombin in HEL cells. *J. Biol. Chem.* **266**, 958–965.

Brass, L., Vassallo, R., Belmonte, E., Ahuja, M., Cichowski, K., and Hoxie, J. (1992). Structure and function of the human platelet thrombin receptor. Studies using monoclonal antibodies directed against a defined domain within the receptor N terminus. *J. Biol. Chem.* **267**, 13795–13798.

Brummel, K., Butenas, S., and Mann, K. G. (1999). An integrated study of fibrinogen during blood coagulation. *J. Biol. Chem.* **274**, 22862–22870.

Bugge, T. H., Xiao, Q., Kombrinck, K., Flock, M., Holmbäck, K., Danton, M., Colbert, M. C., Witte, D., Fujikawa, K., Davie, E., and Degan, J. (1996). Fatal embryonic bleeding events in mice lacking tissue factor, the cell-associated initiator of blood coagulation. *Proc. Natl. Acad. Sci. USA* **93,** 6258–6263.

Camerer, E., Huang, W., and Coughlin, S. R. (2000). Tissue factor- and Factor X-dependent activation of PAR2 by Factor VIIa. *Proc. Natl. Acad. Sci. USA* **97,** 5255–5260.

Carmeliet, P., Mackman, N., Moons, Luther, T., Gressens, P., VanVlaenderen, I., Demunck, Kasper, M., Breier, G., Evrard, P., Muller, M., Risau, W., Edgington, T., and Coleen, D. (1996). Role of tissue factor in embryonic blood vessel development. *Nature* **383,** 73–75.

Carveth, H., Shaddy, R., Whatley, R., McIntyre, T., Prescott, S., and Zimmerman, G. (1992). Regulation of platelet-activating factor (PAF) synthesis and PAF-mediated neutrophil adhesion to endothelial cells activated by thrombin. *Sem. Thromb. Hemost.* **18,** 126–134.

Chen, J., Ishii, M., Wang, L., Ishii, K., and Coughlin, S. R. (1994). Thrombin receptor activation. Confirmation of the intramolecular tethered liganding hypothesis and discovery of an alternative intermolecular liganding mode. *J. Biol. Chem.* **269,** 16041–16045.

Chesebro, J., Zoldhhelyi, P., and Fuster, V. (1991). Pathogenesis of thrombosis in unstable angina. *Am. J. Cardiol.* **68,** 2B–10B.

Cheung, W. M., D'Andrea, M. R., Andrade-Gordon, P., and Damiano, B. P. (1999). Altered vascular injury responses in mice deficient in protease-activated receptor-1. *Thromb. & Vasc. Biol.* **1999,** 3014–3024.

Cirino, G., Cicala, M., Bucci, M., Sorrentino, L., Maraganore, J., and Stove, S. (1996). Thrombin functions as an inflammatory mediator through activation of its receptor. *J. Exp. Med.* **183,** 821–827.

Cocks, T., Fong, B., Chow, J., Anderson, G., Frauman, A., Goldie, R., Henry, P., Carr, M., Hamilton, R., and Moffatt, J. (1999). A protective role for protease-activated receptors in the airways. *Nature* **398,** 156–160.

Connolly, A., Ishihara, H., Kahn, M. L., Farese, R. V., and Coughlin, S. R. (1996). Role of the thrombin receptor in development and evidence for a second receptor. *Nature* **38,** 516–519.

Cook, J., Sitko, G., Bednar, B., Condra, C., Mellott, M., Feng, D., Nutt, R., Shafer, J., Gould, R., and Connolly, T. (1995). An antibody against the exosite of the cloned thrombin receptor inhibits experimental arterial thrombosis in the African green monkey. *Circulation* **91,** 2961–2971.

Cui, J., O'Shea, K. S., Purkayastha, A., Saunders, T. L., and Ginsburg, D. (1996). Fatal haemorrhage and incomplete block to embryogenesis in mice lacking coagulation factor V. *Nature (Lond.)* **384,** 66–68.

Cunningham, M., Rondeau, E., Chen, X., Coughlin, S. R., Holdsworth, S., and Tipping, P. (2000). Protease-activated receptor 1 mediates thrombin-dependent, cell-mediated renal inflammation in crescentic glomemlonephritis. *J. Exp. Med.* **191,** 455–462.

Cupit, L. D., Schmidt, V. A., and Bahou, W. F. (1999). Proteolytically activated receptor-3. A member of an emerging gene family of protease receptors expressed on vascular endothelial cells and platelets. *Trends Cardiovasc. Med.* **9,** 42–48.

D'Andrea, M., Derian, C., Santulli, R. J., and Andrade-Gordon, P. (2001). Differential expression of protease-activated receptors-1 and -2 stromal fibroblasts of normal, benigh, and malignant human tissues. *Am. J. Pathol.* **158,** 2031–2041.

Darrow, A. L., Fung-Leung, W., Ye, R. D., Santulli, R. J., Cheung, W. M., Derian, C. K., Burns, C. L., Damiano, B. P., Zhou, L., Keena, C. M., Peterson, P., and Andrade-Gordon, P. (1996). Biological consequences of thrombin receptor deficiency in mice. *Thromb. Haemost.* **76,** 860–866.

Dewerchin, M., Liang, Z., Moons, L., Carmeliet, P., Castellino, F., Collen, D., and Rosen, E. (2000). Blood coagulation factor X deficiency causes partial embryonic lethality and fatal neonatal bleeding in mice. *Thromb. Haemost.* **83,** 185–190.

Dohlman, H., Thorner, J., Caron, M., and Letkowitz, R. (1991). Model systems for the study of seven-transmembrane-segment receptors. *Annu. Rev. Biochem.* **60,** 653–688.

Downing, M. R., Bloom, J. W., and Maan, K. G. (1978). Comparison of the inhibition of thrombin by three plasma protease inhibitors. *Biochemistry* **17,** 2649.

Ehrlich, J., Boyle, E., Labriola, J., Kovacich, J., Santucci, R., Fearns, C., Morgan, E., Yun, W., Luther, T., Kojikawa, O., Martin, T., Pohlman, T., Vernier, E., and Mackman, N. (2000). Inhibition of the tissue factor-thrombin pathway limits infarct size after myocardial ischemia–reperfusion injury by reducing inflammation. *Am. J. Pathol.* **157,** 1849–1862.

Esmon, C. T. (1987). The regulation of natural anticoagulant pathways. *Science* **235,** 1348–1352.

Esmon, C. T. (2000). Introduction: are natural anticoagulants candidates for modulating the inflammatory response to endotoxin? *Blood* **95,** 1113–1116.

Even-Ram, S., Uziely, B., Cohen, P., Grisaru-Granovsky, S., Maoz, M., Ginzburg, Y., Reich, R., Vlodavsky, I., and Bar-Shavit, R. (1998). Thrombin receptor overexpression in malignant and physiological invasion processes. *Nat. Med.* **4,** 909–914.

Faruqi, T., Weiss, E., Shapiro, M. J., Huang, W., and Coughlin, S. R. (2000). Structure–function analysis of protease-activated receptor 4 tethered ligand peptides. Determinants of specificity and utility in assays of receptor function. *J. Biol. Chem.* **2000,** 19728–19734.

Furman, M. I., Longbin, L., Benoit, S. E., Becker, R. C., Barnard, M. R., and Michelson, A. D. (1998). The cleaved peptide of the thrombin receptor is a strong platelet agonist. *Proc. Natl. Acad. Sci. USA* **95,** 3082–3087.

Gertstzen, R. E., Chen, J. I., Ishii, M., Ishii, K., Wang, L., Nanevicz, T., Turck, C. W., Vu, T.-K., and Coughlin, S. R. (1994). Specificity of the thrombin receptor for agonist peptide is defined by its extracellular surface. *Nature* **368,** 648.

Griffin, C., Srinivasan, Y., Zheng, Y.-W., Huang, W., and Coughlin, S. R. (2001). A role for thrombin receptor signaling in endothelial cells during embryonic development. *Science* **293,** 1666–1669.

Hartwig, J., Bokoch, G., Carpenter, C., Janmey, P., Taylor, L., Toker, A., and Stossel, T. (1995). Thrombin receptor ligation and activated Rac uncap actin filament barbed ends through phosphoinositide synthesis in permeabilized human platelets. *Cell* **82,** 643–653.

Hasan, A., Amenta, S., and Schmaier, A. (1996). Bradykinin and its metabolite, Arg-Pro-Pro-Gly-Phe, and selective inhibitors of alpha-thrombin induced platelet activation. *Circulation* **94,** 517–528.

Hauck, R., Schulz, C., Schomig, A., Hoffman, R., and Panettieri, J. R. (1999). alpha-Thrombin stimulates contraction of human bronchial rings by activation of protease-activated receptors. *Am. J. Physiol.* **277,** L22–L29.

Heras, M., Chesebro, J., Webster, M., Mruk, J., Grill, D., Penny, W., Bowie, E. J. W., Badimon, L., and Fuster, V. (1990). Hirudin, heparin and placebo during deep arterial injury in the pig: the *in vivo* role of thrombin in platelet-mediated thrombosis. *Circulation* **82,** 1476–1484.

Herbert, J., Guy, A., Lamarche, I., Mares, A., Savi, P., and Dol, F. (1997). Intimal hyperplasia following vascular injury is not inhibited by an antisense thrombin receptor oligodeoxynucleotide. *J. Cell. Physiol.* **170,** 106–114.

Hou, L., Ravenall, S., Macey, M., Harriott, P., Kapas, S., and Howells, G. L. (1998). Protease-activated receptors and their role in IL-6 and NF-IL-6 expression in human gingival fibroblasts. *J. Periodontal. Res.* **33,** 205–211.

Hung, D. T., Vu, T.-K., Wheaton, V. I., Ishii, K., and Coughlin, S. R. (1992). Cloned platelet thrombin receptor is necessary for thrombin-induced platelet activation. *J. Clin. Invest.* **89,** 1350–1353.

Ishihara, H., Connolly, A., Zeng, D., Kahn, M. L., Zheng, W., Timmons, C., Tram, T., and Coughlin, S. R. (1997). Protease-activated receptor 3 is a second thrombin receptor in humans. *Nature* **386,** 502–506.

Jenkins, A. L., Chinni, C., DeNiesse, M., and Blackhart, B. D. (2000). Expression of protease-activated receptor-2 during development. *Dev. Dynam.* **218,** 465–471.

Johnson, K., Choi, Y., DeGroot, E., Samuels, I., Creasey, A., and Aarden, L. (1998). Potential mechanisms for a proinflammatory vascular cytokine response to coagulation activation. *J. Immunol.* **160,** 5130–5135.

Kahn, M. L., Nakanishi-Matsui, M., Shapiro, M. J., Ishihara, H., and Coughlin, S. R. (1999). Protease-activated receptors 1 and 4 mediate activation of human platelets by thrombin. *J. Clin. Invest.* **103,** 879–887.

Kahn, M. L., Zheng, Y. W., Huang, W., Bigornia, V., Zeng, D., Moff, S., Farese, Jr., R. V., Tam, C., and Coughlin, S. R. (1998). A dual thrombin receptor system for platelet activation. *Nature* **394,** 690–694.

Kimura, M., Anersen, T., Fenton, J., Bahou, W., and Aviv, A. (1996). Plasmin-platelet interaction involves cleavage of the functional thrombin receptor. *Am. J. Physiol. (Cell Physiol.)* **40,** C54–C60.

Kinlough-Rathbone, R. L., Rand, M., and Packham, M. A. (1993). Rabbit and rat platelets do not respond to thrombin receptor peptides that activate human platelets. *Blood* **82,** 103–106.

Kozasa, T., Jiang, X., Hart, M., Sternweis, P., Singer, W., Gilman, A., Bollag, G., and Sternweis, P. (1998). RhoGEF, a GTPase-activating protein for Galpha2 and Galpha3. *Science* **280,** 2109–2111.

LeBonniec, B., and Esmon, C. T. (1991). Glu-192—Gln substitution in thrombin mimics the catalytic switch induced by thrombomodulin. *Proc. Natl. Acad. Sci. USA* **88,** 7371–7375.

Liu, L.-W., Vu, T.-K., Esmon, C., and Coughlin, S. (1991). The region of the thrombin receptor resembling hirudin binds to thrombin and alters enzyme specificity. *J. Biol. Chem.* **266,** 16977–16980.

Lourbakos, A., Chinni, C., Thompson, P., Potempa, J., Travis, J., Mackie, E., and Pike, R. (1998). Cleavage and activation of proteinase-activated receptor-2 on human neutrophils by gingipain-R from *Porphyromonas gingivalis. FEBS Lett.* **435,** 45–48.

Maraganore, J. M., Bourdon, P., Jablonski, J., Ramachandran, K. L., and Fenton, J. W. (1990). Design and characterization of Hirulogs: A novel class of bivalent peptide inhibitors of thrombin. *Biochemistry* **29,** 7095–7101.

Mathews, I. I., Padmanabhan, K. P., Ganesh, V., Tulinsky, A., Ishii, M., Chen, J., Turck, C. W., Coughlin, S. R., and Fenton, J. W. (1994). Crystallographic structures of thrombin complexed with thrombin receptor peptides: existence of expected and novel binding modes. *Biochemistry* **33,** 3266–3279.

Meloni, F., and Schmaier, A. (1991). Low molecular weight kininogen binds to platelets to modulate thrombin-induced platelet activation. *J. Biol. Chem.* **266,** 6786–6794.

Mirza, H., Schmidt, V., Derian, C., Jesty, J., and Bahou, W. (1997). Mitogenic responses mediated through the proteinase activated receptor-2 are induced by mast cell α- and β- tryptases. *Blood* **90,** 3914–3922.

Mirza, H., Yatsula, V., and Bahou, W. F. (1996). The proteinase activated receptor-2 (PAR-2) mediates mitogenic responses in human vascular endothelial cells. Molecular characterization and evidence for functional coupling to the thrombin receptor. *J. Clin. Invest.* **97,** 1705–1714.

Molino, M., Bainton, D. F., Hoxie, J., Coughlin, S. R., and Brass, L. F. (1997a). Thrombin receptors on human platelets. Initial localization and subsequent redistribution during platelet activation. *J. Biol. Chem.* **272,** 6011–6017.

Molino, M., Blanchard, N., Belmonte, E., Tarver, A. P., Abrams, C., Hoxie, J. A., Cerletti, C., and Brass, L. F. (1995). Proteolysis of the human platelet and endothelial cell thrombin receptor by neutrophil-derived cathepsin. *J. Biol. Chem.* **270,** 11168–11175.

Molino, M., Raghunath, P., Ahuja, M., Hoxie, J. A., Bras, L., and Barnathan, E. S. (1998). Differential expression of functional protease-activated receptor-2 (PAR-2) in human vascular smooth muscle cells. *Arteriosclerosis Thrombosis Vasc. Biol.* **18,** 825–832.

Molino, M., Woolkalis, M. J., Reavey-Cantwell, J., Practico, D., Andrade-Gordon, P., and Brass, L. (1997b). Endothelial cell thrombin receptors and PAR-2: two protease-activated receptors located in a single cellular environment. *J. Biol. Chem.* **272,** 11133–11141.

Mosher, D., Vaheri, A., Choate, J., and Gahmberg, C. (1979). Action of thrombin on surface glycoproteins of human platelets. *Blood* **53,** 437–445.

Mueller, B., Reisfeld, R., Edgington, T. S., and Ruf, W. (1992). Expression of tissue factor by melanoma cells promotes efficient hematogeneous metastasis. *Proc. Natl. Acad. Sci. USA* **11,** 832–836.

Nakanishi-Matsui, M., Zheng, Y. W., Sulciner, D., Weiss, E., Ludeman, M., and Coughlin, S. (2000). PAR3 is a cofactor for PAR4 activation by thrombin. *Nature* **404,** 609–613.

Nanevicz, T., Wang, L., Chen, M., Ishii, M., and Coughlin, S. R. (1996). Thrombin receptor activating mutations. Alteration of an extracellular agonist recognition domain causes constitutive signaling. *J. Biol. Chem.* **271,** 702–706.

Nelken, N. A., Soifer, S. J., O'Keefe, J., Vu, T.-K. H., Charo, I. F., and Coughlin, S. R. (1992). Thrombin receptor expression in normal and atherosclerotic human arteries. *J. Clin. Invest.* **90,** 1614–1621.

Neurath, H. (1984). Evolution of proteolytic enzymes. *Science* **224,** 350–357.

Nicholson, A. C., Nachman, R. L., Altieri, D. C., Summers, B. D., Ruf, W., Edgington, T. S., and Hajjar, D. P. (1996). Effector cell protease receptor-1 is a vascular receptor for coagulation factor Xa*. *J. Biol. Chem.* **271,** 28407–28413.

Nierodzik, M. K. F., and Karpatkin, S. (1992). Effect of thrombin treatment of tumor cells on adhesion of tumor cells to platelets *in vitro* and tumor metastasis *in vivo. Cancer Res.* **52,** 3267–3272.

Nierodzik, M. L., Chen, K., Takeshita, K., Li, J. J., Huang, Y. Q., Feng, X. S., D'Andrea, M. R., Andrade-Gordon, P., and Karpatkin, S. (1998). Protease-activated receptor 1 (PAR-1) is required and rate-limiting for thrombin—enhances experimental pulmonary metastasis. *Blood* **92,** 3694–3700.

Nierodzik, M. L., Plotkin, A., Kajumo, F., and Kartpatkin, S. (1991). Thrombin stimulates tumor–platelet adhesion *in vitro* and metastasis *in vivo. J. Clin. Invest.* **87,** 299–336.

Norton, K. J., Scarborough, R. M., Kutok, J. L., Escobedo, M. A., Nannizzi, L., and Coller, B. S. (1993). Immunologic analysis of the cloned platelet thrombin receptor activation mechanism: evidence supporting receptor cleavage, release of the N-terminal peptide, and insertion of the tethered ligand into a protected environment. *Blood* **82,** 2125–2136.

Nystedt, S., Emilsson, K., Larsson, A. K., Strombeck, B., and Sundelin, J. (1995). Molecular cloning and functional expression of the gene encoding the human proteinase-activated receptor 2. *Eur. J. Biochem.* **232,** 84–89.

Nystedt, S., Emilsson, K., Wahlestedt, C., and Sundelin, J. (1994). Molecular cloning of a potential proteinase activated receptor. *Proc. Natl. Acad. Sci. USA* **91,** 9208–9212.

Nystedt, S., Ramakrishnan, S., and Sundelin, J. (1996). The proteinase-activated receptor 2 is induced by inflammatory mediators in human endothelial cells. *J. Biol. Chem.* **271,** 14910–14915.

O'Brien, P. J., Prevost, N., Molino, M., Hollinger, M., Woolkalis, M., Woulfe, D., and Bras, L. (2000). Thrombin responses in human endothelial cells. Contributions from receptors other than PAR1 include the transactivation of PAR2 by thrombin-cleaved PAR1. *J. Biol. Chem.* **275,** 13502–13509.

Papadaki, M., Ruef, J., Nguyen, K., Li, F., Patterson, C., Eskin, S., McIntire, L., and Runge, M. (1998). Differential regulation of protease activated receptor-1 and tissue plasminogen activator expression by shear stress in vascular smooth muscle cells. *Circ. Res.* **83,** 1027–1034.

Practico, D., Murphy, N., and Fitzgerald, D. (1997). Interaction of a thrombin inhibitor and a platelet GP IIb/IIIa antagonist in vivo: evidence that thrombin mediates platelet aggregation and subsequent thromboxane A2 formation during coronary thrombolysis. *J. Pharmacol. Exp. Ther.* **281,** 1178–1185.

Puri, R., Zhou, F., Hu, C., Colman, R., and Colman, R. (1991). High molecular weight kininogen inhibits thrombin-induced platelet aggregation and cleavage of aggregin by inhibiting binding of thrombin to platelets. *Blood* **77,** 500–507.

Rao, L., and Pendurthi, U. (2001). Factor VIIIa-induced gene expression: potential implications in pathophysiology. *Trends Cardiovasc. Med.* **11,** 14–21.

Renesto, P., Si-Tahar, M., Moniatte, M., Balloy, V., VanDorsselaer, A., Pidard, D., and Chignard, M. (1997). Specific inhibition of thrombin-induced cell activation by the neutrophil proteinases elastase, cathepsin G, and proteinase 3: evidence for distinct cleavage sites within the aminoterminal domain of the thrombin receptor. *Blood* **89,** 1944–1953.

Riewald, M., and Ruf, W. (2001). Mechanistic coupling of protease signaling and initiation of coagulation by tissue factor. *Proc. Natl. Acad. Sci. USA* **98,** 7742–7747.

Riewald, M., Kravchenko, V., Petrovan, R., O'Brien, P., Brass, L., Ulevitch, R., and Ruf, W. (2001). Gene induction by coagulation factor Xa is mediated by activation of protease-activated receptor 1. *Blood* **97,** 3109–3116.

Rosen, E., Chan, J., Idusogie, E., Clotman, F., Vlasuk, G., Luther, T., Jalbert, L., Albrecht, S., Zhong, L., Lissens, A., Schoonjans, L., Moons, L., Collen, D., Castellino, F., and Carmeliet, P. (1997). Mice lacking factor VII develop normally but suffer fatal perinatal bleeding. *Nature* **390,** 290–294.

Rosenberg, R. D. (1996). The absence of the blood clotting regulator thrombomodulin causes embryonic lethality in mice before development of a functional cardiovascular system. *Thromb. Haemost.* **74,** 52–57.

Sambrano, G. R., Huang, W., Faruqi, T., Mahrus, S., Craik, C., and Coughlin, S. (2000). Cathepsin G activates protease-activated receptor-4 in human platelets. *J. Biol. Chem.* **275,** 6819–6823.

Scarborough, R., Naughton, M., Teng, W., Hung, D., Rose, J., Vu, T., Wheaton, V., Turck, C., and Coughlin, S. (1992). Tethered ligand agonist peptides: structural requirements for thrombin receptor activation reveal mechanism of proteolytic unmasking of agonist function. *J. Biol. Chem.* **267,** 13146–14149.

Schlaeger, T., Bartunkova, S., Lawitts, J., Teichmann, G., Risau, W., Deutsch, U., and Saton, T. (1997). Uniform vascular-endothelial-cell-specific gene expression in both embryonic and adult transgenic mice. *Proc. Natl. Acad. Sci. USA* **94,** 3058–3063.

Schmeidler-Sapiro, K. T., Ratnoff, O. D., and Gordon, E. M. (1991). Mitogenic effects of coagulation factor XII and factor XIIa on Hep G_2 cells. *Proc. Natl. Acad. Sci. USA* **88,** 4382–4385.

Schmidt, V., Vitale, E., and Bahou, W. (1996). Genomic cloning and characterization of the human thrombin receptor gene: evidence for a novel gene family that includes PAR-2. *J. Biol. Chem.* **271,** 9307–9312.

Schmidt, V. A., Nierman, W. C., Feldblyum, T. B., Maglott, D. R., and Bahou, W. F. (1997). The human thrombin receptor and proteinase activated receptor-2 genes are tightly linked on chromosome 5q13. *Br. J. Haematol.* **97,** 523–529.

Schmidt, V. A., Nierman, W. C., Maglott, D. R., Cupit, L. D., Moskowitz, K. A., Wainer, J. A., and Bahou, W. F. (1998). The human proteinase-activated receptor-3 (PAR-3) gene. Identification within a PAR gene cluster and characterization in vascular endothelial cells and platelets. *J. Biol. Chem.* **273,** 15061–15068.

Shima, C., Majima, M., and Katori, M. (1992). A stable metabolite Arg-Pro-Pro-Gly-Phe of bradykinin in the degradation pathway in human plasma. *Japan J. Pharmacol.* **60,** 111–119.

Sobey, C., Moffatt, J., and Cocks, T. (1999). Evidence for selective effects of chronic hypertension on cerebral artery vasodilation to protease-activated receptor-2 activation. *Stroke* **30,** 1933–1940.

Soifer, S., Peters, K., O'Keefe, J., and Coughlin, S. R. (1994). Disparate temporal expression of the prothrombin and thrombin receptor genes during mouse development. *Am. J. Pathol.* **144,** 60–69.

Steinberg, S., Robinson, R., Lieberman, H., Stern, D., and Rosen, M. (1991). Thrombin modulates phosphoinositide metabolism, cytosolic calcium, and impulse initiation in the heart. *Circ. Res.* **68,** 1216–1229.

Steinhoff, M., Vergnolle, N., Young, S., Tognetto, M., Amadesi, S., Ennes, H., Trevisani, M., hollenberg, M., Wallace, J. C. G., Mitchell, S., Williams, L., Geppetti, P., Mayer, E., and Bunnett, N. (2000). Agonists of proteinase-activated receptor 2 induce inflammation by a neurogenic mechanism. *Nat. Med.* **6,** 151–158.

Stoyanov, B., Volinia, S., Hanck, T., Rubio, I., Loubtchenkov, M., Malek, D., Stoyanova, S., Vanhaesbroeck, B., Dhand, R., and Nurnberg, B. (1995). Cloning and characterization of a G⁻protein-activated human phosphoinositide-3 kinase. *Science* **269,** 690–693.

Suidan, H. S., Bouvier, J., Schaerer, E., Stone, S. R., Monard, D., and Tschopp, J. (1994). Granzyme A released upon stimulation of cytotoxic T lymphocytes activates the thrombin receptor on neuronal cells and astrocytes. *Proc. Natl. Acad. Sci. USA* **91,** 8112–8116.

Sun, W. Y., Witte, D. P., Degen, J. L., Colbert, M. C., Burkart, M. C., Holmbäck, K., Xiao, Q., Bugge, T. H., and Degen, S. J. (1998). Prothrombin deficiency results in embryonic and neonatal lethality in mice. *Proc. Natl. Acad. Sci. USA* **95,** 7597–7602.

Takada, M., Tanaka, H., Yamada, T., Ito, O., Kogushi, M., Yanagimachi, M., Kawamura, T., Musha, T., Yoshida, F., Ito, M., Kobayashi, H., Yoshitake, S., and Saito, I. (1998). Antibody to thrombin receptor inhibits neointimal smooth muscle cell accumulation without causing inhibition of platelet aggregation or altering hemostatic parameters after angioplasty in rat. *Circ. Res.* **82,** 980–987.

Taylor, J. F., Chang, A., Esmon, C. T., D'Angelo, A., Vigano-D'angelo, S., and Blick, K. (1987). Protein C prevents the coagulopathic and lethal effects of *E. coli* infusion in the baboon. *J. Clin. Invest.* **97,** 918–925.

Taylor, J. F., Chang, A., Ruf, W., Morrissey, J., Hinshaw, L., Catlett, R., Blick, K., and Edginton, T. (1991). Lethal *E. coli* septic shock is prevented by blocking tissue factor with monoclonal antibody. *Circ. Shock* **1991,** 33–127134.

Tiruppathi, C., Lum, H., Andersen, T. T., Fenton, J. W.2., and Malik, A. B. (1992). Thrombin receptor 14-amino acid peptide binds to endothelial cells and stimulates calcium transients. *Am. J. Physiol.* **263,** L595–L601.

Toomey, J., Kratzer, K., Lasky, N., Stanton, J., and Broze, Jr. G. (1996). Targeted disruption of the murine tissue factor gene results in embryonic lethality. *Blood* **88,** 1583–1587.

Turgeon, V. S. N., and Houenou, L. (2001). Thrombin: a neuronal cell modulator. *Thromb. Res.* **99,** 417–427.

Venter, J., Adams, M., Sutton, G., Kerlavage, A., Smith, H., and Hunkapiller, M. (1998). Shotgun sequencing of the human genome. *Science* **280,** 1540–1542.

Vu, T., Hung, D., Wheaton, V., and Coughlin, S. (1991a). Molecular cloning of a functional thrombin receptor reveals a novel proteolytic mechanism of receptor activation. *Cell* **64,** 1057–1068.

Vu, T.-K., Wheaton, V., Hung, D., Charo, I., and Coughlin, S. (1991b). Domains specifying thrombin–receptor interaction. *Nature* **353,** 674–677.

Wilcox, J., Rodriguez, J., Subramanian, R., Ollerenshaw, J., Zhong, C., Hayzer, D., Horaist, C., Hanson, S., Lumdsen, A., and Salam, T. (1994). Characterization of thrombin receptor expression during vascular lesion formation. *Circ. Res.* **75,** 1029–1038.

Xu, W., Andersen, H., Whitmore, T. E., Presnell, S. R., Yee, D. P., Shing, A., Gilbert, T., Davie, E. W., and Foster, D. C. (1998). Cloning and characterization of human protease-activated receptor 4. *Proc. Natl. Acad. Sci. USA* **95,** 6642–6646.

Xue, J., Wu, Q., Westfield, L. A., Tuley, E. A., Lu, D., Zhang, Q., Shim, K., Zheng, X., and Sadler, J. E. (1998). Incomplete embryonic lethality and fatal neonatal hemorrhage caused by prothrombin deficiency in mice. *Proc. Natl. Acad. Sci. USA* **95,** 7603–7607.

Zimmerman, G., McIntyre, T., and Prescott, S. (1985). Thrombin stimulates the adherence of neutrophils to human endothelial cells *in vitro*. *J. Clin. Invest.* **76,** 2235–2246.

Zucker, S., Conner, C., DiMassimo, B., Ende, H., and Bahou, W. (1995). Thrombin induces cell surface activation of gelatinase A in vascular endothelial cells: physiological regulation of angiogenesis. *J. Biol. Chem.* **270,** 23730–23738.

12

Emmprin (CD147), a Cell Surface Regulator of Matrix Metalloproteinase Production and Function

Bryan P. Toole
Department of Anatomy and Cellular Biology
Tufts University School of Medicine
Boston, Massachusetts 02111

I. Introduction

Matrix metalloproteinases (MMPs) play a central role in numerous normal and pathological events. Thus the precise regulation of production and activation of MMPs is important in determining their correct localization and the appropriate extent of their action. Likewise, since MMPs have an incredibly diverse number of substrates and since cleavage of these substrates leads to numerous cellular responses, inappropriate regulation is potentially destructive and, accordingly, MMPs have been implicated in a number of disease processes. Many physiological regulators of MMP production that have been characterized are soluble cytokines and growth factors, but the possibility that cell-surface interactions regulate MMP production and activation has also been appreciated for some time.

Emmprin (basigin, CD147) is an integral plasma membrane glycoprotein of the Ig superfamily that probably has several functions, but one established property of

Current Topics in Developmental Biology, Vol. 54

emmprin is its ability to induce synthesis of various MMPs. In this chapter I will review our current knowledge of emmprin, especially with respect to induction of MMP production in cancer.

II. Regulation of MMP Production by Emmprin

A. Heterotypic Cell Interactions

Emmprin activity was initially discovered in a series of studies performed by Dr. Chitra Biswas and her colleagues, in which they demonstrated stimulation of interstitial collagenase (MMP-1) production in cocultures of tumor cells and fibroblasts (Biswas, 1982, 1984). These studies led to identification of a factor (TCSF, tumor cell-derived collagenase stimulatory factor) that is associated with the surface of tumor cells and that stimulates MMP synthesis in human fibroblasts (Biswas and Nugent, 1987; Ellis *et al.*, 1989; Kataoka *et al.*, 1993; Nabeshima *et al.*, 1991). The factor was subsequently cloned and identified as a transmembrane glycoprotein and member of the Ig superfamily (Biswas *et al.*, 1995; Guo *et al.*, 1997) (Fig. 1). It was also shown to be present in normal epithelia but at a lower level of expression (DeCastro *et al.*, 1996). At this time it was renamed

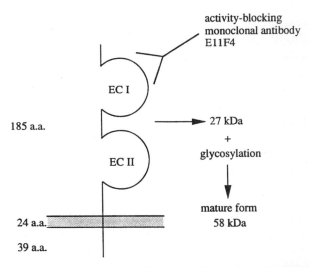

Figure 1 The emmprin molecule. Emmprin is an integral glycoprotein localized on the plasma membrane. It contains an ectodomain, composed of two immunoglobulin loops (ECI and ECII), a transmembrane segment and a cytoplasmic segment. The monoclonal blocking antibody, EIIF4, reacts with ECI, implicating this region in its activity. From Nabeshima *et al.* (1999), with permission.

emmprin (extracellular matrix metalloproteinase inducer). The amino acid sequence of emmprin is identical to those of human basigin (Miyauchi *et al.*, 1991) and M6 antigen, a protein present on membranes of leukocytes from patients with arthritis (Kasinrerk *et al.*, 1992). The functions of basigin and M6 were not then known but it was proposed that emmprin/M6 might stimulate MMP production in arthritis as well as cancer (Biswas *et al.*, 1995). Another antigen, Hab18G, which is present on the surface of human hepatoma cells, has also been shown to be identical in sequence to emmprin (Jiang *et al.*, 2001).

Purified emmprin, derived from human tumor cell membranes, stimulates production of interstitial collagenase (MMP-1), gelatinase A (MMP-2), and stromelysin-1 (MMP-3), but has little effect on production of TIMP-1 or TIMP-2, in fibroblasts (Kataoka *et al.*, 1993) and in endothelial cells (Caudroy *et al.*, 2002). Stimulation of synthesis of membrane-type-MMPs (MT-MMPs) in cocultures of human glioblastoma cells and brain tumor-derived fibroblasts is also dependent on glioma cell-associated emmprin (Sameshima *et al.*, 2000b). Both MT1- and MT2-MMP, but not MT3-MMP, were stimulated in this system. MT1-MMP production was also stimulated in cocultures of human melanoma cells and dermal fibroblasts in an emmprin-dependent manner (Kanekura *et al.*, 2002). Increased activation of MMP-2 by emmprin has also been observed, presumably due to the increased levels of MT-MMPs (Kanekura *et al.*, 2002; Kataoka *et al.*, 1993; Nabeshima *et al.*, 2002; Sameshima *et al.*, 2000b). However, it has been noted that different fibroblast populations differ widely in their response to emmprin (Guo *et al.*, 1997; Kataoka *et al.*, 1993; Lim *et al.*, 1998; Prescott *et al.*, 1989); the basis for this difference has not yet been elucidated but may be related to varying degrees of expression of a putative receptor.

MMP synthesis in fibroblasts in response to stimulation by emmprin is a relatively slow process, taking 24–48 h to reach maximum (Lim *et al.*, 1998; Prescott *et al.*, 1989). Treatment of human lung fibroblasts with purified emmprin leads to phosphorylation of the MAP kinase p38, and induction of MMP-1 synthesis is partially dependent on activity of this kinase but not ERK1/2 or SAPK/JNK (Lim *et al.*, 1998). Emmprin stimulation of MMP production in fibroblasts is also dependent on N-glycosylation of its extracellular domains. Since partially glycosylated emmprin is relatively inactive, it is likely that all three glycosylation sites in the two Ig domains must be occupied for full activity (Guo *et al.*, 1997; Sun and Hemler, 2001). It is not yet clear whether the carbohydrate side chains participate directly in ligand binding to a putative fibroblast receptor or whether they are necessary for preservation of an active molecular configuration. Since it has been shown that the activity-blocking monoclonal antibody, E11F4, recognizes a peptide rather than carbohydrate epitope (Biswas *et al.*, 1995), it is evident that the carbohydrate groups alone cannot be responsible for MMP-inducing activity. This antibody recognizes the distal Ig loop of emmprin, implying that this region of emmprin contains the structure responsible for activity.

B. Homophilic Emmprin Interactions

In recent studies it has been shown that overexpression of emmprin in fibroblasts (Li *et al.*, 2001) or in tumor cells (Caudroy *et al.*, 2002; Zucker *et al.*, 2001) leads to increased MMP production within the same population of cells. When fibroblasts are infected with a recombinant human or mouse emmprin adenovirus, emmprin expression increases severalfold and leads to a corresponding increase in production of MMP-1 and MMP-3 (Li *et al.*, 2001). One possible interpretation of this result is that emmprin on the surface of these fibroblasts may interact in a heterophilic manner with a putative receptor on adjacent fibroblasts. However, treatment of fibroblasts overexpressing emmprin with antibody to emmprin gives rise to even more dramatic increases in MMP production, suggesting that homophilic emmprin interactions within the membrane or between cells may be responsible for or augment induction of MMP synthesis (Li *et al.*, 2001). Overexpression of emmprin in minimally invasive MB-MDA-436 human breast carcinoma cells also leads to an increase in MMP production. In addition, these emmprin-transfected MDA436 cells were found to be more invasive than vector-transfected controls (Caudroy *et al.*, 2002; Zucker *et al.*, 2001).

An important role for homophilic emmprin interactions in stimulation of MMP production is strongly supported by the observation that underglycosylated ectodomain of emmprin is an antagonist of emmprin action (Sun and Hemler, 2001). When the emmprin ectodomain, prepared as an Fc-fusion protein, is added to MDA-435 human breast carcinoma cells, MMP-2 production and MMP-2-dependent invasion of reconstituted basement membrane by these cells are almost completely inhibited (Sun and Hemler, 2001). Furthermore, addition of fully glycosylated emmprin to fibroblasts or to MDA-435 cells, but not addition of underglycosylated emmprin, causes stimulation of MMP synthesis. This inhibition is reversed by addition of underglycosylated emmprin or emmprin ectodomain (Sun and Hemler, 2001). These results imply that emmprin acts via homophilic interaction between emmprin molecules on apposing cells. As stated in Section II,A, the distal Ig loop of the emmprin ectodomain is likely to include the MMP-stimulatory region (Biswas *et al.*, 1995). It has now been shown that homophilic interaction of emmprin molecules is mediated by this region of emmprin (Sun and Hemler, 2001; Yoshida *et al.*, 2000) and thus it is likely that this interaction is crucial to stimulation of MMP production.

The experiments just described provide convincing evidence for the role of homophilic emmprin oligomerization in MMP induction, both in heterotypic interactions between tumor cells and fibroblasts and in homotypic interactions between tumor cells. However, other findings are difficult to reconcile with this conclusion. First, previous studies have shown that homophilic interactions between emmprin molecules do occur (Fadool and Linser, 1996; Yoshida *et al.*, 2000), but evidence was found only for binding in a cis fashion, i.e., within the same cell membrane, and not for trans interactions, i.e., between molecules on apposing cells (Yoshida

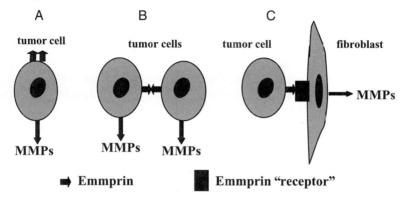

Figure 2 Possible emmprin-mediated interactions stimulating MMP production. (A) Homophilic cis interaction between emmprin molecules within the plasma membrane of a tumor cell. (B) Homophilic trans interaction between emmprin molecules on apposing tumor cells. (C) Heterophilic interactions between emmprin on a tumor cell and a putative emmprin receptor on a fibroblast.

et al., 2000). Second, if emmprin interactions at the surface of tumor cells induce MMP synthesis (irrespective of whether they are cis or trans), then tumor cells expressing high levels of endogenous emmprin should consistently produce MMPs known to be stimulated by emmprin. However, for example, human lung carcinoma, breast carcinoma, and glioblastoma cells express high levels of cell surface emmprin but low levels of MMPs (Kataoka *et al.,* 1993; Lim *et al.,* 1998; Polette *et al.,* 1997; Sameshima *et al.,* 2000b). On the other hand, considerable variations can occur in the levels of expression of MMPs in tumor cells (Heppner *et al.,* 1996; van den Oord *et al.,* 1997; Wright *et al.,* 1994), and thus endogenous emmprin may under some circumstances upregulate MMP production in tumor cells themselves. Third, data published to date indicate that fibroblasts usually express low levels of emmprin (Ellis *et al.,* 1989; Lim *et al.,* 1998; Polette *et al.,* 1997), making it unlikely that heterotypic interactions of fibroblasts with tumor cells would involve homophilic trans interactions between emmprin molecules. However, peritumoral fibroblasts may sometimes express or bind emmprin (Caudroy *et al.,* 1999), allowing these homophilic interactions with tumor cell emmprin to occur. Further work should clarify these important aspects of regulation of MMP production by emmprin.

Some of the potential interactions involved in emmprin regulation of MMP production, as discussed previously, are summarized in Fig. 2.

C. Unidentified MMP-9 Inducing Factors

An unresolved question is whether emmprin induces production of MMP-9 (gelatinase B). Correlations between emmprin and MMP-9 expression have been noted

(Foda *et al.*, 2001; Spinale *et al.*, 2000; van den Oord *et al.*, 1997; Zucker *et al.*, 2001) but no evidence has appeared in the literature showing directly that emm-prin stimulates MMP-9 synthesis (e.g., see Kataoka *et al.*, 1993). Although growth factors and cytokines produced by cancer cells may be responsible for inducing MMP-9 synthesis by stromal cells, the possibility that emmprin-like molecules may function at the cell surface to enhance MMP-9 production has not been fully explored. Experimental coculture systems in which tumor cells stimulate MMP-9 synthesis or activation in fibroblasts have implicated direct cell contact (Himel-stein *et al.*, 1994; Segain *et al.*, 1996) as well as secreted factors. An autocrine mechanism promoting secretion of MMP-9 from tumor cells has also been pro-posed (Hyuga *et al.*, 1994), as well as fibroblast stimulation of cancer cell pro-duction of MMP-9 (Himelstein and Muschel, 1996; Lengyel *et al.*, 1995). The mechanism of regulation of MMP-9 production in cancer clearly requires more investigation.

III. Emmprin Binding Partners

It is obvious, from the preceding discussion, that emmprin binds to itself to form homooligomers and that this interaction may be important in MMP induction. However, it is also clear that emmprin has several other binding partners and that interactions with these proteins are likely to be important in emmprin function. However, their role in MMP induction, if any, is not yet known.

A. Intramembrane Interactions

Structural analyses have demonstrated that the transmembrane and cytoplasmic domains of emmprin are highly conserved among species, suggesting that these regions are of functional importance. The conserved properties of the transmem-brane region, i.e., a central charged residue and a leucine zipper motif, suggest that intramembrane interactions with other proteins are likely to occur (Fossum *et al.*, 1991; Kasinrerk *et al.*, 1992; Miyauchi *et al.*, 1991). Accordingly, interaction of emmprin with several other components of cell membranes has been demon-strated. For example, emmprin interacts with integrins, $\alpha3\beta1$ and $\alpha6\beta1$, within the plasma membrane of HT1080 fibrosarcoma cells (Berditchevski *et al.*, 1997). Emmprin also acts as a chaperone for assembly of lactate transporters in the plasma membrane (Kirk *et al.*, 2000; Wilson *et al.*, 2002). Moreover, emmprin binds to cyclophilins, facilitating HIV virus entry into mammalian cells (Pushkarsky *et al.*, 2001) and promoting neutrophil chemotactic activity (Yurchenko *et al.*, 2002). Fi-nally, emmprin is restricted to different specialized areas of the plasma membrane during maturation of some cell types, e.g., retinal pigment epithelium (Marmorstein *et al.*, 1998) and spermatozoa (Nehme *et al.*, 1993), implying interactions within the membrane that effect specific localization.

All of these interactions are likely to involve the transmembrane and/or cytoplasmic domains of emmprin. However, it is not known whether any of these binding partners are involved in the proteolytic processes stimulated by emmprin. Rather, it seems likely that emmprin has multiple molecular functions, and further work is required to clarify the precise role of emmprin interactions with the above binding partners in physiological and pathological processes.

B. Binding of Emmprin to MMP-1 and Annexin II

Recent work in our laboratory has been directed toward identifying binding partners that might be involved in or result from heterotypic cell interactions leading to MMP induction. These interactions would be trans interactions rather than intramembrane, cis interactions described in the previous section.

Using phage display to probe for emmprin-binding proteins produced by fibroblasts, we have identified MMP-1 as a binding partner. This was confirmed by emmprin-ligand affinity chromatography of fibroblast extracts. In addition, using immunoaffinity chromatography of tumor cell membrane extracts with anti-emmprin antibody, we isolated a complex containing emmprin and MMP-1 in a one-to-one molar ratio. Immunocytochemistry revealed that MMP-1 and emmprin are both present on the surface of human lung carcinoma cells, suggesting that emmprin may be a docking site on the tumor cell surface for MMP-1 produced by fibroblasts (Guo *et al.*, 2000).

We also used reversible cross-linking of emmprin to the fibroblast cell surface and emmprin-ligand chromatography of fibroblast membrane extracts to identify proteins on the fibroblast cell surface that bind emmprin. In this manner we identified annexin II as a binding partner (H.Guo and BPT, unpublished results). Annexin II is a Ca^{2+}-dependent, lipid-binding protein that is commonly found intracellularly but, under some circumstances, is also a prominent cell surface component, e.g. on endothelial cells (Hajjar *et al.*, 1996), neurons (Jacovina *et al.*, 2001), and cancer cells (Tressler *et al.*, 1993). Interestingly, annexin II binds plasminogen activator and plasminogen to the surface of several cell types and may be an important regulator of endothelial cell surface fibrinolysis (Choi *et al.*, 2001; Hajjar and Acharya, 2000; Hajjar *et al.*, 1996). Annexin II usually occurs as a tetramer containing two subunits of a 36-kDa heavy chain and two subunits of an 11-kDa light chain. The 11-kDa component has been shown to bind cathepsin B to the surface of tumor cells (Mai *et al.*, 2000). (See Chapter 10 by Sloane and Cavallo-Medvel for further discussion of the interaction between cathepsin B and annexin II.) Thus annexin II may be an important regulator of protease docking or action at the cell surface. We are currently investigating the role of the emmprin–annexin II interaction in emmprin function.

IV. Role of Emmprin in Tumor Progression

A. Elevated Emmprin Expression in Malignant Tumors

Emmprin is expressed at various levels in a large number of normal cell types, both in the adult and in the embryo (Fossum *et al.,* 1991; Igakura *et al.,* 1998; Nehme *et al.,* 1995). However, emmprin expression in a wide variety of tumors, especially malignant tumors, is usually much greater than in corresponding normal tissue, e.g., in human breast, lung, and kidney carcinomas (Caudroy *et al.,* 1999; Dalberg *et al.,* 2000; Lim *et al.,* 1998; Muraoka *et al.,* 1993; Polette *et al.,* 1997), as well as in malignant melanomas (Kanekura *et al.,* 2002; van den Oord *et al.,* 1997), gliomas (Sameshima *et al.,* 2000a), and lymphomas (Nabeshima *et al.,* 2002).

In one extensive study, the levels of expression of emmprin were analyzed at the mRNA and protein levels in breast and lung tumors from approximately 20 patients each (Caudroy *et al.,* 1999; Polette *et al.,* 1997). The relative distribution of emmprin and MMP-2 mRNAs was compared by *in situ* hybridization in normal lung tissue versus squamous cell carcinomas of the lung and in benign mammary growths versus ductal carcinomas of the breast (Polette *et al.,* 1997). Emmprin mRNA was detected in all breast carcinomas and 20 out of 22 lung carcinomas. Both preinvasive and invasive cancer cells were positive, but tumor stromal cells and peritumoral tissue showed insignificant emmprin mRNA reactivity. Normal and benign epithelia were negative. MMP-2 mRNA, on the other hand, was restricted to stromal cells close to tumor cell clusters. The relative distribution of emmprin mRNA in lung and breast carcinoma cells versus MMP-1 or MMP-2 mRNAs in peritumoral fibroblasts has been confirmed by other groups of investigators (Dalberg *et al.,* 2000; Lim *et al.,* 1998). Expression of emmprin mRNA in this cadre of patients was also analyzed by Northern blotting; the results showed low expression in normal or benign tissues but high levels at all stages of tumor progression in both lung and breast cancers. Analyses of distribution within tumors made by quantitative image cytometry showed that high levels of emmprin mRNA were expressed in preinvasive and invasive nests of tumor cells versus low amounts in normal or peritumoral tissues (Polette *et al.,* 1997). Both normal and tumor epithelia stained with antibody to emmprin but expression of emmprin was much stronger in tumor tissue (Caudroy *et al.,* 1999). Since virtually no signal for emmprin mRNA was seen in the normal epithelia, the presence of emmprin protein implies that turnover of emmprin in normal cells is lower than in tumor cells. Unexpectedly, emmprin protein was also present at a significant level in fibroblasts in the tumor stroma but not in fibroblasts distant from tumor cells. Of interest also was the observation that MMP-2 was associated with the same fibroblasts as was emmprin. Since these cells did not express emmprin mRNA, as visualized by *in situ* hybridization, it was hypothesized that this emmprin was produced by tumor cells, shed, and then became bound to the fibroblasts, inducing synthesis of MMP-2

(Caudroy *et al.*, 1999). (See Chapter 14 by Vitorelli for detailed discussion of plasma membrane shedding events in tumor cells.)

Emmprin levels are also elevated in transitional cell carcinomas of the bladder (Muraoka *et al.*, 1993). In this study elevated emmprin levels in approximately 30 patients with various grades of cancer progression were documented by immunohistochemistry and by Western blotting. In addition, cytology of cells sloughed into the urine revealed elevated emmprin levels in patients with transitional cell carcinoma but not in patients with other tumors or without tumors.

In another study it was found that malignant human glioblastoma cells express much higher levels of emmprin than benign gliomas and normal brain tissue, and that MMP-2 levels were correspondingly high in the glioblastomas (Sameshima *et al.*, 2000a). Of special interest was the observation that emmprin expression in the endothelium of blood vessels in normal brain versus in the gliomas exhibits the opposite relationship to that of rest of the brain tissue. That is, in normal human brain, blood vessel endothelial cells stained strongly for emmprin whereas in angiogenic vessels associated with malignant glioblastomas, the endothelial cells were negative. This transition from positive to negative staining for emmprin on endothelial cells correlates strongly with breakdown of the blood–brain barrier. The association of emmprin with blood–brain barrier endothelium has also been observed in other species (Schlosshauer, 1993; Seulberger *et al.*, 1992), but its physiological significance is not yet understood.

Although emmprin is expressed at a moderate level in normal nonneoplastic keratinocytes (DeCastro *et al.*, 1996), its presence in oral squamous cell carcinoma is associated with MMP production and tumor cell invasion (Bordador *et al.*, 2000). Emmprin expression is also correlated with melanoma invasion and metastasis in human patients (Kanekura *et al.*, 2002; van den Oord *et al.*, 1997).

The finding that emmprin expression is usually elevated in malignant cancer cells has been confirmed in an analysis of disseminated micrometastatic cells derived from the bone marrow of carcinoma patients, using genomic and transcription-based techniques (Klein *et al.*, 2002). Emmprin was found to have among the highest incidence of expression in these cells compared to other proteins. However, little is known about the regulation of emmprin expression in normal or pathological tissues. The promoter region of the emmprin gene has been partially characterized and shown to contain putative Sp1 and AP2 sites (Guo *et al.*, 1998; Liang *et al.*, 2002; Miyauchi *et al.*, 1995). At least one of the Sp1 elements is functional and regulated by both Sp1 and Sp3 transcriptional factors (Liang *et al.*, 2002). However, the regulatory effectors that stimulate transcription of emmprin are not known. A potential negative regulator of emmprin expression is the putative tumor suppressor, pinin (Shi and Sugrue, 2000; Shi *et al.*, 2000, 2001). Pinin expression leads to increased cell adhesion and decreased proliferation in epithelia as a result of altered regulation of several genes that influence cell cycle and motility, including downregulation of emmprin (Shi *et al.*, 2001). Possibly, then, loss of pinin expression, as occurs in several types of tumor cells (Shi *et al.*, 2000), would augment emmprin expression, and consequently lead to increased MMP synthesis and action.

B. Promotion of Tumor Growth and Invasion *In Vivo* by Emmprin

Since malignant tumor cells usually express emmprin *in vivo* and *in vitro* at much higher levels than normal and benign cells, we recently tested whether overexpression of emmprin stimulates tumor progression (Zucker *et al.*, 2001). We used MDA-MB-436 human breast carcinoma cells that produce slow-growing primary tumors in nude mice and express relatively low levels of emmprin. We transfected the cells with emmprin cDNA and selected stable transfectant clones with increased expression of emmprin. The emmprin transfectants grew at similar rates to vector-transfected controls in monolayer cell culture. The transfected tumor cells were injected into the mammary fat pad of groups of 10 nude mice in three separate *in vivo* experiments using different transfectant clones. In all three experiments, the mice injected with emmprin transfectants grew large tumors over a 12-week period whereas vector transfectants developed small tumors that were primarily detectable only at autopsy (Fig. 3, see color insert). In addition, the emmprin transfectants exhibited extensive invasion into surrounding abdominal wall muscle whereas controls did not. Animal survival was markedly decreased in mice carrying the emmprin transfectants compared to controls. We conclude from this study that increased expression of emmprin leads to increases in tumor cell growth and tumor cell invasion *in vivo*.

An important question is whether increased tumor growth and invasion resulting from increased expression of emmprin are due to induction of MMP production by emmprin. In the foregoing study it was found that tumors arising in animals carrying the emmprin transfectants expressed increased levels of MMP-2 and MMP-9, as documented by *in situ* hybridization and by gelatin substrate zymography (Zucker *et al.*, 2001). *In situ* hybridization revealed that these MMPs were produced by the cancer cells and by peritumoral stromal cells. Several studies have shown that, in *de novo* human carcinomas, gelatinases are usually produced by peritumoral stromal cells rather than by the carcinoma cells themselves (DeClerck, 2000; Heppner *et al.*, 1996; Johnsen *et al.*, 1998; Polette *et al.*, 1997). However, some reports have shown that these MMPs are also produced by carcinoma cells *in vivo* (Canete-Soler *et al.*, 1994; Sehgal *et al.*, 1998; Soini *et al.*, 1994). MMP-3 (stromelysin-1) is expressed by stromal cells or carcinoma cells at different stages of progression of skin carcinoma in mice (Wright *et al.*, 1994). Although it is clear that stromal MMPs contribute to tumor progression under some circumstances (Itoh *et al.*, 1998; Masson *et al.*, 1998), the relative importance of these two sources of tumor MMPs in tumor growth and metastasis is not yet established. However, it is likely that promotion of tumor progression by emmprin is due, at least in part, to induction of MMP production. Evidence discussed in Section II indicates that emmprin may be responsible for increased production of MMPs both in cancer cells, as a result of homotypic cell interactions, and in stromal cells, as a result of heterotypic cell interactions.

C. Promotion of Tumor Cell Growth and Invasion *In Vitro* by Emmprin

The effect of emmprin on MMP-2 production and tumor cell invasion has been examined in cocultures of oral squamous cell carcinoma cells and peritumor-derived fibroblasts (Bordador *et al.,* 2000). In this study the tumor cells were plated on a filter coated with reconstituted basement membrane matrix; the fibro-blasts were plated in a well beneath the filter. Tumor cell invasion of the matrix was found to be dependent on emmprin and to result from emmprin stimula-tion of MMP-2 production. Based on these data, it was thought that tumor-cell-derived emmprin stimulated MMP-2 production in the fibroblasts and that this MMP-2 then acted to enhance tumor cell invasion. Another possible interpretation is that emmprin-mediated, homotypic tumor cell interactions caused upregulation of MMP-2 production by the tumor cells themselves, and consequently led to in-creased invasiveness. As discussed in Section II,B, this has been shown to be the case with human breast carcinoma cells (Sun and Hemler, 2001; Zucker *et al.,* 2001). In a similar study of cocultures of human melanoma cells and fibroblasts, however, it was clear that inclusion of fibroblasts enhanced emmprin-dependent invasion by the melanoma cells, thus favoring the former interpretation (Kanekura *et al.,* 2002).

Studies from our laboratory have shown that stimulation of emmprin production leads, not only to increased invasiveness, but also to stimulation of anchorage-independent growth in breast cancer cells (E. Marieb and BPT, unpublished re-sults). The signaling pathways involved in these effects are currently under inves-tigation.

V. Diverse Cellular Functions of Emmprin and Its Homologs in Physiology and Pathology

Several proteins, independently discovered as novel antigens in a wide variety of systems in other species, are homologous to emmprin, e.g., mouse gp42 and basigin (Altruda *et al.,* 1989; Miyauchi *et al.,* 1990), rat OX47 and CE9 (Fossum *et al.,* 1991; Nehme *et al.,* 1993), and chick 5A11, HT7, and neurothelin (Fadool and Linser, 1993; Schlosshauer *et al.,* 1995; Seulberger *et al.,* 1992). However, the functions of these antigens were not identified at that time; the MMP-inducing activity of emmprin, as discovered by Biswas and co-workers, was the first bio-chemical function described. Emmprin and its homologs are now also termed CD147. In addition, emmprin is related to several other Ig superfamily members, e.g., embigin (Miyauchi *et al.,* 1991) and neuroplastin (Langnaese *et al.,* 1997; Smalla *et al.,* 2000).

A knockout mouse has been produced in which basigin, the murine homolog of emmprin, is lacking (Igakura *et al.,* 1998). Basigin-null embryos exhibit numerous defects, including inefficient implantation. Since MMPs are known to be involved in implantation (Alexander *et al.,* 1996; Vu and Werb, 2000), defective implantation may result from misregulation of MMP production due to lack of basigin stimulation. Basigin-null embryos that successfully implant survive past birth but have deficiencies in spermatogenesis (Igakura *et al.,* 1998; Saxena *et al.,* 2002), retinal and photoreceptor development and maintenance (Hori *et al.,* 2000; Ochrietor *et al.,* 2001, 2002), neurological functions (Naruhashi *et al.,* 1997) and lymphocyte responses (Igakura *et al.,* 1996). The relevance of MMP stimulation, if any, to these processes has not been established.

Many physiological functions in which emmprin may participate have also been suggested by other studies. Systems in which emmprin is likely to be important include chaperone functions (Kirk *et al.,* 2000), calcium transport (Jiang *et al.,* 2001), neutrophil chemotaxis (Yurchenko *et al.,* 2002), and blood–brain barrier development (Schlosshauer, 1993). A likely molecular function for emmprin, other than induction of MMP production, is mediation of adhesive cell interactions. For example, the chick homolog of emmprin mediates embryonic retinal cell aggregation and influences glial cell maturation (Fadool and Linser, 1993). Studies of emmprin in hematopoetic cell activation also suggest a role in cell interactions, e.g., during dendritic cell differentiation (Cho *et al.,* 2001; Kasinrerk *et al.,* 1992, 1999; Spring *et al.,* 1997; Stonehouse *et al.,* 1999). Furthermore, emmprin is expressed on erythrocyte lineage cells, including mature erythrocytes. It has been shown that injection of a monoclonal antibody against emmprin into mice causes selective trapping of erythrocytes in the spleen, suggesting that expression of emmprin on erythrocytes plays a critical role in recirculation of mature erythrocytes from the spleen into the general circulation (Coste *et al.,* 2001). It appears unlikely that MMPs are involved in all of these processes, implying that emmprin acts at more than one molecular level. However, it has been shown that emmprin expression precedes the appearance of MMPs during stretching of microvascular endothelial cells, thereby suggesting a causal relationship (Haseneen *et al.,* 2002). Additional studies are required to determine the precise mechanism of action of emmprin in each of these systems.

In addition to its role in numerous physiological events, emmprin expression is associated with several disease processes, including rheumatoid arthritis (Ibrahim *et al.,* 2002; Konttinen *et al.,* 2000a,b; Tomita *et al.,* 2002), atherosclerosis (Liang *et al.,* 2002), heart failure (Li *et al.,* 2000; Spinale *et al.,* 2000), lung injury (Foda *et al.,* 2001), and viral entry into cells (Pushkarsky *et al.,* 2001). Of special interest is the finding that treatment of patients exhibiting acute graft-versus-host disease with antibody to emmprin shows promising efficacy, presumably due to suppression of leukocyte activation (Deeg *et al.,* 2001). The evidence obtained in most of these disease processes, however, is mainly correlative but suggests that emmprin is an important factor in their pathogenesis and, in some cases, may act via induction of MMPs. However, more study is required to establish the specific role of emmprin.

VI. Concluding Remarks

Tumor cells create a pericellular environment in which many MMPs and other proteases become concentrated, thereby enhancing their ability to invade extracellular matrices and to process precursors of factors that promote tumor progression (Vu and Werb, 2000; Werb, 1997). Emmprin, a cell surface glycoprotein and member of the Ig superfamily, most likely participates in regulation of these processes. Current evidence indicates that emmprin stimulates MMP production via homotypic and heterotypic cell interactions, that emmprin expression is greatly increased on the surface of malignant tumor cells, and that increased expression of emmprin leads to augmentation of transformed characteristics in vitro and to marked stimulation of tumor growth and invasion *in vivo*. Under some circumstances stimulation of MMP production may be mediated by homophilic interactions between emmprin molecules on apposing cells, but more work is required to clarify the molecular mechanism of emmprin action and to determine whether other interactions are involved. Several emmprin-binding partners, including MMP-1, have been characterized but their roles in MMP induction and in tumor progression are not clear. Although recent evidence suggests that stromal production of MMPs promotes tumor progression, the relative contributions of stromal versus tumor cell production of MMPs and the role of emmprin in regulating MMP production from either source need to be further clarified. In addition to cancer, emmprin is involved in numerous physiological and pathological events that may or may not involve regulation of MMP synthesis. Emmprin clearly promotes tumor progression but whether or not emmprin serves more than one molecular function in malignant tumor cell behavior also remains to be determined.

References

Alexander, C. M., Hansell, E. J., Behrendtsen, O., Flannery, M. L., Kishnani, N. S., Hawkes, S. P., and Werb, Z. (1996). Expression and function of matrix metalloproteinases and their inhibitors at the maternal–embryonic boundary during mouse embryo implantation. *Development* **122,** 1723–1736.

Altruda, F., Cervella, P., Gaeta, M. L., Daniele, A., Giancotti, F., Tarone, G., Stefanuto, G., and Silengo, L. (1989). Cloning of cDNA for a novel mouse membrane glycoprotein (gp42): shared identity to histocompatibility antigens, immunoglobulins and neural-cell adhesion molecules. *Gene* **85,** 445–451.

Berditchevski, F., Chang, S., Bodorova, J., and Hemler, M. E. (1997). Generation of monoclonal antibodies to integrin-associated proteins. Evidence that alpha3beta1 complexes with EMMPRIN/basigin/OX47/M6. *J. Biol. Chem.* **272,** 29174–29180.

Biswas, C. (1982). Tumor cell stimulation of collagenase production by fibroblasts. *Biochem. Biophys. Res. Commun.* **109,** 1026–1034.

Biswas, C. (1984). Collagenase stimulation in cocultures of human fibroblasts and human tumor cells. *Cancer Lett.* **24,** 201–207.

Biswas, C., and Nugent, M. A. (1987). Membrane association of collagenase stimulatory factor(s) from B-16 melanoma cells. *J. Cell. Biochem.* **35,** 247–258.

Biswas, C., Zhang, Y., DeCastro, R., Guo, H., Nakamura, T., Kataoka, H., and Nabeshima, K. (1995). The human tumor cell-derived collagenase stimulatory factor (renamed EMMPRIN) is a member of the immunoglobulin superfamily. *Cancer Res.* **55,** 434–439.

Bordador, L. C., Li, X., Toole, B., Chen, B., Regezi, J., Zardi, L., Hu, Y., and Ramos, D. M. (2000). Expression of emmprin by oral squamous cell carcinoma. *Int. J. Cancer* **85,** 347–352.

Canete-Soler, R., Litzky, L., Lubensky, I., and Muschel, R. J. (1994). Localization of the 92 kd gelatinase mRNA in squamous cell and adenocarcinomas of the lung using in situ hybridization. *Am. J. Pathol.* **144,** 518–527.

Caudroy, S., Polette, M., Nawrocki-Raby, B., Ciao, J., Toole, B. P., Zucker, S., and Birembaut, P. (2002). EMMPRIN-mediated MMPs in tumor and endothelial cells. *Clin. Exp. Metastasis,* in press.

Caudroy, S., Polette, M., Tournier, J. M., Burlet, H., Toole, B., Zucker, S., and Birembaut, P. (1999). Expression of the extracellular matrix metalloproteinase inducer (EMMPRIN) and the matrix metalloproteinase-2 in bronchopulmonary and breast lesions. *J. Histochem. Cytochem.* **47,** 1575–1580.

Cho, J. Y., Fox, D. A., Horejsi, V., Sagawa, K., Skubitz, K. M., Katz, D. R., and Chain, B. (2001). The functional interactions between CD98, beta1-integrins, and CD147 in the induction of U937 homotypic aggregation. *Blood* **98,** 374–382.

Choi, K. S., Fitzpatrick, S. L., Filipenko, N. R., Fogg, D. K., Kassam, G., Magliocco, A. M., and Waisman, D. M. (2001). Regulation of plasmin-dependent fibrin clot lysis by annexin II heterotetramer. *J. Biol. Chem.* **276,** 25212–25221.

Coste, I., Gauchat, J. F., Wilson, A., Izui, S., Jeannin, P., Delneste, Y., MacDonald, H. R., Bonnefoy, J. Y., and Renno, T. (2001). Unavailability of CD147 leads to selective erythrocyte trapping in the spleen. *Blood* **97,** 3984–3988.

Dalberg, K., Eriksson, E., Enberg, U., Kjellman, M., and Backdahl, M. (2000). Gelatinase A, membrane type 1 matrix metalloproteinase, and extracellular matrix metalloproteinase inducer mRNA expression: correlation with invasive growth of breast cancer. *World J. Surg.* **24,** 334–340.

DeCastro, R., Zhang, Y., Guo, H., Kataoka, H., Gordon, M. K., Toole, B., and Biswas, G. (1996). Human keratinocytes express EMMPRIN, an extracellular matrix metalloproteinase inducer. *J. Invest. Dermatol.* **106,** 1260–1265.

DeClerck, Y. A. (2000). Interactions between tumour cells and stromal cells and proteolytic modification of the extracellular matrix by metalloproteinases in cancer. *Eur. J. Cancer* **36,** 1258–1268.

Deeg, H. J., Blazar, B. R., Bolwell, B. J., Long, G. D., Schuening, F., Cunningham, J., Rifkin, R. M., Abhyankar, S., Briggs, A. D., Burt, R., Lipani, J., Roskos, L. K., White, J. M., Havrilla, N., Schwab, G., and Heslop, H. E. (2001). Treatment of steroid-refractory acute graft-versus-host disease with anti-CD147 monoclonal antibody ABX-CBL. *Blood* **98,** 2052–2058.

Ellis, S. M., Nabeshima, K., and Biswas, C. (1989). Monoclonal antibody preparation and purification of a tumor cell collagenase-stimulatory factor. *Cancer Res.* **49,** 3385–3391.

Fadool, J. M., and Linser, P. J. (1993). 5A11 antigen is a cell recognition molecule which is involved in neuronal–glial interactions in avian neural retina. *Dev. Dynam.* **196,** 252–262.

Fadool, J. M., and Linser, P. J. (1996). Evidence for the formation of multimeric forms of the 5A11/HT7 antigen. *Biochem. Biophys. Res. Commun.* **229,** 280–286.

Foda, H. D., Rollo, E. E., Drews, M., Conner, C., Appelt, K., Shalinsky, D. R., and Zucker, S. (2001). Ventilator-induced lung injury upregulates and activates gelatinases and EMMPRIN: attenuation by the synthetic matrix metalloproteinase inhibitor, Prinomastat (AG3340). *Am. J. Respir. Cell. Mol. Biol.* **25,** 717–724.

Fossum, S., Mallett, S., and Barclay, A. N. (1991). The MRC OX-47 antigen is a member of the immunoglobulin superfamily with an unusual transmembrane sequence. *Eur. J. Immunol.* **21,** 671–679.

Guo, H., Li, R., Zucker, S., and Toole, B. P. (2000). EMMPRIN (CD147), an inducer of matrix metalloproteinase synthesis, also binds interstitial collagenase to the tumor cell surface. *Cancer Res.* **60,** 888–891.

Guo, H., Majmudar, G., Jensen, T. C., Biswas, C., Toole, B. P., and Gordon, M. K. (1998). Characterization of the gene for human EMMPRIN, a tumor cell surface inducer of matrix metalloproteinases. *Gene* **220,** 99–108.

Guo, H., Zucker, S., Gordon, M. K., Toole, B. P., and Biswas, C. (1997). Stimulation of matrix metalloproteinase production by recombinant extracellular matrix metalloproteinase inducer from transfected Chinese hamster ovary cells. *J. Biol. Chem.* **272,** 24–27.

Hajjar, K. A., and Acharya, S. S. (2000). Annexin II and regulation of cell surface fibrinolysis. *Ann. NY Acad. Sci.* **902,** 265–271.

Hajjar, K. A., Guevara, C. A., Lev, E., Dowling, K., and Chacko, J. (1996). Interaction of the fibrinolytic receptor, annexin II, with the endothelial cell surface. Essential role of endonexin repeat 2. *J. Biol. Chem.* **271,** 21652–21659.

Haseneen, N. A., Vaday, G. G., Zucker, S., and Foda, H. D. (2002). Mechanical stretch induces MMP-2 release and activation in lung endothelium: role of emmprin. *Am. J. Physiol.,* in press.

Heppner, K. J., Matrisian, L. M., Jensen, R. A., and Rodgers, W. H. (1996). Expression of most matrix metalloproteinase family members in breast cancer represents a tumor-induced host response. *Am. J. Pathol.* **149,** 273–282.

Himelstein, B. P., and Muschel, R. J. (1996). Induction of matrix metalloproteinase 9 expression in breast carcinoma cells by a soluble factor from fibroblasts. *Clin. Exp. Metastasis* **14,** 197–208.

Himelstein, B. P., Canete-Soler, R., Bernhard, E. J., and Muschel, R. J. (1994). Induction of fibroblast 92 kDa gelatinase/type IV collagenase expression by direct contact with metastatic tumor cells. *J. Cell Sci.* **107,** 477–486.

Hori, K., Katayama, N., Kachi, S., Kondo, M., Kadomatsu, K., Usukura, J., Muramatsu, T., Mori, S., and Miyake, Y. (2000). Retinal dysfunction in basigin deficiency. *Invest. Ophthalmol. Vis. Sci.* **41,** 3128–3133.

Hyuga, S., Nishikawa, Y., Sakata, K., Tanaka, H., Yamagata, S., Sugita, K., Saga, S., Matsuyama, M., and Shimizu, S. (1994). Autocrine factor enhancing the secretion of M(r) 95,000 gelatinase (matrix metalloproteinase 9) in serum-free medium conditioned with murine metastatic colon carcinoma cells. *Cancer Res.* **54,** 3611–3616.

Ibrahim, S. M., Koczan, D., and Thiesen, H. J. (2002). Gene-expression profile of collagen-induced arthritis. *J. Autoimmun.* **18,** 159–167.

Igakura, T., Kadomatsu, K., Kaname, T., Muramatsu, H., Fan, Q. W., Miyauchi, T., Toyama, Y., Kuno, N., Yuasa, S., Takahashi, M., Senda, T., Taguchi, O., Yamamura, K., Arimura, K., and Muramatsu, T. (1998). A null mutation in basigin, an immunoglobulin superfamily member, indicates its important roles in peri-implantation development and spermatogenesis. *Dev. Biol.* **194,** 152–165.

Igakura, T., Kadomatsu, K., Taguchi, O., Muramatsu, H., Kaname, T., Miyauchi, T., Yamamura, K., Arimura, K., and Muramatsu, T. (1996). Roles of basigin, a member of the immunoglobulin superfamily, in behavior as to an irritating odor, lymphocyte response, and blood–brain barrier. *Biochem. Biophys. Res. Commun.* **224,** 33–36.

Itoh, T., Tanioka, M., Yoshida, H., Yoshioka, T., Nishimoto, H., and Itohara, S. (1998). Reduced angiogenesis and tumor progression in gelatinase A-deficient mice. *Cancer Res.* **58,** 1048–1051.

Jacovina, A. T., Zhong, F., Khazanova, E., Lev, E., Deora, A. B., and Hajjar, K. A. (2001). Neuritogenesis and the nerve growth factor-induced differentiation of PC-12 cells requires annexin II-mediated plasmin generation. J. Biol. Chem. 276, 49350–49358.

Jiang, J. L., Zhou, Q., Yu, M. K., Ho, L. S., Chen, Z. N., and Chan, H. C. (2001). The involvement of HAb18G/CD147 in regulation of store-operated calcium entry and metastasis of human hepatoma cells. J. Biol. Chem. 276, 46870–46877.

Johnsen, M., Lund, L. R., Romer, J., Almholt, K., and Dano, K. (1998). Cancer invasion and tissue remodeling: common themes in proteolytic matrix degradation. Curr. Opin. Cell Biol. 10, 667–671.

Kanekura, T., Chen, X., and Kanzaki, T. (2002). Basigin (CD147) is expressed on melanoma cells and induces tumor cell invasion by stimulating production of matrix metalloproteinases by fibroblasts. Int. J. Cancer 99, 520–528.

Kasinrerk, W., Fiebiger, E., Stefanova, I., Baumruker, T., Knapp, W., and Stockinger, H. (1992). Human leukocyte activation antigen M6, a member of the Ig superfamily, is the species homologue of rat OX-47, mouse basigin, and chicken HT7 molecule. J. Immunol. 149, 847–854.

Kasinrerk, W., Tokrasinwit, N., and Phunpae, P. (1999). CD147 monoclonal antibodies induce homotypic cell aggregation of monocytic cell line U937 via LFA-1/ICAM-1 pathway. Immunology 96, 184–192.

Kataoka, H., DeCastro, R., Zucker, S., and Biswas, C. (1993). Tumor cell-derived collagenase-stimulatory factor increases expression of interstitial collagenase, stromelysin, and 72-kDa gelatinase. Cancer Res. 53, 3154–3158.

Kirk, P., Wilson, M. C., Heddle, C., Brown, M. H., Barclay, A. N., and Halestrap, A. P. (2000). CD147 is tightly associated with lactate transporters MCT1 and MCT4 and facilitates their cell surface expression. EMBO J. 19, 3896–3904.

Klein, C. A., Seidl, S., Petat-Dutter, K., Offner, S., Geigl, J. B., Schmidt-Kittler, O., Wendler, N., Passlick, B., Huber, R. M., Schlimok, G., Baeuerle, P. A., and Riethmuller, G. (2002). Combined transcriptome and genome analysis of single micrometastatic cells. Nat. Biotechnol. 20, 387–392.

Konttinen, Y. T., Li, T. F., Hukkanen, M., Ma, J., Xu, J. W., and Virtanen, I. (2000a). Fibroblast biology. Signals targeting the synovial fibroblast in arthritis. Arthritis Res. 2, 348–355.

Konttinen, Y. T., Li, T. F., Mandelin, J., Liljestrom, M., Sorsa, T., Santavirta, S., and Virtanen, I. (2000b). Increased expression of extracellular matrix metalloproteinase inducer in rheumatoid synovium. Arthritis Rheum. 43, 275–280.

Langnaese, K., Beesley, P. W., and Gundelfinger, E. D. (1997). Synaptic membrane glycoproteins gp65 and gp55 are new members of the immunoglobulin superfamily. J. Biol. Chem. 272, 821–827.

Lengyel, E., Gum, R., Juarez, J., Clayman, G., Seiki, M., Sato, H., and Boyd, D. (1995). Induction of M_r 92,000 type IV collagenase expression in a squamous cell carcinoma cell line by fibroblasts. Cancer Res. 55, 963–967.

Li, R., Huang, L., Guo, H., and Toole, B. P. (2001). Basigin (murine EMMPRIN) stimulates matrix metalloproteinase production by fibroblasts. J. Cell Physiol. 186, 371–379.

Li, Y. Y., McTiernan, C. F., and Feldman, A. M. (2000). Interplay of matrix metalloproteinases, tissue inhibitors of metalloproteinases and their regulators in cardiac matrix remodeling. Cardiovasc. Res. 46, 214–224.

Liang, L., Major, T., and Bocan, T. (2002). Characterization of the promoter of human extracellular matrix metalloproteinase inducer (EMMPRIN). Gene 282, 75–86.

Lim, M., Martinez, T., Jablons, D., Cameron, R., Guo, H., Toole, B., Li, J. D., and Basbaum, C. (1998). Tumor-derived EMMPRIN (extracellular matrix metalloproteinase inducer) stimulates collagenase transcription through MAPK p38. FEBS Lett. 441, 88–92.

Mai, J., Finley, R. L. Jr., Waisman, D. M., and Sloane, B. F. (2000). Human procathepsin B interacts with the annexin II tetramer on the surface of tumor cells. J. Biol. Chem. 275, 12806–12812.

Marmorstein, A. D., Gan, Y. C., Bonilha, V. L., Finnemann, S. C., Csaky, K. G., and Rodriguez-Boulan, E. (1998). Apical polarity of N-CAM and EMMPRIN in retinal pigment epithelium resulting from suppression of basolateral signal recognition. *J. Cell. Biol.* **142,** 697–710.

Masson, R., Lefebvre, O., Noel, A., Fahime, M. E., Chenard, M. P., Wendling, C., Kebers, F., LeMeur, M., Dierich, A., Foidart, J. M., Basset, P., and Rio, M. C. (1998). In vivo evidence that the stromelysin-3 metalloproteinase contributes in a paracrine manner to epithelial cell malignancy. *J. Cell Biol.* **140,** 1535–1541.

Miyauchi, T., Jimma, F., Igakura, T., Yu, S., Ozawa, M., and Muramatsu, T. (1995). Structure of the mouse basigin gene, a unique member of the immunoglobulin superfamily. *J. Biochem.* (*Tokyo*) **118,** 717–724.

Miyauchi, T., Kanekura, T., Yamaoka, A., Ozawa, M., Miyazawa, S., and Muramatsu, T. (1990). Basigin, a new, broadly distributed member of the immunoglobulin superfamily, has strong homology with both the immunoglobulin V domain and the beta-chain of major histocompatibility complex class II antigen. *J. Biochem.* (*Tokyo*) **107,** 316–323.

Miyauchi, T., Masuzawa, Y., and Muramatsu, T. (1991). The basigin group of the immunoglobulin superfamily: complete conservation of a segment in and around transmembrane domains of human and mouse basigin and chicken HT7 antigen. *J. Biochem.* (*Tokyo*) **110,** 770–774.

Muraoka, K., Nabeshima, K., Murayama, T., Biswas, C., and Koono, M. (1993). Enhanced expression of a tumor-cell-derived collagenase-stimulatory factor in urothelial carcinoma: its usefulness as a tumor marker for bladder cancers. *Int. J. Cancer* **55,** 19–26.

Nabeshima, K., Kataoka, H., Toole, B. P., and Koono, M. (1999). Activation and induction of collagenases. *In* "Collagenases" (W. Hoeffler, Ed.), pp. 91–113. R. G. Landes, Austin, TX.

Nabeshima, K., Lane, W. S., and Biswas, C. (1991). Partial sequencing and characterization of the tumor cell-derived collagenase stimulatory factor. *Arch. Biochem. Biophys.* **285,** 90–96.

Nabeshima, K., Suzumiya, K., Ohshima, K., Sameshima, T., Shimao, Y., Toole, B. P., Tamura, K., and Kikuchi, M. (2002). Expression of emmprin, a cell surface inducer of matrix metalloproteinases, in T-cell lymphomas. Submitted for publication.

Naruhashi, K., Kadomatsu, K., Igakura, T., Fan, Q. W., Kuno, N., Muramatsu, H., Miyauchi, T., Hasegawa, T., Itoh, A., Muramatsu, T., and Nabeshima, T. (1997). Abnormalities of sensory and memory functions in mice lacking Bsg gene. *Biochem. Biophys. Res. Commun.* **236,** 733–737.

Nehme, C. L., Cesario, M. M., Myles, D. G., Koppel, D. E., and Bartles, J. R. (1993). Breaching the diffusion barrier that compartmentalizes the transmembrane glycoprotein CE9 to the posterior-tail plasma membrane domain of the rat spermatozoon. *J. Cell Biol.* **120,** 687–694.

Nehme, C. L., Fayos, B. E., and Bartles, J. R. (1995). Distribution of the integral plasma membrane glycoprotein CE9 (MRC OX-47) among rat tissues and its induction by diverse stimuli of metabolic activation. *Biochem J.* **310,** 693–698.

Ochrietor, J. D., Moroz, T. M., Kadomatsu, K., Muramatsu, T., and Linser, P. J. (2001). Retinal degeneration following failed photoreceptor maturation in 5A11/basigin null mice. *Exp. Eye Res.* **72,** 467–477.

Ochrietor, J. D., Moroz, T. P., Clamp, M. F., Timmers, A. M., Muramatsu, T., and Linser, P. J. (2002). Inactivation of the Basigin gene impairs normal retinal development and maturation. *Vision Res.* **42,** 447–453.

Polette, M., Gilles, C., Marchand, V., Lorenzato, M., Toole, B., Tournier, J. M., Zucker, S., and Birembaut, P. (1997). Tumor collagenase stimulatory factor (TCSF) expression and localization in human lung and breast cancers. *J. Histochem. Cytochem.* **45,** 703–709.

Prescott, J., Troccoli, N., and Biswas, C. (1989). Coordinate increase in collagenase mRNA and enzyme levels in human fibroblasts treated with the tumor cell factor, TCSF. *Biochem. Int.* **19,** 257–266.

Pushkarsky, T., Zybarth, G., Dubrovsky, L., Yurchenko, V., Tang, H., Guo, H., Toole, B., Sherry, B., and Bukrinsky, M. (2001). CD147 facilitates HIV-1 infection by interacting with virus-associated cyclophilin A. *Proc. Natl. Acad. Sci. USA* **98,** 6360–6365.

Sameshima, T., Nabeshima, K., Toole, B. P., Yokogami, K., Okada, Y., Goya, T., Koono, M., and Wakisaka, S. (2000a). Expression of emmprin (CD147), a cell surface inducer of matrix metalloproteinases, in normal human brain and gliomas. *Int. J. Cancer* **88,** 21–27.

Sameshima, T., Nabeshima, K., Toole, B. P., Yokogami, K., Okada, Y., Goya, T., Koono, M., and Wakisaka, S. (2000b). Glioma cell extracellular matrix metalloproteinase inducer (EMMPRIN) (CD147) stimulates production of membrane-type matrix metalloproteinases and activated gelatinase A in co-cultures with brain-derived fibroblasts. *Cancer Lett.* **157,** 177–184.

Saxena, D. K., Oh-Oka, T., Kadomatsu, K., Muramatsu, T., and Toshimori, K. (2002). Behaviour of a sperm surface transmembrane glycoprotein basigin during epididymal maturation and its role in fertilization in mice. *Reproduction* **123,** 435–444.

Schlosshauer, B. (1993). The blood–brain barrier: morphology, molecules, and neurothelin. *Bioessays* **15,** 341–346.

Schlosshauer, B., Bauch, H., and Frank, R. (1995). Neurothelin: amino acid sequence, cell surface dynamics and actin colocalization. *Eur. J. Cell Biol.* **68,** 159–166.

Segain, J. P., Harb, J., Gregoire, M., Meflah, K., and Menanteau, J. (1996). Induction of fibroblast gelatinase B expression by direct contact with cell lines derived from primary tumor but not from metastases. *Cancer Res.* **56,** 5506–5512.

Sehgal, G., Hua, J., Bernhard, E. J., Sehgal, I., Thompson, T. C., and Muschel, R. J. (1998). Requirement for matrix metalloproteinase-9 (gelatinase B) expression in metastasis by murine prostate carcinoma. *Am. J. Pathol.* **152,** 591–596.

Seulberger, H., Unger, C. M., and Risau, W. (1992). HT7, Neurothelin, Basigin, gp42 and OX-47-many names for one developmentally regulated immuno-globulin-like surface glycoprotein on blood–brain barrier endothelium, epithelial tissue barriers and neurons. *Neurosci. Lett.* **140,** 93–97.

Shi, J., and Sugrue, S. P. (2000). Dissection of protein linkage between keratins and pinin, a protein with dual location at desmosome–intermediate filament complex and in the nucleus. *J. Biol. Chem.* **275,** 14910–14915.

Shi, Y., Ouyang, P., and Sugrue, S. P. (2000). Characterization of the gene encoding pinin/DRS/memA and evidence for its potential tumor suppressor function. *Oncogene* **19,** 289–297.

Shi, Y., Simmons, M. N., Seki, T., Oh, S. P., and Sugrue, S. P. (2001). Change in gene expression subsequent to induction of Pnn/DRS/memA: increase in p21(cip1/waf1). *Oncogene* **20,** 4007–4018.

Smalla, K. H., Matthies, H., Langnase, K., Shabir, S., Bockers, T. M., Wyneken, U., Staak, S., Krug, M., Beesley, P. W., and Gundelfinger, E. D. (2000). The synaptic glycoprotein neuroplastin is involved in long-term potentiation at hippocampal CA1 synapses. *Proc. Natl. Acad. Sci. USA* **97,** 4327–4332.

Soini, Y., Hurskainen, T., Hoyhtya, M., Oikarinen, A., and Autio-Harmainen, H. (1994). 72 KD and 92 KD type IV collagenase, type IV collagen, and laminin mRNAs in breast cancer: a study by in situ hybridization. *J. Histochem. Cytochem.* **42,** 945–951.

Spinale, F. G., Coker, M. L., Heung, L. J., Bond, B. R., Gunasinghe, H. R., Etoh, T., Goldberg, A. T., Zellner, J. L., and Crumbley, A. J. (2000). A matrix metalloproteinase induction/activation system exists in the human left ventricular myocardium and is upregulated in heart failure. *Circulation* **102,** 1944–1949.

Spring, F. A., Holmes, C. H., Simpson, K. L., Mawby, W. J., Mattes, M. J., Okubo, Y., and Parsons, S. F. (1997). The Oka blood group antigen is a marker for the M6 leukocyte activation antigen, the human homolog of OX-47 antigen, basigin and neurothelin, an immunoglobulin superfamily molecule that is widely expressed in human cells and tissues. *Eur. J. Immunol.* **27,** 891–897.

Stonehouse, T. J., Woodhead, V. E., Herridge, P. S., Ashrafian, H., George, M., Chain, B. M., and Katz, D. R. (1999). Molecular characterization of U937-dependent T-cell co-stimulation. *Immunology* **96,** 35–47.

Sun, J., and Hemler, M. E. (2001). Regulation of MMP-1 and MMP-2 production through CD147/extracellular matrix metalloproteinase inducer interactions. *Cancer Res.* **61,** 2276–2281.

Tomita, T., Nakase, T., Kaneko, M., Shi, K., Takahi, K., Ochi, T., and Yoshikawa, H. (2002). Expression of extracellular matrix metalloproteinase inducer and enhancement of the production of matrix metalloproteinases in rheumatoid arthritis. *Arthritis Rheum.* **46,** 373–378.

Tressler, R. J., Updyke, T. V., Yeatman, T., and Nicolson, G. L. (1993). Extracellular annexin II is associated with divalent cation-dependent tumor cell-endothelial cell adhesion of metastatic RAW117 large-cell lymphoma cells. *J. Cell. Biochem.* **53,** 265–276.

van den Oord, J. J., Paemen, L., Opdenakker, G., and de Wolf-Peeters, C. (1997). Expression of gelatinase B and the extracellular matrix metalloproteinase inducer EMMPRIN in benign and malignant pigment cell lesions of the skin. *Am. J. Pathol.* **151,** 665–670.

Vu, T. H., and Werb, Z. (2000). Matrix metalloproteinases: effectors of development and normal physiology. *Genes Dev.* **14,** 2123–2133.

Werb, Z. (1997). ECM and cell surface proteolysis: regulating cellular ecology. *Cell* **91,** 439–442.

Wilson, M. C., Meredith, D., and Halestrap, A. P. (2002). Fluorescence resonance energy transfer studies on the interaction between the lactate transporter MCT1 and CD147 provide information on the topology and stoichiometry of the complex in situ. *J. Biol. Chem.* **277,** 3666–3672.

Wright, J. H., McDonnell, S., Portella, G., Bowden, G. T., Balmain, A., and Matrisian, L. M. (1994). A switch from stromal to tumor cell expression of stromelysin-1 mRNA associated with the conversion of squamous to spindle carcinomas during mouse skin tumor progression. *Mol. Carcinog.* **10,** 207–215.

Yoshida, S., Shibata, M., Yamamoto, S., Hagihara, M., Asai, N., Takahashi, M., Mizutani, S., Muramatsu, T., and Kadomatsu, K. (2000). Homo-oligomer formation by basigin, an immunoglobulin superfamily member, via its N-terminal immunoglobulin domain. *Eur. J. Biochem.* **267,** 4372–4380.

Yurchenko, V., Zybarth, G., O'Connor, M., Dai, W. W., Franchin, G., Hao, T., Guo, H., Hung, H. C., Toole, B., Gallay, P., Sherry, B., and Bukrinsky, M. (2002). Active-site residues of cyclophilin A are crucial for its signaling activity via CD147. *J. Biol. Chem.* **277,** 22959–22965.

Zucker, S., Hymowitz, M., Rollo, E. E., Mann, R., Conner, C. E., Cao, J., Foda, H. D., Tompkins, D. C., and Toole, B. P. (2001). Tumorigenic potential of extracellular matrix metalloproteinase inducer (EMMPRIN). *Am. J. Pathol.* **158,** 1921–1928.

13

The Evolving Roles of Cell Surface Proteases in Health and Disease: Implications for Developmental, Adaptive, Inflammatory, and Neoplastic Processes

Joseph A. Madri
Department of Pathology
Yale University School of Medicine
New Haven, Connecticut, 06510

I. Introduction

Cell surface proteolysis is an important process, thought to be crucial in a growing number of biological functions including development, normal homeostasis and adaptive responses, degenerative diseases, inflammation and repair, and several aspects of the neoplastic process. In addition to a growing appreciation of the general importance of cell surface proteolysis there has also been a growing appreciation of the diversity of enzymes and enzyme systems which are involved in this process (Black and White, 1998; Blobel, 2000; Bourguignon *et al.*, 1998; Kaushal and Shah, 2000; McCawley and Matrisian, 2001; Primakoff and Myles, 2000). In this chapter examples illustrating cell-surface proteolysis in each of these areas will be discussed.

II. Development

Protease activities are crucial during development. In addition to extracellular matrix remodeling, cell surface proteases have been shown to be essential in modulating cell proliferation, migration, and differentiation as well as in signaling cell attraction and repulsion.

Current Topics in Developmental Biology, Vol. 54

Receptor tyrosine kinases (RTKs) comprise a family of surface receptors expressed by and crucial to endothelial development including VEGFR1, 2, and 3 and Tie 1 and 2 (Gale and Yancopoulos, 1999). The spatiotemporal expression of these RTKs has been determined to be essential for normal vasculogenesis and angiogenesis (Gale and Yancopoulos, 1999). Deletion of individual members of this receptor superfamily and deletion/overexpression of their ligands (VEGF, angiopoietin 1, and angiopoietin 2) have uncovered complex interrelationships among family members during embryonic vasculogenesis and angiogenesis (Gale and Yancopoulos, 1999), suggesting that specific, coordinated, temporally controlled activation is required for normal development. In most cases, activation is mediated by ligand engagement while downregulation of RTK activity is thought to occur via phosphotyrosine-specific phosphatase activities and/or internalization and degradation of the receptor complex, and/or internalization and reexpression of the receptor at the cell surface. However, another recently described mechanism of receptor downregulation/modulation at play in several cell types including endothelial cells appears to be involved in modulation of Tie-1 signaling. Similar to what has been noted in several other receptor signaling pathways operational during development, Tie-1 undergoes endoproteolytic cleavage of its ectodomain (extracellular ligand-binding domain) at the cell surface. Endoproteolytic cleavage of cell surface receptors generates soluble ligand-binding domains and membrane-anchored cytoplasmic and transmembrane domains (Cabrera et al., 1996; Vecchi, Baulida, and Carpenter, 1996; Vecchi and Carpenter, 1997). This process has been documented in several cell types and has been found to be associated with protein kinase C activation and with inflammatory cytokine stimulation (Cabrera et al., 1996; Vecchi et al., 1996; Vecchi and Carpenter, 1997; Yabkowitz et al., 1997, 1999). The soluble ligand-binding domain has the capability of functioning as a ligand sink/reservoir, decreasing the effective concentration of its cognate ligand and thus acting as a negative regulator of receptor activity. In contrast, the membrane-anchored intracellular fragment has been observed to function as either a negative or constitutively active form of the receptor, depending upon the cell type studied (Cabrera et al., 1996; Vecchi et al., 1996). Endoproteolytic cleavage of selected receptors is thought to be mediated by metalloproteases, including members of the ADAM family of metalloproteases (Blobel, 2000; Kheradmand and Werb, 2002). Although control of these endoproteolytic activities is currently incompletely understood, it is appreciated that specific receptor activation and activation of protein kinase C is necessary. In the case of endothelial cells, the enzymatic activity is stimulated by inflammatory cytokines and selected growth factors, illustrating the tightly regulated control of this process in particular cell types (Cabrera et al., 1996; Vecchi et al., 1996; Vecchi and Carpenter, 1997; Yabkowitz et al., 1997, 1999).

Investigators (Yabkowitz et al., 1997, 1999) have demonstrated endoproteolytic cleavage of the Tie-1 orphan receptor in human umbilical vein endothelial cell cultures (HUVEC). Although its ligand(s) are not currently known, the Tie-1 receptor has been found to be essential for normal vasculogenesis and angiogenesis as evidenced by embryonic and perinatal lethality in Tie-1 deficient mice (Gale and Yancopoulos, 1999). Its upregulation during angiogenesis is associated with normal physiologic responses, tumor growth, and wound healing. The lack of identifiable ligands for the holoreceptor, its endoproteolytic cleavage, the persistence of its membrane-anchored cytoplasmic domain *in vitro,* its presence *in vivo* in actively remodeling tissues, and its ability to differentially interact with phosphoproteins suggest that the Tie-1 membrane-anchored cytoplasmic domain may be a key participant in signal transduction (McCarthy *et al.,* 1999) and that endoproteolytic processing of this receptor may constitute a major activation pathway. Thus, metalloproteinase generation of a soluble extracellular ligand-binding domain of Tie-1 may cause a downregulation of receptor activation by binding to and sequestering its still undefined ligand(s). Alternatively, the formation of the membrane-anchored cytoplasmic domain may result either in constitutive activation of its cytoplasmic domain or in generating an inactive cytoplasmic domain and reduced expression of the potentially activatable holoreceptor on the cell surface. In addition, endoproteolytic cleavage may result in the generation of a cytoplasmic domain which is capable of binding to distinct adapter and signaling molecules, resulting in a signal which may antagonize the signal elicited by ligand engagement of the holoreceptor or in a positive signal distinct from that elicited by the holoreceptor (Ilan and Madri, 1999; Ilan *et al.,* 2001a).

Another example of cell-surface proteolysis as a modulator of endothelial function in the vasculature is the cleavage and shedding of PECAM-1 (CD31). PECAM-1 is a 130-kDa glycoprotein member of the immunoglobulin superfamily of cell adhesion molecules. It is widely expressed on endothelial cells (Albelda *et al.,* 1990; Muller *et al.,* 1989), platelets, monocytes, granulocytes, subsets of lymphocytes, and bone-marrow hematopoietic cells (Ashman and Aylett, 1991; Goyert *et al.,* 1986; Ohto *et al.,* 1985; Stockinger *et al.,* 1990). On continuous endothelial cell monolayers and most vascular endothelial cells *in vivo,* PECAM-1 expression is concentrated at areas of cell–cell contact (Albelda *et al.,* 1990; Muller *et al.,* 1989). As a cell adhesion molecule, PECAM-1 is capable of mediating homophilic (Albelda *et al.,* 1991; Fawcett *et al.,* 1995; Muller *et al.,* 1992) as well as heterophilic binding (DeLisser *et al.,* 1993; Muller *et al.,* 1992). PECAM-1 is thought to be involved in the transendothelial migration of leukocytes during their extravasation to sites of inflammation and has been found to be an important component in the maintenance of vascular permeability (Graesser *et al.,* 2002). In addition to its adhesive functions, PECAM-1 appears to play important roles as a scaffolding molecule, capable of binding tyrosine phosphorylated β-catenin (Ilan *et al.,* 1999), γ-catenin (Ilan *et al.,* 2000), STAT family members (Ilan *et al.,* 2001), and, when its ITAM domain tyrosines (Y663 and Y686) are phosphorylated, SHP-2 (Lu *et al.,* 1997; Lu, Yan, and Madri, 1996). Its scaffolding functions are thought to, in part, mediate cellular junction dynamics, differentiation state, and apoptosis (Graesser *et al.,* 2002; Ilan *et al.,* 2001, 2001a). *In vivo* and *in vitro* studies have demonstrated that PECAM-1 ectodomain is cleaved and shed, raising the possibility that the shed ectodomain and the retained transmembrane and cyto-plasmic domains can function to alter cellular behaviors (Ilan *et al.,* 2001a; Losy *et al.,* 1999; Minagar *et al.,* 2001; Serebruany *et al.,* 1999). It has been shown that soluble recombinant PECAM–IgG fusion protein can block transendothelial cell migration of neutrophils, monocytes, and NK cells *in vitro* and can inhibit acute inflammation *in vivo* (Serebruany *et al.,* 1999). More recently, constitutive secretion of soluble PECAM-1 into the plasma of transgenic mice has been shown to significantly blunt the acute inflammatory response (Liao, Schenkel, and Muller, 1999). These data suggest that secreted PECAM-1 molecules may exhibit an im-portant function on their own. For example, an increase in the serum levels of soluble PECAM-1 has been observed in multiple sclerosis (MS) patients with ac-tive gadolinium enhancing lesions (Losy *et al.,* 1999). Thus, there is a good deal of evidence for functional significance of shed ectodomain of PECAM-1. There is also evidence being accrued supporting the notion that the cytoplasmic domain of PECAM-1 exhibits unique functional characteristics (Ilan *et al.,* 1999, 2000, 2001, 2001a; Ilan and Madri, 1999; Lu *et al.,* 1996, 1997). We have shown that PECAM-1 is shed from endothelial cell surfaces during apoptosis and accumulates in the culture medium as a 100-kDa soluble protein (Ilan *et al.,* 2001a). The cleav-age mediating the shedding was found to be metalloproteinase (MMP)-dependent, as GM6001, a broad-spectrum MMP inhibitor, inhibited PECAM-1 accumulation in the culture medium in a dose-responsive manner. We also found that PECAM-1 cleavage generated the formation of a truncated (Tr.), 28-kDa molecule, composed of the transmembrane and the cytoplasmic domains, and that transfections of the full-length (Fl) and the truncated PECAM-1 gene constructs into endothelial and nonendothelial cells resulted in significantly more β-catenin and SHP-2 bound to the truncated than to the full-length PECAM-1. In addition, stable expression of the truncated PECAM-1 in SW480 colon carcinoma cells resulted in a signifi-cant decrease in cell proliferation, whereas expression of comparable levels of the full-length PECAM-1 had no effect. Interestingly, the observed decrease in cell proliferation was due, in part, to an increase in programmed cell death (apoptosis) that correlated with continuous caspase 8 cleavage and p38/JNK phosphorylation. These results support the intimate involvement of the metalloproteinase-generated PECAM-1 cytoplasmic domain in signal transduction cascades and also are consis-tent with a role for the retained cytoplasmic domain in the modulation of apoptosis (Ilan *et al.,* 2001a).

Although still incompletely understood, the two examples just given illustrate the potential importance of cell-surface-associated metalloproteinases in affecting the modulation of cellular signaling by endoproteolytic cleavage of a wide variety of cell-surface components.

Other examples of modulation of receptor/ligand expression and activity by endoproteolytic cleavage during development include netrin-1/Deleted in Colon Carcinoma (DCC)-mediated axonal chemoattraction and Ephrin-A2/Eph3A-mediated axon repulsion involving endoproteolytic cleavage of either DCC or Ephrin-A2 by metalloproteases.

In vitro commissural axon outgrowth and *in vivo* guidance of spinal commissural neurons have been found to be dependent upon netrin-1 protein and DCC (Deleted in Colorectal Cancer) interactions (Galko and Tessier-Lavigne, 2000). Further investigations led to the observation that Netrin-1-mediated axonal outgrowth was enhanced by inhibitors of metalloproteinases, which resulted in inhibition of a metalloproteinase-dependent cleavage of DCC in close proximity to its transmembrane domain (Galko and Tessier-Lavigne, 2000). These findings suggest that a tightly controlled balance of proteolytic activities is necessary for proper axonal guidance. The controlled cleavage of DCC has the potential of affecting Netrin-1/DCC interactions and subsequent signaling cascades in several ways. Cleavage of DCC could render the cell less able or unable to interact with Netrin-1. Alternatively, the cleaved ectodomain of DCC may function as a sink for Netrin-1, significantly diminishing its effective local concentration and effectiveness. Additionally, the retained transmembrane and cytoplasmic domain of the cleaved DCC may exhibit functions distinct from those of the full-length molecule, modulating cell apoptosis and/or survival. Elucidation of the cleavage site of DCC will aid in determining which metalloproteinase(s) are responsible for its cleavage.

Cell–cell contact (adhesion), in addition to evoking attractive signaling, has also been found to result in repulsive signaling in the nervous and vascular systems (Hattori, Osterfield, and Flanagan, 2000; Wang, Chen, and Anderson, 1998). Specifically, cell-contact-mediated repulsive signaling has been observed in the process of axon guidance. This ephrin-mediated process has been shown to involve the formation of a stable ephrin-A2-metalloproteinase (ADAM10/*kuzbanian*) complex, which upon binding to its cognate Eph receptor (EphA3) triggers Ephrin-A2 cleavage and axonal repulsion. Similarly, in the development of the vascular system a number of Ephrin–Eph interactions have been shown to be operational (Ilan and Madri, 1999; Wang *et al.*, 1998).

Another example of Eph/Ephrin-mediated modulation of attractive and repulsive forces during development is the influence of Ephrin-B2 and Eph-B4 on the interdigitation and segregation of arterial and venous structures throughout the vascular system and on the process of myocardial trabeculation during cardiac development (Gale and Yancopoulos, 1999; Wang *et al.*, 1998). Using a knockout strategy Wang *et al.* found that Ephrin-B2, a transmembrane ligand of the Ephrin-B family, marks arterial endothelial cells while Eph-B4, a transmembrane receptor of the Eph-B family, marks venous endothelium and that engagement of and reciprocal signaling generated by this ligand–receptor pair was required for proper development of the cardiovascular system. Interdigitation, segregation, and adhesive interactions between the arterial and venous components are critical for the development and maintenance of proper structure and function of the vascular system. Development of the primary capillary plexus, which develops from fusion of blood islands and initial angiogenic sprouts, requires complex adhesion events consisting of cadherin and Ig family members as well as Ephrin/Eph interactions between arterial and venous endothelial cell populations (DeLisser *et al.*, 1997; Ilan *et al.*, 1999; Wang *et al.*, 1998). Following this, the primary capillary plexus undergoes remodeling and angiogenesis, resulting in an arborizing, interdigitating network of arterial and venous vessels with connections at the level of the capillary. These attractive "cis" adhesions are stabilizing and provide a link between the arterial and venous components of the vascular system at the capillary level (Yancopoulos, Klagsbrun, and Folkman, 1998). In contrast, during the interdigitation/arborization phase of vascular tree development, arterial and venous segments of the vasculature are required to move past one another without making long-term connections. These "trans" interactions (Yancopoulos *et al.,* 1998) are similar to the repulsive interactions noted during the process of axonal guidance. Since specific Ephrins and Ephs appear to be involved in both these processes (axonal guidance and vessel interdigitation and arborization), it is attractive to speculate that in the processes of vascular interdigitation and arborization, the function(s) of the Ephrin/Eph adhesive pair(s) involved may be modulated by endoproteolytic cleavage by an ADAM or other cell-surface protease family members. Similarly, since other families of adhesion receptors (cadherins and Ig family members) are present on the vascular cells involved in these "en passant" maneuvers, it is not unreasonable to investigate the potential roles of ADAMs and other cell-surface endoproteolytic enzymes (MT-MMPs) as mediators of as yet undefined endoproteolytic events occurring on the cell surfaces involving these adhesion receptors and their potential dysregulation in a variety of vascular malformations.

III. Normal Homeostasis and Adaptive Responses

Maintenance of homeostasis entails a continuous, dynamic response to the environment that sometimes may include controlled proteolytic responses crucial for modulation of local growth factor levels, optimal functioning of the clotting and complement systems, and remodeling of our connective tissues and vascular system (Kheradmand and Werb, 2002; McCawley and Matrisian, 2001). Adaptation to external stimuli can result in a range of cellular and tissue responses including hypertrophy, hyperplasia, and metaplasia (Cotran, Kumar, and Collins, 1999). In this section of the chapter the involvement of cell-surface proteolysis in the processes of cardiac hypertrophy (Asakura *et al.,* 2002), adaptive angiogenesis (Haas *et al.,* 2000), and airway epithelial cell responsiveness to pathogens (Lemjabbar and Basbaum, 2002) will be considered.

Cardiac hypertrophy is a well-documented and studied adaptive/compensatory response to increased blood pressure that preserves cardiac function but also is a risk factor for the development of cardiovascular disease and often leads to heart failure (Cotran *et al.,* 1999). The process of cardiac hypertrophy is known to be initiated by activation of G-protein-coupled receptors by a variety of ligands including Angiotensin II, phenylepherine, and endothelin-1, while the downstream signaling cascades involved in mediating the development of the hypertrophy are only incompletely understood. In one report, investigators demonstrated the involvement of ADAM12 in the development of cardiac hypertrophy (Asakura *et al.,* 2002). Using murine models of cardiac hypertrophy generated by transverse aortic banding and selected G-protein-coupled receptor agonists and tissue cultures of cardiac myocytes treated with G-protein-coupled receptor agonists, the investigators illustrated that the resultant hypertrophy was mediated by the activation of a particular metalloproteinase (ADAM12) cell surface cleavage of heparin-binding epidermal growth factor and subsequent activation of the epidermal growth factor receptor. Although the exact mechanism(s) and signaling pathway(s) involved in the activation of ADAM12 were not elucidated, this finding clearly illustrates the importance of controlled cell-surface proteolysis in an important adaptive mechanism and may lead to new and novel therapeutic approaches to treating pathological cardiac hypertrophy. These findings also raise the possibility that specific ADAMs may be involved in the maintenance of myocyte size and atrophy by tightly regulating the release of surface-bound growth factors. This report also leads one to consider the possibility of ADAMS (or other cell surface proteolytic activities) functioning as modulators of adaptive (physiologic) and pathologic hypertrophy/metaplasia in other tissues and organs. Cell surface proteolytic enzymes may be proven to function as markers/discriminators/initiators or enhancers of adaptive (exercise-induced) and/or pathological cardiac hypertrophy associated with coronary artery atherosclerosis or idiopathic hypertrophic cardiac hypertrophy. In addition, it will be of interest to investigate the roles of ADAMS (and other cell-surface proteolytic enzymes) as long-term downstream modulators of the effects of G-protein-coupled receptor agonists and antagonists on vascular endothelial and vascular smooth-muscle cell migration and proliferation, important processes in the maintenance of vascular integrity and responsiveness to injury (Bell *et al.,* 1992; Bell and Madri, 1989, 1990).

In addition to the cardiac and skeletal muscle hypertrophy associated with exercise, it is well recognized that there is a proportional microvascular response to exercise consisting of an increased capillary-to-muscle-fiber ratio (Haas *et al.,* 2000). This increase in capillary:fiber ratio involves a process of controlled angiogenesis that is responsive to the workloads and oxygen demands of the muscle tissues involved. Studies have demonstrated a transient, coordinated, and localized upregulation of MT1-MMP and MMP-2 in response to chronic skeleton muscle stimulation (Haas *et al.,* 2000). In this model of adaptive angiogenesis, controlled, precise, discrete breakdown of the investing basement membrane occurs (Fig. 1) which allows endothelial cell processes to migrate into the surrounding interstitial tissue and participate in the development of new vessels. This process correlates with the coordinate upregulation of MT1-MMP and MMP-2. Both enzymes are localized to the endothelial cells undergoing angiogenesis, and inhibition of these enzymes by hydroxamic-acid-based inhibitors abrogates the exercise-induced angiogenesis. Prior to the upregulation of MT1-MMP and MMP-2, soluble growth factors are known to be expressed (Hudlicka, 1998). In light of findings demonstrating ADAM-mediated release of growth factors during cardiac hypertrophy (Asakura

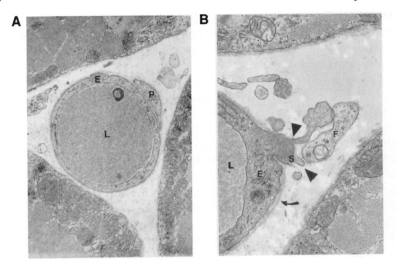

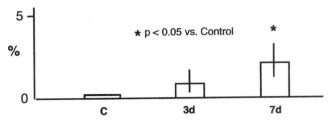

Figure 1 Basement membrane disturbances in chronically stimulated muscle. (A) Electron micrograph of a capillary from a control unstimulated extensor hallucis proprius (EHP) muscle showing a thin endothelial cell layer (E) with smooth luminal (L) and abluminal surfaces surrounded by an intact basement membrane within which is located the cellular process of a pericyte (P). Bar $= 0.5 \, \mu$m. (B) Capillary from 3-day stimulated EHP muscle illustrating the lumen (L) of the vessel and showing an abluminal endothelial cell (E) cytoplasmic projection (S) into the interstitium. The basement membrane is intact around the capillary (arrow) but absent at the tip of the sprout (S) (arrowheads). The adjacent cell (F) likely is a fibroblast. Bar $= 0.2 \, \mu$m. (C) Graphic representation of the percentage of capillary basement membrane disrupted after 3 and 7 days of chronic electrical stimulation of the EHP muscle. Based on data from Haas *et al.* (2000).

et al., 2002) it is reasonable to consider the possibility that induction/activation of ADAM family members may, in part, be responsible for the initiation of this local growth factor mediated process. In addition, ADAM family members may also participate in the activation and modulation of the induced MT1-MMP and MMP-2 activities.

Higher vertebrate species are endowed with a formidable armamentarium to ward off pathogens. This includes the range of host responses that comprise the innate immune system and the adaptive immune system (Cotran *et al.*, 1999; Janeway and Travers, 1997). One component of the innate immune system is the ability of subsets of host epithelial cells (airway epithelium) to upregulate their production of mucus, which entraps pathogens and is swept out of the bronchial tree by the action of cilia and expectoration (Cotran *et al.*, 1999; Lemjabbar and Basbaum, 2002). In a study using both *in vitro* and *in vivo* models, investigators found that gram-positive organisms (specifically a component of the organisms' cell walls known as lipoteichoic acid) elicit mucin production by epithelial cells via a novel signaling pathway. Elucidation of the signaling pathway revealed that the gram-positive organism (or lipoteichoic acid) binds to and activates platelet-activating-factor receptor, a G-protein-coupled receptor, present on the epithelial cell surface (Lemjabbar and Basbaum, 2002). Similar to what has been described following activation of other G-protein-coupled receptors (Asakura *et al.*, 2002), engagement of the platelet-activating-factor receptor resulted in activation of the disintegrin and metalloproteinase ADAM10/*kuzbanian*. This activation of ADAM10 then resulted in cleavage of cell-surface-tethered heparin-binding epidermal growth factor, allowing it to bind to and activate epidermal growth factor receptor. The resultant mucin induction was found to be signaled via a Ras, Mek, pp90rsk, NFκB-mediated pathway independent of CD14 and Toll receptor signaling (Lemjabbar and Basbaum, 2002).

Thus, in light of the growing appreciation of the importance of ADAMs in a wide variety of signaling processes, the control of specificity should be addressed. For example, in the case of ADAM10/*kuzbanian*, it appears that ADAMs can function as sheddases mediating repulsive forces during axonal guidance, while playing pivotal roles in growth factor signaling (release) during cardiac hypertrophy and responses to pathogen challenge. Several important questions remain to be addressed: Is specificity determined by the presence of potential ADAM10 substrates on the cell surface; by the presence of receptors capable of activating ADAMs; by the presence of particular ADAMs on the cell surface; or by a combination of some or all of the foregoing putative potential control mechanisms?

IV. Inflammation and Repair

During the processes of homing and extravasation at sites of inflammation, circulating leukocytes become tissue-resident cells for variable periods of time. The transmigration of leukocytes from the bloodstream into perivascular and interstitial tissues is a crucial event in these processes. During homing, lymphocytes bind via specific adhesion receptors expressed on high endothelial venules (HEV) as sites of entry into the secondary lymphoid organs (with the exception of the spleen where lymphocyte entry is via blood sinusoids in the marginal zones). Although granulocytes and monocytes do not home, along with lymphocytes, they will adhere to the microvascular endothelium in inflamed tissues and, in response to chemoattractant or adhesive gradients, migrate across endothelial cell layers and accumulate at the sites of injury or infection. Both processes of homing and extravasation during inflammation are dynamic and involve attachments to, and detachments from, a variety of cell types and extracellular matrix components before the transmigrating immune cells encounter antigen or antigen-presenting cells in the involved tissues and deliver appropriate immunological responses (Janeway and Travers, 1997; Madri and Graesser, 2000). The general mechanisms involved in both processes are similar (Springer, 1995, 1995a). The extravasating leukocytes are first tethered to the endothelial cells and roll along the blood vessel wall. Integrins on the cell surface are then activated and used to adhere tightly to the endothelial cells. Leukocytes will then transmigrate across the endothelial layer after undergoing release from their contacts with the luminal surface of the endothelium, traverse the subendothelial basement membrane, and finally migrate into the perivascular/interstitial tissue. All these steps are mediated by dynamic binding of adhesion molecules on the extravasating leukocytes to counterreceptors on endothelial cells lining the vessel wall (Madri and Graesser, 2000).

It is wellknown that proteolytic processes play important roles in the processes of inflammation and repair. Leukocyte proteases have been shown to facilitate transmigration of polymorphonuclear leukocytes during the acute phase of inflammatory responses, and similarly, proteases have been implicated in monocytic and lymphocytic transmigration in chronic inflammatory conditions (Madri and Graesser, 2000). Studies on cell lines from patients with lymphoproliferative diseases such as Burkitt's lymphoma, B-cell lymphoblastic leukemia, and multiple myeloma have found that these lymphoid cells express elevated levels of MMP-2 and/or MMP-9 or (Stetler-Stevenson *et al.*, 1997; Vacca *et al.*, 1998). In several infectious processes MMP levels have been documented to be increased. For example, MMP-9 levels were found to be increased in the cerebral spinal fluid of rabbits with induced bacterial meningitis, and the increase was found to be associated with leukocytes that had infiltrated the CNS (Azeh *et al.*, 1998). Patients with viral meningitis, characterized by invasion of neutrophils, monocytes, and lymphocytes into the subarachnoidal space, were found to exhibit elevated MMP-2 levels (Kolb *et al.*, 1998). In addition, studies in several laboratories have shown that MMPs play important roles in experimental rodent models of multiple sclerosis (Anthony *et al.*, 1998; Gijbels, Galardy, and Steinman, 1994; Graesser *et al.*, 1998; Graesser, Mahooti, and Madri, 2000; Hewson *et al.*, 1995; Romanic and Madri, 1994).

The signals that regulate expression of MMPs in leukocytes are the subject of ongoing research. Since the engagement of integrin receptors is known to trigger a number of intracellular signaling events in T cells (Davis *et al.*, 1990; Graesser *et al.*, 1998; Hynes, 1992; Nojima *et al.*, 1990, 1992; Romanic *et al.*, 1997; Romanic and Madri, 1994; Shimuzu and Shaw, 1991; Yamada *et al.*, 1991), it was reasoned that signals mediated by the engagement of leukocyte adhesion receptors during the initial stages of transendothelial cell migration may lead to secretion and/or activation of MMPs in the migrating cells (Graesser *et al.*, 1998, 2000; Madri and Graesser, 2000; Romanic and Madri, 1994). Indeed, recent data suggest that integrin-mediated signals regulate MMP secretion and/or activity in a number of cell types and for each cell type examined, there appears to be a degree of specificity, i.e., certain integrin engagements will induce particular MMPs (Chintala *et al.*, 1996; Kubota *et al.*, 1997; Nakahara *et al.*, 1997; Remy *et al.*, 1999; Seftor *et al.*, 1992, 1993; Tremble, Chiquet-Ehrismann, and Werb, 1994; Tremble, Damsky, and Werb, 1995). In addition, several studies in our and other laboratories have demonstrated that particular MMPs induced via integrin-mediated signals (including $\alpha4\beta1$) play significant roles in both *in vivo* and *in vitro* models for T-cell transmigration (Baron *et al.*, 1993; Graesser *et al.*, 1998, 2000; Romanic *et al.*, 1997; Romanic and Madri, 1994). In antigen-specific, activated T lymphocytes, $\alpha4\beta1$ integrin engagement has been shown to affect the coordinated induction and activation state of MT1-MMP and MMP-2 (Fig. 2) (Graesser *et al.*, 1998, 2000). In these studies, MT-1-MMP (membrane-type matrix metalloproteinase, MMP-14), a transmembrane metalloproteinase that has been implicated as the molecule that both tethers and activates MMP-2 at the cell surface (Butler *et al.*, 1998; Cao *et al.*, 1995; Kinoshita *et al.*, 1998; Sato *et al.*, 1994, 1996; Strongin *et al.*, 1995), was found to form a cell surface tertiary complex comprising MT1-MMP, MMP-2, and TIMP-2 (Fig. 3; see color insert), which facilitates T lymphocyte transmigration in areas of inflammation (Graesser *et al.*, 1998, 2000). In addition, following adhesion and the induction and assembly of this tertiary proteolytic complex, the complex is clustered at the putative invading podosome (invadopodia) of the migrating T cells (Fig. 3) (Madri and Graesser, 2000).

These findings are consistent with observations made in other model systems (Gilles *et al.*, 1998; Thompson *et al.*, 1994). In one such study, the activity of MMP-2 was increased in human breast carcinoma cells upon adhesion to collagen I, but not collagen IV, laminin, or fibronectin (Thompson *et al.*, 1994). In another study, simultaneous induction of MMP-2 and its activator, MT1-MMP, were noted during *in vitro* three-dimensional collagen matrix culture of endothelial cells as an *in vitro* model of angiogenesis (Haas, Davis, and Madri, 1998). The coordinate induction of MMP-2 and MT1-MMP was observed when cell were plated on or in type I collagen gels, but not when cells were plated on coatings of type I collagen or Matrigel or in Matrigel gels, suggesting that specific integrin and mechanosensor engagement and signaling are required. Further, in this model, MMP-2 activation was blocked with matrix metalloproteinase inhibitors that also inhibited endothelial-cell tube formation. The induction of MT1-MMP was further clarified with the sequencing of the MT1-MMP promoter and the elucidation of a tandem Sp1and consensus Egr-1 transcription factor binding site in the promoter (Haas *et al.*, 1999). Accrued data indicate that stimuli of angiogenesis (chronic

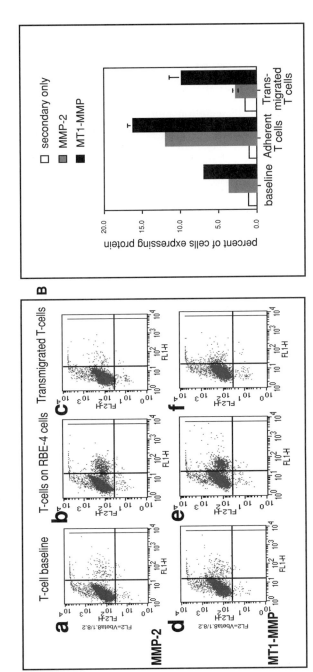

Figure 2 Coordinate induction of MT1-MMP and MMP-2 in activated, antigen-specific T lymphocytes following adhesion to and transmigration through an endothelial cell monolayer. (A) FACS analysis of T cells prior to adhesion onto endothelial cells (panels a and d, baseline), T cells adherent to the EC monolayer (panels b and e, adherent T cells), and T cells following transmigration through the EC monolayer and the underlying subendothelial ECM (panels c and f, transmigrated T cells). Panels b and e are representative FACS analyses of the transient MMP-2 and MT1-MMP induction following adhesion to brain microvascular endothelial cells (RBE4 cells). Return to near-baseline levels was noted in transmigrated T cells, following their transmigration through the EC monolayer and the subendothelial ECM and collection in the lower chamber of the transwell (panels c and f). Compare the densities in the right upper quadrants. (B) Graphic representation of the coordinate induction of T cell MT1-MMP (black boxes) and MMP-2 (gray boxes) from one of four experiments each comprising triplicate samples illustrating a transient 2.4-fold induction of MT1-MMP and a 3.4-fold induction of MMP-2 following adhesion to VCAM-1 positive brain microvascular endothelial cells. Following transmigration through the EC monolayer and the subendothelial ECM (20 h after the start of the experiment) expression of both enzymes has returned to near baseline levels.

exercise, culture in type I collagen gels) and remodeling (cyclic strain) elicit induction of Egr-1, which displaces Sp1 from the site on the MT1-MMP .promoter, resulting in induction of MT1-MMP (Haas *et al.,* 1999; Yamaguchi *et al.,* 2002). Interestingly, shear stress, while it induces increased expression of Egr-1, does not elicit induction of MT1-MMP (Yun, Madri, and Sumpio, 2002). Recent data indicate that in this instance, while Egr-1 is induced, Sp1 becomes phosphorylated, increasing its affinity for the site on the MT1-MMP promoter, thus preventing its displacement by Egr-1 and MT1-MMP induction (Yun *et al.,* 2002). Thus, the modulation of MT1-MMP activity appears to be regulated at several levels by integrin and mechanosensor receptors. The regulation of this cell surface protease is complex, involving differential induction of specific transcription factors (Sp1 and Egr-1) as well as their modification (Sp1 phosphorylation) (Fig. 4). This multilevel regulation in addition to differential induction following engagement of specific cell-matrix adhesion molecules (β1 integrins) and possibly cell–cell adhesion molecules (cadherins and Ig family members) allows for a tightly controlled expression of this cell surface protease. This broad range of control mechanisms allows for a finely tuned, tightly regulated response.

V. Neoplastic Processes

It is currently widely accepted that cell-surface proteases play important, if not pivotal roles in tumor invasion and metastasis, functioning as disruptors of tissue barriers, as liberators of ECM-bound growth factors, and/or as sheddases of cell-surface-tethered growth factors, receptors, and ligands (Black and White, 1998; Kheradmand and Werb, 2002; McCawley and Matrisian, 2001). The diverse capacities of cell-surface proteases are exemplified by the diverse activities exhibited by membrane-type matrix metalloproteinase-1, MT1-MMP. Since its discovery as a cell surface metalloproteinase expressed on carcinoma cells and capable of tethering and activating pro-MMP-2, MT1-MMP (Sato *et al.,* 1994) has been found capable of proteolyzing gelatin, fibronectin, vitronectin, collagen, aggregan, and fibrin (McCawley and Matrisian, 2001) and has been associated with an invasive phenotype (Ellenrieder *et al.,* 2000). MT1-MMP also has been found to interact with and cleave the cell-surface hyaluronic acid- and MMP-9-binding protein CD44 and cell-surface-bound tissue transglutaminase (Belkin *et al.,* 2001; Kajita *et al.,* 2001). In these instances, too, the MT1-MMP mediated cleavages were found to affect cell migration, with CD44 cleavage associated with increased cell migration and tissue transglutaminase cleavage associated with decreased tumor cell migration.

The recently described CD44-MMP-9–hyaluronic acid cell surface complex is another example of a biologically relevant proteolytic activity found tethered to cell surfaces. Investigators have shown that in mouse mammary carcinoma and human melanoma cells, CD44-associated cell surface MMP-9 mediates tumor-cell type IV

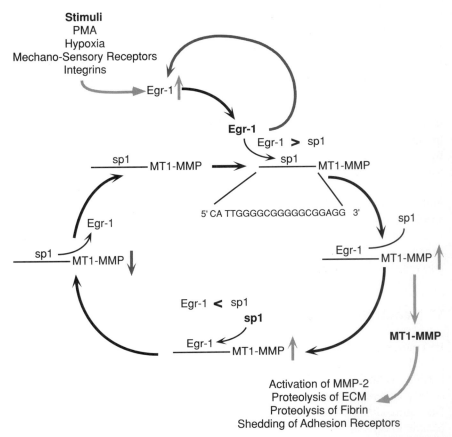

Figure 4 Schematic depicting the complex roles of Sp1 and Egr-1 transcription factors in modulating the baseline and stimulated induction of MT1-MMP. In several *in vitro* and *in vivo* settings, induction of Egr-1 expression by a number of stimuli will cause the displacement of Sp1 from the MT1-MMP promoter, leading to a significantly greater rate of transcription of MT1-MMP. This, in turn, leads to an activation of pro-MMP-2 and selective proteolysis of ECM components and shedding of adhesion receptors. Interestingly, shear stress was found to inhibit MT1-MMP transcription in spite of Egr-1 up-regulation, indicating that Egr-1 is *not* a key transcriptional factor in this circumstance in capillary endothelial cells. During shear stress Sp1 is phosphorylated, increasing its affinity for the MT1-MMP promoter, thus preventing its displacement by Egr-1 and maintaining MT1-MMP expression at baseline levels. Thus, regulation of MT1-MMP in endothelial cells appears to be complex, involving differential induction of specific transcription factors (Sp1 and Egr-1) as well as their modification (phosphorylation of Sp1). This multilevel regulation in addition to differential induction following engagement of specific cell-matrix adhesion molecules (beta 1 integrins) and cell–cell adhesion molecules (cadherins and Ig family members) allows for a tightly controlled expression of this cell surface protease. Based on data from Haas *et al.* (1998, 1999, 2000); Yamaguchi *et al.* (2002); Yun *et al.* (2002).

collagen degradation and tumor-cell invasion of myoblast monolayers *in vitro*. Further, hyaluronic acid was found to induce clustering of CD44/MMP-9 complexes and disruption of cluster formation inhibited tumor invasiveness *in vivo* (Yu and Stamenkovic, 1999). The CD44/MMP-9 cell surface complex has also been implicated in the activation of TGFβ1, a growth factor crucial in the regulation of tissue remodeling, angiogenesis, and tumor invasion (Yu and Stamenkovic, 2000). This and other studies (Kajita *et al.*, 2001) illustrating CD44/MMP-9/hyaluronic acid interactions, and MT1-MMP/MMP-2 interactions, and MT1-MMP/CD44 and MT1-MMP/tissue transglutaminase interactions are indicative of the diverse range of mechanisms utilized by cells to tether, activate, and cluster a variety of proteolytic activities on their surfaces for a variety of different purposes (Black and White, 1998; Bourguignon *et al.*, 1998; Ellenrieder *et al.*, 2000; Kheradmand and Werb, 2002; McCawley and Matrisian, 2001; Sounni *et al.*, 2002).

VI. Conclusions

The findings discussed in this chapter are illustrative of the intense interest and broad range of investigations currently underway in the areas of cell surface protease biology and pathobiology. Future investigations utilizing transgenic, knockout, and knockin technologies as well as cDNA, protein, and tissue arrays will likely expand our knowledge of the numbers and kinds of cell-surface proteases and protease complexes and lead to a more complete understanding of the importance of these moieties in development, homeostasis, adaptation, and disease.

Acknowledgments

I thank all the postdoctoral fellows, students, and technicians who have contributed to the laboratory over the years, especially Anne Romanic, Tara Haas, Donnasue Graesser, Neta Ilan, Sandra Davis, Adeline Tucker, and Anna Soloweij. This work was supported by USPHS grants R01-HL-51018, R37-HL-28373, and P01-DK-38979.

References

Albelda, S., Muller, W., Buck, C., and Newman, P. (1991). Molecular and cellular properties of PECAM-1 (endoCAM/CD31): a novel vascular cell–cell adhesion molecule. *J. Cell Biol.* **114,** 1059–1068.

Albelda, S., Oliver, P., Romer, L., and Buck, C. (1990). EndoCAM: a novel endothelial cell-cell adhesion molecule. *J. Cell Biol.* **110,** 1227–1237.

Anthony, D. C., Miller, K. M., Fearn, S., Townsend, M. J., Opdenaker, G., Wells, G. M., Clements, J. M., Chandler, S., Gearing, A. J., and Pery, V. H. (1998). Matrix metalloproteinase expression in an experimentallyinduced DTH model of multiple sclerosis in the rat CNS. *J. Neuroimmunol.* **87,** 62–72.

Asakura, M., Kitakaze, M., Takashima, S., Liao, Y., Ishikura, F., Yoshinaka, T., Ohmoto, H., Node, K., Yoshino, K., Ishiguro, H., Asanuma, H., Sanada, S., Matsumura, Y., Takeda, H., Beppu, S., Tada, M., Hori, M., and Higashiyama, S. (2002). Cardiac hypertrophy is inhibited by antagonism of ADAM12 processing of HB-EGF: metalloproteinase inhibitors as a new therapy. *Nat. Med.* **8,** 35–40.

Ashman, L., and Aylett, G. (1991). Expression of CD31 epitopes on human lymphocytes: CD31 monoclonal antibodies differentiate between naive (CD45RA+) and memory (CD45RA-) CD4-positive T cells. *Tiss. Ant.* **38,** 208–212.

Azeh, I., Mader, M., Smirnov, A., Beuche, W., Nau, R., and Weber, F. (1998). Experimental pneumococcal meningitis in rabbits: the increase of matrix metalloproteinase-9 in cerebrospinal fluid correlates with leukocyte invasion. *Neurosci. Lett.* **256,** 127–130.

Baron, J., Madri, J., Ruddle, N., Hashinm, G., and Janeway, C. (1993). Surface expression of alpha4 integrin by CD4 T cells is required for their entry into brain parenchyma. *J. Exp. Med.* **177,** 57–68.

Belkin, A., Akimov, S., Zaritskaya, L., Ratnikov, B., Deryugina, E., and Strongin, A. (2001). Matrix-dependent proteolysis of surface transglutaminase by membrane-type metalloproteinase regulates cancer cell adhesion and locomotion. *J. Biol. Chem.* **276,** 18415–18422.

Bell, L., and Madri, J. (1989). Effect of platelet factors on migration of cultured bovine aortic endothelial and smooth muscle cells. *Circ. Res.* **65,** 1057–1065.

Bell, L., and Madri, J. (1990). Influence of the angiotensin system on endothelial and smooth muscle cell migration. *Am. J. Pathol.* **137,** 7–12.

Bell, L., Luthringer, D., Madri, J., and Warren, S. (1992). Autocrine angiotensin system regulation of bovine aortic endothelial cell migration and plasminogen activator involves modulation of proto-oncogene pp60c-src expression. *J. Clin. Invest.* **89,** 315–320.

Black, R., and White, J. (1998). ADAMs: focus on the protease domain. *Curr. Opin. Cell Biol.* **10,** 654–659.

Blobel, C. (2000). Remarkable roles of proteolysis on and beyond the cell surface. *Curr. Opin. Cell Biol.* **12,** 606–612.

Bourguignon, L., Gunja-Smith, Z., Iida, N., Zhu, H., Young, L., Muller, W., and Cardiff, R. (1998). CD44v(3,8–10) is involved in cytoskeleton-mediated tumor cell migration and matrix metalloproteinase (MMP-9) association in metastatic breast cancer cells. *J. Cell Physiol.* **176,** 206–215.

Butler, G., Butler, M., Atkinson, S., Will, H., Tamura, T., vanWestrum, S., Crabbe, T., Clements, J., d'Orthos, M., and Murphy, G. (1998). The TIMP-2 membrane type-1 metalloproteinase "receptor" regulates the concentration and efficient activation of progelatinase A: a kinetic study. *J. Biol. Chem.* **273,** 871–880.

Cabrera, N., Diaz-Rodriguez, E., Becker, E., Martin-Zanca, D., and Pandiella, A. (1996). TrkA receptor ectodomain cleavage generates a tyrosine-phosphorylated cell-associated fragment. *J. Cell Biol.* **132,** 427–436.

Cao, J., Sato, H., Takino, T., and Seiki, M. (1995). The C-terminal region of membrane type matrix metalloproteinase is a functional transmembrane domain required for pro-gelatinase A activation. *J. Biol. Chem.* **270,** 801–805.

Chintala, S., Sawaya, R., Gokaslan, Z., and Rao, J. (1996). Modulation of matrix metalloproteinase-2 and invasion in human glioma cells by $\alpha 3\beta 1$ integrin. *Cancer Lett.* **103,** 201–208.

Cotran, R., Kumar, V., and Collins, T. (1999). "Pathologic Basis of Disease." W. B. Saunders Co., Philadelphia.

Davis, L., Oppenheimer-Marks, N., Bednarczyk, J., McIntyre, B., and Lipsky, P. (1990). Fibronectin promotes proliferation of naive and memory T cells by signaling through both the VLA-4 and VLA-5 integrin molecules. *J. Immunol.* **145,** 785–793.

DeLisser, H., Christofidou-Solomidou, M., Strieter, R., Burdick, M., Robinson, C., Wexler, R., Kerr, J., Garlanda, C., Merwin, J., Madri, J., and Albelda, S. (1997). Involvement of endothelial PECAM-1/CD31 in angiogenesis. *Am. J. Pathol.* **151,** 671–677.

DeLisser, H., Yan, H., Newman, P., Muller, W., Buck, C., and Albelda, S. (1993). Platelet/endothelial cell adhesion molecule-1 (CD31)-mediated cellular aggregation involves surface glycosaminoglycans. *J. Biol. Chem.* **268,** 16037–16046.

Ellenrieder, V., Alber, B., Lacher, U., Hendler, S., Menke, A., Boeck, W., Wagner, M., Wilda, M., Friess, H., Buchler, M., Adler, G., and Gress, T. (2000). Role of MT-MMPs and MMP-2 in pancreatic cancer progression. *Int. J. Cancer* **85,** 14–20.

Fawcett, J., Buckley, C., Holness, C., Bird, I., Spragg, J., Saunders, J., Harris, A., and Simmons, D. (1995). Mapping the homotypic binding sites in CD31 and the role of CD31 adhesion in the formation of interendothelial cell contacts. *J. Cell Biol.* **128,** 1229–1241.

Gale, N. W., and Yancopoulos, G. D. (1999). Growth factors acting via endothelial cell-specific receptor tyrosine kinases: VEGFs, Angiopoietins, and ephrins in vascular development. *Genes Dev.* **13,** 1055–1066.

Galko, M., and Tessier-Lavigne, M. (2000). Function of an axonal chemoattractant modulated by metalloprotease activity. *Science* **289,** 1365–1367.

Gijbels, K., Galardy, R., and Steinman, L. (1994). Reversal of experimental autoimmune encephalomyeltis with a hydroxmate inhibitor of matrix metalloproteases. *J. Clin. Invest.* **94,** 2177–2182.

Gilles, C., Bassuk, J., Pulyaeva, H., Sage, E., Foidart, J., and Thompson, E. (1998). SPARC/osteonectin induces metrix metalloproteinase-2 activation in human breast cancer cell lines. *Cancer Res.* **58,** 5529–5536.

Goyert, S., Ferrero, E., Seremetis, S., Winchester, R., Silver, J., and Mattison, A. (1986). Biochemistry and expression of myelomonocytic antigens. *J. Immunol.* **137,** 3909–3914.

Graesser, D., Mahooti, S., Haas, T., Davis, S., Clark, R., and Madri, J. (1998). The interrelationship of alpha4 integrin and matrix metalloproteinase-2 in the pathogenesis of experimental autoimmune encephalomyelitis. *Lab. Invest.* **78,** 1445–1458.

Graesser, D., Mahooti, S., and Madri, J. (2000). Distinct roles for matrix metalloproteinase-2 and alpha4 integrin in autoimmune T cell extravasation and residency in brain parenchyma during experimental autoimmune encephalomyelitis. *J. Neuroimmunol.* **109,** 121–131.

Graesser, D., Solowiej, A., Bruckner, M., Osterweil, E., Juedes, A., Davis, S., Ruddle, N., Engelhardt, B., and Madri, J. (2002). Altered vascular permeability and early onset of experimental autoimmune encephalomyelitis in PECAM-1-deficient mice. *J. Clin. Invest.* **109,** 383–392.

Haas, T., Davis, S., and Madri, J. (1998). Three-dimensional type I collagen lattices induce coordinate expression of matrix metalloproteinases MT1-MMP and MMP-2 in microvascular endothelial cells. *J. Biol. Chem.* **273,** 3604–3610.

Haas, T., Milkiewicz, M., Davis, S., Zhou, A., Egginton, S., Brown, M., Madri, J., and Hudlicka, O. (2000). Matrix metalloproteinase activity is required for activity-induced angiogenesis in rat skeletal muscle. *Am. J. Physiol. Heart Circ. Physiol.* **279,** H1540–H1547.

Haas, T., Stitelman, D., Davis, S., Apte, S., and Madri, J. (1999). Egr-1 mediates extracellular matrix-driven transcription of membrane type 1 matrix metalloproteinase in endothelium. *J. Biol. Chem.* **274,** 22679–22685.

Hattori, M., Osterfield, M., and Flanagan, J. (2000). Regulated cleavage of a contact-mediated axon repellent. *Science* **289,** 1360–1365.

Hewson, A., Smith, T., Leonard, J., and Cuzner, M. (1995). Suppression of experimental allergic encephalomyelitis in the Lew rat by the matrix metalloproteinase inhibitor RO31-9790. *Inflamm. Res.* **44,** 345–349.

Hudlicka, O. (1998). Growth of arterioles in chronically stimulated adult rat skeletal muscle. *Microcirculation* **5,** 7–23.

Hynes, R. (1992). Integrins: versitality, modulation, and signaling in cell adhesion. *Cell* **69,** 11–25.

Ilan, N., and Madri, J. (1999). New paradigms of signaling in the vasculature: ephrins and metalloproteases. *Curr. Opin. Biotechnol.* **10,** 536–540.

Ilan, N., Cheung, L., Miller, S., Mohsenin, A., Tucker, A., and Madri, J. (2001). PECAM-1 is a modulator of STAT family member phosphorylation and localization: lessons from a transgenic mouse. *Dev. Biol.* **232,** 219–232.

Ilan, N., Cheung, L., Pinter, E., and Madri, J. (2000). PECAM-1 (CD31): A scaffolding molecule for selected catenin family members whose binding is mediated by different tyrosine and serine/threonine phosphorylation. *J. Biol. Chem.* **275,** 21435–21443.

Ilan, N., Mahooti, S., Rimm, D., and Madri, J. (1999). PECAM-1 (CD31) functions as a reservoir for and a modulator of tyrosine-phosphorylated β-catenin. *J. Cell Sci.* **112,** 3005–3014.

Ilan, N., Mohsenin, A., Cheung, L., and Madri, J. (2001a). PECAM-1 shedding during apoptosis generates a membrane-anchored truncated molecule with unique signaling characteristics. *FASEB J.* **15,** 362–372.

Janeway, C., and Travers, P. (1997). "Immunobiology—The Immune System in Health and Disease." Current Biology Limited, London & Garland Publishing Inc., New York.

Kajita, M., Itoh, Y., Chiba, T., Mori, H., Okada, A., Kinoh, H., and Seiki, M. (2001). Membrane-type 1 matrix metalloproteinase cleaves CD44 and promotes cell migration. *J. Cell Biol.* **153,** 893–904.

Kaushal, G., and Shah, S. (2000). The new kids on the block: ADAMTSs, potentially multifunctional metalloproteinases of the ADAM family. *J. Clin. Invest.* **105,** 1335–1337.

Kheradmand, F., and Werb, Z. (2002). Shedding light on sheddases: role in growth and development. *BioEssays* **24,** 8–12.

Kinoshita, T., Sato, H., Okada, A., Ohuchi, E., Imain, K., Okada, Y., and Seiki, M. (1998). TIMP-2 promotes activation of progelatinase A by membrane-type 2 matrix-metalloproteinase immobilized on agarose beads. *J. Biol. Chem.* **273,** 16098–16103.

Kolb, S., Lahrtz, F., Paul, R., Leppert, D., Nadal, D., Pfistet, H., and Fontana, A. (1998). Matrix metalloproteinases and tissue inhibitors of metalloproteinases in viral meningitis: upregulation of MMP-9 and TIMP-1 in cerebrospinal fluid. *J. Neuroimmunol.* **84,** 143–150.

Kubota, S., Ito, H., Ishibashi, Y., and Seyama, Y. (1997). Anti-α3 integrin antibody induces the activated form of matrix metalloproteinase-2 (MMP-2) with concomitant stimulation of invasion through matrigel by human rhabdomyosarcoma cells. *Int. J. Cancer* **70,** 106–111.

Lemjabbar, H., and Basbaum, C. (2002). Platelet-activating factor receptor and ADAM10 mediate responses to *Staphylococcus aureus* in epithelial cells. *Nat. Med.* **8,** 41–46.

Liao, F., Schenkel, A., and Muller, W. (1999). Transgenic mice expressing different levels of soluble platelet/endothelial cell adhesion molecule-IgG display distinct inflammatory phenotypes. *J. Immunol.* **163,** 5640–5648.

Losy, J., Niezgoda, A., and Wender, M. (1999). Increased serum levels of soluble PECAM-1 in multiple sclerosis patients with brain gadolinium-enhancing lesions. *J. Neuroimmunol.* **99,** 169–172.

Lu, T., Barreuther, M., Davis, S., and Madri, J. (1997). Platelet endothelial cell adhesion molecule-1 is phosphorylatable by c-Src, binds Src-Src homology 2 domain, and exhibits immunoreceptor tyrosine-based activation motif-like properties. *J. Biol. Chem.* **272,** 14442–14446.

Lu, T., Yan, L., and Madri, J. (1996). Integrin engagement mediates tyrosine dephosphorylation on platelet-endothelial cell adhesion molecule 1. *Proc. Natl. Acad. Sci. USA* **93,** 11808–11813.

Madri, J., and Graesser, D. (2000). Cell migration in the immune system: the evolving inter-related roles of adhesion molecules and proteinases. *Dev. Immunol.* **7,** 103–116.

McCarthy, M., Burrows, R., Bell, S., Christie, G., Bell, P., and Brindle, N. (1999). Potential roles of metalloprotease mediated ectodomain cleavage in signalling by the endothelial receptor tyrosine kinase Tie-1. *Lab. Invest.* **79,** 889–896.

McCawley, L., and Matrisian, L. (2001). Matrix metalloproteinases: they're not just for matrix anymore. *Curr. Opin. Cell Biol.* **13,** 534–540.

Minagar, A., Jy, W., Jimenez, J., Sheremata, W., Mauro, L., Mao, W., Horstman, L., and Ahn, Y. (2001). Elevated plasma endothelial microparticles in multiple sclerosis. *Neurology* **56,** 1319–1324.

Muller, W., Berman, M., Newman, P., DeLisser, H., and Albelda, S. (1992). A heterophilic adhesion mechanism for platelet/endothelial cell adhesion molecule-1. *J. Exp. Med.* **175,** 1401–1404.

Muller, W., Ratti, C., McDonnell, S., and Cohn, Z. (1989). A human endothelial cell-restricted, externally disposed plasmolemmal protein enriched in intercellular junctions. *J. Exp. Med.* **170,** 399–414.

Nakahara, H., Howard, L., Thompson, E., Sato, H., Seiki, M., Yeh, Y., and Chen, W. (1997). Transmembrane/cytoplasmic domain-mediated membrane type-1 matrix metalloprotease docking to invadopodia is required for cell invasion. *Proc. Natl. Acad. Sci. USA* **94,** 7959–7964.

Nojima, Y., Humphries, M., Mould, A., Yamada, K., Schlossman, S., and Morimoto, C. (1990). VLA-4 mediates CD3-dependent CD4+ T cell activation via the CS1 alternatively spliced domain of fibronectin. *J. Exp. Med.* **172,** 1185–1192.

Nojima, Y., Rothstein, D., Sugita, K., Schlossman, S., and Morimoto, C. (1992). Ligation of VLA-4 on T cells stimulates tyrosine phosphorylation of a 150 kDa protein. *J. Exp. Med.* **175,** 1045–1053.

Ohto, H., Maeda, H., Shibata, Y., Chen, R., Ozaki, Y., Higashihara, M., Takeuchi, A., and Tohyama, H. (1985). A novel leukocyte differentiation antigen: tow monoclonal antibodies TM2 and TM3 define a 120-kd molecule present on neutrophils, monocytes, platelets, and activated lymphoblasts. *Blood* **66,** 873–881.

Primakoff, P., and Myles, D. (2000). The ADAM gene family-surfqace proteins with adhesion and protease activity. *Trends Genet.* **16,** 83–87.

Remy, L., Bellaton, C., Pourreyron, C., Anderson, W., Pereira, A., Deumortier, J., and Jacquier, M. (1999). Integrins and metalloproteinases: an efficient collaboration in the invasive process. *Bull. Cancer* **86,** 154–158.

Romanic, A., and Madri, J. (1994). The induction of 72-kD gelatinase in T cells upon adhesion to endothelial cells is VCAM-1 dependent. *J. Cell Biol.* **125,** 1165–1178.

Romanic, A., Graesser, D., Visintin, I., Baron, J., Janeway, C., and Madri, J. (1997). T cell adhesion to endothelial cell adhesion molecules and extracellular matrix is modulated upon transendothelial cell migration *Lab. Invest.* **76,** 11–23.

Sato, H., Kinoshita, T., Takino, T., Nakayma, K., and Seiki, M. (1996). Activation of a recombinant membrane type-1 matrix metalloproteinase (MT1-MMP) by furin and its interaction with tissue inhibitor of metalloproteinases (TIMP-2). *FEBS Lett.* **393,** 101–104.

Sato, H., Takino, T., Okada, Y., Cao, J., Shinagawa, A., Yamamoto, E., and Seiki, M. (1994). A matrix metalloproteinase expressed on the surface of invasive tumour cells. *Nature* **370,** 61–65.

Seftor, R., Seftor, E., Gehlsen, K., Stetler-Stevenson, W., Brown, P., Ruoslahti, E., and Hendrix, M. (1992). Role of $\alpha v \beta 3$ integrin in human melanoma cell invasion. *Proc. Nat. Acad. Sci. USA* **89,** 1557–1561.

Seftor, R., Seftor, E., Stetler-Stevenson, W., and Hendrix, M. (1993). The 72 KDa type IV collagenase is modulated via differential expression of $\alpha v \beta 3$ and $\alpha 5 \beta 1$ integrins during human melanoma cell invasion. *Cancer Res.* **53,** 3411–3415.

Serebruany, V., Murugesan, S., Pothula, A., Semaan, H., and Gurbel, P. (1999). Soluble PECAM-1, but not P-selectin, nor osteonectin identify acute myocardial infarction in patients presenting with chest pain. *Cardiology* **91,** 50–55.

Shimuzu, Y., and Shaw, S. (1991). Lymphocyte interactions with extracellular matrix. *FASEB J.* **5,** 2292–2299.

Sounni, N., Baramova, E., Munaut, C., Maquoi, E., Frankenne, F., Foidart, J., and Noel, A. (2002). Expression of membrane type 1 matrix metalloproteinase (MT1-MMP) in A2058 melanoma cells is associated with MMP-2 activation and increased tumor growth and vascularization. *Int. J. Cancer* **98,** 23–28.

Springer, T. (1995). Traffic signals on endothelium for lymphocyte recirculation and leukocyte emigration. *Annu. Rev. Physiol.* **57,** 827–872.

Springer, T. (1995a). Signals on endothelium for lymphocyte recirculation and leukocyte emigration: the area code paradigm. *Harvey Lect. Series* **89,** 53–103.

Stetler-Stevenson, M., Mansoor, A., Lim, M., Fukushima, P., Kehrl, J., Marti, G., Ptaszynski, K., Wang, J., and Stetler-Stevenson, W. (1997). Expression of matrix metalloproteinases and tissue inhibitors of metalloproteinases in reactive and neoplastic lymphoid cells. *Blood* **89**, 1708–1715.

Stockinger, S., Gadd, S., Ehere, R., Majdic, O., Schreiber, W., Kasinrerk, W., Strass, B., Schnabl, E., and Knapp, W. (1990). Molecular characterization and functional analysis of the leukocyte surface protein CD31. *J. Immunol.* **145**, 3889–3897.

Strongin, A., Collier, I., Bannikov, G., Marmer, B., Grant, G., and Goldbert, G. (1995). Mechanism of cell surface activation of 72-kDa type IV collagenase. *J. Biol. Chem* **10**, 5331–5338.

Thompson, E., Yu, M., Bueono, J., Jin, L., Maiti, S., Palao-Marco, F., Pulyaeva, H., Tamborlane, J., Tirgari, R., and Wapnir, R. (1994). Collagen induced MMP-2 activation in human breast cancer. *Breast Cancer Res. Treat.* **31**, 357–370.

Tremble, P., Chiquet-Ehrismann, R., and Werb, Z. (1994). The extracellular matrix ligands fibronectin and tenascin collaborate in regulating collagenase gene expression in fibroblasts *Mol. Biol. Cell* **5**, 439–453.

Tremble, P., Damsky, C., and Werb, Z. (1995). Components of the nuclear signaling cascade that regulate collagenase gene expression in response to integrin-driven signals. *J. Cell Biol.* **129**, 1707–1720.

Vacca, A., Ribatti, D., Iurlaro, M., Albini, A., Minischetti, M., Bussolino, R., Pellegrino, A., Ria, R., Rusnati, M., Presta, M., Vincenti, V., Persico, M., and Damacco, F. (1998). Human lymphoblastoid cells produce extracellular matrix-degrading enzymes and induce endothelial cell proliferation, migration, morphogenesis, and angiogenesis. *Int. J. Clin. Lab. Res.* **28**, 55–68.

Vecchi, M., and Carpenter, G. (1997). Constitutive proteolysis of the ErbB-4 receptor tyrosine kinase by a unique sequential mechanism. *J. Cell Biol.* **139**, 995–1003.

Vecchi, M., Baulida, J., and Carpenter, G. (1996). Selective cleavage of the heregulin receptor ErbB-4 by protein kinase C activation. *J. Biol. Chem.* **271**, 18989–18995.

Wang, H., Chen, Z.-F., and Anderson, D. (1998). Molecular distinction and angiogenic interaction between embryonic arteries and veins revealed by ephrin-B2 and its receptor Eph-B4. *Cell* **93**, 741–753.

Yabkowitz, R., Meyer, S., Black, T., Elliot, G., Merewether, L.A., and Yamane, H.K. (1999). Inflammatory cytokines and vascular endothelial growth factor stimulate the release of soluble tie receptor from human endothelial cells via metalloprotease activation. *Blood* **93**, 1969–1979.

Yabkowitz, R., Meyer, S., Yanagihara, D., Brankow, D., Staley, T., Elliot, G., Hu, S., and Ratzkin, B. (1997). Regulation of tie receptor expression on human endothelial cells by protein kinase C-mediated release of soluble tie. *Blood* **90**, 706–715.

Yamada, A., Nikaido, T., Nojima, Y., Schlossman, S., and Morimoto, C. (1991). Activation of human CD4 T lymphocytes: interaction of fibronectin with VLA-5 receptor on CD4 cells induces the AP-1 transcription factor. *J. Immunol.* **146**, 53–56.

Yamaguchi, S., Yamaguchi, M., Yatsuyanagi, E., Yun, S.-S., Nakajima, N., Madri, J., and Sumpio, B. (2002). Cyclic strain stimulates Egr-1-mediated expression of MT1-MMP in endothelium. *Lab. Invest.* **82**, 949–956.

Yancopoulos, G., Klagsbrun, M., and Folkman, J. (1998). Vasculogenesis, angiogenesis and growth factors: Ephrins enter the fray at the border. *Cell* **93**, 661–664.

Yu, Q., and Stamenkovic, I. (1999). Localization of matrix metalloproteinase 9 to the cell surface provides a mechanism for CD44-mediated tumor invasion. *Genes Dev.* **13**, 35–48.

Yu, Q., and Stamenkovic, I. (2000). Cell surface-localized matrix metalloproteinase-9 proteolytically activates TGF-beta and promotes tumor invasion and angiogenesis. *Genes Dev.* **14**, 163–176.

Yun, S., Dardik, A., Haga, M., Yamashita, A., Yamaguchi, S., Koh, Y., Madri, J. A., and Sumpio, B. E. (2002). Shear stress stimulates phosphorylation of the Sp1 transcription factor in microvascular endothelium. *J. Biol. Chem.* **277**, 34808–34814.

14

Shed Membrane Vesicles and Clustering of Membrane-Bound Proteolytic Enzymes

M. Letizia Vittorelli
Dipartimento di Biologia cellulare e dello Sviluppo
Viale delle Scienze, Parco D'Orleans II
90128-Palermo, Italy

I. Introduction

Eukaryotic cells appear to release into the extracellular medium several populations of exovesicles which are suggested to have different origins and functions and are identified by different names. This report specifically deals with vesicles believed to originate from the cell membrane and named membrane vesicles. These, in fact, are structures in which membrane-bound proteolytic enzymes are clustered and are suggested to play important roles in matrix remodeling. This report will begin with an introductory description of the specific characteristics reported for each exovesicle population.

A. Membrane Vesicles

The first described population of exovesicles appears to originate directly from the plasma membrane with a mechanism apparently similar to that of viral budding

Current Topics in Developmental Biology, Vol. 54

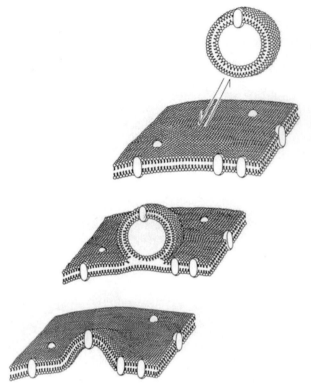

Figure 1 A schematic view of membrane vesicle release. From Dainiak, N. (1991). Surface membrane-associated regulation of cell assembly, differentiation, and growth. *Blood* **78,** 264–276 (Fig. 2). Copyright American Society of Hematology, used by permission.

(Fig. 1). Freeze-fracture electron micrographs (Fig. 2) and electron-microscopic analyses (Fig. 3) show that vesicles shed by circulating human lymphocytes (Dainiak, 1991) and by *in vitro* cultured human breast carcinoma cells (Dolo *et al.,* 1994) are relatively large and heterogeneous in respect to overall size; their diameters range from about 100 to more than 1000 nm. Shed vesicles belonging to this population are produced by several kinds of cells; they are also called "exfoliated microvesicles" (Trams *et al.,* 1981), "microvesicles" (Heijnen *et al.,* 1999), and "matrix vesicles" (the name used specifically for vesicles shed by hypertrophic chondriocytes). Most reports concern their role in matrix remodeling, cell migration, and tumor progression mechanisms. In several instances, however, they are also known to mediate cell–cell interactions.

B. Exosomes

Exovesicles having a smaller diameter of 40–80 nm are called exosomes. These particles are reported to be released by exocytosis from cellular compartments

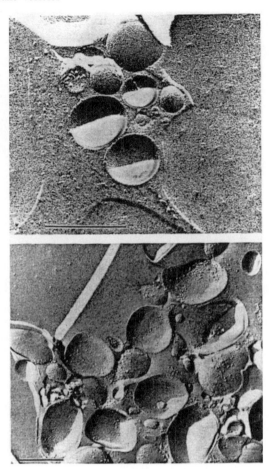

Figure 2 Freeze-fracture electronmicrographs of extracellular vesicles shed from the surface of circulating human lymphocytes. Bars, 0.25 μm. From: Dainiak, N. (1991). Surface membrane-associated regulation of cell assembly, differentiation, and growth. *Blood* **78,** 264–276 (Fig. 3). Copyright American Society of Hematology, used by permission.

displaying intraluminal membrane-bound vesicles which are collectively named multivesicular bodies (MVBs). They are part of the endosomal system which consists of early endosomes, late endosomes, and lysosomes. Early endosomes are the major entry site for endocytosed material, whereas late endosomes are thought to receive newly synthesized lysosomal hydrolases directly from the trans-Golgi network. The distinction between late endosomes and lysosomes, however, is not very sharp (Geuze, 1998). MVBs are considered to belong to the late endosomes/ lysosomal category and are probably formed by inward budding of the vesicle membrane (Fig. 4), which creates a membrane-enclosed compartment in which the lumen is topologically equivalent to the cytoplasm (Gruenberg and Maxfield,

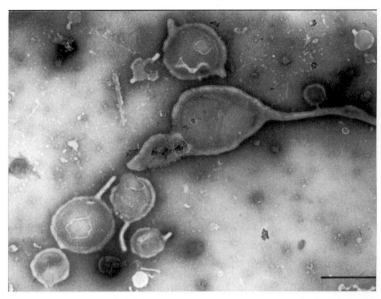

Figure 3 Isolated vesicles observed by electron microscopy with negative staining. Bar, 0.5 μm. From: Dolo, V., Pizzurro, P., Ginestra, A., and Vittorelli, M. L. (1995b). Inhibitory effects of vesicles shed by human breast carcinoma cells on lymphocyte ^{3}H-thymidine incorporation, are neutralised by anti TGF-beta antibodies. *J. Submicrosc. Cytol. Pathol.* **27,** 535–541 (Fig. 2). Copyright *J. Submicrosc. Cytol. Pathol.* Editorial Office, used by permission.

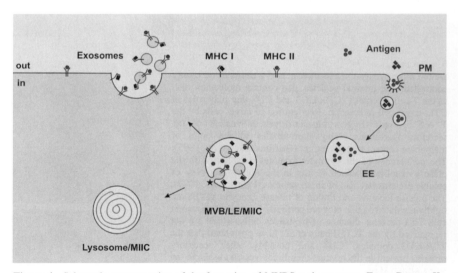

Figure 4 Schematic representation of the formation of MVBS and exosomes. From: Denzer, K., Kleijmeer, M. J., Heijnen, H. F., Stoorvogel, W., and Geuze, H. J. (2000a). Exosome: from internal vesicle of the multivesicular body to intercellular signaling device. *J. Cell Sci.* **113 Pt19,** 3365–3374 (Fig. 1). Copyright The Company of Biologists LTD, used by permission.

1995; Denzer *et al.,* 2000a). MVBs seem to follow two distinct pathways: (1) they fuse with lysosomes, which results in the degradation of their contents, or (2) they fuse with the plasma membrane, which results in the exocytosis of the internal vesicles into extracellular space. The possibility that these processes involve distinct categories of MVBs cannot be excluded, and the final destination of MVBs might be determined by the interaction of different SNARE proteins (Luzio *et al.,* 2000).

The release pathway of exosomes was described almost 20 years ago in differentiating red blood cells (Pan and Johnstone, 1983), but more recently its presence has been reported in most cells of the hemopoietic lineage where exosomes appear to play important roles in cell–cell signaling events (Denzer *et al.,* 2000a). Exosomes produced by professional antigen presenting cells (APCs), such as B cells (Raposo *et al.,* 1996) and dendritic cells (Zitvogel *et al.,* 1998), have been found to be rich in MHC class I and class II molecules. The majority of intracellular MHC class II molecules reside in late endocytic compartments and lysosomes. These endocytic compartments are therefore collectively named MHC class II enriched compartments (MIICs). MIICs are MVB-like; it is therefore suggested that exosomes derive from MIICs compartments. Internal vesicles of MIICs and exosomes are enriched in members of the tetraspan protein superfamily such as CD37, CD53, CD63, CD81, CD82 (Escola *et al.,* 1998), and CD9 (Thery *et al.,* 1999). These proteins are able to form oligomeric complexes with other tetraspan proteins and may function cooperatively with each other in complex formation with cell-surface proteins. They also can interact with integrins and with T-cell coreceptors CD4 and CD8. In B lymphocytes, tetraspan proteins and costimulatory molecule CD86 (B7.2) are confined to the internal vesicles of MIICs and are enriched in exosomes (Escola *et al.,* 1998). Exosomes therefore may have the potential to bind and excite signaling at a distance from the site of their production, carrying exogenously derived peptides bound to MHC class II molecules which can simulate CD4$^+$ T cell proliferation *in vitro*. Other MIIC markers, such as Lamp-1, Lamp-2, and HLA-DM, reside primarily in MIIC-limiting membrane and are absent from exosomes (Escola *et al.,* 1998).

Exosomes play important roles in eliciting immune responses. Cytolytic T cells produce exosomes that deliver killing reagents specifically to their targets (Peters *et al.,* 1989, 1990). Antigen presenting cells release exosomes which carry peptide-MHC molecules and are capable of activating specific subsets of T cells *in vitro* (Raposo *et al.,* 1996; Zitvogel *et al.,* 1998). Exosomes released by B lymphocytes bind specifically to follicular dendritic cells (FDCs) which are accessory cells of the immune system essential for germinal center reaction. FDCs play an important role in the presentation of processed antigens bound to MHC class II molecules to specific T lymphocytes. FDCs do not express MHC class II molecules; they are only able to acquire them by passively binding to exosomes released by B lymphocytes and possibly by other cells of the immune system (Denzer *et al.,* 2000b).

A potential use for exosomes in immunotherapy has been recently demonstrated. Murine dendritic cells (DC) pulsed with tumor-specific peptides were administrated to mice with established P815 tumors. At day 60, approximately 20% of the mice were tumor-free. The administration of exosomes derived from DC resulted in a more effective response with tumor eradication in 60% of mice. The tumor clearance was a MHC class I restricted response, as specific CD8$^+$ T lymphocytes were detected in the spleen of cured animals (Zitvogel et al., 1998).

C. Argosomes

Very recently a third kind of exovesicles, called argosomes, was described in *Drosophila* embryos. These vesicles are thought to be involved in the release and spreading of membrane-bound morphogens. In developing tissues, a pattern forms in response to the graded distribution of signaling molecules (morphogens) that modulate the transcriptional activity of responding cells in a concentration-dependent manner. Morphogens specify cell fates; therefore, during development their concentrations are carefully controlled by both morphogen producing cells and the tissue through which they travel. Despite their ability to travel long distances, many morphogens are tightly associated with plasma membranes. Morphogen spreading was thought to be dependent upon their release from the cell membrane; however, it was frequently impossible to detect free morphogen protein throughout the entire range over which they can activate cellular responses. Wingless (Wg) *Drosophila* membrane-bound morphogen, needed for wing development and homolog to mammalian Wnt, in normal embryos was detected by immunostaining inside vesicles two to three cells away from its source. In embryos that are endocytosis defective, the morphogen was found to be restricted to the synthesizing cells (Dierick and Bejsovec, 1999; Christian, 2000). Argosomes seem to be responsible for the spreading of this morphogen (Greco et al., 2001). The Wg protein, in fact, colocalizes with argosomes, produced by the morphogen expressing cells. Argosomes derived from the basolateral membrane of the *Drosophila* imaginal disk epithelium and which travel through adjacent tissue were found in specific areas.

Argosomes are thought to represent a new kind of exovesicle. They are found in endosomes and are suggested to move in tissues traveling prevalently in endosomes; shed vesicles and exosomes, in contrast, are released in the extracellular medium. As shown by Fig. 5, two models are proposed to explain argosome production. Model 1 suggests that argosomes, like exosomes, are generated from multivesicular endosomes. The observation that Wg protein is found on the plasma membranes of the cells that synthesize it and also within the cells MVBs validates this hypothesis (van den Heuvel et al., 1989). Model 2, on the contrary, suggests that exosomes are formed by endocytosis of entire regions of the plasma membrane by a neighboring cell. Long range movements of argosomes would therefore require the subsequent release of these endocytosed vesicles. This second mechanism has

Model 1   ## Model 2

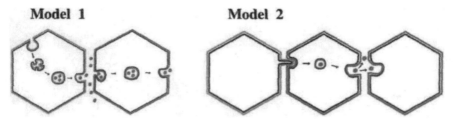

Figure 5 Two models for the generation of argosomes. From: Greco, V., Hannus, M., and Eaton, S. (2001). Argosomes: A potential vehicle for the spread of morphogens through epithelia. *Cell* **106,** 633–645 (Fig. 4 C). Copyright *Cell* Editorial Office, used by permission.

been suggested to occur in *Drosophila* where endocytosis of transmembrane ligands such as Notch occurs (Klueg and Muskavitch, 1999); it also seems to occur between MDCK epithelial cells and to be exploited by *Listeria* as it spreads through the epithelium (Robbins *et al.,* 1999).

II. Membrane Vesicle Release

Shedding of relatively large membrane vesicles originating from the plasma membrane was first described by Tarin (1967, 1969). Tarin reported that shed membrane vesicles were released *in vivo* by murine breast carcinoma cells and bound to collagen fibers in areas where fibril degradation was observed. Around this time, shed vesicles were also observed in the cartilage matrix and were suggested to play a role in matrix mineralization (Anderson, 1969). Some years later, vesicle shedding was reported to occur in *in vitro* cultured leukemic and melanoma cells and other tumor cells such as breast, colon, and ovary carcinomas and glioma and fibrosarcoma cells. Vesicle shedding was also observed with highly motile nontumor cells, including embryonic cells, activated lymphocytes, and macrophages (Taylor and Black, 1986; Dainiak, 1991), and in primary cultures of avian chondrocytes (Hale and Wuthier, 1987). Vesicle shedding occurs *in vitro* at measurable rates in all tumor cells analyzed thus far, in primary cultures of human fibroblast (Ginestra and Vittorelli, unpublished results), and of murine fetal cerebral neurons (Ginestra and Cestelli, personal communication). However, shedding is not detected in MCF-10A, an immortalized, nontransformed cell line of epithelial origin (Ginestra *et al.,* 1998).

Shedding is an energy-requiring, vital phenomenon, for which the synthesis of RNAs and proteins is essential (Dainiak and Sorba, 1991). Characterization of the shed vesicles does not provide evidence for the presence within the vesicle of cytosolic proteins, or of nuclear fragments similar to those observed in apoptotic cells (Taylor *et al.,* 1983; Albanese *et al.,* 1998). Vesicle membranes carry most of the cell-surface antigens expressed by the producing cells (Dolo *et al.,* 1995a), but have a specific composition for both membrane lipid (van Blitterswijk *et al.*, 1982; Armstrong *et al.*, 1988) gangliosides (Dolo *et al.*, 2000) and protein components (Trams *et al.*, 1981; Lerner *et al.*, 1983; Dolo *et al.*, 1998). Some molecules clustered in membranes of shed vesicles appear to be selected for specific vesicle function (Dolo *et al.*, 1998).

Shedding mechanisms are not well understood. Among the proposed mechanisms, disruption or remodeling of cytoskeletal components received the most extensive attention. Liepins (1983) demonstrated that shedding of membrane vesicles by P815 tumor cells could be induced by low temperature, colchicines, and vinblastine, all of which disrupt microtubules. Contractile proteins were reported to participate in vesicle shedding (Dainiak *et al.*, 1988). We also observed that membrane vesicle shedding is stimulated by concanavalin A and by wheat germ agglutinin (Cassarà *et al.*, unpublished results). Since concanavalin A inhibits endocytosis of cell surface proteins, these observations link the shedding process with this pathway. In breast carcinoma cells, shedding is inhibited by quercetin (Cassarà *et al.*, 1998). Quercetin is a plant flavonoid inhibiting 1-phosphatidylinositol 4-kinase (EC 2.7.1.67, PI kinase) and phosphatidylinositol 4-phosphate 5-kinases (EC 2.7.1.68, PIP kinase) and therefore causes a decrease of inositol 1,4,5-triphosphate (IP_3) concentration, which in turn decreases the release of calcium from intracellular sources (Singhal *et al.*, 1995).

As shown by Fig. 6, shedding occurs with high intensity when subconfluent carcinoma cells are cultured in the presence of fetal calf serum and it is morphologically similar to virus budding. Serum stimulatory effects on the rate of vesicle shedding have been reported in several different cell lines (Chiba *et al.*, 1989; Dolo *et al.*, 1994; Ginestra *et al.*, 1997).

Early reports suggested involvement of enhanced proteolysis or lipolysis at the cell surface in vesicle production (Taylor and Black, 1986). A partial confirmation of proteinases involved in the shedding phenomenon was obtained more recently. We observed that when fetal calf serum is preadsorbed by gelatin–Sephrose affinity chromatography, it partially loses its stimulatory effect on shedding. This loss of activity seems to be due to adsorption of serum metalloproteinases. This conclusion is based on the result that the addition of bathophenantroline, a metalloproteinase inhibitor unable to cross the cell membrane, to non-preadsorbed serum has a similar inhibitory effect on membrane vesicle shedding (Cassara *et al.*, 1998).

In murine melanoma (Taylor *et al.*, 1988) and human breast carcinoma cells (Cassarà *et al.*, 1998), shedding is inhibited by treatment with the differentiating agent retinoic acid. In breast carcinoma cells, shedding is also inhibited by 8-Cl-cAMP, a site selective cAMP analogous (Ally *et al.*, 1988). In contrast, treatments of 8701-BC breast carcinoma cells with inhibitors of intracellular vesicular trafficking such as brefeldin A (Misumi *et al.*, 1986) and monensin (Tartakoff, 1983) do not modify the shedding rate (Cassarà *et al.*, 1998). Methylamine, an inhibitor of exocytosis (Maxfield *et al.*, 1979), has only a moderate inhibitory affect. As

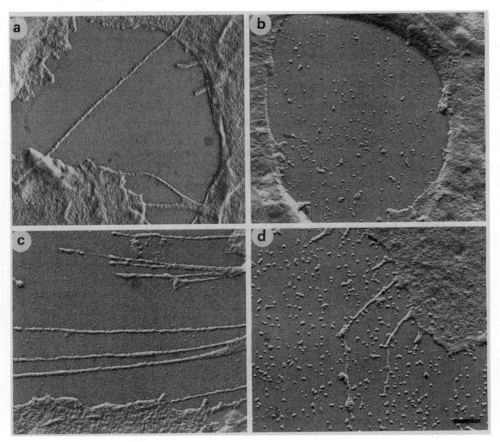

Figure 6   TEM analysis of 8701-BC breast carcinoma cells cultured in serum-free (a and c) and in 10% FCS (b and d) showing release of vesicles occurring in complete media. Bar, 0.1 μm. From: Dolo, V., Ginestra, A., Cassara, D., Violini, S., Lucania, G., Torrisi, M. R., Nagase, H., Canevari, S., Pavan, A., and Vittorelli, M. L. (1998). Selective localization of matrix metalloproteinase 9, beta1 integrins, and human lymphocyte antigen class I molecules on membrane vesicles shed by 8701-BC breast carcinoma cells. *Cancer Res.* **58,** 4468–4474 (Fig. 1). Copyright American Association for Cancer Research, used by permission.

shown by results from several treatments, the rate of membrane vesicle shedding is not directly related to the rate of cell proliferation (Cassarà *et al.,* 1998).

Rab3c, a small GTP-binding protein of the Rab subfamily, was found to be highly expressed by chondrocytes in the hypertrophic zone. Since its pattern of expression corresponds with the genesis of matrix vesicles during endocondrial ossification, Rab3c is suggested to be involved in matrix vesicle trafficking during skeletal development (Pavlos *et al.,* 2001).

III. Role of Shed Vesicles in Cell–Cell Interactions

A. Shed Vesicles and Tumor Escape from Immune Rejection

In several instances, vesicles shed by tumor cells were reported to play a relevant role in tumor escape from immune surveillance. The first indication that vesicles shed by a growing tumor may assist tumor "escape" comes from studies showing that nodes draining a tumor have a large content of shed vesicles and that, while highly stimulated, they are functionally impaired. Moreover, serum of tumor bearing rats was found to inhibit the activity of immune lymph node cytotoxic lymphocytes against the specific tumor. This phenomenon was attributed to the spontaneous shedding of antigen-rich membrane fragments and membrane vesicles by tumor cells (Alexander, 1974). Taylor and Black (1985) later showed that membrane vesicles shed by murine melanoma cells inhibit the expression of MHC class II molecules in macrophages. More recently, we observed that vesicles shed by breast carcinoma cells inhibited proliferation of peripheral blood lymphocytes and that their inhibitory effect was completely abolished by the addition of anti-TGFβ antibodies (Dolo et al., 1995b). In vesicles shed by these breast carcinoma cell lines, we also observed the presence of tumor-associated antigens and of HLA class I molecules (Dolo et al., 1995a). HLA class I molecules were shown to be specifically clustered in areas of the cell membrane which are shed as membrane vesicles (Dolo et al., 1998). Several studies have shown the immunosuppressive potential of TGFβ is an important promoter of malignant cell growth because of the TGFβ induced interference with the generation of tumor-specific cytotoxic T lymphocytes (de Visser and Kast, 1999). TGFβ is in fact the most potent immunosuppressor described to date. We propose that the immunosuppressive potential of TGFβ can be targeted against tumor-specific cytotoxic T lymphocytes by vesicle associated antigens. Subsets of lymphocytes in fact, may adhere to membrane vesicles released from tumor cells by recognizing tumor specific antigens (TSA). Such interactions may hamper activation of anti-TSA cytotoxic T lymphocytes.

TGFβ is, however, a multifunctional cytokine suggested to facilitate tumor progression by several different mechanisms. TGFβ's localization in shed vesicles may have other important consequences in addition to those concerning the cytokine immunosuppressive role. Shed membrane vesicles are rich in matrix metalloproteinases (see Section IV). It is tempting to hypothesize that in shed vesicles the coordinated metalloproteinases and TGFβ functions could provide a physiological mechanism of tissue remodeling adopted by malignant cells to promote tumor growth and invasion through formation of a permissive stroma in peritumoral areas. TGFβ has also emerged as a major modulator of angiogenesis. In ras-transformed mammary epithelial cells, for example, TGFβ was shown to cause the production of vascular endothelial growth factor (VEGF) and VEGF receptor-2, thus activating a signaling system centrally involved in tumor angiogenesis (Breier et al.,

2002). At the same time the permissive stroma created in peritumoral areas could facilitate endothelial cell migration and tumor vascularization.

Other mechanisms based on shedding of membrane-associated molecules could help tumor cells to escape from induced apoptosis. For instance, biologically active Fas antigen was found to be present in vesicles shed by human colorectal carcinoma cell lines. Shedding of the Fas antigen is likely to decrease the susceptibility of tumor cells to Fas ligand (Fas-L) induced apoptosis. Moreover, Fas-bearing vesicles were capable of interacting with Fas-L, efficiently neutralizing its ability to induce apoptosis. In contrast, Fas-L was present in vesicles shed by an activated human T-cell line and Fas-L-bearing vesicles were capable of inducing apoptosis of Fas expressing cells (Albanese *et al.*, 1998). Monleon *et al.* (2001) showed by immunoelectron microscopy that in human T cells, preformed Fas-L and APO2L/TRAIL are stored inside cytoplasmic compartments, approximately 500 nm in diameter, with characteristics of MVBs. Upon PHA activation, the mobilization of these particles toward the plasma membrane resulting in the secretion of the internal microvesicles loaded with FasL and APO2L/TRAIL was evident. The rapid release of these molecules in their bioactive form is involved in T-cell activation-induced cell death.

B. Shed Vesicles and Growth Factor Release

Shed membrane vesicles are suggested to be involved also in other cell–cell signaling mechanisms. Many lymphokines and several other signaling molecules are transmembrane cell-surface proteins. In some instances, shed vesicles are known to transport these regulatory factors. Granulopoietic colony-stimulating activity (CSA) was found to be present in high-speed centrifugation pellets of media conditioned by peripheral blood leukocytes (Rubin and Cowan, 1973). Membrane-bound burst-promoting activity (mBPA) was detected in vesicles shed by B lymphocytes (Feldman *et al.*, 1987), and it was found to be concentrated in shed vesicle membranes compared to different areas of the cell plasma membrane (Guha *et al.*, 1993, 1994). A similar phenomenon was reported for human macrophage colony-stimulating factor (M-CSF) (Tuck *et al.*, 1994).

Vesicles are also involved in the secretion of some molecules found in the extracellular matrix, but devoid of the classical signal sequence which normally allows vectorial transfer of the neosynthesized protein across the membrane of the endoplasmic reticulum. The first documented reports concern Lectin 14 (L-14), a soluble lactose-binding lectin, which interacts with extracellular glycoconjugates. Externalization of L-14, which is a signaling molecule highly expressed in muscle cells, is developmentally regulated. As myogenic cells progressively differentiate, cytosolic L-14 is concentrated in cellular ectoplasm beneath regions of the plasma membrane, which then appear to be evaginated to form extracellular vesicles (Cooper and Barondes, 1990). Mackenzie *et al.* (2001) reported that interleukin 1β

(IL-1β) is also secreted by microvesicle shedding. IL-1β is released from activated immune cells after a secondary stimulus such as extracellular ATP acting on P2X(7) receptors. Human THP-1 monocytes shed vesicles from their plasma membranes within 2–5 s after the activation of P2X(7) receptors; 2 min later the released vesicles contained IL-1β, which only later appeared in the vesicle-free supernatant. Phosphatidylserine translocation preceded microvesicle shedding as measured by FITC-conjugated annexin-V binding to the plasma membrane (Mackenzie *et al.*, 2001). Shedding of membrane vesicles seems to be also involved in secretion of basic fibroblast growth factor. Vesicles shed by Sk-Hep1 human hepatoma cells which spontaneously produce FGF-2, and FGF-2 transfected NIH 3T3 cells, were found to carry the factor and were able to promote chemotaxis and uPA expression in GM7373 endothelial cells. FGF-2 release and vesicle shedding were induced in parallel by serum addition (unpublished results).

In addition to membrane-bound factors and leaderless secretory proteins, other signaling molecules are probably released in shed vesicles. As already mentioned TGFβ was found in vesicles shed by breast carcinoma cells (Dolo *et al.*, 1995b); the factor was also found in vesicles shed by hypertrophic chondrocytes where it appears to be involved in modulating matrix mineralization (D'Angelo *et al.*, 2001; Derfus *et al.*, 2001). Other signaling molecules found in shed vesicles include two ectoprotein kinases, expressed by human leukemic cells on their surfaces; one of them phosphorylates serines and threonines, and the other, which is markedly enhanced by Mn^{2+}, is specific for tyrosines (Paas and Fishelson, 1995).

IV. Role of Shed Vesicles in Cell Matrix Interactions

A. Shed Vesicles and Tumor Invasion

Shed vesicles are thought to play a role in matrix remodeling occurring during cell migration and tumor invasion. Early electron microscopic studies of invasive murine skin and mammary carcinomas *in vivo* showed the destruction of connective tissues by membrane vesicles, in the immediate vicinity of the invading epithelial cells (Tarin, 1967, 1969). Poste and Nicolson (1980) reported that fusion of the membrane vesicles shed from a highly metastatic variant of murine melanoma cells with a poorly metastatic clone increased the invasive potential of the latter.

The ability of membrane vesicles to promote invasive activity is most likely related to the association within vesicle membranes of proteolytic enzymes that participate in extracellular matrix (ECM) degradation. One of the first reports concerning vesicle associated proteolytic enzymes was published by Zucker *et al.* They reported that *in vitro* cultured mouse melanoma cells released a component of membrane vesicles having metalloproteinase activity (M_r 59, 000) (Zucker *et al.*, 1987). The presence of collagenolytic activity was also reported in the vesicles shed by human rectal carcinoma cells (Murayama *et al.*, 1991). We showed that vesicles shed by breast carcinoma cells carry MMP-9 (gelatinase B) (Dolo *et al.*, 1994). More recently, we observed that MMP-9 is clustered in the vesicle membranes but is not easily detected in other areas of the cell plasma membrane. Most of the vesicle membrane-bound MMP-9 is in the proenzyme form, as a component of a high-molecular weight complex in which it appears to be bound to its inhibitor TIMP-1 (Tissue Inhibitor of Metalloproteinases 1) (Dolo *et al.*, 1998). Vesicles shed by HT-1080 fibrosarcoma cells, in addition to MMP-9, carry MMP-2 (gelatinase A) and uPA (urokinase type of plasminogen activator), which is bound to its membrane receptor. The addition of physiological concentrations of plasminogen to vesicles shed in serum-free medium increases ^{3}H type IV collagen degradation by vesicles and converts vesicle-associated progelatinases into their active forms, showing that plasmin, generated by vesicle-bound uPA, activates the MMPs (Ginestra *et al.*, 1997).

The presence of a variety of proteolytic enzymes in the vesicle membrane indicates that these structures are capable of promoting the proteolytic cascade required for the localized degradation of extracellular matrix. Vesicles may provide a large membrane surface for binding of proteinases and thus greatly facilitate their activation. In addition, vesicles are rich in integrins (Dolo *et al.*, 1995a, 1998) and therefore bind specifically to their substrates. Vesicle shedding, therefore, may represent an important mechanism by which tumor cells induce ECM degradation during invasion and metastasis.

Vesicle involvement in tumor invasion and dissemination is also suggested by relationships between metastatic potential and the amount of shed vesicles described in variants of the murine melanoma B-16 cell line. Highly metastatic variants were found to shed more membrane vesicles than did poorly metastatic ones (Taylor *et al.*, 1988). The same relationship between invasiveness measured *in vitro* on a reconstituted basement membrane (Matrigel) substrate and both the amount and the proteolytic content of shed vesicle was also observed comparing MCF-10, a nontumor immortalized cell line derived from human mammary breast epithelium, three human breast carcinoma cell lines having a different invasive potential, and HT-1080 fibrosarcoma cells. MCF-10 cells did not release a measurable amount of vesicles and were completely unable to invade Matrigel (Ginestra *et al.*, 1998).

Urokinase type of plasminogen activator was also found to be associated with vesicles shed by the aggressive prostatic carcinoma cell line PC3 (Angelucci *et al.*, 2000). Vesicles shed by PC3 cells were found to adhere and degrade both collagen IV and Matrigel. The addition of plasminogen enhanced the degradation in a dose-dependent manner. Addition of membrane vesicles shed by PC3 cells to cultures of the poorly invasive prostate cancer cell line LnCaP enhanced the adhesive and invasive capabilities of the latter.

Vesicles shed *in vivo* by tumor cells have been detected in biological fluids such as the pleural effusion of a mammary carcinoma patient and in serum (Dolo *et al.*, 1995a). Large amounts of vesicles were also found in malignant ovarian tumor fluids (Dolo *et al.*, 1999). A positive correlation between tumor malignancy and both vesicle-amount and vesicle-associated MMP-2 activity was noted (Ginestra *et al.,* 1999). These findings suggest that vesicle content in biological fluids could represent a useful marker of tumor aggressiveness and tumor progression.

B. Shed Vesicles and Endothelial Cell Migration

Other studies confirming the role of shed vesicles in matrix remodeling are still in progress. Some analyses concern evaluation of lytic enzymes in vesicles shed by GM7373 cells grown in different conditions. GM7373 is an immortalized line of fetal endothelial cells obtained from bovine aorta. When cultured in nonconfluent conditions these cells acquire an invasive phenotype. Vesicles shed by migrating and by confluent GM7373 cells have been compared for their quantity and for their gelatinolytic pattern. Migrating cells shed approximately twice as much vesicle as confluent ones. Moreover, vesicles shed by migrating cells carry several bands of gelatinolytic activities that were not detected or had a lower concentration in vesicles shed by confluent ones. Proteolytic enzymes, identified in GM7373 vesicles, include MMP-2, MMP-9, and MT1-MMP. The concentration of all three metalloproteinases was found to be increased in vesicles shed by migrating cells. A gelatinolytic band, sensitive to the serine protease inhibitor PMSF and having the typical electrophoretic migration of Seprase, and uPA were found only in vesicles shed by migrating (and not by resting) GM7373 cells. Western blotting with anti-MT1-MMP antibodies showed that gelatinolytic enzymes present in vesicle membranes are associated in large complexes. Digestion assays performed on purified ECM components (collagen types I and IV, laminin, and fibronectin) confirmed that vesicles shed by migrating cells are much more active in promoting matrix digestion as compared to vesicles shed by resting ones (Ghersi *et al.,* manuscript in preparation).

An increased amount of shed vesicle and an increased activity of vesicle-bound lytic enzymes were also observed after treatment of cultured human umbilical cells (HUVEC) with angiogenic factors such as FGF-2, VEGF, and thrombospondin-1. The addition of shed vesicles to HUVEC resulted in autocrine stimulation of invasion through Matrigel and in cord formation (Taraboletti *et al.,* 2002).

C. Shed Vesicles and Blood Coagulation

In some instances, vesicles shed by tumor cells were reported to increase emboli formation (Gasic *et al.,* 1978) and to have procoagulant activity (Dvorak *et al.,* 1981). A procoagulant effect was also reported for vesicles shed by lipopolysaccharide activated monocytes (Satta *et al.,* 1994). The procoagulant effect is, however, more specifically promoted by activated platelet vesicles.

Activated platelets were shown to release both membrane vesicles and exosomes, and in this system the differences in composition and function between these two populations of exovesicles were analyzed in some detail (Heijnen *et al.,* 1999). Relatively large vesicles (>0.2 μm in diameter) referred to as "microvesicles," and exosomes (40–80 nm in diameter) shed by activated platelets were collected separately and analyzed for the presence of GP1b, a plasma membrane marker which is absent in MVBs, and for the presence of the tetraspan protein CD63, which is a marker for late endocytic compartments. Exosomes are enriched in CD63 (the protein was, however, present also in microvesicles); in contrast, GP1b was prevalent in microvesicles. Platelet microvesicles were found to contain several other platelet surface glycoproteins such as the integrin chains α_{IIb}, β_3, and β_I and also P-selectin, which is translocated from intracellular compartments to the cell surface after activation. Phosphatidylserine was found to be present on the outer membrane leaflet of microvesicles, where it provides an anionic surface for the binding of coagulation factors VIII, Va, and Xa. Thus, microvesicles shed by platelets have the potential to exert a procoagulant function at a distance from the site of platelet activation. In the presence of 5 mM CaCl$_2$, binding sites for FITC-conjugated annexin-V were mainly present on microvesicles and not on exosomes. These results indicate that the outer leaflet of platelet exosomes is probably not enriched in phosphatidylserine. Microvesicles therefore are the sites in which procoagulant activity predominates.

The shedding of platelet microvesicles has been correlated with the hydrolysis of actin-binding protein and the disruption of actin–membrane interactions. Following incubation of platelets with the physiologic agonists collagen or thrombin, the extent of activation of the Ca^{2+}-dependent protease calpain correlated positively with the amount of microvesicles formed; furthermore, the shedding of procoagulant-containing microvesicles was inhibited by inhibitors of calpain. Fox *et al.* (1991) hypothesized that association of the platelet membrane skeleton with the plasma membrane maintained the integrity of the plasma membrane and prevented shedding of microvesicles. A physiologic role for calpain in microvesicle shedding was proposed.

D. Shed Vesicles and Bone Formation

Relatively large membrane vesicles are released, both *in vivo* and *in vitro,* from the plasma membrane of hypertrophic chondrocytes; these vesicles, generally named matrix vesicles, are suggested to be involved in the onset of calcification (Anderson, 1969; Hale and Wuthier, 1987). During mineralization of the growth plate, the first crystal phase forms and grows inside the vesicle lumen. Vesicles shed by hypertrophic chondrocytes contain several specific proteins forming a nucleational core complex rich in Ca^{2+} and P$_i$, moreover, they contain annexins II, V, and VI and alkaline phosphates. The annexins are Ca^{2+} channel-forming proteins which enable Ca^{2+} influx inside the vesicle (Kirsch *et al.,* 2000); annexin V also binds directly to types II and X collagen, thereby anchoring the vesicle to the ECM. Alkaline phosphates are enzymes which release inorganic phosphate from organic phosphate compounds (Anderson, 1995). Matrix vesicle composition varies according to changes in the cell physiology. Using cultures of chick embryonic hypertrophic chondrocytes in which mineralization was triggered by treatment with vitamin C and phosphate, Kirsch *et al.* (1997) showed that both control and treated chondrocytes produced vesicles of similar shape and dimensions. However, they observed that only vesicles shed by mineralizing chondrocytes had a very strong alkaline phosphate activity and contained annexin V.

Once crystals grown inside vesicle lumen have reached a certain size, they rupture the vesicle membrane and grow out into the extracellular matrix. It is suggested that vesicle-associated MMPs could be involved in this process and in degradation and remodeling of ECM for subsequent mineralization. Matrix metalloproteinases extracted from matrix vesicles released by rat costochondral chondrocytes (MMP-2 and MMP-3) were found to extensively degrade proteoglycans and this activity was found to be increased in vesicles shed in the growth zone (Dean *et al.*, 1992; Schmitz *et al.*, 1996a). MMP-2, MMP-9, and MMP-13 were shown to be associated with vesicles isolated from avian growth plate cartilage; these vesicles also carry TGFβ in a small latent complex which appears to be an integral membrane component (D'Angelo *et al.*, 2001). According to D'Angelo *et al.*, vesicle-associated MMP-2 is involved in MMP-13 activation and active MMP-13 in turn is responsible for latent TGFβ activation occurring in ECM in parallel with mineral appearance. According to Maeda *et al.* (2001a), in matrix vesicles released by rat costochondral chondrocytes the vesicle associated metalloproteinase involved in TGFβ1 activation is, however, MMP-3. The addition of anti-MMP-3 antibodies to matrix vesicle extracts, in fact, blocked activation of recombinant human latent TGFβ1. The activated factor seems to play a relevant role in the ossification process since it is capable of increasing the capacity of chondrocyte-derived matrix vesicles to generate inorganic pyrophosphate from ATP by stimulating nucleoside triphosphate pyrophospho-hydrolase, and to increase the vesicle biomineralization capability (Derfus *et al.*, 2001). Metalloproteinase activity in growth plate chondrocyte cultures is found to be modulated by vesicle and plasma membrane-associated protein kinase C (PKC) activity, which in turn is modulated by vitamin D metabolites (Maeda *et al.*, 2001b; Schmitz *et al.*, 1996b; Schwartz *et al.*, 2002).

V. Fate of Shed Membrane Vesicles

Few data are available concerning the final destination of shed vesicles. Vesicles shed *in vivo* by tumor cells, recognized after labeling with antibodies against tumor associated antigens, are found, often in association with fibrin fiber networks, in biological fluids such as ascitic and pleural effusions (Dolo *et al.*, 1995a) and in serum (Dolo *et al.*, 1999). Vesicles shed *in vitro* appear to have a rather short half-life. In fact, the amount of vesicles recovered from conditioned media of 8701-BC cells, in which shedding occurs almost exclusively after serum stimulation, reaches the maximum value 3 h after serum addition and decreases afterwards (Cassarà *et al.*, 1998). This result indicates that the rate of vesicle release induced by serum addition, after a burst, slows down rather quickly; moreover, it indicates that vesicles shed into the culture medium do not remain unmodified for long. We propose that shed vesicles, rich in integrins (Dolo *et al.*, 1995a, 1998), can bind to matrix components where they are slowly degraded and deliver their content into the cell matrix. It is, however, likely that vesicles can also fuse with the cell membranes of surrounding cells. This hypothesis is supported by reports describing passive transfer of vesicle membrane proteins between cells (Tabibzadeh *et al.*, 1994; Hewett and Murray, 1996).

VI. Concluding Remarks and Perspective

Relatively large membrane vesicles (diameters ranging from 100 nm to 1 μm) are shed from plasma membranes through unidentified budding mechanisms. These membrane structures are specifically enriched in selected plasma-membrane components including integrins, HLA molecules, and cell surface-associated proteolytic enzymes. MMP-9, MMP-2, MMP-13, MT1-MMP, MMP-1, uPA, and seprase have been detected in vesicles shed by one or more cell lines. Highly motile cells and cells actively involved in bone formation release more vesicles, and these vesicles are richer in membrane-bound proteolytic enzymes. Vesicle membranes appear capable of promoting the activation of proteolytic cascades in the pericellular space, facilitating cell motility and tumor invasion. In hypertrophic chondrocytes, shed vesicles promote the first steps of bone mineralization and the following steps of matrix remodeling needed for bone formation. Vesicles shed by platelets and by some tumor cells exert a procoagulant activity. Shed vesicles appear also to be involved in the release of several signaling molecules and in tumor escape from immune rejection.

Many studies indicate that cells can produce not only these relatively large membrane vesicles, but also other populations of exovesicles having specific characteristics and functions (exosomes and argosomes). Since most of the data on shed vesicles were collected prior to the characterization of these newly identified particles, it is conceivable that previous information may need to be reevaluated.

References

Albanese, J., Meterissian, S., Kontogiannea, M., Dubreuil, C., Hand, A., Sorba, S., and Dainiak, N. (1998). Biologically active Fas antigen and its cognate ligand are expressed on plasma membrane-derived extracellular vesicles. *Blood* **91,** 3862–3874.

Alexander, P. (1974). Proceedings: Escape from immune destruction by the host through shedding of surface antigens: is this a characteristic shared by malignant and embryonic cells. *Cancer Res.* **34,** 2077–2082.

Ally, S., Tortora, G., Clair, T., Grieco, D., Merlo, G., Katsaros, D., Ogreid, D., Doskeland, S. O., Jahnsen, T., and Cho-Chung, Y. S. (1988). Selective modulation of protein kinase isozymes by the site-selective analog 8-chloroadenosine 3′,5′-cyclic monophosphate provides a biological means for control of human colon cancer cell growth. *Proc. Natl. Acad. Sci. USA* **85,** 6319–6322.

Anderson, H. C. (1969). Vesicles associated with calcification in the matrix of epiphyseal cartilage. *J. Cell Biol.* **41,** 59–72.

Anderson, H. C. (1995). Molecular biology of matrix vesicles. *Clin. Orthop.* 266–280.

Angelucci, A., D'Ascenzo, S., Festuccia, C., Gravina, G. L., Bologna, M., Dolo, V., and Pavan, A. (2000). Vesicle-associated urokinase plasminogen activator promotes invasion in prostate cancer cell lines. *Clin. Exp. Metastasis* **18,** 163–170.

Armstrong, M. J., Storch, J., and Dainiak, N. (1988). Structurally distinct plasma membrane regions give rise to extracellular membrane vesicles in normal and transformed lymphocytes. *Biochim. Biophys. Acta* **946,** 106–112.

Breier, G., Blum, S., Peli, J., Groot, M., Wild, C., Risau, W., and Reichmann, E. (2002). Transforming growth factor-beta and Ras regulate the VEGF/VEGF-receptor system during tumor angiogenesis. *Int. J. Cancer* **97,** 142–148.

Cassarà, D., Ginestra, A., Dolo, V., Miele, M., Caruso, G., Lucania, G., and Vittorelli, M. L. (1998). Modulation of vesicle shedding in 8701 BC human breast carcinoma cells. *J. Submicrosc. Cytol. Pathol.* **30,** 45–53.

Chiba, I., Jin, R., Hamada, J., Hosokawa, M., Takeichi, N., and Kobayashi, H. (1989). Growth-associated shedding of a tumor antigen (CE7) detected by a monoclonal antibody. *Cancer Res.* **49,** 3972–3975.

Christian, J. L. (2000). BMP, Wnt and Hedgehog signals: how far can they go. *Curr. Opin. Cell Biol.* **12,** 244–249.

Cooper, D. N., and Barondes, S. H. (1990). Evidence for export of a muscle lectin from cytosol to extracellular matrix and for a novel secretory mechanism. *J. Cell Biol.* **110,** 1681–1691.

D'Angelo, M., Billings, P. C., Pacifici, M., Leboy, P. S., and Kirsch, T. (2001). Authentic matrix vesicles contain active metalloproteases (MMP). A role for matrix vesicle-associated MMP-13 in activation of transforming growth factor-beta. *J. Biol. Chem.* **276,** 11347–11353.

Dainiak, N. (1991). Surface membrane-associated regulation of cell assembly, differentiation, and growth. *Blood* **78,** 264–276.

Dainiak, N., and Sorba, S. (1991). Intracellular regulation of the production and release of human erythroid-directed lymphokines. *J. Clin. Invest.* **87,** 213–220.

Dainiak, N., Riordan, M. A., Strauss, P. R., Feldman, L., and Kreczko, S. (1988). Contractile proteins participate in release of erythroid growth regulators from mononuclear cells. *Blood* **72,** 165–171.

Dean, D. D., Schwartz, Z. V., Muniz, O. E., Gomez, R., Swain, L. D., Howell, D. S., and Boyan, B. D. (1992). Matrix vesicles contain metalloproteinases that degrade proteoglycans. *Bone Miner.* **17,** 172–176.

Denzer, K., Kleijmeer, M. J., Heijnen, H. F., Stoorvogel, W., and Geuze, H. J. (2000a). Exosome: from internal vesicle of the multivesicular body to intercellular signaling device. *J. Cell Sci.* **113 Pt19,** 3365–3374.

Denzer, K., van Eijk, M., Kleijmeer, M. J., Jakobson, E., de Groot, C., and Geuze, H. J. (2000b). Follicular dendritic cells carry MHC class II-expressing microvesicles at their surface. *J. Immunol.* **165,** 1259–1265.

Derfus, B. A., Camacho, N. P., Olmez, U., Kushnaryov, V. M., Westfall, P. R., Ryan, L. M., and Rosenthal, A. K. (2001). Transforming growth factor beta-1 stimulates articular chondrocyte elaboration of matrix vesicles capable of greater calcium pyrophosphate precipitation. *Osteoarthritis Cartilage* **9**, 189–194.

de Visser, K. E., and Kast, W. M. (1999). Effects of TGF-beta on the immune system: implications for cancer immunotherapy. *Leukemia* **13**, 1188–1199.

Dierick, H., and Bejsovec, A. (1999). Cellular mechanisms of wingless/Wnt signal transduction. *Curr. Top. Dev. Biol.* **43**, 153–190.

Dolo, V., Adobati, E., Canevari, S., Picone, M. A., and Vittorelli, M. L. (1995a). Membrane vesicles shed into the extracellular medium by human breast carcinoma cells carry tumor-associated surface antigens. *Clin. Exp. Metastasis* **13**, 277–286.

Dolo, V., D'Ascenzo, S., Violini, S., Pompucci, L., Festuccia, C., Ginestra, A., Vittorelli, M. L., Canevari, S., and Pavan, A. (1999). Matrix-degrading proteinases are shed in membrane vesicles by ovarian cancer cells in vivo and in vitro. *Clin. Exp. Metastasis* **17**, 131–140.

Dolo, V., Ginestra, A., Cassara, D., Violini, S., Lucania, G., Torrisi, M. R., Nagase, H., Canevari, S., Pavan, A., and Vittorelli, M. L. (1998). Selective localization of matrix metalloproteinase 9, beta1 integrins, and human lymphocyte antigen class I molecules on membrane vesicles shed by 8701-BC breast carcinoma cells. *Cancer Res.* **58**, 4468–4474.

Dolo, V., Ginestra, A., Ghersi, G., Nagase, H., and Vittorelli, M. L. (1994). Human breast carcinoma cells cultured in the presence of serum shed membrane vesicles rich in gelatinolytic activities. *J. Submicrosc. Cytol. Pathol.* **26**, 173–180.

Dolo, V., Li, R., Dillinger, M., Flati, S., Manela, J., Taylor, B. J., Pavan, A., and Ladisch, S. (2000). Enrichment and localization of ganglioside G(D3). and caveolin-1 in shed tumor cell membrane vesicles. *Biochim. Biophys. Acta* **1486**, 265–274.

Dolo, V., Pizzurro, P., Ginestra, A., and Vittorelli, M. L. (1995b). Inhibitory effects of vesicles shed by human breast carcinoma cells on lymphocyte 3H-thymidine incorporation, are neutralised by anti TGF-beta antibodies. *J. Submicrosc. Cytol. Pathol.* **27**, 535–541.

Dvorak, H. F., Dickersin, G. R., Dvorak, A. M., Manseau, E. J., and Pyne, K. (1981). Human breast carcinoma: fibrin deposits and desmoplasia. Inflammatory cell type and distribution. Microvasculature and infarction. *J. Natl. Cancer Inst.* **67**, 335–345.

Escola, J. M., Kleijmeer, M. J., Stoorvogel, W., Griffith, J. M., Yoshie, O., and Geuze, H. J. (1998). Selective enrichment of tetraspan proteins on the internal vesicles of multivesicular endosomes and on exosomes secreted by human B-lymphocytes. *J. Biol. Chem.* **273**, 20121–20127.

Feldman, L., Cohen, C. M., Riordan, M. A., and Dainiak, N. (1987). Purification of a membrane-derived human erythroid growth factor. *Proc. Natl. Acad. Sci. USA* **84**, 6775–6779.

Fox, J. E., Austin, C. D., Reynolds, C. C., and Steffen, P. K. (1991). Evidence that agonist-induced activation of calpain causes the shedding of procoagulant-containing microvesicles from the membrane of aggregating platelets. *J. Biol. Chem.* **266**, 13289–13295.

Gasic, G. J., Boettiger, D., Catalfamo, J. L., Gasic, T. B., and Stewart, G. J. (1978). Aggregation of platelets and cell membrane vesiculation by rat cells transformed in vitro by Rous sarcoma virus. *Cancer Res.* **38**, 2950–2955.

Geuze, H. J. (1998). The role of endosomes and lysosomes in MHC class II functioning. *Immunol. Today* **19**, 282–287.

Ginestra, A., La Placa, M. D., Saladino, F., Cassara, D., Nagase, H., and Vittorelli, M. L. (1998). The amount and proteolytic content of vesicles shed by human cancer cell lines correlates with their in vitro invasiveness. *Anticancer Res.* **18**, 3433–3437.

Ginestra, A., Miceli, D., Dolo, V., Romano, F. M., and Vittorelli, M. L. (1999). Membrane vesicles in ovarian cancer fluids: a new potential marker. *Anticancer Res.* **19**, 3439–3445.

Ginestra, A., Monea, S., Seghezzi, G., Dolo, V., Nagase, H., Mignatti, P., and Vittorelli, M. L. (1997). Urokinase plasminogen activator and gelatinases are associated with membrane vesicles shed by human HT1080 fibrosarcoma cells. *J. Biol. Chem.* **272**, 17216–17222.

Greco, V., Hannus, M., and Eaton, S. (2001). Argosomes: a potential vehicle for the spread of morphogens through epithelia. *Cell* **106**, 633–645.

Gruenberg, J., and Maxfield, F. R. (1995). Membrane transport in the endocytic pathway. *Curr. Opin. Cell Biol.* **7,** 552–563.

Guha, A., Mason, R. P., Chen, L., Tuck, D. P., and Dainiak, N. (1994). Modulation of hematopoiesis by lymphocyte membrane-derived components. *Leukemia* **8 Suppl 1,** S227–S230.

Guha, A., Tuck, D., Sorba, S., and Dainiak, N. (1993). Expression of membrane-bound burst-promoting activity is mediated by allogeneic effector cells. *Exp. Hematol.* **21,** 1335–1341.

Hale, J. E., and Wuthier, R. E. (1987). The mechanism of matrix vesicle formation. Studies on the composition of chondrocyte microvilli and on the effects of microfilament-perturbing agents on cellular vesiculation. *J. Biol. Chem.* **262,** 1916–1925.

Heijnen, H. F., Schiel, A. E., Fijnheer, R., Geuze, H. J., and Sixma, J. J. (1999). Activated platelets release two types of membrane vesicles: microvesicles by surface shedding and exosomes derived from exocytosis of multivesicular bodies and alpha-granules. *Blood* **94,** 3791–3799.

Hewett, P. W., and Murray, C. (1996). Modulation of human endothelial cell procoagulant activity in tumour models in vitro. *Int. J. Cancer* **66,** 784–789.

Kirsch, T., Harrison, G., Golub, E. E., and Nah, H. D. (2000). The roles of annexins and types II and X collagen in matrix vesicle-mediated mineralization of growth plate cartilage. *J. Biol. Chem.* **275,** 35577–35583.

Kirsch, T., Nah, H. D., Shapiro, I. M., and Pacifici, M. (1997). Regulated production of mineralization-competent matrix vesicles in hypertrophic chondrocytes. *J. Cell Biol.* **137,** 1149–1160.

Klueg, K. M., and Muskavitch, M. A. (1999). Ligand–receptor interactions and trans-endocytosis of Delta, Serrate and Notch: members of the Notch signalling pathway in *Drosophila. J. Cell Sci.* **112(Pt 19),** 3289–3297.

Lerner, M. P., Lucid, S. W., Wen, G. J., and Nordquist, R. E. (1983). Selected area membrane shedding by tumor cells. *Cancer Lett.* **20,** 125–130.

Liepins, A. (1983). Possible role of microtubules in tumor cell surface membrane shedding, permeability, and lympholysis. *Cell Immunol.* **76,** 120–128.

Luzio, J. P., Rous, B. A., Bright, N. A., Pryor, P. R., Mullock, B. M., and Piper, R. C. (2000). Lysosome-endosome fusion and lysosome biogenesis. *J. Cell Sci.* **113(Pt 9),** 1515–1524.

Mackenzie, A., Wilson, H. L., Kiss-Toth, E., Dower, S. K., North, R. A., and Surprenant, A. (2001). Rapid secretion of interleukin-1beta by microvesicle shedding. *Immunity* **15,** 825–835.

Maeda, S., Dean, D. D., Gay, I., Schwartz, Z., and Boyan, B. D. (2001a). Activation of latent transforming growth factor beta1 by stromelysin 1 in extracts of growth plate chondrocyte-derived matrix vesicles. *J. Bone Miner. Res.* **16,** 1281–1290.

Maeda, S., Dean, D. D., Sylvia, V. L., Boyan, B. D., and Schwartz, Z. (2001b). Metalloproteinase activity in growth plate chondrocyte cultures is regulated by 1,25-(OH)(2)D(3) and 24,25-(OH)(2)D(3) and mediated through protein kinase C. *Matrix Biol.* **20,** 87–97.

Maxfield, F. R., Willingham, M. C., Davies, P. J., and Pastan, I. (1979). Amines inhibit the clustering of alpha2-macroglobulin and EGF on the fibroblast cell surface. *Nature* **277,** 661–663.

Misumi, Y., Misumi, Y., Miki, K., Takatsuki, A., Tamura, G., and Ikehara, Y. (1986). Novel blockade by brefeldin A of intracellular transport of secretory proteins in cultured rat hepatocytes. *J. Biol. Chem.* **261,** 11398–11403.

Monleon, I., Martinez-Lorenzo, M. J., Monteagudo, L., Lasierra, P., Taules, M., Iturralde, M., Pineiro, A., Larrad, L., Alava, M. A., Naval, J., and Anel, A. (2001). Differential secretion of Fas li. *J. Immunol.* **167,** 6736–6744.

Murayama, T., Kataoka, H., Koita, H., Nabeshima, K., and Koono, M. (1991). Glycocalyceal bodies in a human rectal carcinoma cell line and their interstitial collagenolytic activities. *Virchows Arch. B Cell Pathol. Incl. Mol. Pathol.* **60,** 263–270.

Paas, Y., and Fishelson, Z. (1995). Shedding of tyrosine and serine/threonine ecto-protein kinases from human leukemic cells. *Arch. Biochem. Biophys.* **316,** 780–788.

Pan, B. T., and Johnstone, R. M. (1983). Fate of the transferrin receptor during maturation of sheep reticulocytes in vitro: selective externalization of the receptor. *Cell* **33**, 967–978.

Pavlos, N. J., Xu, J., Papadimitriou, J. M., and Zheng, M. H. (2001). Molecular cloning of the mouse homologue of Rab3c. *J. Mol. Endocrinol.* **27**, 117–122.

Peters, P. J., Geuze, H. J., van der Donk, H. A., and Borst, J. (1990). A new model for lethal hit delivery by cytotoxic T lymphocytes. *Immunol. Today* **11**, 28–32.

Peters, P. J., Geuze, H. J., van der Donk, H. A., Slot, J. W., Griffith, J. M., Stam, N. J., Clevers, H. C., and Borst, J. (1989). Molecules relevant for T cell-target cell interaction are present in cytolytic granules of human T lymphocytes. *Eur. J. Immunol.* **19**, 1469–1475.

Poste, G., and Nicolson, G. L. (1980). Arrest and metastasis of blood-borne tumor cells are modified by fusion of plasma membrane vesicles from highly metastatic cells. *Proc. Natl. Acad. Sci. USA* **77**, 399–403.

Raposo, G., Nijman, H. W., Stoorvogel, W., Liejendekker, R., Harding, C. V., Melief, C. J., and Geuze, H. J. (1996). B lymphocytes secrete antigen-presenting vesicles. *J. Exp. Med.* **183**, 1161–1172.

Robbins, J. R., Barth, A. I., Marquis, H., de Hostos, E. L., Nelson, W. J., and Theriot, J. A. (1999). Listeria monocytogenes exploits normal host cell processes to spread from cell to cell. *J. Cell Biol.* **146**, 1333–1350.

Rubin, S. H., and Cowan, D. H. (1973). Assay of granulocytic progenitor cells in human peripheral blood. *Exp. Hematol.* **1**, 127–131.

Satta, N., Toti, F., Feugeas, O., Bohbot, A., Dachary-Prigent, J., Eschwege, V., Hedman, H., and Freyssinet, J. M. (1994). Monocyte vesiculation is a possible mechanism for dissemination of membrane-associated procoagulant activities and adhesion molecules after stimulation by lipopolysaccharide. *J. Immunol.* **153**, 3245–3255.

Schmitz, J. P., Dean, D. D., Schwartz, Z., Cochran, D. L., Grant, G. M., Klebe, R. J., Nakaya, H., and Boyan, B. D. (1996a). Chondrocyte cultures express matrix metalloproteinase mRNA and immunoreactive protein; stromelysin-1 and 72 kDa gelatinase are localized in extracellular matrix vesicles. *J. Cell Biochem.* **61**, 375–391.

Schmitz, J. P., Schwartz, Z., Sylvia, V. L., Dean, D. D., Calderon, F., and Boyan, B. D. (1996b). Vitamin D3 regulation of stromelysin-1 (MMP-3) in chondrocyte cultures is mediated by protein kinase C. *J. Cell Physiol.* **168**, 570–579.

Schwartz, Z., Sylvia, V. L., Larsson, D., Nemere, I., Casasola, D., Dean, D. D., and Boyan, B. D. (2002). 1α25(OH2)D3 regulates chondrocyte matrix vesicle protein kinase C (PKC) directly via G-protein-dependent mechanisms and indirectly via incorporation of PKC during matrix vesicle biogenesis. *J. Biol. Chem.* **277**, 11828–11837.

Singhal, R. L., Yeh, Y. A., Praja, N., Olah, E., Sledge, G. W. Jr., and Weber, G. (1995). Quercetin down-regulates signal transduction in human breast carcinoma cells. *Biochem. Biophys. Res. Commun.* **208**, 425–431.

Tabibzadeh, S. S., Kong, Q. F., and Kapur, S. (1994). Passive acquisition of leukocyte proteins is associated with changes in phosphorylation of cellular proteins and cell–cell adhesion properties. *Am. J. Pathol.* **145**, 930–940.

Taraboletti, G., D'Ascenzo, S., Borsotti, P., Giavazzi, R., Pavan, A., and Dolo, V. (2002). Shedding of the matrix metalloproteinases MMP-2, MMP-9, and MT1-MMP as membrane vesicle-associated components by endothelial cells. *Am. J. Pathol.* **160**, 673–680.

Tarin, D. (1967). Sequential electron microscopical study of experimental mouse skin carcinogenesis. *Int. J. Cancer* **2**, 195–211.

Tarin, D. (1969). Fine structure of murine mammary tumours: the relationship between epithelium and connective tissue in neoplasms induced by various agents. *Br. J. Cancer* **23**, 417–425.

Tartakoff, A. M. (1983). Perturbation of vesicular traffic with the carboxylic ionophore monensin. *Cell* **32**, 1026–1028.

Taylor, D. D., and Black, P. H. (1985). Inhibition of macrophage Ia antigen expression by shed plasma membrane vesicles from metastatic murine melanoma lines. *J. Natl. Cancer Inst.* **74,** 859–867.

Taylor, D. D., and Black, P. H. (1986). Shedding of plasma membrane fragments. Neoplastic and developmental importance. *Dev. Biol. (NY 1985)* **3,** 33–57.

Taylor, D. D., Chou, I. N., and Black, P. H. (1983). Isolation of plasma membrane fragments from cultured murine melanoma cells. *Biochem. Biophys. Res. Commun.* **113,** 470–476.

Taylor, D. D., Taylor, C. G., Jiang, C. G., and Black, P. H. (1988). Characterization of plasma membrane shedding from murine melanoma cells. *Int. J. Cancer* **41,** 629–635.

Thery, C., Regnault, A., Garin, J., Wolfers, J., Zitvogel, L., Ricciardi-Castagnoli, P., Raposo, G., and Amigorena, S. (1999). Molecular characterization of dendritic cell-derived exosomes. Selective accumulation of the heat shock protein hsc73. *J. Cell Biol.* **147,** 599–610.

Trams, E. G., Lauter, C. J., Salem, N. Jr., and Heine, U. (1981). Exfoliation of membrane ecto-enzymes in the form of micro-vesicles. *Biochim. Biophys. Acta* **645,** 63–70.

Tuck, D. P., Cerretti, D. P., Hand, A., Guha, A., Sorba, S., and Dainiak, N. (1994). Human macrophage colony-stimulating factor is expressed at and shed from the cell surface. *Blood* **84,** 2182–2188.

van Blitterswijk, W. J., De Veer, G., Krol, J. H., and Emmelot, P. (1982). Comparative lipid analysis of purified plasma membranes and shed extracellular membrane vesicles from normal murine thymocytes and leukemic GRSL cells. *Biochim. Biophys. Acta* **688,** 495–504.

van den Heuvel, M., Nusse, R., Johnston, P., and Lawrence, P. A. (1989). Distribution of the wingless gene product in Drosophila embryos: a protein involved in cell-cell communication. *Cell* **59,** 739–749.

Zitvogel, L., Regnault, A., Lozier, A., Wolfers, J., Flament, C., Tenza, D., Ricciardi-Castagnoli, P., Raposo, G., and Amigorena, S. (1998). Eradication of established murine tumors using a novel cell-free vaccine: dendritic cell-derived exosomes. *Nat. Med.* **4,** 594–600.

Zucker, S., Wieman, J. M., Lysik, R. M., Wilkie, D. P., Ramamurthy, N., and Lane, B. (1987). Metastatic mouse melanoma cells release collagen-gelatin degrading metalloproteinases as components of shed membrane vesicles. *Biochim. Biophys. Acta* **924,** 225–237.

Index

Contents of Previous Volumes

447